TIME SERIES PREDICTION:
FORECASTING THE FUTURE
AND UNDERSTANDING THE PAST

T0173970

TIME SERIES PREDICTION:
FORECASTING THE FUTURE
AND UNDERSTANDING THE PAST

Proceedings of the NATO Advanced Research
Workshop on Comparative Time Series Analysis
held in Santa Fe, New Mexico, May 14–17, 1992

Editors

Andreas S. Weigend
Xerox PARC
Palo Alto, CA and
University of Colorado
Boulder, CO

Neil A. Gershenfeld
MIT Media Laboratory
Cambridge, MA

Proceedings Volume XV

Santa Fe Institute
Studies in the Sciences of Complexity

 *Advanced Book Program*

 Routledge
Taylor & Francis Group
New York London

Director of Publications, Santa Fe Institute: *Ronda K. Butler-Villa*
Publications Assistant, Santa Fe Institute: *Della L. Ulibarri*

First published 1994 by Westview Press

Published 2018 by Routledge
711 Third Avenue, New York, NY 10017, USA
2 Park Square, Milton Park, Abingdon, Oxon OX14 4RN

Routledge is an imprint of the Taylor & Francis Group, an informa business

ISBN 13: 978-0-201-62602-5 (pbk)

This volume was typeset using T$_{\mathrm{E}}$Xtures on a Macintosh II computer.

About the Santa Fe Institute

The *Santa Fe Institute* (SFI) is a multidisciplinary graduate research and teaching institution formed to nurture research on complex systems and their simpler elements. A private, independent institution, SFI was founded in 1984. Its primary concern is to focus the tools of traditional scientific disciplines and emerging new computer resources on the problems and opportunities that are involved in the multidisciplinary study of complex systems—those fundamental processes that shape almost every aspect of human life. Understanding complex systems is critical to realizing the full potential of science, and may be expected to yield enormous intellectual and practical benefits.

All titles from the *Santa Fe Institute Studies in the Sciences of Complexity* series will carry this imprint which is based on a Mimbres pottery design (circa A.D. 950–1150), drawn by Betsy Jones. The design was selected because the radiating feathers are evocative of the outreach of the Santa Fe Institute Program to many disciplines and institutions.

Santa Fe Institute
Studies in the Sciences of Complexity

Lectures Volumes

Vol.	Editor	Title
I	D. L. Stein	Lectures in the Sciences of Complexity, 1989
II	E. Jen	1989 Lectures in Complex Systems, 1990
III	L. Nadel & D. L. Stein	1990 Lectures in Complex Systems, 1991
IV	L. Nadel & D. L. Stein	1991 Lectures in Complex Systems, 1992
V	L. Nadel & D. L. Stein	1992 Lectures in Complex Systems, 1993

Lecture Notes Volumes

Vol.	Author	Title
I	J. Hertz, A. Krogh, & R. Palmer	Introduction to the Theory of Neural Computation, 1990
II	G. Weisbuch	Complex Systems Dynamics, 1990
III	W. D. Stein & F. J. Varela	Thinking About Biology, 1993

Reference Volumes

Vol.	Author	Title
I	A. Wuensche & M. Lesser	The Global Dynamics of Cellular Automata: Attraction Fields of One-Dimensional Cellular Automata, 1992

Proceedings Volumes

Vol.	Editor	Title
I	D. Pines	Emerging Syntheses in Science, 1987
II	A. S. Perelson	Theoretical Immunology, Part One, 1988
III	A. S. Perelson	Theoretical Immunology, Part Two, 1988
IV	G. D. Doolen et al.	Lattice Gas Methods for Partial Differential Equations, 1989
V	P. W. Anderson, K. Arrow, D. Pines	The Economy as an Evolving Complex System, 1988
VI	C. G. Langton	Artificial Life: Proceedings of an Interdisciplinary Workshop on the Synthesis and Simulation of Living Systems, 1988
VII	G. I. Bell & T. G. Marr	Computers and DNA, 1989
VIII	W. H. Zurek	Complexity, Entropy, and the Physics of Information, 1990
IX	A. S. Perelson & S. A. Kauffman	Molecular Evolution on Rugged Landscapes: Proteins, RNA and the Immune System, 1990
X	C. G. Langton et al.	Artificial Life II, 1991
XI	J. A. Hawkins & M. Gell-Mann	The Evolution of Human Languages, 1992
XII	M. Casdagli & S. Eubank	Nonlinear Modeling and Forecasting, 1992
XIII	J. E. Mittenthal & A. B. Baskin	Principles of Organization in Organisms, 1992
XIV	D. Friedman & J. Rust	The Double Auction Market: Institutions, Theories, and Evidence, 1993
XV	A. S. Weigend & N. A. Gershenfeld	Time Series Prediction: Forecasting the Future and Understanding the Past

List of Authors for Correspondence

Casdagli, Martin
    First Boston Corporation, Equity Financial Strategies, 55 E. 52nd St., 7th floor,
    New York, NY 10055
    e-mail: mcasda@equity.fbc.com, phone: (212)909-4157, fax: (212)909-1431
Clemens, Chris
    Astronomy Department, RLM 15.308, University of Texas at Austin, Austin,
    TX 78712
    e-mail: cclemens@astro.as.utexas.edu, phone: (512)471-4419
Dirst, Matthew
    Department of Music, Stanford University, Stanford, CA 94305
    e-mail: mdirst@leland.standford.edu, phone: (415)321-9593
Fraser, Andrew M.
    Systems Science Ph.D. Program, Portland State University, P. O. Box 751,
    Portland, Oregon 97207-0751
    e-mail: andy@ee.pdx.edu, phone: (503)725-4989
Gershenfeld, Neil
    MIT Media Lab, Room E15-425, 20 Ames Street, Cambridge, MA 02139
    e-mail: neilg@media.mit.edu, phone: (617)253-7680, fax: (617)258-6264
Glass, Leon
    Department of Physiology, McGill University, 3655 Drummond St., Montreal,
    Quebec H3G 1Y6, Canada
    e-mail: glass@cnd.mcgill.ca, phone (514)398-4338
Granger, Clive
    Economics Department D-008, University of California at San Diego, La Jolla,
    CA 92093-0508
    e-mail: mbacci@weber.ucsd.edu, phone: (619)534-3856
Hübner, Udo
    Phys.-Techn. Budesanstalt, Lab. 4.42, Bundesallee 100, D-3300 Braunschweig,
    Germany
    e-mail: huebn405@rz.braunschweig.ptb.d400.de, phone: 49(531)592-4422,
    fax: 49(531)592-4006
Kantz, Holger
    Department of Theoretical Physics, University of Wuppertal, Gauss-str. 20,
    D-42097 Wuppertal 1, Germany
    e-mail: kantz@wptu0.physik.uni-wuppertal.de, phone: 49(202)439-2739
Kaplan, Daniel
    Department of Physiology, McGill University, 3655 Drummond St., Montreal,
    Quebec H3G 1Y6, Canda
    e-mail: danny@cnd.mcgill.ca, phone: (514) 398-8092
Kostelich, Eric
    Department of Mathematics, Box 871804, Arizona State University, Tempe,
    AZ 85287-1804
    e-mail: kostelich@asu.edu, phone: (602)965-5006, fax: (602)965-8119
LeBaron, Blake
    Department of Economics, 1180 Observatory Drive, University of Wisconsin,
    Madison, WI 53706
    e-mail: blakel@vms.macc.wisc.edu, phone: (608)263-2516
Lequarré, Jean
    Union Bank of Switzerland, LHIS Lausanne Development Center, 6, Avenue de
    Provence, CH-1007 Lausanne, Switzerland
    e-mail: jean.lequarre@zh001.ubs.ubs.arcom.ch, phone: 41(21)318.50.36,
    fax: 41(21)25.68.96

Lewis, Peter
    Code OR, USN Postgraduate School, Monterey, CA 93943
    e-mail: 1526P@navpgs.bitnet, phone: (408)646-2283
Mozer, Mike
    Department of Computer Science and Institute of Cognitive Science,
    University of Colorado, Boulder, CO 80309-0430
    e-mail: mozer@cs.colorado.edu, phone: (303)492-4103, fax: (303)492-2844
Paluš, Milan
    Santa Fe Institute, 1660 Old Pecos Trail, Suite A, Santa Fe, NM 87501
    e-mail: mp@santafe.edu, phone (505)984-8800; fax (505)982-0565
Pineda, Fernando J.
    Room 2-254, Applied Physics Laboratory, The John Hopkins University, Laurel,
    MD 20723-6099
    e-mail: fernando@aplcomm.jhuapl.edu, phone: (410)792-5480, fax: (301)953-6904
Press, William
    Harvard College Observatory, 60 Garden Street, Cambridge, MA 02138
    e-mail: wpress@cfa.harvard.edu, phone: (617)495-4908
Rigney, David
    Beth Israel Hospital, Room KB-26, 330 Brookline Avenue, Boston, MA 02215
    e-mail: david@astro.bih.harvard.edu, phone: (617)753-5121
Sauer, Tim
    Department of Mathematics, George Mason University, Fairfax, VA 22030
    e-mail: tsauer@gmu.edu, phone: (703)993-1471 or (703)993-1460,
    fax: (703)993-1491
Smith, Leonard
    Mathematics Institute, Oxford University, OX1 3LB Oxford, United Kingdom
    e-mail: lenny@maths.ox.ac.uk, phone: 44(865)270 506, fax: 44(865)270 515
Swinney, Harry
    Department of Physics, University of Texas at Austin, Austin, TX 78712-1081
    e-mail: swinney@chaos.utexas.edu, phone: (512)471-4619, fax: (512)471-1558
Theiler, James
    MS-B213, Los Alamos National Laboratory, Los Alamos, NM 87545
    e-mail: jt@t13.lanl.gov, phone: (505)665-5682, fax: (505)665-3003
Wan, Eric
    Department of Electrical Engineering, Durand, Rm. 104, Stanford University,
    Stanford, CA 94305
    e-mail: wan@isl.stanford.edu, phone: (415)723-4769, fax: (415)723-8473
Weigend, Andreas
    Department of Computer Science and Institute of Cognitive Science,
    University of Colorado, Boulder, CO 80309-0430
    e-mail: weigend@cs.colorado.edu, phone: (303)492-2524
    or [1-0-ATT] 0-700-WEIGEND, fax: (303)492-2844
Zhang, Xiru
    Thinking Machines Corporation, 245 First Street, Cambridge, MA 02142
    e-mail: xiru@think.com, phone (617)234-2026, fax: (617)234-4444
Zheleznyak, Alexander
    Institute of Applied Physics, Russian Academy of Science, Uljanov 46, Nizhny
    Novgorod, 603600 Russia
    e-mail: alzhel@appl.nnov.su, phone: 7 8312 384283, fax: 7 8312 369717

Contents

Preface
Andreas S. Weigend and Neil A. Gershenfeld   xv

The Future of Time Series: Learning and Understanding
Neil A. Gershenfeld and Andreas S. Weigend   1

Section I. Description of the Data Sets   71

Lorenz-Like Chaos in NH_3-FIR Lasers (Data Set A)
Udo Hübner, Carl-Otto Weiss, Neal Broadus Abraham,
and Dingyuan Tang   73

Multi-Channel Physiological Data: Description and
Analysis (Data Set B)
David R. Rigney, Ary L. Goldberger, Wendell C. Ocasio,
Yuhei Ichimaru, George B. Moody, and Roger G. Mark   105

Foreign Currency Dealing: A Brief Introduction
(Data Set C)
Jean Y. Lequarré   131

Whole Earth Telescope Observations of the White Dwarf
Star (PG1159-035) (Data Set E)
J. Christopher Clemens   139

Times Series Prediction: Forecasting the Future and
Understanding the Past, Eds. A. S. Weigend and N. A. Gershenfeld, SFI Studies
in the Sciences of Complexity, Proc. Vol. XV, Addison-Wesley, 1993   **xi**

Baroque Forecasting: On Completing J. S. Bach's Last
Fugue (Data Set F)
  Matthew Dirst and Andreas S. Weigend   151

Section II. Time Series Prediction   173

Time Series Prediction by Using Delay Coordinate
Embedding (Data Set A)
  Tim Sauer   175

Time Series Prediction by Using a Connectionist Network
with Internal Delay Lines (Data Set A)
  Eric A. Wan   195

Simple Architectures on Fast Machines: Practical Issues in
Nonlinear Time Series Prediction (Data Sets C,D,F)
  Xiru Zhang and Jim Hutchinson   219

Neural Net Architectures for Temporal Sequence
Processing (Data Set C)
  Michael C. Mozer   243

Forecasting Probability Densities by Using Hidden Markov
Models with Mixed States (Data Set D)
  Andrew M . Fraser and Alexis Dimitriadis   265

Time Series Prediction by Using the Method of Analogues
(Data Set A)
  Eric J. Kostelich and Daniel P. Lathrop   283

Modeling Time Series by Using Multivariate Adaptive
Regression Splines (MARS) (Data Sets A,B,C)
  P. A. W. Lewis, B. K. Ray, and J. G. Stevens   297

Visual Fitting and Extrapolation (Data Set A)
  George G. Lendaris and Andrew M. Fraser   319

Does a Meeting in Santa Fe Imply Chaos? (Data Set A)
  Leonard A. Smith   323

Contents

Section III. Time Series Analysis and Characterization 345

Exploring the Continuum Between Deterministic and
Stochastic Modeling (Data Sets A,B,D)
Martin Casdagli and Andreas S. Weigend 347

Estimating Generalized Dimensions and Choosing Time
Delays: A Fast Algorithm (Data Sets A,D)
Fernando J. Pineda and John C. Sommerer 367

Identifying and Quantifying Chaos by Using Information-
Theoretic Functionals (Data Sets A,D,E)
Milan Paluš 387

A Geometrical Statistic for Detecting Deterministic
Dynamics (Data Sets A,B,C,D)
Daniel T. Kaplan 415

Detecting Nonlinearity in Data with Long Coherence Times
(Data Set E)
James Theiler, Paul S. Linsay, and David M. Rubin 429

Nonlinear Diagnostics and Simple Trading Rules for High-
Frequency Foreign Exchange Rates (Data Set C)
Blake LeBaron 457

Noise Reduction by Local Reconstruction of the Dynamics
(Data Set A)
Holger Kantz 475

Section IV. Practice and Promise 491

Large-Scale Linear Methods for Interpolation, Realization,
and Reconstruction of Noisy, Irregularly Sampled Data
William H. Press and George B. Rybicki 493

Complex Dynamics in Physiology and Medicine
Leon Glass and Daniel T. Kaplan 513

Forecasting in Economics
Clive W. J. Granger 529

Finite-Dimensional Spatial Disorder: Description and
Analysis
  V. S. Afraimovich, M. I. Rabinovich, and A. L. Zheleznyak   **539**

Spatio-Temporal Patterns: Observations and Analysis
  Harry L. Swinney   **557**

Appendix: Accessing the Server   **569**

Bibliography   **571**

Index   **631**

Preface

This book is the result of an unsuccessful joke. During the summer of 1990, we were both participating in the Complex Systems Summer School of the Santa Fe Institute. Like many such programs dealing with "complexity," this one was full of exciting examples of how it can be possible to recognize when apparently complex behavior has a simple understandable origin. However, as is often the case in young disciplines, little effort was spent trying to understand how such techniques are interrelated, how they relate to traditional practices, and what the bounds on their reliability are. Addressing these issues is necessary for suggestive results to grow into a mature discipline. These problems were particularly apparent in time series analysis, an area that we arrived at through our physics theses. Out of frustration with the fragmented and anecdotal literature, we made what we thought was a humorous suggestion: run a competition. Much to our surprise, no one laughed and, to our further surprise, the Santa Fe Institute promptly agreed to support it. The rest is history (630 pages worth).

Reasons why a competition might be a bad idea abound: science is a thoughtful activity, not a simple race; the relevant disciplines are too dissimilar and the questions too difficult to permit meaningful comparisons; and the required effort might be prohibitively large in return for potentially misleading results. On the other hand, regardless of the very different techniques and language games of the different disciplines that study time series (physics, biology, economics,...), very

Times Series Prediction: Forecasting the Future and
Understanding the Past, Eds. A. S. Weigend and N. A. Gershenfeld, SFI Studies
in the Sciences of Complexity, Proc. Vol. XV, Addison-Wesley, 1993   **XV**

similar questions are asked: What will happen next? What kind of system produced the time series? How can it be described? How much can we know about the system? These questions can have quantitative answers that permit direct comparisons. And with the growing penetration of computer networks, it had become feasible to announce a competition, to distribute the data (withholding the continuations), and subsequently to collect and analyze the results. We began to realize that a competition might not be such a crazy idea.

The Santa Fe Institute seemed ideally placed to support such an undertaking. It spans many disciplines and addresses broad questions that do not easily fall within the purview of a single academic department. Following its initial commitment, we assembled a group of advisors[1] to represent many of the relevant disciplines in order to help us decide if and how to proceed. These initial discussions progressed to the collection of a large library of candidate data sets, the selection of a representative small subset, the specification of the competition tasks, and finally the publicizing and then running of the competition (which was remotely managed by Andreas in Bangkok and Neil in Cambridge, Massachusetts). After its close, we ran a NATO Advanced Research Workshop to bring together the advisory board, representatives of the groups that had provided the data, successful participants, and interested observers. This heterogeneous group was able to communicate using the common reference of the competition data sets; the result is this book. It aims to provide a snapshot of the range of new techniques that are currently used to study time series, both as a reference for experts and as a guide for novices.

Scanning the contents, we are struck by the variety of routes that lead people to study time series. This subject, which has a rather dry reputation from a distance (we certainly thought that), lies at the heart of the scientific enterprise of building models from observations. One of our goals was to help clarify how new time series techniques can be broadly applicable beyond the restricted domains within which they evolved (such as simple chaos experiments), and equally well how theories of everything can be applicable to nothing given the limitations of real data.

We had another hidden agenda in running this competition. Any one such study can never be definitive, but our hope was that the real result would be planting a seed for an ongoing process of using new technology to share results in what is, in effect, a very large collective research project. The many papers in this volume that use the competition tasks as starting points for the broader and deeper study of these common data sets suggests that our hope might be fulfilled. This survey of what is possible is in no way meant to suggest that better results are impossible. We will be pleased if the Santa Fe data sets and results become common reference benchmarks, and even more pleased if they are later discarded and replaced by more worthy successors.

---

[1]The advisors were Leon Glass (biology), Clive Granger (economics), Bill Press (astrophysics and numerical analysis), Maurice Priestley (statistics), Itamar Procaccia (dynamical systems), T. Subba Rao (statistics), and Harry Swinney (experimental physics).

An undertaking such as this requires the assistance of more friends (and thoughtful critics) than we knew we had. We thank the members of the advisory board, the providers of the data sets, and the competition entrants for participating in a quixotic undertaking based on limited advance information. We thank the Santa Fe Institute (SFI) and NATO for support.[2] We are grateful for the freedom provided by Stanford, Harvard, Chulalongkorn, MIT, and Xerox PARC. We thank Ronda Butler-Villa and Della Ulibarri for the heroic job of helping us assemble this book, and we thank the one hundred referees for their critical comments. We also thank our friends for not abandoning us despite all the trouble they had with this enterprise.

Finally, we must thank each other for tolerating and successfully filtering each other's occasionally odd ideas about how to run a time series competition, which neither of us would have been able to do (or understand) alone.

Neil Gershenfeld	Andreas Weigend
Cambridge, MA	San Francisco, CA

July 1993

[2]Core funding for SFI is provided by the John D. and Catherine T. MacArthur Foundation, the National Science Foundation, grant PHY-8714918, and the U.S. Department of Energy, grant ER-FG05-88ER25054.

TIME SERIES PREDICTION:
FORECASTING THE FUTURE
AND UNDERSTANDING THE PAST

Neil A. Gershenfeld† and Andreas S. Weigend‡
†MIT Media Laboratory, 20 Ames Street, Cambridge, MA 02139;
e-mail: neilg@media.mit.edu.
‡Xerox PARC, 3333 Coyote Hill Road, Palo Alto, CA 94304;
e-mail: weigend@cs.colorado.edu.
Address after August 1993: Andreas Weigend, Department of Computer Science
and Institute of Cognitive Science, University of Colorado, Boulder, CO 80309-0430.

The Future of Time Series: Learning and Understanding

Throughout scientific research, measured time series are the basis for characterizing an observed system and for predicting its future behavior. A number of new techniques (such as state-space reconstruction and neural networks) promise insights that traditional approaches to these very old problems cannot provide. In practice, however, the application of such new techniques has been hampered by the unreliability of their results and by the difficulty of relating their performance to those of mature algorithms. This chapter reports on a competition run through the Santa Fe Institute in which participants from a range of relevant disciplines applied a variety of time series analysis tools to a small group of common data sets in order to help make meaningful comparisons among their approaches. The design and the results of this competition are described, and the historical and theoretical backgrounds necessary to understand the successful entries are reviewed.

Times Series Prediction: Forecasting the Future and
Understanding the Past, Eds. A. S. Weigend and N. A. Gershenfeld, SFI Studies
in the Sciences of Complexity, Proc. Vol. XV, Addison-Wesley, 1993   **1**

1. INTRODUCTION

The desire to predict the future and understand the past drives the search for laws that explain the behavior of observed phenomena; examples range from the irregularity in a heartbeat to the volatility of a currency exchange rate. If there are known underlying deterministic equations, in principle they can be solved to forecast the outcome of an experiment based on knowledge of the initial conditions. To make a forecast if the equations are not known, one must find both the rules governing system evolution and the actual state of the system. In this chapter we will focus on phenomena for which underlying equations are not given; the rules that govern the evolution must be inferred from regularities in the past. For example, the motion of a pendulum or the rhythm of the seasons carry within them the potential for predicting their future behavior from knowledge of their oscillations without requiring insight into the underlying mechanism. We will use the terms "understanding" and "learning" to refer to two complementary approaches taken to analyze an unfamiliar time series. *Understanding* is based on explicit mathematical insight into how systems behave, and *learning* is based on algorithms that can emulate the structure in a time series. In both cases, the goal is to explain observations; we will not consider the important related problem of using knowledge about a system for controlling it in order to produce some desired behavior.

Time series analysis has three goals: forecasting, modeling, and characterization. The aim of *forecasting* (also called *predicting*) is to accurately predict the short-term evolution of the system; the goal of *modeling* is to find a description that accurately captures features of the long-term behavior of the system. These are not necessarily identical: finding governing equations with proper long-term properties may not be the most reliable way to determine parameters for good short-term forecasts, and a model that is useful for short-term forecasts may have incorrect long-term properties. The third goal, system *characterization*, attempts with little or no *a priori* knowledge to determine fundamental properties, such as the number of degrees of freedom of a system or the amount of randomness. This overlaps with forecasting but can differ: the complexity of a model useful for forecasting may not be related to the actual complexity of the system.

Before the 1920s, forecasting was done by simply extrapolating the series through a global fit in the time domain. The beginning of "modern" time series prediction might be set at 1927 when Yule invented the autoregressive technique in order to predict the annual number of sunspots. His model predicted the next value as a weighted sum of previous observations of the series. In order to obtain "interesting" behavior from such a linear system, outside intervention in the form of external shocks must be assumed. For the half-century following Yule, the reigning paradigm remained that of linear models driven by noise.

However, there are simple cases for which this paradigm is inadequate. For example, a simple iterated map, such as the logistic equation (Eq. (11), in Section

3.2), can generate a broadband power spectrum that cannot be obtained by a linear approximation. The realization that apparently complicated time series can be generated by very simple equations pointed to the need for a more general theoretical framework for time series analysis and prediction.

Two crucial developments occurred around 1980; both were enabled by the general availability of powerful computers that permitted much longer time series to be recorded, more complex algorithms to be applied to them, and the data and the results of these algorithms to be interactively visualized. The first development, state-space reconstruction by time-delay embedding, drew on ideas from differential topology and dynamical systems to provide a technique for recognizing when a time series has been generated by deterministic governing equations and, if so, for understanding the geometrical structure underlying the observed behavior. The second development was the emergence of the field of machine learning, typified by neural networks, that can adaptively explore a large space of potential models. With the shift in artificial intelligence from rule-based methods towards data-driven methods,[1] the field was ready to apply itself to time series, and time series, now recorded with orders of magnitude more data points than were available previously, were ready to be analyzed with machine-learning techniques requiring relatively large data sets.

The realization of the promise of these two approaches has been hampered by the lack of a general framework for the evaluation of progress. Because time series problems arise in so many disciplines, and because it is much easier to describe an algorithm than to evaluate its accuracy and its relationship to mature techniques, the literature in these areas has become fragmented and somewhat anecdotal. The breadth (and the range in reliability) of relevant material makes it difficult for new research to build on the accumulated insight of past experience (researchers standing on each other's toes rather than shoulders).

Global computer networks now offer a mechanism for the disjoint communities to attack common problems through the widespread exchange of data and information. In order to foster this process and to help clarify the current state of time series analysis, we organized the Santa Fe Time Series Prediction and Analysis Competition under the auspices of the Santa Fe Institute during the fall of 1991. The goal was not to pick "winners" and "losers," but rather to provide a structure for researchers from the many relevant disciplines to compare quantitatively the results of their analyses of a group of data sets selected to span the range of studied problems. To explore the results of the competition, a NATO Advanced Research Workshop was held in the spring of 1992; workshop participants included members of the competition advisory board, representatives of the groups that had collected the data, participants in the competition, and interested observers. Although the participants came from a broad range of disciplines, the discussions were framed by

[1]Data sets of hundreds of megabytes are routinely analyzed with massively parallel supercomputers, using parallel algorithms to find near neighbors in multidimensional spaces (K. Thearling, personal communication, 1992; Bourgoin et al., 1993).

the analysis of common data sets and it was (usually) possible to find a meaningful common ground. In this overview chapter we describe the structure and the results of this competition and review the theoretical material required to understand the successful entries; much more detail is available in the articles by the participants in this volume.

2. THE COMPETITION

The planning for the competition emerged from informal discussions at the Complex Systems Summer School at the Santa Fe Institute in the summer of 1990; the first step was to assemble an advisory board to represent the interests of many of the relevant fields.[2] With the help of this group we gathered roughly 200 megabytes of experimental time series for possible use in the competition. This volume of data reflects the growth of techniques that use enormous data sets (where automatic collection and processing is essential) over traditional time series (such as quarterly economic indicators, where it is possible to develop an intimate relationship with each data point).

In order to be widely accessible, the data needed to be distributed by ftp over the Internet, by electronic mail, and by floppy disks for people without network access. The latter distribution channels limited the size of the competition data to a few megabytes; the final data sets were chosen to span as many of a desired group of attributes as possible given this size limitation (the attributes are shown in Figure 2). The final selection was:

A. **A clean physics laboratory experiment.** 1,000 points of the fluctuations in a far-infrared laser, approximately described by three coupled nonlinear ordinary differential equations (Hübner et al., this volume).

B. **Physiological data from a patient with sleep apnea.** 34,000 points of the heart rate, chest volume, blood oxygen concentration, and EEG state of a sleeping patient. These observables interact, but the underlying regulatory mechanism is not well understood (Rigney et al., this volume).

C. **High-frequency currency exchange rate data.** Ten segments of 3,000 points each of the exchange rate between the Swiss franc and the U.S. dollar. The average time between two quotes is between one and two minutes (Lequarré, this volume). If the market was efficient, such data should be a random walk.

[2] The advisors were Leon Glass (biology), Clive Granger (economics), Bill Press (astrophysics and numerical analysis), Maurice Priestley (statistics), Itamar Procaccia (dynamical systems), T. Subba Rao (statistics), and Harry Swinney (experimental physics).

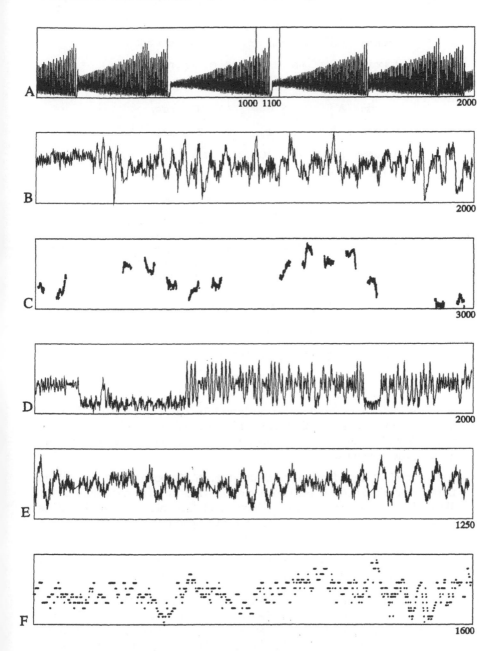

FIGURE 1 Sections of the competition data sets.

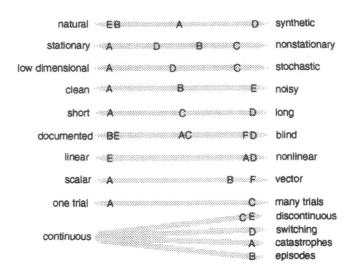

natural ‐‐‐EB‐‐‐‐‐‐‐‐‐‐‐‐A‐‐‐‐‐‐‐‐‐‐‐‐D‐‐‐ synthetic
stationary ‐‐‐A‐‐‐‐‐‐D‐‐‐‐‐B‐‐‐‐‐C‐‐‐‐‐ nonstationary
low dimensional ‐‐‐A‐‐‐‐‐‐‐‐‐‐‐‐D‐‐‐‐‐‐‐‐‐‐‐‐C‐‐‐ stochastic
clean ‐‐‐A‐‐‐‐‐‐‐‐‐‐‐B‐‐‐‐‐‐‐‐‐E‐‐‐ noisy
short ‐‐‐A‐‐‐‐‐‐‐‐‐‐‐‐‐C‐‐‐‐‐‐‐‐‐‐‐‐D‐‐‐ long
documented ‐‐‐BE‐‐‐‐‐‐‐‐‐‐‐AC‐‐‐‐‐‐‐‐‐‐‐FD‐‐‐ blind
linear ‐‐‐E‐‐‐‐‐‐‐‐‐‐‐‐‐‐‐‐‐‐‐‐‐‐‐‐‐AD‐‐‐ nonlinear
scalar ‐‐‐A‐‐‐‐‐‐‐‐‐‐‐‐‐‐‐‐‐‐‐‐‐‐‐B‐‐‐F‐‐‐ vector
one trial ‐‐‐A‐‐‐‐‐‐‐‐‐‐‐‐‐‐‐‐‐‐‐‐‐‐‐C‐‐‐ many trials
continuous CE discontinuous
D switching
A catastrophes
B episodes

can dynamics make money? C

can dynamics save lives? B

FIGURE 2   Some attributes spanned by the data sets.

D. **A numerically generated series designed for this competition.** A driven particle in a four-dimensional nonlinear multiple-well potential (nine degrees of freedom) with a small nonstationarity drift in the well depths. (Details are given in the Appendix.)

E. **Astrophysical data from a variable star.** 27,704 points in 17 segments of the time variation of the intensity of a variable white dwarf star, collected by the *Whole Earth Telescope* (Clemens, this volume). The intensity variation arises from a superposition of relatively independent spherical harmonic multiplets, and there is significant observational noise.

F. **A fugue.** J. S. Bach's final (unfinished) fugue from *The Art of the Fugue*, added after the close of the formal competition (Dirst and Weigend, this volume).

The amount of information available to the entrants about the origin of each data set varied from extensive (Data Sets B and E) to blind (Data Set D). The original files will remain available. The data sets are graphed in Figure 1, and some of the characteristics are summarized in Figure 2. The appropriate level of description for models of these data ranges from low-dimensional stationary dynamics to stochastic processes.

After selecting the data sets, we next chose **competition tasks** appropriate to the data sets and research interests. The participants were asked to:

- predict the (withheld) continuations of the data sets with respect to given error measures,
- characterize the systems (including aspects such as the number of degrees of freedom, predictability, noise characteristics, and the nonlinearity of the system),
- infer a model of the governing equations, and
- describe the algorithms employed.

The data sets and competition tasks were made publicly available on August 1, 1991, and competition entries were accepted until January 15, 1992. Participants were required to describe their algorithms. (Insight in some previous competitions was hampered by the acceptance of proprietary techniques.) One interesting trend in the entries was the focus on prediction, for which three motivations were given: (i) because predictions are falsifiable, insight into a model used for prediction is verifiable; (ii) there are a variety of financial incentives to study prediction; and (iii) the growth of interest in machine learning brings with it the hope that there can be universally and easily applicable algorithms that can be used to generate forecasts. Another trend was the general failure of simplistic "black-box" approaches—in all successful entries, exploratory data analysis preceded the algorithm application.[3]

It is interesting to compare this time series competition to the previous state of the art as reflected in two earlier competitions (Makridakis & Hibon, 1979; Makridakis et al., 1984). In these, a very large number of time series was provided (111 and 1001, respectively), taken from business (forecasting sales), economics (predicting recovery from the recession), finance, and the social sciences. However, all of the series used were very short, generally less than 100 values long. Most of the algorithms entered were fully automated, and most of the discussion centered around linear models.[4] In the Santa Fe Competition all of the successful entries were fundamentally nonlinear and, even though significantly more computer power was used to analyze the larger data sets with more complex models, the application of the algorithms required more careful manual control than in the past.

---

[3] The data, analysis programs, and summaries of the results are available by anonymous ftp from ftp.santafe.edu, as described in the Appendix to this volume. In the competition period, on average 5 to 10 people retrieved the data per day, and 30 groups submitted final entries by the deadline. Entries came from the U.S., Europe (including former communist countries), and Asia, ranging from junior graduate students to senior researchers.

[4] These discussions focused on issues such as the order of the linear model. Chatfield (1988) summarizes previous competitions.

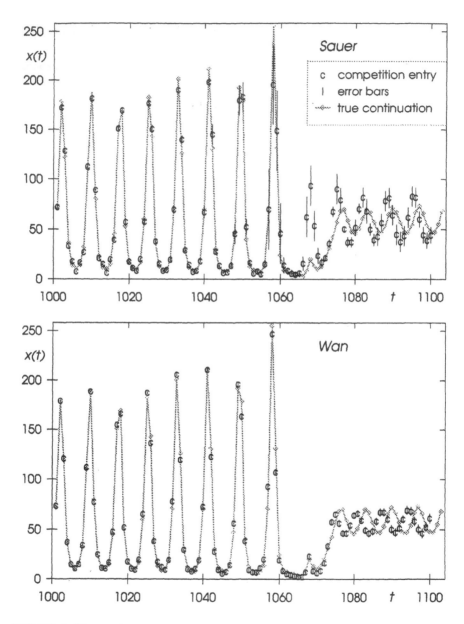

FIGURE 3 The two best predicted continuations for Data Set A, by Sauer and by Wan. Predicted values are indicated by "c," predicted error bars by vertical lines. The true continuation (not available at the time when the predictions were received) is shown in grey (the points are connected to guide the eye).

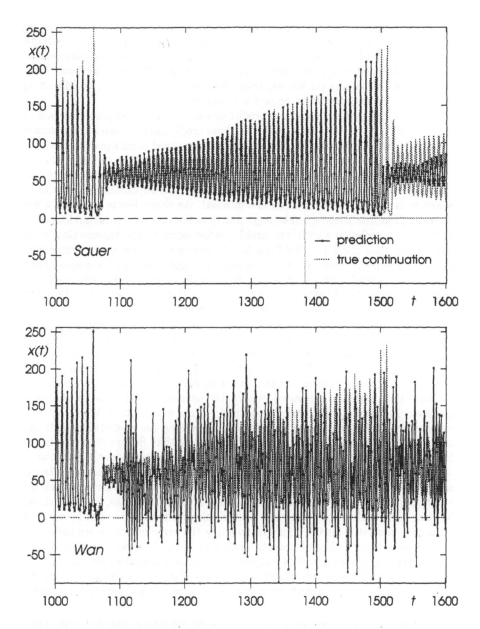

FIGURE 4 Predictions obtained by the same two models as in the previous figure, but continued 500 points further into the future. The solid line connects the predicted points; the grey line indicates the true continuation.

As an example of the results, consider the intensity of the laser (Data Set A; see Figure 1). On the one hand, the laser can be described by a relatively simple "correct" model of three nonlinear differential equations, the same equations that Lorenz (1963) used to approximate weather phenomena. On the other hand, since the 1,000-point training set showed only three of four collapses, it is difficult to predict the next collapse based on so few instances.

For this data set we asked for predictions of the next 100 points as well as estimates of the error bars associated with these predictions. We used two measures to evaluate the submissions. The first measure (normalized mean squared error) was based on the predicted values only; the second measure used the submitted error predictions to compute the likelihood of the observed data given the predictions. The Appendix to this chapter gives the definitions and explanations of the error measures as well as a table of all entries received. We would like to point out a few interesting features. Although this single trial does not permit fine distinctions to be made between techniques with comparable performance, two techniques clearly did much better than the others for Data Set A; one used state-space reconstruction to build an explicit model for the dynamics and the other used a connectionist network (also called a neural network). Incidentally, a prediction based solely on visually examining and extrapolating the training data did much worse than the best techniques, but also much better than the worst.

Figure 3 shows the two best predictions. Sauer (this volume) attempts to understand and develop a *representation* for the geometry in the system's state space, which is the best that can be done without knowing something about the system's governing equations, while Wan (this volume) addresses the issue of *function approximation* by using a connectionist network to learn to emulate the input-output behavior. Both methods generated remarkably accurate predictions for the specified task. In terms of the measures defined for the competition, Wan's squared errors are one-third as as large as Sauer's, and—taking the predicted uncertainty into account—Wan's model is four times more likely than Sauer's.[5] According to the competition scores for Data Set A, this puts Wan's network in the first place.

A different picture, which cautions the hurried researcher against declaring one method to be universally superior to another, emerges when one examines the evolution of these two prediction methods further into the future. Figure 4 shows the same two predictors, but now the continuations extend 500 points beyond the 100 points submitted for the competition entry (no error estimates are shown).[6] The neural network's class of potential behavior is much broader than what can be generated from a small set of coupled ordinary differential equations, but the state-space model is able to reliably forecast the data much further because its explicit description can correctly capture the character of the long-term dynamics.

---

[5]The likelihood ratio can be obtained from Table 2 in the Appendix as $\exp(-3.5)/\exp(-4.8)$.

[6]Furthermore, we invite the reader to compare Figure 5 by Sauer (this volume, p. 191) with Figure 13 by Wan (this volume, p. 213). Both entrants start the competition model at the same four (new) different points. The squared errors are compared in the Table on p.192 of this book.

In order to understand the details of these approaches, we will detour to review the framework for (and then the failure of) linear time series analysis.

3. LINEAR TIME SERIES MODELS

Linear time series models have two particularly desirable features: they can be understood in great detail and they are straightforward to implement. The penalty for this convenience is that they may be entirely inappropriate for even moderately complicated systems. In this section we will review their basic features and then consider why and how such models fail. The literature on linear time series analysis is vast; a good introduction is the very readable book by Chatfield (1989), many derivations can be found (and understood) in the comprehensive text by Priestley (1981), and a classic reference is Box and Jenkins' book (1976). Historically, the general theory of linear predictors can be traced back to Kolmogorov (1941) and to Wiener (1949).

Two crucial assumptions will be made in this section: the system is assumed to be linear and stationary. In the rest of this chapter we will say a great deal about relaxing the assumption of linearity; much less is known about models that have coefficients that vary with time. To be precise, unless explicitly stated (such as for Data Set D), we assume that the underlying equations do not change in time, i.e., *time invariance* of the system.

3.1 ARMA, FIR, AND ALL THAT

There are two complementary tasks that need to be discussed: understanding how a given model behaves and finding a particular model that is appropriate for a given time series. We start with the former task. It is simplest to discuss separately the role of external inputs (moving average models) and internal memory (autoregressive models).

3.1.1 PROPERTIES OF A GIVEN LINEAR MODEL.

Moving average (MA) models. Assume we are given an external input series $\{e_t\}$ and want to modify it to produce another series $\{x_t\}$. Assuming linearity of the system and causality (the present value of x is influenced by the present and N past values of the input series e), the relationship between the input and output is

$$x_t = \sum_{n=0}^{N} b_n e_{t-n} = b_0 e_t + b_1 e_{t-1} + \cdots + b_N e_{t-N}. \tag{1}$$

This equation describes a convolution filter: the new series x is generated by an Nth-order filter with coefficients $b_0, \cdots, b_n$ from the series e. Statisticians and econometricians call this an Nth-order *moving average* model, MA(N). The origin of this (sometimes confusing) terminology can be seen if one pictures a simple smoothing filter which averages the last few values of series e. Engineers call this a *finite impulse response* (FIR) filter, because the output is guaranteed to go to zero at N time steps after the input becomes zero.

Properties of the output series x clearly depend on the input series e. The question is whether there are characteristic features independent of a specific input sequence. For a linear system, the response of the filter is independent of the input. A characterization focuses on properties of the system, rather than on properties of the time series. (For example, it does not make sense to attribute linearity to a time series itself, only to a system.)

We will give three equivalent characterizations of an MA model: in the time domain (the impulse response of the filter), in the frequency domain (its spectrum), and in terms of its autocorrelation coefficients. In the first case, we assume that the input is nonzero only at a single time step t_0 and that it vanishes for all other times. The response (in the time domain) to this "impulse" is simply given by the b's in Eq. (1): at each time step the impulse moves up to the next coefficient until, after N steps, the output disappears. The series $b_N, b_{N-1}, \cdots, b_0$ is thus the impulse response of the system. The response to an arbitrary input can be computed by superimposing the responses at appropriate delays, weighted by the respective input values ("convolution"). The transfer function thus completely describes a linear system, i.e., a system where the superposition principle holds: the output is determined by impulse response and input.

Sometimes it is more convenient to describe the filter in the frequency domain. This is useful (and simple) because a convolution in the time domain becomes a product in the frequency domain. If the input to a MA model is an impulse (which has a flat power spectrum), the discrete Fourier transform of the output is given by $\sum_{n=0}^{N} b_n \exp(-i2\pi n f)$ (see, for example, Box & Jenkins, 1976, p.69). The power spectrum is given by the squared magnitude of this:

$$\left| 1 + b_1 e^{-i2\pi 1 f} + b_2 e^{-i2\pi 2 f} + \cdots + b_N e^{-i2\pi N f} \right|^2 . \tag{2}$$

The third way of representing yet again the same information is, in terms of the autocorrelation coefficients, defined in terms of the mean $\mu = \langle x_t \rangle$ and the variance $\sigma^2 = \langle (x_t - \mu)^2 \rangle$ by

$$\rho_\tau \equiv \frac{1}{\sigma^2} \langle (x_t - \mu)(x_{t-\tau} - \mu) \rangle . \tag{3}$$

The angular brackets $\langle \cdot \rangle$ denote expectation values, in the statistics literature often indicated by $E\{\cdot\}$. The autocorrelation coefficients describe how much, on average, two values of a series that are τ time steps apart co-vary with each other. (We will later replace this linear measure with mutual information, suited also to describe nonlinear relations.) If the input to the system is a stochastic process with input values at different times uncorrelated, $\langle e_i e_j \rangle = 0$ for $i \neq j$, then all of the cross terms will disappear from the expectation value in Eq. (3), and the resulting autocorrelation coefficients are

$$
\rho_\tau =
\begin{cases}
\dfrac{1}{\sum_{n=0}^{N} b_n^2} \displaystyle\sum_{n=\tau}^{N} b_n b_{n-|\tau|} & |\tau| \leq N \,, \\
0 & |\tau| > N \,.
\end{cases}
\tag{4}
$$

Autoregressive (AR) models. MA (or FIR) filters operate in an open loop without feedback; they can only transform an input that is applied to them. If we do not want to drive the series externally, we need to provide some feedback (or memory) in order to generate internal dynamics:

$$
x_t = \sum_{m=1}^{M} a_m x_{t-m} + e_t \,.
\tag{5}
$$

This is called an Mth-order *autoregressive model* (AR(M)) or an *infinite impulse response* (IIR) filter (because the output can continue after the input ceases). Depending on the application, e_t can represent either a controlled input to the system or noise. As before, if e is white noise, the autocorrelations of the output series x can be expressed in terms of the model coefficients. Here, however—due to the feedback coupling of previous steps—we obtain a set of linear equations rather than just a single equation for each autocorrelation coefficient. By multiplying Eq. (5) by $x_{t-\tau}$, taking expectation values, and normalizing (see Box & Jenkins, 1976, p.54), the autocorrelation coefficients of an AR model are found by solving this set of linear equations, traditionally called the *Yule-Walker equations*,

$$
\rho_\tau = \sum_{m=1}^{M} a_m \rho_{\tau-m}, \qquad \tau > 0 \,.
\tag{6}
$$

Unlike the MA case, the autocorrelation coefficient need not vanish after M steps. Taking the Fourier transform of both sides of Eq. (5) and rearranging terms shows that the output equals the input times $(1 - \sum_{m=1}^{M} a_m \exp(-i2\pi mf))^{-1}$. The power spectrum of output is thus that of the input times

$$
\frac{1}{|1 - a_1 e^{-i2\pi 1 f} - a_2 e^{-i2\pi 2 f} - \cdots - a_M e^{-i2\pi M f}|^2} \,.
\tag{7}
$$

To generate a specific realization of the series, we must specify the initial conditions, usually by the first M values of series x. Beyond that, the input term e_t is crucial for the life of an AR model. If there was no input, we might be disappointed by the series we get: depending on the amount of feedback, after iterating it for a while, the output produced can only decay to zero, diverge, or oscillate periodically.[7]

Clearly, the next step in complexity is to allow both AR and MA parts in the model; this is called an ARMA(M, N) model:

$$x_t = \sum_{m=1}^{M} a_m x_{t-m} + \sum_{n=0}^{N} b_n e_{t-n} \,. \tag{8}$$

Its output is most easily understood in terms of the *z-transform* (Oppenheim & Schafer, 1989), which generalizes the discrete Fourier transform to the complex plane:

$$X(z) \equiv \sum_{t=-\infty}^{\infty} x_t z^t \,. \tag{9}$$

On the unit circle, $z = \exp(-i2\pi f)$, the z-transform reduces to the discrete Fourier transform. Off the unit circle, the z-transform measures the rate of divergence or convergence of a series. Since the convolution of two series in the time domain corresponds to the multiplication of their z-transforms, the z-transform of the output of an ARMA model is

$$\begin{aligned} X(z) &= A(z)X(z) + B(z)E(z) \\ &= \frac{B(z)}{1 - A(z)}\, E(z) \end{aligned} \tag{10}$$

(ignoring a term that depends on the initial conditions). The input z-transform $E(z)$ is multiplied by a transfer function that is unrelated to it; the transfer function will vanish at zeros of the MA term $(B(z) = 0)$ and diverge at poles $(A(z) = 1)$ due to the AR term (unless cancelled by a zero in the numerator). As $A(z)$ is an Mth-order complex polynomial, and $B(z)$ is Nth-order, there will be M poles and N zeros. Therefore, the z-transform of a time series produced by Eq. (8) can be decomposed into a rational function and a remaining (possibly continuous) part due to the input. The number of poles and zeros determines the number of *degrees of freedom* of the system (the number of previous states that the dynamics retains). Note that since only the ratio enters, there is no unique ARMA model. In the extreme cases, a finite-order AR model can always be expressed by an infinite-order MA model, and vice versa.

ARMA models have dominated all areas of time series analysis and discrete-time signal processing for more than half a century. For example, in speech recognition and synthesis, Linear Predictive Coding (Press et al., 1992, p.571) compresses

---

[7]In the case of a first-order AR model, this can easily be seen: if the absolute value of the coefficient is less than unity, the value of x exponentially decays to zero; if it is larger than unity, it exponentially explodes. For higher-order AR models, the long-term behavior is determined by the locations of the zeroes of the polynomial with coefficients a_i.

speech by transmitting the slowly varying coefficients for a linear model (and possibly the remaining error between the linear forecast and the desired signal) rather than the original signal. If the model is good, it transforms the signal into a small number of coefficients plus residual white noise (of one kind or another).

3.1.2 FITTING A LINEAR MODEL TO A GIVEN TIME SERIES

Fitting the coefficients. The Yule-Walker set of linear equations (Eq. (6)) allowed us to express the autocorrelation coefficients of a time series in terms of the AR coefficients that generated it. But there is a second reading of the same equations: they also allow us to estimate the coefficients of an AR(M) model from the observed correlational structure of an observed signal.[8] An alternative approach views the estimation of the coefficients as a regression problem: expressing the next value as a function of M previous values, i.e., linearly regress x_{t+1} onto $\{x_t, x_{t-1}, \ldots, x_{t-(M-1)}\}$. This can be done by minimizing squared errors: the parameters are determined such that the squared difference between the model output and the observed value, summed over all time steps in the fitting region, is as small as possible. There is no comparable conceptually simple expression for finding MA and full ARMA coefficients from observed data. For all cases, however, standard techniques exist, often expressed as efficient recursive procedures (Box & Jenkins, 1976; Press et al., 1992).

Although there is no reason to expect that an arbitrary signal was produced by a system that can be written in the form of Eq. (8), it is reasonable to attempt to approximate a linear system's true transfer function (z-transform) by a ratio of polynomials, i.e., an ARMA model. This is a problem in function approximation, and it is well known that a suitable sequence of ratios of polynomials (called Padé approximants; see Press et al., 1992, p.200) converges faster than a power series for arbitrary functions.

Selecting the (order of the) model. So far we have dealt with the question of how to estimate the coefficients from data for an ARMA model of order (M, N), but have not addressed the choice for the order of the model. There is not a unique best choice for the values or even for the number of coefficients to model a data set—as the order of the model is increased, the fitting error decreases, but the test error of the forecasts beyond the training set will usually start to increase at some point because the model will be fitting extraneous noise in the system. There are several heuristics to find the "right" order (such as the Akaike Information Criterion (AIC), Akaike, 1970; Sakomoto et al., 1986)—but these heuristics rely heavily on the linearity of the model and on assumptions about the distribution from which the errors are drawn. When it is not clear whether these assumptions hold, a simple approach (but wasteful in terms of the data) is to hold back some of

---

[8] In statistics, it is common to emphasize the difference between a given model and an estimated model by using different symbols, such as $\hat{a}$ for the estimated coefficients of an AR model. In this paper, we avoid introducing another set of symbols; we hope that it is clear from the context whether values are theoretical or estimated.

the training data and use these to evaluate the performance of competing models. Model selection is a general problem that will reappear even more forcefully in the context of nonlinear models, because they are more flexible and, hence, more capable of modeling irrelevant noise.

3.2 THE BREAKDOWN OF LINEAR MODELS

We have seen that ARMA coefficients, power spectra, and autocorrelation coefficients contain the same information about a linear system that is driven by uncorrelated white noise. Thus, *if and only if* the power spectrum is a useful characterization of the relevant features of a time series, an ARMA model will be a good choice for describing it. This appealing simplicity can fail entirely for even simple nonlinearities if they lead to complicated power spectra (as they can). Two time series can have very similar broadband spectra but can be generated from systems with very different properties, such as a linear system that is driven stochastically by external noise, and a deterministic (noise-free) nonlinear system with a small number of degrees of freedom. One the key problems addressed in this chapter is how these cases can be distinguished—linear operators definitely will not be able to do the job.

Let us consider two nonlinear examples of discrete-time maps (like an AR model, but now nonlinear):

■ The first example can be traced back to Ulam (1957): the next value of a series is derived from the present one by a simple parabola

$$x_{t+1} = \lambda\, x_t\, (1 - x_t)\,. \tag{11}$$

Popularized in the context of population dynamics as an example of a "simple mathematical model with very complicated dynamics" (May, 1976), it has been found to describe a number of controlled laboratory systems such as hydrodynamic flows and chemical reactions, because of the universality of smooth unimodal maps (Collet, 1980). In this context, this parabola is called the *logistic map* or *quadratic map*. The value x_t deterministically depends on the previous value x_{t-1}; λ is a parameter that controls the qualitative behavior, ranging from a fixed point (for small values of λ) to deterministic chaos. For example, for $\lambda = 4$, each iteration destroys one bit of information. Consider that, by plotting x_t against x_{t-1}, each value of x_t has two equally likely predecessors or, equally well, the average slope (its absolute value) is two: if we know the location within ϵ before the iteration, we will on average know it within 2ϵ afterwards. This exponential increase in uncertainty is the hallmark of deterministic chaos ("divergence of nearby trajectories").

■ The second example is equally simple: consider the time series generated by the map

$$x_t = 2x_{t-1} \pmod 1\,. \tag{12}$$

The action of this map is easily understood by considering the position x_t written in a binary fractional expansion (i.e., $x_t = 0.d_1 d_2 \ldots = (d_1 \times 2^{-1}) + (d_2 \times 2^{-2}) + \ldots$): each iteration shifts every digit one place to the left ($d_i \leftarrow d_{i+1}$). This means that the most significant digit d_1 is discarded and one more digit of the binary expansion of the initial condition is revealed. This map can be implemented in a simple physical system consisting of a classical billiard ball and reflecting surfaces, where the x_t are the successive positions at which the ball crosses a given line (Moore, 1991).

Both systems are completely deterministic (their evolutions are entirely determined by the initial condition x_0), yet they can easily generate time series with broadband power spectra. In the context of an ARMA model a broadband component in a power spectrum of the output must come from external noise input to the system, but here it arises in two one-dimensional systems as simple as a parabola and two straight lines. Nonlinearities are essential for the production of "interesting" behavior in a deterministic system, the point here is that even simple nonlinearities suffice.

Historically, an important step beyond linear models for prediction was taken in 1980 by Tong and Lim (see also Tong, 1990). After more than five decades of approximating a system with *one* globally linear function, they suggested the use of *two* functions. This *threshold autoregressive model* (TAR) is globally nonlinear: it consists of choosing one of two local linear autoregressive models based on the value of the system's state. From here, the next step is to use many local linear models; however, the number of such regions that must be chosen may be very large if the system has even quadratic nonlinearities (such as the logistic map). A natural extension of Eq. (8) for handling this is to include quadratic and higher order powers in the model; this is called a Volterra series (Volterra, 1959).

TAR models, Volterra models, and their extensions significantly expand the scope of possible functional relationships for modeling time series, but these come at the expense of the simplicity with which linear models can be understood and fit to data. For nonlinear models to be useful, there must be a process that exploits features of the data to guide (and restrict) the construction of the model; lack of insight into this problem has limited the use of nonlinear time series models. In the next sections we will look at two complementary solutions to this problem: building explicit models with state-space reconstruction, and developing implicit models in a connectionist framework. To understand why both of these approaches exist and why they are useful, let us consider the nature of scientific modeling.

4. UNDERSTANDING AND LEARNING

Strong models have strong assumptions. They are usually expressed in a few equations with a few parameters, and can often explain a plethora of phenomena. In weak models, on the other hand, there are only a few domain-specific assumptions. To compensate for the lack of explicit knowledge, weak models usually contain many more parameters (which can make a clear interpretation difficult). It can be helpful to conceptualize models in the two-dimensional space spanned by the axes data-poor↔data-rich and theory-poor↔theory-rich. Due to the dramatic expansion of the capability for automatic data acquisition and processing, it is increasingly feasible to venture into the theory-poor and data-rich domain.

Strong models are clearly preferable, but they often originate in weak models. (However, if the behavior of an observed system does not arise from simple rules, they may not be appropriate.) Consider planetary motion (Gingerich, 1992). Tycho Brahe's (1546–1601) experimental observations of planetary motion were accurately described by Johannes Kepler's (1571–1630) phenomenological laws; this success helped lead to Isaac Newton's (1642–1727) simpler but much more general theory of gravity which could derive these laws; Henri Poincaré's (1854–1912) inability to solve the resulting three-body gravitational problem helped lead to the modern theory of dynamical systems and, ultimately, to the identification of chaotic planetary motion (Sussman & Wisdom, 1988, 1992).

As in the previous section on linear systems, there are two complementary tasks: discovering the properties of a time series generated from a given model, and inferring a model from observed data. We focus here on the latter, but there has been comparable progress for the former. Exploring the behavior of a model has become feasible in interactive computer environments, such as Cornell's dstool,[9] and the combination of traditional numerical algorithms with algebraic, geometric, symbolic, and artificial intelligence techniques is leading to automated platforms for exploring dynamics (Abelson, 1990; Yip, 1991; Bradley, 1992). For a nonlinear system, it is no longer possible to decompose an output into an input signal and an independent transfer function (and thereby find the correct input signal to produce a desired output), but there are adaptive techniques for controlling nonlinear systems (Hübler, 1989; Ott, Grebogi & Yorke, 1990) that make use of techniques similar to the modeling methods that we will describe.

The idea of weak modeling (data-rich and theory-poor) is by no means new— an ARMA model is a good example. What is new is the emergence of weak models (such as neural networks) that combine broad generality with insight into how to manage their complexity. For such models with broad approximation abilities and few specific assumptions, the distinction between memorization and generalization becomes important. Whereas the signal-processing community sometimes uses the

---

[9] Available by anonymous ftp from macomb.tn.cornell.edu in pub/dstool.

term *learning* for any adaptation of parameters, we need to contrast learning without generalization from learning with generalization. Let us consider the widely and wildly celebrated fact that neural networks can learn to implement the exclusive OR (XOR). But—what kind of learning is this? When four out of four cases are specified, no generalization exists! Learning a truth table is nothing but rote memorization: learning XOR is as interesting as memorizing the phone book. More interesting—and more realistic—are real-world problems, such as the prediction of financial data. In forecasting, nobody cares how well a model fits the training data—only the quality of future predictions counts, i.e., the performance on novel data or the *generalization* ability. Learning means extracting regularities from training examples that do transfer to new examples.

Learning procedures are, in essence, statistical devices for performing inductive inference. There is a tension between two goals. The immediate goal is to fit the training examples, suggesting devices as general as possible so that they can learn a broad range of problems. In connectionism, this suggests large and flexible networks, since networks that are too small might not have the complexity needed to model the data. The ultimate goal of an inductive device is, however, its performance on cases it has not yet seen, i.e., the quality of its predictions outside the training set. This suggests—at least for noisy training data—networks that are not too large since networks with too many high-precision weights will pick out idiosyncrasies of the training set and will not generalize well.

An instructive example is polynomial curve fitting in the presence of noise. On the one hand, a polynomial of too low an order cannot capture the structure present in the data. On the other hand, a polynomial of too high an order, going through all of the training points and merely interpolating between them, captures the noise as well as the signal and is likely to be a very poor predictor for new cases. This problem of fitting the noise in addition to the signal is called *overfitting*. By employing a regularizer (i.e., a term that penalizes the complexity of the model) it is often possible to fit the parameters and to select the relevant variables at the same time. Neural networks, for example, can be cast in such a Bayesian framework (Buntine & Weigend, 1991).

To clearly separate memorization from generalization, the true continuation of the competition data was kept secret until the deadline, ensuring that the continuation data could not be used by the participants for tasks such as parameter estimation or model selection.[10] Successful forecasts of the withheld *test set* (also called *out-of-sample predictions*) from the provided *training set* (also called *fitting set*) were produced by two general classes of techniques: those based on state-space reconstruction (which make use of explicit understanding of the relationship between the internal degrees of freedom of a deterministic system and an observable of the system's state in order to build a model of the rules governing the measured behavior of the system), and connectionist modeling (which uses potentially rich

---

[10]After all, predictions are hard, particularly those concerning the future.

models along with learning algorithms to develop an implicit model of the system). We will see that neither is uniquely preferable. The domains of applicability are not the same, and the choice of which to use depends on the goals of the analysis (such as an understandable description vs. accurate short-term forecasts).

4.1 UNDERSTANDING: STATE-SPACE RECONSTRUCTION

Yule's original idea for forecasting was that future predictions can be improved by using immediately preceding values. An ARMA model, Eq.(8), can be rewritten as a dot product between vectors of the time-lagged variables and coefficients:

$$x_t = \mathbf{a} \cdot \mathbf{x}_{t-1} + \mathbf{b} \cdot \mathbf{e}_t, \tag{13}$$

where $\mathbf{x}_t = (x_t, x_{t-1}, \ldots, x_{t-(d-1)})$, and $\mathbf{a} = (a_1, a_2, \ldots, a_d)$. (We slightly change notation here: what was M (the order of the AR model) is now called d (for dimension).) Such lag vectors, also called tapped delay lines, are used routinely in the context of signal processing and time series analysis, suggesting that they are more than just a typographical convenience.[11]

In fact, there is a deep connection between time-lagged vectors and underlying dynamics. This connection was was proposed in 1980 by Ruelle (personal communication), Packard et al. (1980), and Takens (1981; he published the first proof), and later strengthened by Sauer et al. (1991). Delay vectors of sufficient length are not just a representation of the state of a linear system—it turns out that delay vectors can recover the full geometrical structure of a nonlinear system. These results address the general problem of inferring the behavior of the intrinsic degrees of freedom of a system when a function of the state of the system is measured. If the governing equations and the functional form of the observable are known in advance, then a Kalman filter is the optimal linear estimator of the state of the system (Catlin, 1989; Chatfield, 1989). We, however, focus on the case where there is little or no *a priori* information available about the origin of the time series.

There are four relevant (and easily confused) spaces and dimensions for this discussion:[12]

1. The *configuration space* of a system is the space "where the equations live." It specifies the values of all of the potentially accessible physical degrees of freedom of the system. For example, for a fluid governed by the Navier-Stokes

---

[11]For example, the spectral test for random number generators is based on looking for structure in the space of lagged vectors of the output of the source; these will lie on hyperplanes for a linear congruential generator $x_{t+1} = ax_t + b \pmod{c}$ (Knuth, 1981, p.90).

[12]The first point (configuration space and potentially accessible degrees of freedom) will not be used again in this chapter. On the other hand, the dimension of the solution manifold (the actual degrees of freedom) will be important both for characterization and for prediction.

partial differential equations, these are the infinite-dimensional degrees of freedom associated with the continuous velocity, pressure, and temperature fields.
2. The *solution manifold* is where "the solution lives," i.e., the part of the configuration space that the system actually explores as its dynamics unfolds (such
   as the support of an attractor or an integral surface). Due to unexcited or correlated degrees of freedom, this can be much smaller than the configuration
   space; the dimension of the solution manifold is the number of parameters that
   are needed to uniquely specify a distinguishable state of the overall system. For
   example, in some regimes the infinite physical degrees of freedom of a convect-
   ing fluid reduce to a small set of coupled ordinary differential equations for a
   mode expansion (Lorenz, 1963). Dimensionality reduction from the configuration space to the solution manifold is a common feature of dissipative systems:
   dissipation in a system will reduce its dynamics onto a lower dimensional subspace (Temam, 1988).
3. The *observable* is a (usually) one-dimensional function of the variables of configuration, an example is Eq. (51) in the Appendix. In an experiment, this might
   be the temperature or a velocity component at a point in the fluid.
4. The *reconstructed state space* is obtained from that (scalar) observable by combining past values of it to form a lag vector (which for the convection case
   would aim to recover the evolution of the components of the mode expansion).

Given a time series measured from such a system—and no other information
about the origin of the time series—the question is: What can be deduced about
the underlying dynamics?

Let y be the state vector on the solution manifold (in the convection example
the components of y are the magnitude of each of the relevant modes), let $dy/dt =$
$f(y)$ be the governing equations, and let the measured quantity be $x_t = x(y(t))$
(e.g., the temperature at a point). The results to be cited here also apply to systems
that are described by iterated maps. Given a delay time τ and a dimension d, a lag
vector x can be defined,

$$\text{lag vector}: \quad x_t = (x_t, x_{t-\tau}, \ldots, x_{t-(d-1)\tau}). \tag{14}$$

The central result is that the behavior of x and y will differ only by a smooth local
invertible change of coordinates (i.e., the mapping between x and y is an embedding, which requires that it be diffeomorphic) for almost every possible choice of
$f(y), x(y)$, and τ, as long as d is large enough (in a way that we will make precise),
x depends on at least some of the components of y, and the remaining components of y are coupled by the governing equations to the ones that influence x.
The proof of this result has two parts: a local piece, showing that the linearization
of the embedding map is almost always nondegenerate, and a global part, showing

1000 points

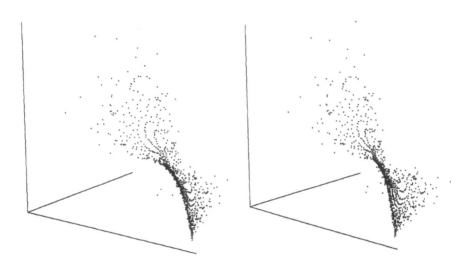

25000 points

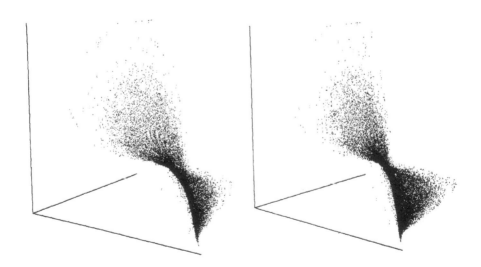

FIGURE 5   Stereo pairs for the three-dimensional embedding of Data Set A. The shape of the surface is apparent with just the 1,000 points that were given.

the diagonal of the embedding space and, as τ is increased, it sets a length scale for the reconstructed dynamics. There can be degenerate choices for τ for which the embedding fails (such as choosing it to be exactly equal to the period of a periodic system), but these degeneracies almost always will be removed by an arbitrary perturbation of τ. The intrinsic noise in physical systems guarantees that these results hold in all known nontrivial examples, although in practice, if the coupling between degrees of freedom is sufficiently weak, then the available experimental resolution will not be large enough to detect them (see Casdagli et al., 1991, for further discussion of how noise constrains embedding).[13]

Data Set A appears complicated when plotted as a time series (Figure 1). The simple structure of the system becomes visible in a figure of its three-dimensional embedding (Figure 5). In contrast, high-dimensional dynamics would show up as a structureless cloud in such a stereo plot. Simply plotting the data in a stereo plot allows to guess a value of the dimension of the manifold of around two, not far from the submitted computed values of 2.0–2.2. In Section 6, we will discuss in detail the practical issues associated with choosing and understanding the embedding parameters.

Time-delay embedding differs from traditional experimental measurements in three fundamental respects:

1. It provides detailed information about the behavior of degrees of freedom other than the one that is directly measured.
2. It rests on probabilistic assumptions and—although it has been routinely and reliably used in practice—it is not guaranteed to be valid for any system.
3. It allows precise questions only about quantities that are invariant under such a transformation, since the reconstructed dynamics have been modified by an unknown smooth change of coordinates.

This last restriction may be unfamiliar, but it is surprisingly unimportant: we will show how embedded data can be used for forecasting a time series and for characterizing the essential features of the dynamics that produced it. We close this section by presenting two extensions of the simple embedding considered so far.

Filtered embedding generalizes simple time-delay embedding by presenting a linearly transformed version of the lag vector to the next processing stage. The lag vector $\mathbf{x}$ is trivially equal to itself times an identity matrix. Rather than using the identity matrix, the lag vector can be multiplied by any (not necessarily square) matrix. The resulting vector is an embedding if the rank of the matrix is equal to or larger than the desired embedding dimension. (The window of lags can be larger than the final embedding dimension, which allows the embedding procedure

---

[13]The Whitney embedding theorem from the 1930s (see Guillemin & Pollack, 1974, p. 48) guarantees that the number of independent observations d required to embed an arbitrary manifold (in the absence of noise) into a Euclidean embedding space will be no more than twice the dimension of the manifold. For example, a two-dimensiona Möbius strip can be embedded in a three-dimensional Euclidean space, but a two-dimensional Klein bottle requires a four-dimensional space.

to include additional signal processing.) A specific example, used by Sauer (this volume), is embedding with a matrix produced by multiplying a discrete Fourier transform, a low-pass filter, and an inverse Fourier transform; as long as the filter cut-off is chosen high enough to keep the rank of the overall transformation greater than or equal to the required embedding dimension, this will remove noise but will preserve the embedding. There are a number of more sophisticated linear filters that can be used for embedding (Oppenheim & Schafer, 1989), and we will also see that connectionist networks can be interpreted as sets of nonlinear filters.

A final modification of time-delay embedding that can be useful in practice is **embedding by expectation values**. Often the goal of an analysis is to recover not the detailed trajectory of $\mathbf{x}(t)$, but rather to estimate the probability distribution $p(\mathbf{x})$ for finding the system in the neighborhood of a point $\mathbf{x}$. This probability is defined over a measurement of duration T in terms of an arbitrary test function $g(\mathbf{x})$ by

$$\frac{1}{T}\int_0^T g(\mathbf{x}(t))dt = \langle g(\mathbf{x}(t))\rangle_t$$
$$= \int g(\mathbf{x})p(\mathbf{x})\ d\mathbf{x}\,. \tag{15}$$

Note that this is an empirical definition of the probability distribution for the observed trajectory; it is not equivalent to assuming the existence of an invariant measure or of ergodicity so that the distribution is valid for all possible trajectories (Petersen, 1989). If a complex exponential is chosen for the test function

$$\langle e^{i\mathbf{k}\cdot\mathbf{x}(t)}\rangle = \langle e^{i\mathbf{k}\cdot(x_t,x_{t-\tau},\dots,x_{t-(d-1)\tau})}\rangle$$
$$= \int e^{i\mathbf{k}\cdot\mathbf{x}}p(\mathbf{x})\ d\mathbf{x}\,, \tag{16}$$

we see that the time average of this is equal to the Fourier transform of the desired probability distribution (this is just a characteristic function of the lag vector). This means that, if it is not possible to measure a time series directly (such as for very fast dynamics), it can still be possible to do time-delay embedding by measuring a set of time-average expectation values and then taking the inverse Fourier transform to find $p(\mathbf{x})$ (Gershenfeld, 1993a). We will return to this point in Section 6.2 and show how embedding by expectation values can also provide a useful framework for distinguishing measurement noise from underlying dynamics.

We have seen that time-delay embedding, while appearing similar to traditional state-space models with lagged vectors, makes a crucial link between behavior in the reconstructed state space and the internal degrees of freedom. We will apply this insight to forecasting and characterizing deterministic systems later (in Sections 5.1 and 6.2). Now, we address the problem of what can be done if we are unable to understand the system in such explicit terms. The main idea will be to learn to emulate the behavior of the system.

4.2 LEARNING: NEURAL NETWORKS

In the competition, the majority of contributions, and also the best predictions for each set used connectionist methods. They provide a convenient language game for nonlinear modeling. Connectionist networks are also known as neural networks, parallel distributed processing, or even as "brain-style computation"; we use these terms interchangeably. Their practical application (such as by large financial institutions for forecasting) has been marked by (and marketed with) great hope and hype (Schwarz, 1992; Hammerstrom, 1993).

Neural networks are typically used in pattern recognition, where a collection of features (such as an image) is presented to the network, and the task is to assign the input feature to one or more classes. Another typical use for neural networks is (nonlinear) regression, where the task is to find a smooth interpolation between points. In both these cases, all the relevant information is presented simultaneously. In contrast, time series prediction involves processing of patterns that evolve over time—the appropriate response at a particular point in time depends not only on the current value of the observable but also on the past. Time series prediction has had an appeal for neural networkers from the very beginning of the field. In 1964, Hu applied Widrow's adaptive linear network to weather forecasting. In the post-backpropagation era, Lapedes and Farber (1987) trained their (nonlinear) network to emulate the relationship between output (the next point in the series) and inputs (its predecessors) for computer-generated time series, and Weigend, Huberman and Rumelhart (1990, 1992) addressed the issue of finding networks of appropriate complexity for predicting observed (real-world) time series. In all these cases, temporal information is presented spatially to the network by a time-lagged vector (also called tapped delay line).

A number of ingredients are needed to specify a neural network:

- its interconnection architecture,
- its activation functions (that relate the output value of a node to its inputs),
- the cost function that evaluates the network's output (such as squared error),
- a training algorithm that changes the interconnection parameters (called weights) in order to minimize the cost function.

The simplest case is a network **without hidden units**: it consists of one output unit that computes a weighted linear superposition of d inputs, $\text{out}^{(t)} = \sum_{i=1}^{d} w_i x_i^{(t)}$. The superscript (t) denotes a specific "pattern"; $x_i^{(t)}$ is the value of the ith input of that pattern.[14] w_i is the weight between input i and the output. The network output can also be interpreted as a dot-product $\mathbf{w} \cdot \mathbf{x}^{(t)}$ between the weight vector $\mathbf{w} = (w_1, \cdots, w_d)$ and an input pattern $\mathbf{x}^{(t)} = (x_1^{(t)}, \ldots, x_d^{(t)})$.

---

[14]In the context of time series prediction, $x_i^{(t)}$ can be the ith component of the delay vector, $x_i^{(t)} = x_{t-i}$.

Given such an input-output relationship, the central task in learning is to find a way to change the weights such that the actual output $\text{out}^{(t)}$ gets closer to the desired output or $\text{target}^{(t)}$. The closeness is expressed by a cost function, for example, the squared error $E^{(t)} = (\text{out}^{(t)} - \text{target}^{(t)})^2$. A learning algorithm iteratively updates the weights by taking a small step (parametrized by the learning rate η) in the direction that decreases the error the most, i.e., following the negative of the local gradient.[15] The "new" weight $\widetilde{w}_i$, after the update, is expressed in terms of the "old" weight w_i as

$$\widetilde{w}_i = w_i - \eta \frac{\partial E^{(t)}}{\partial w_i} = w_i + 2\eta \underbrace{x_i}_{\text{activation}} \underbrace{(\text{out}^{(t)} - \text{target}^{(t)})}_{\text{error}}. \qquad (17)$$

The weight-change $(\widetilde{w}_i - w_i)$ is proportional to the product of the activation going into the weight and the size of error, here the deviation $(\text{out}^{(t)} - \text{target}^{(t)})$. This rule for adapting weights (for linear output units with squared errors) goes back to Widrow and Hoff (1960).

If the input values are the lagged values of a time series and the output is the prediction for the next value, this simple network is equivalent to determining an $AR(d)$ model through least squares regression: the weights at the end of training equal the coefficients of the AR model.

Linear networks are very limited—exactly as limited as linear AR models. The key idea responsible for the power, potential, and popularity of connectionism is the insertion of one of more layers of **nonlinear hidden units** (between the inputs and output). These nonlinearities allow for interactions between the inputs (such as products between input variables) and thereby allow the network to fit more complicated functions. (This is discussed further in the subsection on neural networks and statistics below.)

The simplest such nonlinear network contains only one hidden layer and is defined by the following components:

- There are d *inputs*.
- The inputs are fully connected to a layer of nonlinear *hidden units*.
- The hidden units are connected to the one linear *output unit*.
- The output and hidden units have adjustable offsets or *biases b*.
- The *weights w* can be positive, negative, or zero.

The response of a unit is called its activation value or, in short, *activation*. A common choice for the nonlinear *activation function* of the hidden units is a

---

[15]Eq. (17) assumes updates after each pattern. This is a stochastic approximation (also called "on-line updates" or "pattern mode") to first averaging the errors over the entire training set, $E = 1/N \sum_t E^{(t)}$ and then updating (also called "batch updates" or "epoch mode"). If there is repetition in the training set, learning with pattern updates is faster.

composition of two operators: an affine mapping followed by a sigmoidal function. First, the inputs into a hidden unit h are linearly combined and a bias b_h is added:

$$\xi_h^{(t)} = \sum_{i=1}^{d} w_{hi} x_i^{(t)} + b_h \, . \tag{18}$$

Then, the output of the unit is determined by passing $\xi_h^{(t)}$ through a sigmoidal function ("squashing function") such as

$$S(\xi_h^{(t)}) = \frac{1}{1 + e^{-a\xi_h^{(t)}}} = \frac{1}{2}\left(1 + \tanh \frac{a}{2}\xi_h^{(t)}\right) , \tag{19}$$

where the slope a determines the steepness of the response.

In the introductory example of a linear network, we have seen how to change the weights when the activations are known at both ends of the weight. How do we update the weights to the hidden units that do not have a target value? The revolutionary (but in hindsight obvious) idea that solved this problem is the chain rule of differentiation. This idea of **error backpropagation** can be traced back to Werbos (1974), but only found widespread use after it was independently invented by Rumelhart et al. (1986a, 1986b) at a time when computers had become sufficiently powerful to permit easy exploration and successful application of the backpropagation rule.

As in the linear case, weights are adjusted by taking small steps in the direction of the negative gradient, $-\partial E/\partial w$. The weight-change rule is still of the same form (activation into the weight) $\times$ (error signal from above). The activation that goes into the weight remains unmodified (it is the same as in the linear case). The difference lies in the error signal. For weights between hidden unit h and the output, the error signal for a given pattern is now $(\text{out}^{(t)} - \text{target}^{(t)}) \times S'(\xi_h^{(t)})$; i.e., the previous difference between prediction and target is now multiplied with the derivative of the hidden unit activation function taken at $\xi_h^{(t)}$. For weights that do not connect directly to the output, the error signal is computed recursively in terms of the error signals of the units to which it directly connects, and the weights of those connections. The weight change is computed locally from incoming activation, the derivative, and error terms from above multiplied with the corresponding weights.[16] Clear derivations of backpropagation can be found in Rumelhart, Hinton, and Williams (1986a) and in the textbooks by Hertz, Krogh, and Palmer (1991, p.117) and Kung (1993, p.154). The theoretical foundations of backpropagation are laid out clearly by Rumelhart et al. (1993).

---

[16] All update rules are local. The flip side of the locality of the update rules discussed above is that learning becomes an iterative process. Statisticians usually do not focus on the emergence of iterative local solutions.

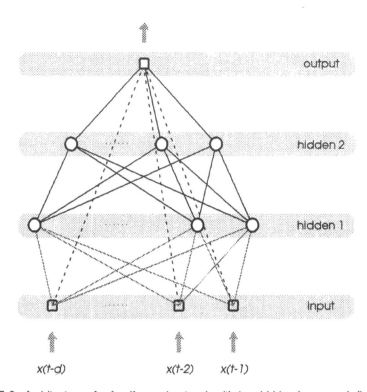

FIGURE 6   Architecture of a feedforward network with two hidden layers and direct connections from the input to the output. The lines correspond to weight values. The dashed lines represent direct connections from the inputs to the (linear) output unit. Biases are not shown.

Figure 6 shows a typical network; activations flow from the bottom up. In addition to a second layer of (nonlinear) hidden units, we also include direct (linear) connections between each input and the output. Although not used by the competition entrants, this architecture can extract the linearly predictable part early in the learning process and free up the nonlinear resources to be employed where they are really needed. It can be advantageous to choose different learning rates for different parts of the architecture, and thus not follow the gradient exactly (Weigend, 1991). In Section 6.3.2 we will describe yet another modification, i.e., sandwiching a bottleneck hidden layer between two additional (larger) layers.

NEURAL NETWORKS AND STATISTICS. Given that feedforward networks with hidden units implement a nonlinear regression of the output onto the inputs, what features do they have that might give them an advantage over more traditional methods? Consider polynomial regression. Here the components of the input vector $(x_1, x_2, \ldots, x_d)$ can be combined in pairs $(x_1 x_2, x_1 x_3, \ldots)$, in triples $(x_1 x_2 x_3, x_1 x_2 x_4, \ldots)$, etc., as well as in combinations of higher powers. This vast number of possible terms can approximate any desired output surface. One might be tempted to conjecture that feedforward networks are able to represent a larger function space with fewer parameters. This, however, is not true: Cover (1965) and Mitchison and Durbin (1989) showed that the "capacity" of both polynomial expansions and networks is proportional to the number of parameters. The real difference between the two representations is in the kinds of constraints they impose. For the polynomial case, the number of possible terms grows rapidly with the input dimension, making it sometimes impossible to use even all of the second-order terms. Thus, the necessary selection of which terms to include implies a decision to permit only specific pairwise or perhaps three-way interactions between components of the input vector. A layered network, rather than limiting the *order* of the interactions, limits only the total *number* of interactions and learns to select an appropriate combination of inputs. Finding a simple representation for a complex signal might require looking for such simultaneous relationships among many input variables. A small network is already potentially fully nonlinear. Units are added to increase the number of features that can be represented (rather than to increase the model order in the example of polynomial regression).

NEURAL NETWORKS AND MACHINE LEARNING. Theoretical work in connectionism ranges from reassuring proofs that neural networks with sigmoid hidden units can essentially fit any well-behaved function and its derivative (Irie & Miyake, 1988; Cybenko, 1989; Funahashi, 1989; White, 1990; Barron, 1993) to results on the ability to generalize (Haussler, personal communication, 1993).[17] Neural networks have found their place in (and helped develop) the broader field of machine learning which studies algorithms for learning from examples. For time series prediction, this has included genetic algorithms (Packard, 1990; Meyer & Packard, 1992; see also Koza, 1993, for a recent monograph on genetic programming), Boltzmann machines (Hinton & Sejnowski, 1986), and conventional AI techniques (Laird & Saul, 1993). The increasing number of such techniques that arrive with strong claims about their performance is forcing the machine learning community to pay greater attention to methodological issues. In this sense, the comparative evaluations of the Santa Fe Competition can also be viewed as a small stone in the mosaic of machine learning.

[17]This paper by Haussler (on PAC [probably approximately correct] learning) will be published in the collection edited by Smolensky, Mozer, and Rumelhart (1994). That collection also contains theoretical connectionist results by Vapnik (on induction principles), Judd (on complexity of learning), and Rissanen (on information theory and neural networks).

5. FORECASTING

In the previous section we have seen that analysis of the geometry of the embedded data and machine learning techniques provide alternative approaches to discovering the relationship between past and future points in a time series. Such insight can be used to forecast the unknown continuation of a given time series. In this section we will consider the details of how prediction is implemented, and in the following section we will step back to look at the related problem of characterizing the essential properties of a system.

5.1 STATE-SPACE FORECASTING

If an experimentally observed quantity arises from deterministic governing equations, it is possible to use time-delay embedding to recover a representation of the relevant internal degrees of freedom of the system from the observable. Although the precise values of these reconstructed variables are not meaningful (because of the unknown change of coordinates), they can be used to make precise forecasts because the embedding map preserves their geometrical structure. In this section we explain how this is done for a time series that has been generated by a deterministic system; in Section 6.2 we will consider how to determine whether or not this is the case (and, if so, what the embedding parameters should be) and, in Section 5.2, how to forecast systems that are not simply deterministic.

Figure 5 is an example of the structure that an embedding can reveal. Notice that the surface appears to be single-valued; this, in fact, must be the case if the system is deterministic and if the number of time lags used is sufficient for an embedding. Differential equations and maps have unique solutions forward in time; this property is preserved under a diffeomorphic transformation and so the first component of an embedded vector must be a unique function of the preceding values $x_{t-\tau}, ..., x_{t-(d-1)\tau}$ once d is large enough. Therefore, the points must lie on a single-valued hypersurface. Future values of the observable can be read off from this surface if it can be adequately estimated from the given data set (which may contain noise and is limited in length).

Using embedding for forecasting appears—at first sight—to be very similar to Yule's original AR model: a prediction function is sought based on time-lagged vectors. The crucial difference is that understanding embedding reduces forecasting to recognizing and then representing the underlying geometrical structure, and once the number of lags exceeds the minimum embedding dimension, this geometry will not change. A global linear model (AR) must do this with a single hyperplane. Since this may be a very poor approximation, there is no fundamental insight into how to choose the number of delays and related parameters. Instead, heuristic rules such as the AIC[18] are used (and vigorously debated). The ease of producing and exploring pictures, such as Figure 5, with modern computers has helped clarify the importance of this point.

Early efforts to improve global linear AR models included systematically increasing the order of interaction ("bilinear" models[19]; Granger & Anderson, 1978), splitting the input space across one variable and allowing for two AR models (threshold autoregressive models, Tong & Lim, 1980), and using the nonlinearities of a Volterra expansion. A more recent example of the evolutionary improvements are adaptive fits with local splines (Multivariate Adaptive Regression Splines, MARS; Friedman, 1991c; Lewis et al., this volume). The "insight" gained from a model such as MARS describes what parameter values are used for particular regions of state space, but it does not help with deeper questions about the nature of a system (such as how many degrees of freedom there are, or how trajectories evolve). Compare this to forecasting based on state-space embedding, which starts by testing for the presence of identifiable geometrical structure and then proceeds to model the geometry, rather than starting with (often inadequate) assumptions about the geometry. This characterization step (to be discussed in detail in Section 6.2) is crucial: simple state-space forecasting becomes problematic if there is a large amount of noise in the system, or if there are nonstationarities on the time scale of the sampling time. Assuming that sufficiently low-dimensional dynamics has been detected, the next step is to build a model of the geometry of the hypersurface in the embedding space that can interpolate between measured points and can distinguish between measurement noise and intrinsic dynamics. This can be done by both local and global representations (as well as by intermediate hybrids).

Farmer and Sidorowich (1987) introduced *local linear models* for state-space forecasting. The simple idea is to recognize that any manifold is locally linear (i.e., locally a hyperplane). Furthermore, the constraint that the surface must be single-valued allows noise transverse to the surface (the generic case) to be recognized and eliminated. Broomhead and King (1986) use *Singular Value Decomposition* (SVD) for this projection; the distinction between local and global SVD is crucial. (Fraser, 1989a, points out some problems with the use of SVD for nonlinear systems.) In this volume, Smith discusses the relationship between local linear and nonlinear models,

---

[18] For linear regression, it is sometimes possible to "correct" for the usually over-optimistic estimate. An example is to multiply the fitting error with $(N + k)/(N - k)$, where N is the number of data points and k is the number of parameters of the model (Akaike, 1970; Sakomoto et al., 1986). Moody (1992) extended this for nonlinear regression and used a notion of effective number of parameters for a network that has convergenced. Weigend and Rumelhart (1991a, 1991b) focused on the increase of the effective network size (expressed as the effective number of hidden units) as a function on training time.

[19] A bilinear model contains second-order interactions between the inputs, i.e., $x_i x_j$. The term "bilinear" comes from the fact that two inputs enter linearly into such products.

as well as the relationship between local and global approaches. Finally, Casdagli and Weigend (this volume) specifically explore the continuum between local and global models by varying the size of the local neighborhood used in the local linear fit. Before returning to this issue in Section 6.3 where we will show how this variation relates to (and characterizes) a system's properties, we now summarize the method used by Tim Sauer in his successful entry. (Details are given by Sauer, this volume.)

In his competition entry, shown in Figure 3, Sauer used a careful implementation of local-linear fitting that had five steps:

1. Low-pass embed the data to help remove measurement and quantization noise. This low-pass filtering produces a smoothed version of the original series. (We explained such filtered embedding at the end of Section 4.1.)
2. Generate more points in embedding space by (Fourier-) interpolating between the points obtained from Step 1. This is to increase the coverage in embedding space.
3. Find the k nearest neighbors to the point of prediction (the choice of k tries to balance the increasing bias and decreasing variance that come from using a larger neighborhood).
4. Use a local SVD to project (possibly very noisy) points onto the local surface. (Even if a point is very far away from the surface, this step forces the dynamics back on the reconstructed solution manifold.)
5. Regress a linear model for the neighborhood and use it to generate the forecast.

Because Data Set A was generated by low-dimensional smooth dynamics, such a local linear model is able to capture the geometry remarkably well based on the relatively small sample size. The great advantage of local models is their ability to adhere to the local shape of an arbitrary surface; the corresponding disadvantage is that they do not lead to a compact description of the system. Global expansions of the surface reverse this tradeoff by providing a more manageable representation at the risk of larger local errors. Giona et al. (1991) give a particularly nice approach to global modeling that builds an orthogonal set of basis functions with respect to the natural measure of the attractor rather than picking a fixed set independent of the data. If $x_{t+1} = f(\mathbf{x}_t)$, $\rho(\mathbf{x})$ is the probability distribution for the state vector $\mathbf{x}$, and $\{p_i\}$ denotes a set of polynomials that are orthogonal with respect to this distribution:

$$\langle p_i(\mathbf{x})p_j(\mathbf{x}) \rangle = \int p_i(\mathbf{x})p_j(\mathbf{x})\rho(\mathbf{x}) \, d\mathbf{x} = \delta_{ij} \, , \tag{20}$$

then the expansion coefficients

$$f(\mathbf{x}) = \sum_i a_i p_i(\mathbf{x}) \tag{21}$$

can be found from the time average by the orthogonality condition:

$$a_i = \langle f(\mathbf{x}_t)p_i(\mathbf{x}_t) \rangle = \langle x_{t+1}p_i(\mathbf{x}_t) \rangle = \lim_{N \to \infty} \frac{1}{N} \sum_{t=1}^{N} x_{t+1}p_i(\mathbf{x}_t) \, . \tag{22}$$

The orthogonal polynomials can be found from Gram-Schmidt orthogonalization on the moments of the time series. This expansion is similar in spirit to embedding by expectation values presented in Eq. (16).

In between global and local models lie descriptions such as **radial basis functions** (Powell, 1987; Broomhead & Lowe, 1988; Casdagli, 1989; Poggio & Girosi, 1990; L. A. Smith, this volume). A typical choice is a mixture of (spherically symmetric) Gaussians, defined by

$$f(\mathbf{x}) = \sum_i w_i e^{-(\mathbf{x} - \mathbf{c}_i)^2/(2\sigma_i^2)} . \tag{23}$$

For each basis function, three quantities have to be determined: its center, $\mathbf{c}_i$; its width, σ_i; and the weight, w_i. In the simplest case, all the widths and centers are fixed. (The centers are, for example, placed on the observed data points.) In a weaker model, these assumptions are relaxed: the widths can be made adaptive, the constraint of spherical Gaussians can be removed by allowing for a general covariance matrix, and the centers can be allowed to adapt freely.

An important issue in function approximation is whether the adjustable parameters are all "after" the nonlinearities (for radial basis functions this corresponds to fixed centers and widths), or whether some of them are also "before" the nonlinearities (i.e., the centers and/or widths can be adjusted). The advantage of the former case is that the only remaining free parameters, the weights, can be estimated by matrix inversion. Its disadvantage is an exponential "curse of dimensionality." [20] In the latter case of adaptive nonlinearities, parameter estimation is harder, but can always be cast in an error backpropagation framework, i.e., solved with gradient descent. The surprising—and promising—result is that in this case when the adaptive parameters are "before" the nonlinearity, the curse of dimensionality is only linear with the dimension of the input space (Barron, 1993).

A goal beyond modeling the geometry of the manifold is to find a set of differential equations that might have produced the time series. This is often a more compact (and meaningful) description, as shown for simple examples by Cremers and Hübler (1987), and by Crutchfield and McNamara (1987). An alternative goal, trying to characterize symbol sequences (which might be obtained by a coarse quantization of real-valued data where each bin has a symbol assigned to it), is suggested by Crutchfield and Young (1989), who try to extract the rules of an automaton that could have generated the observed symbol sequence. Such approaches are interesting but are not yet routinely applicable.

---

[20] The higher the dimension of a data set of a given size, the more sparse the data set appears. If the average distance ϵ to the nearest point is to remain constant, the number of points N needed to cover the space increases exponentially with the dimension of the space d, $N \propto (1/\epsilon)^d$.

5.2 CONNECTIONIST FORECASTING

State-space embedding "solves" the forecasting problem for a low-dimensional deterministic system. If there is understandable structure in the embedding space, it can be detected and modeled; the open questions have to do with finding good representations of the surface and with estimating the reliability of the forecast. This approach of reconstructing the geometry of the manifold will fail if the system is high-dimensional, has stochastic inputs, or is nonstationary, because in these cases, there is no longer a simple surface to model.

Neural networks do not build an explicit description of a surface. On the one hand, this makes it harder to interpret them even for simple forecasting tasks. On the other hand, they promise to be applicable (and mis-applicable) to situations where simpler, more explicit approaches fail: much of their promise comes from the hope that they can learn to emulate unanticipated regularities in a complex signal. This broad attraction leads to results like those seen in Table 2 in the Appendix: the best as well as many of the worst forecasts of Data Set A were obtained with neural networks. The purpose of this section is to examine how neural networks can be used to forecast relatively simple time series (Data Set A, laser) as well as more difficult time series (Data Set D, high-dimensional chaos; Data Set E, currency exchange rates). We will start with *point predictions*, i.e., predictions of a single value at each time step (such as the most likely one), and then turn to the additional estimation of error bars, and eventually to the prediction of the full probability distribution.

A network that is to predict the future must know about the past. The simplest approach is to provide time-delayed samples to its input layer. In Section 4.2 we discussed that, on the one hand, a network without (nonlinear) hidden units is equivalent to an AR model (one linear filter). On the other hand, we showed that with nonliner hidden units, the network combines a number of "squashed" filters.

Eric Wan (at the time a graduate student at Stanford) used a somewhat more difficult architecture for his competition entry. The key modification to the network displayed in Figure 6 is that each connection now becomes an AR filter (tapped delay line). Rather than displaying the explicit buffer of the input units, it suffices then to draw only a single input unit and conceptually move the weight vectors into the tapped delay lines from the input to each hidden unit. This architecture is also known as "time-delay neural network" by Lang, Waibel, and Hinton (1990) or (in the spatial domain) as a network with "linked weights," as suggested by le Cun (1989).

We would like to emphasize that these architectures are all examples of *feedforward* networks: they are trained in "open loop," and there is no feedback of activations in training. (For iterated predictions, however, predictions must be used for the input: the network is run in "closed loop" mode. A more consistent scheme is to train the network in the mode eventually used in prediction; we return to this point at the end of this section.)

Wan's network had 1 input unit, two layers of 12 hidden units each, and 1 output unit. The "generalized weights" of the first layer were tapped delay lines with 25 taps; the second and third layers had 5 taps each. These values are not the result of a simple quantitative analysis, but rather the result of evaluating the performance of a variety of architectures on some part of the available data that Wan had set aside for this purpose. Such careful exploration is important for the successful use of neural networks.

At first sight, selecting an architecture with 1,105 parameters to fit 1,000 data points seems absurd. How is it possible not to overfit if there are more parameters than data points? The key is knowing when to stop. At the outset of training, the parameters have random values, and so changing any one coefficient has little impact on the quality of the predictions. As training progresses and the fit improves, the *effective number of parameters* grows (Weigend & Rumelhart, 1991a, 1991b). The overall error in predicting points out of the training set will initially decrease as the network learns to do something, but then will begin to increase once the network learns to do too much; the location of the minimum of the "cross validation" error determines when the effective network complexity is right.[21]

The competition entry was obtained by a network that was stopped early. It did very well over the prediction interval (Figure 3), but notice how it fails dramatically after 100 time steps (Figure 4) while the local linear forecast continues on. The reason for this difference is that the local linear model with local SVD (described in Section 5.1) is constrained to stay on the reconstructed surface in the embedding space (any point gets projected onto the reconstructed manifold before the prediction is made), whereas the network does not have a comparable constraint.

Short-term predictors optimize the parameters for forecasting the next step. The same architecture (e.g., Figure 6) can also be used to follow the trajectory more closely in the longer run, at the expense of possibly worse single-step predictions, by using the *predicted* value at the input in training, rather than the *true* value.[22] There is a continuum between these two extremes: in training, the errors from both sources can be combined in a weighted way, e.g., $\lambda \times$ (short-term error) + $(1 - \lambda) \times$ (long-term error). This mixed error is then used in backpropagation. Since the network produces very poor predictions at the beginning of the training process, it can be advantageous to begin with $\lambda = 1$ (full teacher forcing) and then anneal to the desired value.

[21]Other ideas, besides early stopping, for arriving at networks of an appropriate size include (i) penalizing network complexity by adding a term to the error function, (ii) removing unimportant weights with the help of the Hessian (the second derivative of the error with respect to the weights), and (iii) expressing the problem in a Bayesian framework.

[22]There are a number of terms associated with this distinction. Engineers use the expression "open loop" when the inputs are set to true values, and "closed loop" for the case when the input are given the predicted values. In the connectionist community, the term "teacher forcing" is used when the inputs are set to the true values, and the term "trajectory learning" (Principe et al., 1993) when the predicted output is fed back to the inputs.

We close the section on point-predictions with some remarks on the question whether predictions for several steps in the future should be made by iterating the single-step prediction T times or by making a noniterated, direct prediction. Farmer and Sidorowich (1988) argue that for deterministic chaotic systems, iterated forecasts lead to better predictions than direct forecasts.[23] However, for noisy series the question "iterated vs. direct forecasts?" remains open. The answer depends on issues such as the sampling time (if the sampling is faster than the highest frequency in the system, one-step predictions will focus on the noise), and the complexity of the input-output map (even the simple parabola of the logistic map becomes a polynomial of order 2^T when direct T step predictions are attempted). As Sauer (this volume) points out, the dichotomy between iterated and direct forecasts is not necessary: combining both approaches leads to better forecasts than using either individually.

5.3 BEYOND POINT-PREDICTIONS

So far we have discussed how to predict the continuation of a time series. It is often desirable and important to also know the confidence associated with a prediction. Before giving some suggestions, we make explicit the two assumptions that minimizing sum squared errors implies in a maximum likelihood framework (which we need not accept):

- The errors of different data points are independent of each other: statistical independence is present if the joint probability between two events is precisely the product between the two individual probabilities. After taking the logarithm, the product becomes a sum. Summing errors thus assumes statistical independence of the measurement errors.
- The errors are Gaussian distributed: i.e., the likelihood of a data point given the prediction is a Gaussian. Taking the logarithm of a Gaussian transforms it into a squared difference. This squared difference can be interpreted as a squared error. Summing squared errors with the same weight for each data point assumes that the uncertainty is the same for each data point (i.e., that the size of the error bar is independent of the location in state space).

The second assumption is clearly violated in the laser data: the errors made by the predictors on Data Set A depend strongly on the location in state space (see, for example, Smith, this volume). This is not surprising, since the local properties of the attractor (such as the rate of divergence of nearby trajectories) vary with the location. Furthermore, different regions are sampled unequally since the training set is finite.

The second assumption can be relaxed: in our own research, we have used maximum likelihood networks that successfully estimate the error bars as a function of

[23]Fraser (personal communication, 1993) points out that this argument can be traced back to Rissanen and Langdon (1981).

the network input. These networks had two output units: the first one predicts the value, the second the error bar. This model of a single Gaussian (where mean and width are estimated depending on the input) is easily generalized to a mixture of Gaussians. It is also possible to explore more general models with the activation distributed over a set of output units, allowing one to predict the probability density.[24] All these models fall within the elegant and powerful probabilistic framework laid out for connectionist modeling by Rumelhart et al. (1993). Another approach to estimating the output errors is to use "noisy weights": rather than characterizing each weight only by its value, Hinton and van Camp (1993) also encode the precision of the weight. In addition to facilitating the application of the minimum description length principle, this formulation yields a probability distribution of the output for a given input.

To force competition entrants to address the issue of the reliability of their estimates, we required them to submit, for Data Set A, both the predicted values and their estimated accuracies. We set as the error measure the likelihood of the true (withheld) data, computed from the submitted predictions and confidence intervals under a Gaussian model. In the Appendix we motivate this measure and list its values for the entries received. Analyzing the competition entries, we were quite surprised by how little effort was put into estimating the error bars. In all cases, the techniques used to generate the submitted error estimates were much less sophisticated than the models used for the point-predictions.

Motivated by this lack of sophistication, following the close of the competition, Fraser and Dimitriadis (this volume) used a hidden Markov model (HMM; see, e.g., Rabiner & Juang, 1986) to predict the evolution of the entire probability density for the computer-generated Data Set D, with 100,000 points the longest of the data sets. Nonhidden Markov models fall in the same class as feedforward networks: their only notion of state is what is presented to them in the moment—there is no implicit memory or stack. They thus implement a regular (although probabilistic) grammar. Hidden Markov models introduce "hidden states" that are not directly observable but are built up from the past. In this sense, they resemble recurrent networks which also form a representation of the past in internal memory that can influence decisions beyond what is seen by the immediate inputs. Both HMMs and recurrent networks thus implement context-free grammars.

---

[24]On the one hand, a localist representation is suited for the prediction of symbols where we want to avoid imposing a metric. For example, in composing music, it is undesirable to inherit a metric obtained from the pitch value (see Dirst & Weigend, this volume). On the other hand, if neighborhood is a sensible distance measure, then higher accuracy can be obtained by smearing out the activation over several units (e.g., Saund, 1989).

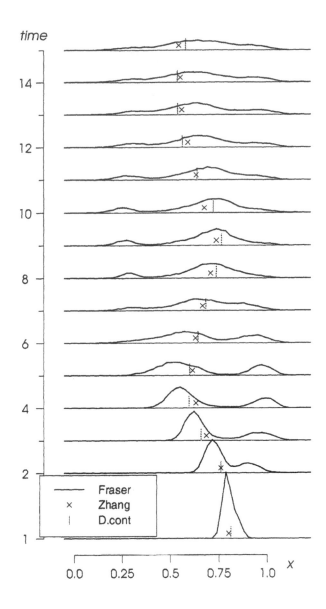

FIGURE 7 Continuations of Data Set D, the computer-generated data. The curves are the predictions for the probability density function from a hidden Markov model (Fraser & Dimitriadis, this volume). The ×'s indicate the point predictions from a neural network (Zhang & Hutchinson, this volume). The "true" continuation is indicated by vertical lines.

The states of a hidden Markov model can be discrete or continuous; Fraser and Dimitriadis use a mixture of both: they generate their predictions from a 20-state model using eighth-order linear autoregressive filters. The model parametrizes multivariate normal distributions.

Figure 7 shows the evolution of the probability density function according to Fraser and Dimitriadis (this volume). They first estimated the roughly 6,000 parameters of their model from the 100,000 points of Data Set D, and then generated several million continuations from their model. We plot the histograms of these continuations (normalized to have the same areas) along with the values obtained by continuing the generation of the data from the differential equations (as described in the Appendix). Unfortunately, the predicted probability density functions by Fraser and Dimitriadis are a lot wider than what the uncertainty due to the dynamics of the system and the stochasticity used in generating the series requires: on the time scale of Figure 7, an ensemble of continuations of Data Set D spreads only about 1%, a lot less than the uncertainty predicted by the mixed-state HMM.

In the competition, we received several sets of point-predictions for Data Set D. They are listed and plotted in the Appendix. In Figure 7 we have included the best of these, obtained by Zhang and Hutchinson (this volume). They trained 108 simple feedforward networks in a brute force approach: the final run alone—after all initial experimentation and architecture selection—used 100 hours on a Connection Machine CM-2 with 8,192 floating point processors. Each network had between 20 and 30 inputs, either one hidden layer with 100 units or two layers with 30 units each, and up to 5 outputs. The complex, unstructured architecture unfortunately does not allow a satisfying interpretation of the resulting networks.

Beyond physical systems: Financial data. We close this section with some remarks about the forecasts for the exchange rate time series (Data Set C). (No prediction tasks were specified for the biological, astronomical, and musical sets.) The market is quite large: in 1989 the daily volume of the currency markets was estimated to be U.S. $650 billion, and in 1993 the market exceeded U.S. $1 trillion (= 10^{12}) on busy days. Lequarré (this volume) reports that 97% of this is speculation— i.e., only 3% of the trades are "actual" transactions.

The simplest model of a market indicator is the efficient market hypothesis: "you can't do better than predicting that tomorrow's rate is the same as today's." Diebold and Nason (1990), for example, review attempts to forecast foreign exchange rates and conclude that none of the methods succeeded in beating the random walk hypothesis out-of-sample. However, all these academic findings about the unpredictability of the vicissitudes of financial markets are based on daily or weekly rates, the only data available until a few years ago. The great deal of activity in foreign exchange trading suggests that it must be possible to make money with better data.

Recently, high-frequency data have become available, and we were fortunate enough to be able to provide a set of such "tick-by-tick" data for the competition.

Data Set C consists of quotes on the time scale of one to two minutes for the exchange rate between the Swiss franc and the U.S. dollar. The market is based on bids and asks to buy and sell. (There is no central market for currency transactions.) Prices are given for a trade of U.S. $10 million, and an offer is good for five seconds(!). In addition to the quote, we included the day of the week and the time after the opening that day (to allow for modeling of intra- and inter-day effects).

In order to balance the desire for statistically significant results and the need to keep the competition prediction task manageable, we asked for forecasts for 10 episodes of 3,000 points each, taken from the period between August 7, 1990 to April 18, 1991.[25] This assessment provided a basic sanity test of the submitted predictors (some of them were worse than chance by a factor of 16), and afterwards the best two groups were invited to analyze a much larger sample. The quality of the predictions is expressed in terms of the following ratio of squared errors:

$$\frac{\sum_t \left(\text{observation}_t - \text{prediction}_t\right)^2}{\sum_t \left(\text{observation}_t - \text{observation}_{t-1}\right)^2} \,. \tag{24}$$

The denominator simply predicts the last observed value—which is the best that can be done for a random walk. A ratio above 1.0 thus corresponds to a prediction that is worse than chance; a ratio below 1.0 is an improvement over a random walk.[26]

In this second round of evaluation, predictions were made for all the points in the gaps between the training segments (still using the competition data for training). The out-of-sample performance on this extended test set, as reported

---

[25] For each trial we asked for six forecasts: 1 minute after the last tick, 15 minutes after the last tick, 60 minutes after the last tick, the closing value of the day of the last tick, the opening value of the next trading day, and the closing value of the fifth trading day (usually one week) after the day of the last tick. One example evaluation is given in the following table. The numbers are the ratio of the sum of squared errors of the submitted predictions by Zhang and Hutchinson (this volume, p. 235) divided by the sum of the squared errors obtained by using the last observation as the prediction.

	1 minute	15 minutes	1 hour
training set ("in-sample")	0.889	0.891	0.885
test set ("out-of-sample")	0.697	1.04	0.988

This table shows the crucial difference between training set and test set performances, and suggests that the uncertainty in the numbers is large. Hence, we proposed the evaluation on larger data sets, as described in the main text.

[26] The random walk model used here for comparison is a weak null hypothesis. LeBaron (this volume) reports statistically significant autocorrelations on Data Set C (see Table 2 in his paper on p. 462 of this volume). To the degree that the in-sample autocorrelations generalize out-of-sample, this justifies a low-order AR model as a stronger null hypothesis. However, to avoid additional assumptions in our comparison here (such as the order of the model), we decided simply to compare to a random walk.

TABLE 1 Performance on exchange rate predictions expressed as the squared error of the predictor divided by the squared error from predicting no change, as defined in Eq. (24). The numbers are as reported by Mozer (this volume, p. 261) for his recurrent networks, and Zhang and Hutchinson (this volume, p. 236) for their feedforward networks.

	1 minute ($N = 18,465$)	15 minutes ($N = 7,246$)	1 hour ($N = 3,334$)
Mozer	0.9976	0.9989	0.9965
Zhang & Hutchinson	1.090	1.103	1.098

by Mozer (this volume) and by Zhang and Hutchinson (this volume), is collected in Table 1. Please refer to their articles for more details and further evaluation.

These results—in particular the last row in Table 1 where all out-of-sample predictions are on average worse than chance—make clear that a naive application of the techniques that worked so well for Data Set A (the laser had less than 1% measurement noise added to deterministic dynamics) and to some degree for Data Set D fails for a data set so close to pure randomness as this financial data set. Future directions include (Weigend, 1991)

- using additional information from other sources (such as other currencies, interest rates, financial indicators, as well as information automatically extracted from incoming newswires, "topic spotting"),
- splitting the problem of predicting returns to the two separate tasks of predicting the squared change (volatility) and its direction,
- employing architectures that allow for stretching and compressing of the time series, as well as enhancing the input with features that collapse time in ways typically done by traders,
- implementing trading strategies (i.e., converting predictions to recommendations for actions), and subsequently improving them by backpropagating the actual loss or profit through a pay-off matrix (taking transaction costs into account).

Financial predictions can also serve as good vehicles to stimulate research in areas such as subset selection (finding relevant variables) and capacity control (avoiding overfitting). The full record of 329,112 quotes (bid and ask) from May 20, 1985 to April 12, 1991 is available as a benchmark data set via anonymous ftp.[27] We encourage the reader to experiment (and inform the authors of positive results).

[27]It corresponds to 11.5 MB. Like the data sets used in the competition, it is available via anonymous ftp to ftp.santafe.edu.

6. CHARACTERIZATION

Simple systems can produce time series that appear to be complicated; complex systems can produce time series that are complicated. These two extremes have different goals and require different techniques; what constitutes a successful forecast depends on where the underlying system falls on this continuum. In this section we will look at characterization methods that can be used to extract some of the essential properties that lie behind an observed time series, both as an end in itself and as a guide to further analysis and modeling.

Characterizing time series through their frequency content goes back to Schuster's "periodogram" (1898). For a simple linear system the traditional spectral analysis is very useful (peaks = modes = degrees of freedom), but different nonlinear systems can have similar featureless broadband power spectra. Therefore, a broadly useful characterization of a nonlinear system cannot be based on its frequency content.

We will describe two approaches that parallel the earlier discussion of forecasting: (1) an explicit analysis of the structure in an embedding space (in Section 6.2 we will introduce the information-theoretic measure of redundancy as an example of understanding by "opening up the box"), and (2) an implicit approach based on analyzing properties of an emulation of the system arrived at through learning (in Section 6.3.1 we will show how local linear methods can be used for characterization, and in Section 6.3.2 we will discuss how neural networks can be used to estimate dimensions, the amount of nonlinearity, and Lyapunov coefficients). Before turning to these more sophisticated analyses, we discuss some simple tests.

6.1 SIMPLE TESTS

This section begins with suggestions for exploring data when no additional information is available. We then turn to time series whose frequency spectra follow a power law where low-frequency structure can lead to artifacts. We then show how surrogate data can be generated with identical linear but different nonlinear structure. Finally, we suggest some ways of analyzing the residual errors.

EXPLORATORY DATA ANALYSIS. The importance of exploring a given data set with a broad range of methods cannot be overemphasized. Besides the many traditional techniques (Tukey, 1977), modern methods of interactive exploration range from interactive graphics with linked plots and virtual movements in a visual space[28] to examination in an auditory space (data sonification). The latter method uses the temporal abilities of our auditory systems—after all, analyzing temporal sequences as static plots has not been a prime goal in human evolution.

[28] A good example is the visualization package xgobi. To obtain information about it, send the one line message send index to statlib@lib.stat.cmu.edu.

LINEAR CORRELATIONS. We have seen in Section 3.2 that nonlinear structure can be missed by linear analysis. But that is not the only problem: linear structure that is present in the data can confuse nonlinear analysis. Important (and notorious) examples are processes with power spectra proportional to $|\omega|^{-\alpha}$, which arise routinely in fluctuating transport processes. At the extreme, white noise is defined by a flat spectrum; i.e., its spectral coefficient is $\alpha = 0$. When white noise is integrated (summed up over time), the result is a random walk process. After squaring the amplitude (in order to arrive at the power spectrum), a random walk yields yields a spectral exponent $\alpha = 2$. The intermediate value of $\alpha = 1$ is seen in everything from electrical resistors to traffic on a highway to music (Dutta & Horn, 1981). This intermediate spectral coefficient of $\alpha = 1$ implies that all time-scales (over which the $1/\omega$ behavior holds) are equally important.

Consider the cloud of points obtained by plotting x_t against $x_{t-\tau}$ for such a series with a lag time τ. (This plot was introduced in Section 3.2 in the context of the logistic map whose phase portrait was a parabola.) In general, a distribution can be described by its moments. The first moments (means) give the coordinates of the center of the point cloud; the second moments (covariance matrix) contain the information about how elongated the point cloud is, and how it is rotated with respect to the axes.[29] We are interested here in this elongation of the point cloud; it can be described in terms of the eigenvalues of its correlation matrix. The larger eigenvalue, λ_+, characterizes the extension in the direction of the larger principal axis along the diagonal, and the smaller eigenvalue, λ_-, measures the extension transverse to it. The ratio of these two eigenvalues can be expressed in terms of the autocorrelation function ρ at the lag τ as

$$\frac{\lambda_-}{\lambda_+} = \frac{1 - \rho(\tau)}{1 + \rho(\tau)}. \tag{25}$$

For a power-law spectrum, the autocorrelation function can be evaluated analytically in terms of the spectral exponent α and the exponential integral Ei (Gershenfeld, 1992). For example, for a measurement bandwidth of 10^{-3} to 10^3 Hz and a lag time τ of 1 sec, this ratio is 0.51 for $\alpha = 1$, 0.005 for $\alpha = 2$, and 0.0001 for $\alpha = 3$. As the spectrum drops off more quickly, i.e., as the spectral exponent gets larger, the autocorrelation function decays more slowly. (In the extreme of a delta function in frequency space, the signal is constant in time.) Large spectral exponents thus imply that the shape of the point cloud (an estimate of the probability distribution in the embedding space) will be increasingly long and skinny, regardless of the detailed dynamics and of the value of τ. If the width becomes small compared to

---

[29]We here consider only first- and second-order moments. They completely characterize a Gaussian distribution, and it is easy to relate them to the conventional (linear) correlation coefficient; see Duda and Hart (1973). A nonlinear relationship between x_t and $x_{t-\tau}$ is missed by an analysis in terms of the first- and second-order moments. This is why a restriction to up to second-order terms is sometimes called linear.

the available experimental resolution, the system will erroneously appear to be one-dimensional. There is a simple test for this artifact: whether or not the quantities of interest change as the lag time τ is varied. This effect is discussed in more detail by Theiler (1991).

SURROGATE DATA. Since the autocorrelation function of a signal is equal to the inverse Fourier transform of the power spectrum, any transformation of the signal that does not change the power spectrum will not change the autocorrelation function. It is therefore possible to take the Fourier transform of a time series, randomize the phases (symmetrically, so that the inverse transform remains real), and then take the inverse transform to produce a series that by construction has the same autocorrelation function but will have removed any nonlinear ordering of the points. This creation of sets of surrogate data provides an important test for whether an algorithm is detecting nonlinear structure or is fooled by linear properties (such as we saw for low frequency signals): if the result is the same for the surrogate data, then the result cannot have anything to do with deterministic rules that depend on the specific sequence of the points. This technique has become popular in recent years; Fraser (1989c), for example, compares characteristics of time series from the Lorenz equation with surrogate versions. Kaplan (this volume) and Theiler et al. (this volume) apply the method of surrogate data to Data Sets A, B, D, and E; the idea is also the basis of Paluš's comparison (this volume) between "linear redundancy" and "redundancy" (we will introduce the concept of redundancy in the next section).

SANITY CHECKS AND SMOKE ALARMS. Once a predictor has been fitted, a number of sanity checks should be applied to the resulting predictions and their residual errors. It can be useful to look at a distribution of the errors sorted by their size (see, e.g., Smith, this volume, Figures 5 and 8 on p. 331 and 336). Such plots distinguish between forecasts that have the same mean squared error but very different distributions (uniformly medium-sized errors versus very small errors along with a few large outliers). Other basic tests plot the prediction errors against the true (or against the predicted) value. This distribution should be flat if the errors are Gaussian distributed, and proportional to the mean for a Poissonian error distribution. The time ordering of the errors can also contain information: a good model should turn the time series into structureless noise for the residual errors; any remaining structure indicates that the predictor missed some features.

Failure to use common sense was readily apparent in many of the entries in the competition. This ranged from forecasts for Data Set A that included large negative values (recall that the training set was strictly positive) to elaborate "proofs" that Data Set D was very low dimensional (following a noise reduction step that had the effect of removing the high-dimensional dynamics). Distinguishing fact, fiction and fallacies in forecasts is often hard: carrying out simple tests is crucial particularly in the light of readily available sophisticated analysis and prediction algorithms that can swiftly and silently produce nonsense.

6.2 DIRECT CHARACTERIZATION VIA STATE SPACE

It is always possible to define a time-delayed vector from a time series, but this certainly does not mean that it is always possible to identify meaningful structure in the embedded data. Because the mapping between a delay vector and the system's underlying state is not known, the precise value of an embedded data point is not significant. However, because an embedding is diffeomorphic (smooth and invertible), a number of important properties of the system will be preserved by the mapping. These include local features such as the number of degrees of freedom, and global topological features such as the linking of trajectories (Melvin & Tufillaro, 1991). The literature on characterizing embedded data in terms of such invariants is vast, motivated by the promise of obtaining deep insight into observations, but plagued by the problem that plausible algorithms will always produce a result—whether or not the result is significant. General reviews of this area may be found in Ruelle and Eckmann (1985), Gershenfeld (1989), and Theiler (1990).[30]

Just as state-space forecasting leaps over the systematic increase in complexity from linear models to bilinear models, etc., these characterization ideas bypass the traditional progression from ordinary spectra to higher order spectra.[31] They are predated by similar efforts to analyze signals in terms of dimensionality (Trunk, 1968) and near-neighbor scaling (Pettis et al., 1979), but they could not succeed until the relationship between observed and unobserved degrees of freedom was made explicit by time-delay embedding. This was done to estimate degrees of freedom by Russel et al. (1980), and implemented efficiently by Grassberger and Procaccia (1983a). Brock, Dechert, and Scheinkman later developed this algorithm into a statistical test, the BDS test (Brock et al., 1988), with respect to the null hypothesis of an iid sequence (which was further refined by Green & Savit, 1991).

We summarize here an information-based approach due to Fraser (1989b) that was successfully used in the competition by Paluš (this volume) (participating in the competition over the network from Czechoslovakia). Although the connection between information theory and ergodic theory has long been appreciated (see, e.g., Petersen, 1989), Shaw (1981) helped point out the connection between dissipative dynamics and information theory, and Fraser and Swinney (1986) first used information-theoretic measures to find optimal embedding lags. This example of

---

[30]The term "embedding" is used in the literature in two senses. In its wider sense, the term denotes any lag-space representation, whether there is a unique surface or not. In its narrower (mathematical) sense used here, the term applies if and only if the resulting surface is unique, i.e., if a diffeomorphism exists between the solution manifold in configuration space and the manifold in lag space.

[31]Chaotic processes are analyzed in terms of bispectra by Subba Rao (1992).

the physical meaning of information (Landauer, 1991) can be viewed as an application of information theory back to its roots in dynamics: Shannon (1948) built his theory of information on the analysis of the single-molecule Maxwell Demon by Szilard in 1929, which in turn was motivated by Maxwell and Boltzmann's effort to understand the microscopic dynamics of the origin of thermodynamic irreversibility (circa 1870).

Assume that a time series $x(t)$ has been digitized to integer values lying between 1 and N. If a total of n_T points have been observed, and a particular value of x is recorded n_x times, then the probability of seeing this value is estimated to be $p_1(x) = n_x/n_T$.[32] (The subscript of the probability indicates that we are at present considering one-dimensional distributions (histograms). It will soon be generalized to d-dimensional distributions.) In terms of this probability, the **entropy** of this distribution is given by

$$H_1(N) = -\sum_{x=1}^{N} p_1(x) \log_2 p_1(x). \tag{26}$$

This is the average number of bits required to describe an isolated observation, and can range from 0 (if there is only one possible value for x) to $\log_2 N$ (if all values of x are equally likely and hence the full resolution of x is required).

In the limit $N \to 1$, there is only one possible value and so the probability of seeing it is unity, thus $H_1(1) = 0$. As N is increased, the entropy grows as $\log N$ if all values are equally probable; it will reach an asymptotic value of $\log M$ independent of N if there are M equally probable states in the time series; and if the probability distribution is more complicated, it can grow as $D_1 \log N$ where D_1 is a constant ≤ 1 (the meaning of D_1 will be explained shortly). Therefore, the dependence of H_1 on N provides information about the resolution of the observable.

The probability of seeing a specific lag vector $\mathbf{x}_t = (x_t, x_{t-\tau}, \ldots, x_{t-(d-1)\tau})$ (see Eq. (14)) in d-dimensional lag space is similarly estimated by counting the relative population of the corresponding cell in the d-dimensional array: $p_d(\mathbf{x}) = n_{\mathbf{x}}/n_T$. The probability of seeing a particular *sequence* of D embedded vectors $(\mathbf{x}_t, \ldots, \mathbf{x}_{t-(D-1)\tau})$ is just $p_{d+D}(x_t, \ldots, x_{t-(d+D-1)\tau})$ because each successive vector is equal to the preceding one with the coordinates shifted over one place and a new observation added at the end. This means that the joint probability of d delayed observations, p_d, is equivalent to the probability of seeing a single point in the d-dimensional embedding space (or the probability of seeing a sequence of $1 + d - n$

---

[32]Note that there can be corrections to such estimates if one is interested in the expectation value of functions of the probability (Grassberger, 1988).

points in a smaller n-dimensional space). In terms of p_d, the **joint entropy** or **block entropy** is

$$H_d(\tau, N) =$$

$$-\sum_{x_t=1}^{N} \cdots \sum_{x_{t-(d-1)\tau}=1}^{N} p_d(x_t, x_{t-\tau}, \ldots, x_{t-(d-1)\tau}) \log_2 p_d(x_t, x_{t-\tau}, \ldots, x_{t-(d-1)\tau}).$$

$$(27)$$

This is the average number of bits needed to describe a sequence. (The range of the sum might seem strange at first sight, but keep in mind that we are assuming that x_t is quantized to integers between 1 and N.) In the limit of small lags, we obtain

$$\lim_{\tau \to 0} p_d(x_t, x_{t-\tau}, \ldots, x_{t-(d-1)\tau}) = p_1(x)$$

$$\Rightarrow H_d(0, N) = H_1(N).$$

$$(28)$$

In the opposite limit, if successive measurements become uncorrelated at large times, the probability distribution will factor:

$$\lim_{\tau \to \infty} p_d(x_t, x_{t-\tau}, \ldots, x_{t-(d-1)\tau}) = p_1(x_t)p_1(x_{t-\tau}), \ldots, p_1(x_{t-(d-1)\tau})$$

$$\Rightarrow \lim_{\tau \to \infty} H_d(\tau, N) = dH_1(N).$$

$$(29)$$

We will return to the τ dependence later; for now assume that the delay time τ is small but nonzero.

We have already seen that $\lim_{N \to 1} H_d(\tau, N) = 0$. The limit of large N is best understood in the context of the **generalized dimensions** D_q (Hentschel & Procaccia, 1983). These are defined by the scaling of the moments of the d-dimensional probability distribution $p_d(\mathbf{x})$ as the number of bins N tends to infinity (i.e., the bin sizes become very small, corresponding to an increasing resolution):

$$D_q = \lim_{N \to \infty} \frac{1}{q-1} \frac{\log_2 \sum_{\mathbf{x}_i} p_d(\mathbf{x}_i)^q}{-\log_2 N}.$$

$$(30)$$

For simple geometrical objects such as lines or surfaces, the D_q's are all equal to the integer topological dimension (1 for a line, 2 for a surface,...). For fractal distributions they need not be an integer, and the q dependence is related to how singular the distribution is. The values of the D_q's are typically similar, and there are strong bounds on how much they can differ (Beck, 1990). D_2 measures the scaling of pairs of points (it is the Grassberger-Procaccia correlation dimension (Grassberger & Procaccia, 1983a); see also Kantz, this volume, and Pineda & Sommerer, this volume), and D_1 provides the connection with entropy:

$$\lim_{q \to 1} D_q = \lim_{N \to \infty} \frac{\sum_{\mathbf{x}_i} p_d(\mathbf{x}_i) \log_2 p_d(\mathbf{x}_i)}{-\log_2 N}$$

$$= \lim_{N \to \infty} \frac{H_d(\tau, N)}{\log_2 N}.$$

$$(31)$$

As N is increased, the prefactor to the logarithmic growth of the entropy is the generalized dimension D_1 of the probability distribution. If the system is deterministic, so that its solutions lies on a low-dimensional attractor, the measured dimension D_1 will equal the dimension of the attractor if the number of time delays used is large enough. If the number of lags is too small, or if the successive observations in the time series are uncorrelated, then the measured dimension will equal the number of lags. The dimension of an attractor measures the number of local directions available to the system and so it (or the smallest integer above it if the dimension is fractal) provides an estimate of the number of degrees of freedom needed to describe a state of the system. If the underlying system has n degrees of freedom, the minimum embedding dimension to recover these dynamics can be anywhere between n and $2n$, depending on the geometry.

The d-dependence of the entropy can be understood in terms of the concept of **mutual information**. The mutual information between two samples is the difference between their joint entropy and the sum of their scalar entropies:

$$
\begin{aligned}
I_2(\tau, N) = &-\sum_{x_t=1}^{N} p_1(x_t) \log_2 p_1(x_t) - \sum_{x_{t-\tau}=1}^{N} p_1(x_{t-\tau}) \log_2 p_1(x_{t-\tau}) \\
&+ \sum_{x_t=1}^{N} \sum_{x_{t-\tau}=1}^{N} p_2(x_t, x_{t-\tau}) \log_2 p_2(x_t, x_{t-\tau}) \\
= &\, 2H_1(\tau, N) - H_2(\tau, N).
\end{aligned}
\tag{32}
$$

If the samples are statistically independent (this means by definition that the probability distribution factors, i.e., $p_2(x_t, x_{t-\tau}) \equiv p_1(x_t) p_1(x_{t-\tau})$), then the mutual information will vanish: no knowledge can be gained for the second sample by knowing the first. On the other hand, if the first sample uniquely determines the second sample ($H_2 = H_1$), the mutual information will equal the scalar entropy $I_2 = H_1$. In between these two cases, the mutual information measures in bits the degree to which knowledge of one variable specifies the other.

The mutual information can be generalized to higher dimensions either by the **joint mutual information**

$$
I_d(\tau, N) = dH_1(\tau, N) - H_d(\tau, N)
\tag{33}
$$

or by the **incremental mutual information** or **redundancy** of one sample

$$
R_d(\tau, N) = H_1(\tau, N) + H_{d-1}(\tau, N) - H_d(\tau, N).
\tag{34}
$$

The redundancy measures the average number of bits about an observation that can be determined by knowing $d-1$ preceding observations. Joint mutual information and redundancy are related by $R_d = I_d - I_{d-1}$.

For systems governed by differential equations or maps, given enough points and enough resolution, the past must uniquely determine the future[33] (to a certain horizon, depending on the finite resolution).

If d is much less than the minimum embedding dimension, then the $d-1$ previous observations do not determine the next one, and so the value of the redundancy will approach zero:

$$p_d(x_t, x_{t-\tau}, \ldots, x_{t-(d-1)\tau}) = p_1(x_t)p_{d-1}(x_{t-\tau}, \ldots, x_{t-(d-1)\tau})$$
$$\Rightarrow H_d = H_1 + H_{d-1} \Rightarrow R_d = 0. \tag{35}$$

On the other hand, if d is much larger than the required embedding dimension, then the new observation will be entirely redundant:[34]

$$p_d(x_t, x_{t-\tau}, \ldots, x_{t-(d-1)\tau}) = p_{d-1}(x_{t-\tau}, \ldots, x_{t-(d-1)\tau})$$
$$\Rightarrow H_d = H_{d-1} \Rightarrow R_d = H_1. \tag{36}$$

The minimum value of d for which the redundancy converges (if there is one) is equal to the minimum embedding dimension at that resolution and delay time, i.e., the size of the smallest Euclidean space that can contain the dynamics without trajectories crossing. Before giving the redundancy for Data Set A of the competition (in Figure 8), we provide relations to other quantities, such as the Lyapunov exponents and the prediction horizon.

The **source entropy** or **Kolmogorov-Sinai entropy**, $h(\tau, N)$, is defined to be the asymptotic rate of increase of the information with each additional measurement given unlimited resolution:

$$h(\tau, N) = \lim_{N \to \infty} \lim_{d \to \infty} H_d(\tau, N) - H_{d-1}(\tau, N). \tag{37}$$

The limit of infinite resolution is usually not needed in practice: the source entropy reaches its maximum asymptotic value once the resolution is sufficiently fine to produce a generating partition (Petersen, 1989, p. 243). The source entropy is

---

[33] Note that the converse remains true for differential equations but need not be true for maps: for example, neither of the two maps introduced in Section 3.2 has a unique preimage in time—they cannot be inverted. Given that most computer simulations approximate the continuous dynamics of differential equations by discrete dynamics, this is a potentially rich source for artifacts. Lorenz (1989) shows how discretization in time can create dynamical features that are impossible in the continuous-time system; see also Grebogi et al. (1990). Rico-Martinez, Kevrekidis, and Adomaitis (1993) discuss noninvertibility in the context of neural networks.

[34] To be precise, $H_d = H_{d-1}$ is only valid for (1) the limit of short times τ, (2) discrete measurements, and (3) the noise-free case.

important because the *Pesin identity* relates it to the sum of the positive Lyapunov exponents (Ruelle & Eckmann, 1985):

$$h(\tau) = \tau h(1) = \tau \sum_i \lambda_i^+ . \tag{38}$$

The Lyapunov exponents λ_i are the eigenvalues of the local linearization of the dynamics (i.e., a local linear model), measuring the average rate of divergence of the principal axes of an ensemble of nearby trajectories. They can be found from the Jacobian (if it is known) or by following trajectories (Brown et al., 1991). Diverging trajectories reveal information about the system which is initially hidden by the measurement quantization. The amount of this information is proportional to the expansion rate of the volume, which is given by the sum of the positive exponents.

If the sequence of a time series generated by differential equations is reversed, the positive exponents will become negative ones and vice versa, and so the sum of negative exponents can be measured on a reversed series (Parlitz, 1992). If the time series was produced by a discrete map rather than a continuous flow (from differential equations), then the governing equations need not be reversible; for such a system the time-reversed series will no longer be predictable. Therefore, examining a series backward as well as forward is useful for determining whether a dynamical system is invertible and, if it is, the rate at which volumes contract. Note that reversing a time series is very different from actually running the dynamics backwards; there may not be a natural measure for the backwards dynamics and, if there is one, it will usually not be the same as that of the forwards dynamics. Since Lyapunov exponents are defined by time averages (and hence with respect to the natural measure), they will also change.

If the embedding dimension d is large enough, the redundancy is just the difference between the scalar entropy and an estimate of the source entropy:

$$R_d(\tau, N) \approx H_1(\tau, N) - h(\tau, N) . \tag{39}$$

In the limit of small lags,

$$H_{d-1}(0, N) = H_d(0, N) \Rightarrow R_d(0, N) = H_1(N) , \tag{40}$$

and for long lags

$$\lim_{\tau \to \infty} H_d(\tau, N) = d H_1(\tau, N) \Rightarrow R_d(\infty, N) = 0 . \tag{41}$$

The value of τ where the redundancy vanishes provides an estimate of the limit of predictability of the system at that resolution (prediction horizon) and will be very short if d is less than the minimum embedding dimension. Once d is larger than the embedding dimension (if there is one), then the redundancy will decay much more slowly, and the slope for small τ will be the source entropy:

$$R_d(\tau, N) = H_1(N) - \tau h(1). \tag{42}$$

We have seen that it is possible from the block entropy (Eq. (27)) and the redundancy (Eq. (34)) to estimate the resolution, minimum embedding dimension, information dimension D_1, source entropy, and the prediction horizon of the data, and hence learn about the number of degrees of freedom underlying the time series and the rate at which it loses memory of its initial conditions. It is also possible to test for nonlinearity by comparing the redundancy with a linear analog defined in terms of the correlation matrix (Paluš, this volume). The redundancy can be efficiently computed with an $O(N)$ algorithm by sorting the measured values on a simple fixed-resolution binary tree (Gershenfeld, 1993b). This tree sort is related to the frequently rediscovered fact that box-counting algorithms (such as are needed for estimating dimensions and entropies) can be implemented in high-dimensional space with an $O(N \log N)$ algorithm requiring no auxiliary storage by sorting the appended indices of lagged vectors (Pineda & Sommerer, this volume). Equal-probability data structures can be used (at the expense of computational complexity) to generate more reliable unbiased entropy estimates (Fraser & Swinney, 1986).

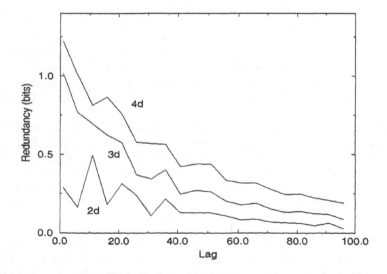

FIGURE 8 The redundancy (incremental mutual information) of Data Set A as a function of the number of time steps. The figure indicates that three past values are sufficient to retrieve most of the predictable structure and that the system has lost the memory of its initial conditions after roughly 100 steps.

Figure 8 shows the results of our redundancy calculation for Data Set A. This figure was computed using 10,000 data points sorted on the three most significant bits. Note that three lags are sufficient to retrieve most of the predictable structure. This is in agreement with the exploratory stereo pairs (Figure 5), where the three dimensions plotted appear to be sufficient for an embedding. Furthermore, we can read off from Figure 8 that the system has lost memory of its initial condition after about 100 steps.

There are many other approaches to characterizing embedded data (and choosing embedding parameters): Liebert and Schuster (1989) comment on a good choice for the delay time by relating the first minimum of mutual information plotted as a function of the lag time (suggested by Fraser & Swinney, 1986) to the generalized correlation integral. Aleksić (1991) plots the distances between images of close points as a function of the number of lags: at the minimum embedding dimension the distance suddenly becomes small. Savit and Green (1991), and Pi and Peterson (1993) exploit the notion of continuity of a function (of the conditional probabilities in the lag space as the number of lags increases). Kennel, Brown, and Abarbanel (1992) look for "false neighbors" (once the space is large enough for the attractor to unfold, they disappear). Further methods are presented by Kaplan, and by Pineda and Sommerer in this volume. Gershenfeld (1992) discusses how high-dimensional probability distributions limit the most complicated deterministic system that can be distinguished from a stochastic one. (The expected fraction of "typical" points in the interior of a distribution is increased; this is true because the ratio of the volume of a thin shell near the surface to the overall volume tends to 1 with increasing dimension.) Work needs to be done to understand the relations, strengths, and weaknesses of these algorithms. However, regardless of the algorithm employed, it is crucial to understand the nature of the errors in the results (both statistical uncertainty and possible artifacts), and to remember that there is no "right" answer for the time delay τ and the number of lags d—the choices will always depend on the goal.

The entries for the analysis part of the competition showed good agreement in the submitted values of the correlation dimension for Data Set A (2.02, 2.05, 2.06, 2.07, 2.2), but the estimates of the the positive Lyapunov exponent (either directly or from the source entropy) were more scattered (.024, .037, .07, .087, .089 bits/step). There were fewer estimates of these quantities for the other data sets (because they were more complicated) and they were more scattered; estimates of the degrees of freedom of Data Set D ranged from 4 to 8.

Insight into embedding can be used for more than characterization; it can also be used to distinguish between measurement noise and intrinsic dynamics. If the system is known, this can be done with a Wiener filter, a two-sided linear filter that estimates the most likely current value (rather than a future value). This method, however, requires advance knowledge of the power spectra of both the desired signal and of the noise, and the recovery will be imperfect if these spectra overlap (Priestley, 1981, p. 775; Press, 1992, p. 574). State-space reconstruction through time-average expectations (Eq. (16)) provides a method for signal separation that requires information only about the noise. If the true observable $x(t)$ is corrupted by additive measurement noise $n(t)$ that is uncorrelated with the system, then the expectation will factor:

$$\langle e^{i\mathbf{k}\cdot(\mathbf{x}(t)+\mathbf{n}(t))} \rangle = \langle e^{i\mathbf{k}\cdot\mathbf{x}(t)} \rangle \langle e^{i\mathbf{k}\cdot\mathbf{n}(t)} \rangle \; ; \tag{43}$$

$\mathbf{k}$ is the wave vector indexing the Fourier transform of the state-space probability density. The noise produces a k-dependent correction in the embedding space; if the noise is uncorrelated with itself on the time scale of the lag time τ (as for additive white noise), then this correction can be estimated and removed solely from knowledge of the probability distribution for the noise (Gershenfeld, 1993a). This algorithm requires sufficient data to accurately estimate the expectation values; Marteau and Abarbanel (1991), Sauer (1992), and Kantz (this volume) describe more data-efficient alternatives based on distinguishing between the low-dimensional system's trajectory and the high-dimensional behavior associated with the noise. Much less is known about the much harder problem of separating noise that enters into the dynamics (Guckenheimer, 1982).

6.3 INDIRECT CHARACTERIZATION: UNDERSTANDING THROUGH LEARNING

In Section 3.1 we showed that a linear time system is fully characterized by its Fourier spectrum (or equivalently by its ARMA coefficients or its autocorrelation function). We then showed how we have to go beyond that in the case of nonlinear systems and focused in Section 6.2 on the properties of the observed points in embedding space. As with forecasting, we move from the direct approach to the case where the attempt to understand the system directly does not succeed: we now show examples of how time series can be characterized through forecasting. The price for this appealing generality will be less insight into the meaning of the results. Both classes of algorithms that were successful in the competition will be put to work for characterization; in Section 6.3.1 we use local linear models to obtain DVS plots ("deterministic vs. stochastic") and, in Section 6.3.2, we explain how properties of the system that are not directly accessible can be extracted from connectionist networks that were trained to emulate the system.

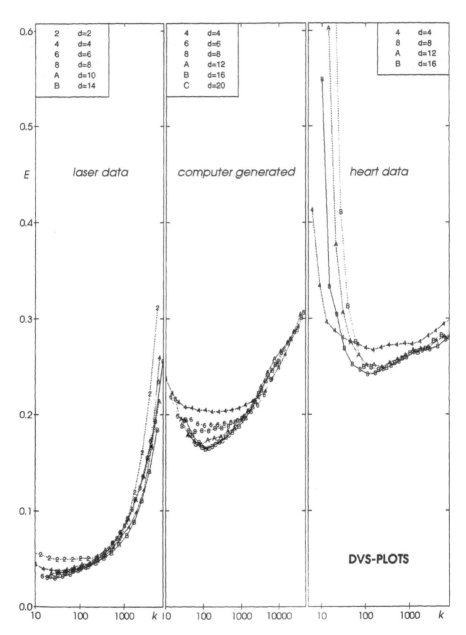

FIGURE 9 Deterministic vs. stochastic plots: The normalized out-of-sample error E is shown as function of the number of neighbors k used to construct a local linear model of order d.

6.3.1 CHARACTERIZATION VIA LOCAL LINEAR MODELS: DVS PLOTS. Forecasting models often possess some "knobs" that can be tuned. The dependence of the prediction error on the settings of these knobs can sometimes reveal—indirectly—some information about properties of the system. In local linear modeling, examples of such knobs are the number of delay values d and the number of neighbors k used to construct the local linear model. Casdagli (1991) introduces the term "deterministic vs. stochastic modeling" (DVS) for the turning of these knobs, and Casdagli and Weigend (this volume) apply the idea to the competition data.

In Figure 9 we show the out-of-sample performance for a local linear model on three of the Santa Fe data sets (laser data A.con, computer-generated data D1.dat, and the heart data B2.dat) as a function of the number of neighbors k used for the linear fit. The left side of each plot corresponds to a simple look-up of the neighbor closest in lag space; the right corresponds to a global linear model that fits a hyperplane through all points. In all three panels the scale of the y-axis (absolute errors) is the same. (Before applying the algorithm, all series were normalized to unit variance.)

The first observation is the overall size of the out-of-sample errors:[35] The laser data are much more predictable than the computer-generated data, which in turn are more predictable than the heart data. The second observation concerns the shape of the three curves. The location of the minimum shifts from the left extreme for the laser (next-neighbor look-up), through a clear minimum for the computer-generated data (of about one hundred neighbors), to a rather flat behavior beyond a sharp drop for the heart data.

So far, we have framed this discussion along the axis (local linear)↔(global linear). We now stretch the interpretation of the fit in order to infer properties of the generating system. Casdagli (1991) uses the term "deterministic" for the left extreme of local linear models, and "stochastic" for the right extreme of global linear models. He motivates this description with computer-generated examples as well as with experimental data from cellular flames, EEG, fully developed turbulence, and the sunspot series. Deterministic nonlinear systems are often successfully modeled by local linear models: the error is small for small neighborhoods, but increases as more and more neighbors are included in the fit. When the neighborhood size is too large, the hyperplane does not accurately approximate the (nonlinear) manifold of the data, and the out-of-sample error increases. This is indeed the case for the laser. The local linear forecasts are 5 to 10 times more accurate than global linear

---

[35] We give average absolute errors since they are more robust than squared errors, but the qualitative behavior is the same for squared errors. The numerical values are obtained by dividing the sum of these "linear" errors by the size of the test example (500 points in all three series). Furthermore, the delay between each past sample ("lag time") is chosen to be one time step in all three series, but the prediction time ("lead time") is chosen to be one time step for the laser, two time steps for the computer-generated data, and four time steps for the heart rate predictions to reflect the different sampling rates with respect to the timescale of the dynamics. The predictions for the heart data were obtained from a bivariate model, using both the heart rate and the chest volume as input. The details are given in the article by Casdagli and Weigend (this volume).

ones, suggesting the interpretation of the laser as a nonlinear deterministic chaotic system. On the other hand, if a system is indeed linear and stochastic, then using smaller neighborhoods makes the out-of-sample predictions worse, due to overfitting of the noise. This is apparent for the heart data, which clearly shows overfitting for small neighborhood sizes and therefore rules out simple deterministic chaos. The DVS plot alone is not powerful enough to decide whether nonlinearities are present in the system or not.[36] Finally, the middle example of computer-generated data falls between these two cases. Nonlinearity, but not low-dimensional chaos, is suggested here since the short-term forecasts at the minimum are between 50% to 100% more accurate than global linear models.

Apart from the number of neighbors, the other knob to turn is the order of the linear model, i.e., the number of time delays d. The order of the AR model that makes successful predictions provides an upper bound on the minimum embedding dimension. For the laser, $d = 2$ is clearly worse than $d = 4$, and the lowest out-of-sample errors are reached for $d = 6$. This indeed is an upper estimate for $d = 3$ (compare to Figures 5 and 8). For the computer-generated data, the quality of the predictions continually increases from $d = 4$ to $d = 12$ and saturates at $d = 16$. For the heart data—since the DVS plots give no indications of low-dimensional chaos—it does not make sense to give an embedding dimension into which the geometry can be disambiguated.

We close this section with a statistical perspective on DVS plots: although their interpretation is necessarily somewhat qualitative, they nicely reflect the trade-off between bias and variance. A weak local linear model has a low bias (it is very flexible), but the parameters have a high variance (since there are only a few data points for estimating each parameter). A strong global linear model has a large model bias, but the parameters can be estimated with a small variance since many data points are available to determine the fit (see, e.g., Geman, Bienenstock, & Doursat, 1992). The fact that the out-of-sample error is a combination of "true" noise and model mismatch is not limited to DVS plots but should be kept in mind as a necessary limitation of any error-based analysis.

6.3.2 CHARACTERIZATION VIA CONNECTIONIST MODELS. Connectionist models are more flexible than local linear models. We first show how it is possible to extract characteristic properties such as the minimal embedding dimension or the manifold dimension from the network, and then indicate how the network's emulation of the system can be used to estimate Lyapunov coefficients.

For simple feedforward networks (i.e., no direct connections between input and output, and no feedback loops), it is relatively easy to see how the hidden units can be used to discover hidden dimensions:

[36] A comparison to DVS plots for financial data (not shown here) suggests that there are more nonlinearities in the heart than in financial data.

■ **Vary network size.** In the early days of backpropagation, networks were trained with varying numbers of hidden units and the "final" test error (when the training "had converged") was plotted as a function of the number of hidden units: it usually first drops and then reaches a minimum; the number of hidden units when the minimum is reached can be viewed as a kind of measure of the degrees of freedom of the system. A similar procedure can determine a kind of embedding dimension by systematically varying the number of input units.

Problems with this approach are that these numbers can be strongly influenced by the *choice of the activation function* and the search algorithm employed. Ignoring the search issue for the moment: if, for example, sigmoids are chosen as the activation function, we obtain the manifold dimension *as expressed by sigmoids,* which is an upper limit to the true manifold dimension. Saund (1989) suggested, in the context of nonlinear dimensionality reduction, to sandwich the hidden layer (let us now call it the "central" hidden layer) between two (large) additional hidden layers. An interpretation of this architecture is that the time-delay input representation is transformed nonlinearly by the first "encoding" layer of hidden units; if there are more hidden units than inputs, it is an expansion into a higher dimensional space. The goal is to find a representation that makes it easy for the network subsequently to parametrize the manifold with as few parameters as possible (done by the central hidden layer). The prediction is obtained by linearly combining the activations of the final "decoding" hidden layer that follows the central hidden layer.[37]

Although an expansion with the additional sandwich layers reduces the dependence on the specific choice of the activation function, a small size of the bottleneck layer can make the *search* (via gradient descent in backpropagation) hard: overfitting even occurs for small networks, before they have reached their full potential (Weigend, in preparation). There are two approaches to this problem: to penalize network complexity, or to use an oversized network and analyze it.

■ **Penalize network complexity or prune.** Most of the algorithms that try to produce small networks have been applied to time series prediction, e.g., "weight elimination" (Weigend, Huberman, & Rumelhart, 1990), "soft

---

[37]This approach can also be used for cleaning and for compressing time series. In cleaning, we use a filter network that tries to remove noise in the series: the output hopefully corresponds to a noise-reduced version of the signal corresponding to a time at the center of the input time window, rather than a prediction beyond the window. In compression, the network is to reproduce the entire input vector at the output after piping it through a bottleneck layer with a small number of hidden units. The signal at the hidden units is a compressed version of the larger input vector. The three cases of prediction, cleaning, and compression are just different parametrizations of the same manifold, allocating more resources to the areas appropriate for the specific task.

weight sharing" (Nowlan & Hinton, 1992), and "optimal brain damage" (developed by le Cun, Denker, & Solla, 1990, and applied to the time series by Svarer, Hansen, & Larsen, 1993). All of these researchers apply their algorithms to the sunspot series and end up with networks of three hidden units. Finding appropriate parameters for the regularizer can be tricky; we now give a method for dimension estimation that does not have such parameters but does require post-training analysis.

- **Analyze oversized networks.** The idea here is to use a large network that easily reaches the training goal (and also easily overfits).[38] The spectrum of the eigenvalues of the *covariance matrix of the (central) hidden unit activations* is computed as a function of training time. The covariance $C_{ij} = \langle (S_i - \overline{S}_i)(S_j - \overline{S}_j) \rangle$ describes the two-point interaction between the activations of the two hidden units i and j. ($\overline{S}_i = \langle S_i \rangle$ is the average activation of hidden unit i.) The number of significantly sized eigenvalues of the covariance matrix (its effective rank) serves as a measure of the effective dimension of the hidden unit space (Weigend & Rumelhart, 1991a, 1991b). It expresses the number of parameters needed to parametrize the solution manifold of the dynamical system in terms of the primitives. Using the sandwich-expansion idea (described on the previous page), the effect of the specific primitives can be reduced. We have also used mutual information to capture dependencies between hidden units; this measure is better suited than linear correlation or covariance if the hidden units have nonlinear functional relations.

All of these approaches have to be used with caution as estimates of the true dimension of the manifold. We have pointed out above that the estimate can be too large (for example, if the sigmoid basis functions are not suitable for the manifold, or if the network is overfitting). But it can also be too small (for example, if the network has essentially learned nothing), as often is the case for financial data (either because there is nothing to be emulated or because the training procedure or the architecture was not suited to the data).

In addition to dimension, networks can be used to extract other properties of the generating system. Here we point to a few possibilities that help locate a series within the space of attributes outlined in Figure 2.

Nonlinearity. DVS plots analyze the error as a function of the nonlinearity of the model (smaller neighborhoods ⇒ more nonlinear). Rather than basing the analysis on the errors, we can use a property of the network to characterize the amount of

---

[38] The network is initialized with very small weights—large enough to break the symmetry but small enough to keep the hidden sigmoids in their linear range. The weights grow as the network learns. In this sense, training time can be viewed in a regularization framework as a complexity term that penalizes weights according their size, strongly at first, and later relaxes.

nonlinearity. Weigend et al. (1990) analyze the distribution of the activations S of sigmoidal hidden units: they show that the ratio of the quadratic part of the Taylor expansion of a sigmoid with respect to the linear part, i.e., $|f''(\xi)|/|f'(\xi)|$, can be expressed in terms of network parameters (the activation S, the net-input ξ, the activation function f, and the slope a are defined in Section 4.2) as $(a|1-2S|)$. The distribution of this statistic (averaged over patterns and hidden units) can be used in addition to the simple comparison of the out-of-sample error of the network to the out-of-sample error of a linear model.

Lyapunov exponents. It is notoriously difficult to estimate Lyapunov exponents from short time records of noisy systems. The hope is that if a network has reliably learned how to emulate such a system, the exponents can be found through the network. This can be done by looking at the out-of-sample errors as a function of prediction time (Weigend et al., 1990), by using the Jacobian of the map implemented by the network (Gencay & Dechert, 1992; Nychka et al., 1992), or by using the trained network to generate time series of any length needed for the application of standard techniques (Brown et al., 1991).

This section on characterization started with important simple tests that apply to any data set and algorithm, continued with redundancy as an example of the detailed information than can be found by analyzing embedded data, and closed with learning algorithms that are more generally applicable but less explicitly understandable. Connectionist approaches to characterization, which throw a broad model (and a lot of computer time) at the data, should be contrasted with the traditional statistical approach of building up nonlinearities by systematically adding terms to a narrow model (which can be estimated much faster than a neural network can be trained) and hoping that the system can be captured by such extensions. This is another example of the central theme of the trade-off that must be made between model flexibility and specificity. The blind (mis)application of these techniques can easily produce meaningless results, but taken together and used thoughtfully, they can yield deep insights into the behavior of a system of which a time series has been observed.

These themes have recurred throughout our survey of new techniques for time series forecasting and characterization. We have seen results that go far beyond what is possible within the canon of linear systems analysis, but we have also seen unprecedented opportunities for the analysis to go astray. We have shown that it can be possible to anticipate, detect, and prevent such errors, and to relate new algorithms to traditional practice, but these steps, as necessary as they are, require significantly more effort. The possibilities that we have presented will hopefully help motivate such an effort. We close this chapter with some thoughts about future directions.

7. THE FUTURE

We have surveyed the results of what appears to be a steady progress of insight over ignorance in analyzing time series. Is there a limit to this development? Can we hope for the discovery of a universal forecasting algorithm that will predict everything about all time series? The answer is emphatically "no!" Even for completely deterministic systems, there are strong bounds on what can be known. The search for a universal time series algorithm is related to Hilbert's vision of reducing all of mathematics to a set of axioms and a decision procedure to test the truth of assertions based on the axioms (*Entscheidungsproblem*); this culminating dream of mathematical research was dramatically dashed by Gödel (1931)[39] and then by Turing. The most familiar result of Turing is the undecidability of the halting problem: it is not possible to decide in advance whether a given computer program will eventually halt (Turing, 1936). But since Turing machines can be implemented with dynamical systems (Fredkin, 1982; Moore, 1991), a universal algorithm that can directly forecast the value of a time series at any future time would need to contain a solution to the halting problem, because it would be able to predict whether a program will eventually halt by examining the program's output. Therefore, there cannot be a universal forecasting algorithm.

The connection between computability and time series goes deeper than this. The invention of Turing machines and the undecidability of the halting problem were side results of Turing's proof of the existence of uncomputable real numbers. Unlike a number such as π, for which there is a rule to calculate successive digits, he showed that there are numbers for which there cannot be a rule to generate their digits. If one was unlucky enough to encounter a deterministic time series generated by a chaotic system with an initial condition that was an uncomputable real number, then the chaotic dynamics would continuously reveal more and more digits of this number. Correctly forecasting the time series would require calculating unseen digits from the observed ones, which is an impossible task.

Perhaps we can be more modest in our aspirations. Instead of seeking complete future knowledge from present observations, a more realistic goal is to find the best model for the data, and a natural definition of "best" is the model that requires the least amount of information to describe it. This is exactly the aim of *Algorithmic Information Theory*, independently developed by Chaitin (1966), Kolmogorov (1965), and Solomonoff (1964). Classical information theory, described in Section 6.2, measures information with respect to a probability distribution of an ensemble of observations of a string of symbols. In algorithmic information theory, information is measured within a single string of symbols by the number of bits needed to specify the shortest algorithm that can generate them. This has led to significant extensions of Gödel and Turing's results (Chaitin, 1990) and, through

---

[39] Hofstadter paraphrases Gödel's Theorem as: "All consistent axiomatic formulations of number theory include undecidable propositions." (Hofstadter, 1979, p. 17).

the *Minimum Description Length* principle, it has been used as the basis for a general theory of statistical inference (Wallace & Boulton, 1968; Rissanen, 1986, 1987). Unfortunately, here again we run afoul of the halting problem. There can be no universal algorithm to find the shortest program to generate an observed sequence because we cannot determine whether an arbitrary candidate program will continue to produce symbols or will halt (e.g., see Cover & Thomas, 1991, p.162).

Although there are deep theoretical limitations on time series analysis, the constraints associated with specific domains of application can nevertheless permit strong useful results (such as those algorithms that performed well in the competition), and can leave room for significant future development. In this chapter, we have ignored many of the most important time series problems that will need to be resolved before the theory can find widespread application, including:

- *Building parametrized models for systems with varying inputs.* For example, we have shown how the stationary dynamics of the laser that produced Data Set A can be correctly inferred from an observed time series, but how can a family of models be built as the laser pump energy is varied in order to gain insight into how this parameter enters into the governing equations? The problem is that for a linear system it is possible to identify an internal transfer function that is independent of the external inputs, but such a separation is not possible for nonlinear systems.

- *Controlling nonlinear systems.* How can observations of a nonlinear system and access to some of its inputs be used to build a model that can then be used to guide the manipulation of the system into a desired state? The control of nonlinear systems has been an area of active research; approaches to this problem include both explicit embedding models (Hübler, 1989; Ott, Grebogi & Yorke, 1990; Bradley, 1992) and implicit connectionist strategies (Miller, Sutton, & Werbos, 1990; White & Sofge, 1992).

- *The analysis of systems that have spatial as well as temporal structure.* The transition to turbulence in a large aspect-ratio convection cell is an experimental example of a spatio-temporal structure (Swinney, this volume), and cellular automata and coupled map lattices have been extensively investigated to explore the theoretical relationship between temporal and spatial ordering (Gutowitz, 1990). A promising approach is to extend time series embedding (in which the task is to find a rule to map past observations to future values) to spatial embedding (which aims to find a map between one spatial, or spatio-temporal, region and another). Unfortunately, the mathematical framework underlying time-delay embedding (such as the uniqueness of state-space trajectories) does not simply carry over to spatial structures. Afraimovich et al. (this volume) explore the prospects for developing such a theory of spatial embedding. A related problem is the combination of data from different sources, ranging from financial problems (using multiple market indicators) to medical data (such as Data Set B).

- *The analysis of nonlinear stochastic processes.* Is it possible to extend embedding to extract a model for the governing equations for a signal generated by a stationary nonlinear stochastic differential equation? Probabilistic approaches to embedding such as the use of expectation values (Eq. (16)) and hidden Markov models (Fraser & Dimitriadis, this volume) may point toward an approach to this problem.

- *Understanding versus learning.* How does understanding (explicitly extracting the geometrical structure of a low-dimensional system) relate to learning (adaptively building models that emulate a complex system)? When a neural network correctly forecasts a low-dimensional system, it has to have formed a representation of the system. What is this representation? Can it be separated from the network's implementation? Can a connection be found between the entrails of the internal structure in a possibly recurrent network, the accessible structure in the state-space reconstruction, the structure in the time series, and ultimately the structure of the underlying system?

The progress in the last decade in analyzing time series has been remarkable and is well witnessed by the contributions to this volume. Where once time series analysis was shaped by linear systems theory, it is now possible to recognize when an apparently complicated time series has been produced by a low-dimensional nonlinear system, to characterize its essential properties, and to build a model that can be used for forecasting. At the opposite extreme, there is now a much richer framework for designing algorithms that can learn the regularities of time series that do not have a simple origin. This progress has been inextricably tied to the arrival of the routine availability of significant computational resources (making it possible to collect large time series, apply complex algorithms, and interactively visualize the results), and it may be expected to continue as the hardware improves.

General access to computer networks has enabled the widespread distribution and collection of common material, a necessary part of the logistics of the competition, thereby identifying interesting results where they once might have been overlooked (such as from students, and from researchers in countries that have only recently been connected to international computer networks). This synchronous style of research is complementary to the more common asynchronous mode of relatively independent publication, and may be expected to become a familiar mechanism for large-scale scientific progress. We hope that, in the short term, the data and results from this competition can continue to serve as basic reference benchmarks for new techniques. We also hope that, in the long term, they will be replaced by more worthy successors and by future comparative studies building on this experience.

In summary, in this overview chapter we started with linear systems theory and saw how ideas from fields such as differential topology, dynamical systems, information theory, and machine learning have helped solve what had appeared to be very difficult problems, leading to both fundamental insights and practical applications. Subject to some ultimate limits, there are grounds to expect significant further extensions of these approaches for handling a broader range of tasks. We predict that a robust theory of nonlinear time series prediction and analysis (and nonlinear signal processing in general) will emerge that will join spectrum analysis and linear filters in any scientist's working toolkit.

ACKNOWLEDGMENTS

An international interdisciplinary project such as this requires the intellectual and financial support of a range of institutions; we would particularly like to thank the Santa Fe Institute, the NATO Science and Technology Directorate, the Harvard Society of Fellows, the MIT Media Laboratory, the Xerox Palo Alto Research Center, and, most importantly, all of the researchers who contributed to the Santa Fe Time Series Competition.

APPENDIX

We describe the prediction tasks and competition results in more detail. We have included only those entries that were received by the close of the competition; many people have since refined their analyses. This appendix is not intended to serve as an exhaustive catalog of what is possible; it is just a basic guide to what has been done.

DATA SET A: LASER

Submissions were evaluated in two ways. First, using the predicted values $\hat{x}_k$ only (in addition to the observed values x_k), we compute the *normalized mean squared error*:

$$\text{NMSE}(N) = \frac{\sum_{k \in \mathcal{T}} \left(\text{observation}_k - \text{prediction}_k\right)^2}{\sum_{k \in \mathcal{T}} \left(\text{observation}_k - \text{mean}_{\mathcal{T}}\right)^2} \approx \frac{1}{\hat{\sigma}_{\mathcal{T}}^2} \frac{1}{N} \sum_{k \in \mathcal{T}} (x_k - \hat{x}_k)^2 ,$$

(44)

where $k = 1 \cdots N$ enumerates the points in the withheld test set $\mathcal{T}$, and $\text{mean}_{\mathcal{T}}$ and $\hat{\sigma}_{\mathcal{T}}^2$ denote the sample average and sample variance of the observed values (targets) in $\mathcal{T}$. A value of NMSE = 1 corresponds to simply predicting the average.

Second, the submitted error bars $\hat{\sigma}_k$ were used to compute the likelihood of the observed data, given the predicted values and the predicted error bars, based

TABLE 2 Entries received before the deadline for the prediction of Data Set A (laser). We give the normalized mean squared error (NMSE), and the negative logarithm of the likelihood of the data given the predicted values and predicted errors. Both scores are averaged over the prediction set of 100 points.

code	method	type	computer	time	NMSE(100)	-log(lik.)
W	conn	1-12-12-1; lag 25,5,5	SPARC 2	12 hrs	0.028	3.5
Sa	loc lin	low-pass embd, 8 dim, 4nn	DEC 3100	20 min	0.080	4.8
McL	conn	feedforward, 200-100-1	CRAY Y-MP	3 hrs	0.77	5.5
N	conn	feedforward, 50-20-1	SPARC 1	3 weeks	1.0	6.1
K	visual	look for similar stretches	SG Iris	10 sec	1.5	6.2
L	visual	look for similar stretches			0.45	6.2
M	conn	feedforward, 50-350-50-50	386 PC	5 days	0.38	6.4
Can	conn	recurrent, 4-4c-1	VAX 8530	1 hr	1.4	7.2
U	tree	k-d tree; AIC	VAX 6420	20 min	0.62	7.3
A	loc lin	21 dim, 30 nn	SPARC 2	1 min	0.71	10.
P	loc lin	3 dim time delay	Sun	10 min	1.3	–
Sw	conn	feedforward	SPARC 2	20 hrs	1.5	–
Y	conn	feedforward, weight-decay	SPARC 1	30 min	1.5	–
Car	linear	Wiener filter, width 100	MIPS 3230	30 min	1.9	–

on an assumption of independent Gaussian errors. Although this assumption may not be justified, it provides a simple form that captures some desirable features of error weighting. Since the original real-valued data is quantized to integer values, the probability of seeing a given point x_k is found by integrating the Gaussian distribution over a unit interval (corresponding to the rounding error of 1 bit of the analog-to-digital converter) centered on x_k:

$$p(x_k|\widehat{x}_k, \widehat{\sigma}_k) = \frac{1}{\sqrt{2\pi\widehat{\sigma}_k^2}} \int_{x_k-0.5}^{x_k+0.5} \exp\left(\frac{-(\xi - \widehat{x}_k)^2}{2\widehat{\sigma}_k^2}\right) d\xi. \quad (45)$$

If the predicted error is large, then the computed probability will be relatively small, independent of the value of the predicted point. If the predicted error is small and the predicted point is close to the observed value, then the probability will be large, but if the predicted error is small and the prediction is not close to the observed value, then the probability will be very small. The potential reward, as well as the risk, is greater for a confident prediction (small error bars). Under the assumption of independent errors, the likelihood of the whole test for the the observed data given the submitted model is then

$$p(\mathrm{D}|\mathrm{M}) = \prod_{k=1}^{N} p(x_k|\widehat{x}_k, \widehat{\sigma}_k) \,. \tag{46}$$

Finally, we take the logarithm of this probability of the data given the model (this turns products into sums and also avoids numerical problems), and then scale the result by the size of the data set N. This defines the *negative average log likelihood*:

$$-\frac{1}{N} \sum_{k=1}^{N} \log p(x_k|\widehat{x}_k, \widehat{\sigma}_k) \,. \tag{47}$$

We give these two statistics for the submitted entries for Data Set A in Table 2 along with a brief summary of the entries. The best two entries (W=Wan, Sa=Sauer) are shown in Figures 3 and 4 in the main text. Figure 10 displays all predictions received for Data Set A. The symbols ($\times$) correspond to the predicted values, the vertical bars to the submitted error bars. The true continuation points are connected by a grey line (to guide the eye).

DATA SET D: COMPUTER-GENERATED DATA

In order to provide a relatively long series of known high-dimensional dynamics (between the extremes of Data Set A and Data Set C) with weak nonstationarity, we generated 100,000 points by numerically integrating the equations of motion for a damped, driven particle

$$\frac{d^2 x}{dt^2} + \gamma \frac{dx}{dt} + \nabla V(\mathbf{x}) = \mathbf{F}(t) \tag{48}$$

TABLE 3 Entries received before the deadline for the prediction of Data Set D (computer-generated data).

index	method	type	computer	time	NMSE(15)	NMSE(30)	NMSE(50)
ZH	conn	...-30-30-1 & 30-100-5	CM-2 (16k)	8 days	0.086	0.57	0.87
U	tree	k-d tree; AIC	VAX 6420	30 min	1.3	1.4	1.4
C	conn	recurrent, 4-4c-1	VAX 8530	n/a	6.4	3.2	2.2
W	conn	1-30-30-1; lags 20,5,5	SPARC 2	1 day	7.1	3.4	2.4
Z	linear	36 AR(8), last 4k pts.	SPARC	10 min	4.8	5.0	3.2
S	conn	feedforward	SPARC 2	20 hrs	17.	9.5	5.5

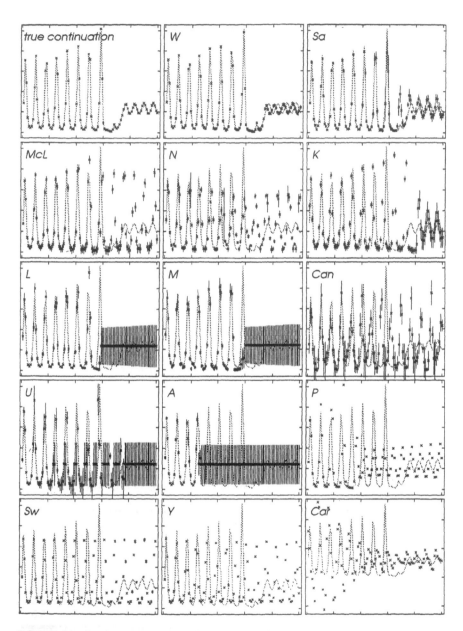

FIGURE 10   Continuations of Data Set A (laser). The letters correspond to the code of the entrant. Grey lines indicate the true continuation, × the predicted values, and vertical bars the predicted uncertainty.

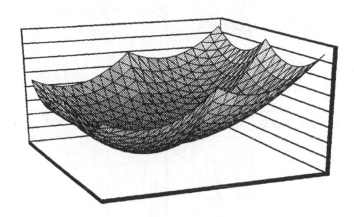

FIGURE 11   The potential $V(\mathbf{x})$ for Data Set D, plotted above the (x_1, x_2) plane.

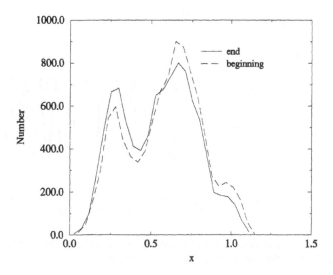

FIGURE 12   Histogram of the probability distribution at the beginning and end of Data Set D, indicating three observable states and the small drift in the relative probabilities.

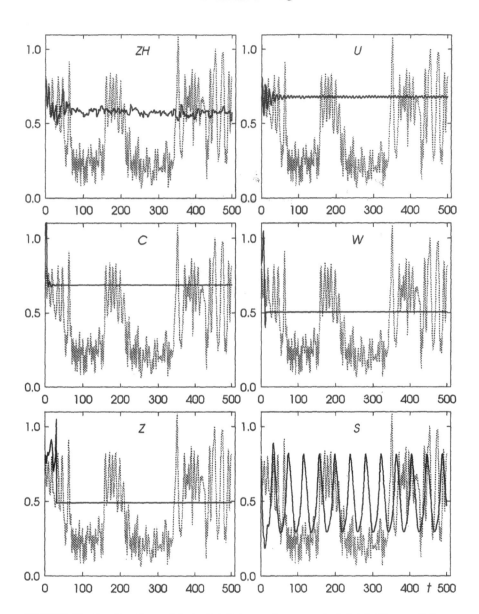

FIGURE 13 Continuations of Data Set D (computer-generated data). The predictions are shown as black lines, the true continuation is indicated in grey.

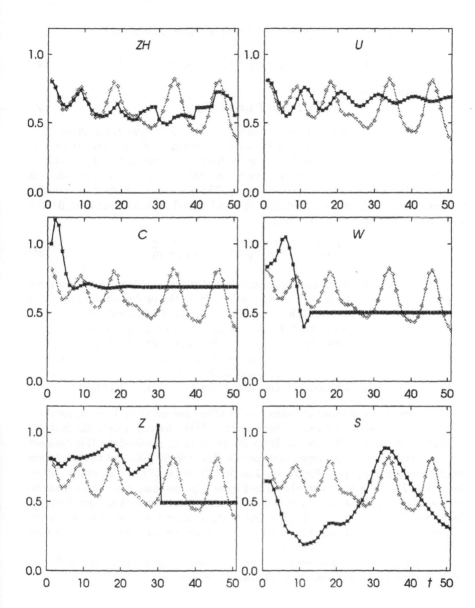

FIGURE 14 Continuations of first 50 points of Data Set D. The predictions are shown as black lines, the true continuation is indicated in grey.

in an asymmetrical four-dimensional four-well potential (Figure 11)

$$V(\mathbf{x}) \; = \; A_4 \left(x_1^2 + x_2^2 + x_3^2 + x_4^2 \right)^2 - A_2 \, |x_1 x_2| - A_1 x_1 \tag{49}$$

with periodic forcing

$$\mathbf{F}(t) = F \sin(\omega t) \, \widehat{\mathbf{x}}_3 \,. \tag{50}$$

$\mathbf{x} = (x_1, x_2, x_3, x_4)$ denotes the location of the particle. This system has nine degrees of freedom (four position, four velocity, and one forcing time). The equations were integrated with 64-bit real numbers using a fixed-step fourth-order Runge-Kutta algorithm (to eliminate possible coupling to an active stepper). The potential has four wells that are tilted by the parameter A_1. This parameter slowly drifted during the generation of the data according to a small biased random walk from 0.02 to 0.06. As a scalar observable we chose

$$\sqrt{(x_1 - 0.3)^2 + (x_2 - 0.3)^2 + x_3^2 + x_4^2} \,. \tag{51}$$

This observable projects the four wells onto three distinguishable states. These three states, and the effect of the drift of A_1 can be seen in the probability distributions at the beginning and the end of the data set (Figure 12). The magnitude of the drift is chosen to be small enough such that the nature of the dynamics does not change, but large enough such that the relative probabilities are different at the beginning and end of the data set. Data Set D was generated with $A_4 = 1$, $A_2 = 1$, A_1 drifting from 0.02 to 0.06, $\gamma = 0.01$, $F = 0.135$, and with $\omega = 0.6$ (for these parameters, in the absence of the weak forcing by the drift, this system is not chaotic).

The contributions received before the deadline are listed in Table 3; they are evaluated by normalized mean squared error, NMSE, averaged over the first 15, 30, and 50 predictions. Figure 13 shows the predictions received for this data set over the entire prediction range of 500 points. The first 50 points of the prediction range are plotted again in Figure 14. The first 15 steps of the best entry (ZH = Zhang and Hutchinson) is also shown in Figure 7 in the main text, in addition to the prediction of the evolution of the probability density function, submitted by Fraser and Dimitriadis after the deadline of the competition. On this time scale, the spread of an ensemble of continuations due to the stochasticity of the algorithm used to generate Dat Set D is small ($\sim 1\%$).

The program and parameter file that we used to generate Data Set D are available through anonymous ftp to `ftp.santafe.edu` .

I. Description of the Data Sets

Udo Hübner,† **Carl-Otto Weiss,**† **Neal Broadus Abraham,**†‡ **and Dingyuan Tang**†

†Phys.-Techn. Bundesanstalt, D-3300 Braunschweig, Germany.
‡Permanent address: Department of Physics, Bryn Mawr College, Bryn Mawr, PA 19010.

Lorenz-Like Chaos in NH₃-FIR Lasers (Data Set A)

I. INTRODUCTION

Spontaneous periodic and chaotic pulsations have been observed in many fields of physics. At least two of those fields, the physics of fluids and lasers, may even show exactly the same dynamics (i.e., obeying the same nonlinear differential equations) if the relevant physical parameters are chosen appropriately. A compact and minimal set of equations for convective fluid dynamics was developed by Lorenz (1963) for the case of the Rayleigh-Bénard convection. Later, Haken (1975) found that the equations of a single-mode laser field interacting with a homogeneously broadened two-level medium are exactly the same as those so-called Lorenz equations. Of course, variables and parameters must be interpreted differently in the case of the laser.

Far-infrared (FIR) lasers have been proposed (Weiss & Klische, 1984) as examples of a realization of the Lorenz-Haken model. Some experiments on such lasers for certain operating conditions (Klische & Weiss, 1985; Weiss & Brock, 1986; Weiss et al., 1988) have shown remarkable similarities with the predictions of that model, including characteristic time-dependent behavior and appropriate instability thresholds.

Times Series Prediction: Forecasting the Future and
Understanding the Past, Eds. A. S. Weigend and N. A. Gershenfeld, SFI Studies
in the Sciences of Complexity, Proc. Vol. XV, Addison-Wesley, 1993   **73**

The obvious correspondence between the dynamics observed on NH_3 laser transitions and the Lorenz-Haken model produced some controversy as to whether such optically pumped lasers can be properly modeled as incoherently pumped two-level lasers (Dupertuis et al., 1987; Khandokhin et al., 1988; Moloney et al., 1986; Ryan & Lawandy, 1987; Uppal et al., 1987). The principal objection is that the laser excitation leads to coherences in the three-level molecular system which cannot be adiabatically eliminated by reducing the equations to an equivalent two-level formulation. The three-level models differ from the Lorenz-Haken model in that they have lower instability thresholds, supercritical bifurcations to periodic pulsations, and (more typical) asymmetric periodic and chaotic attractors.

However, the actual laser systems are more complex than simple coherently coupled three-level systems. Complications arise from (a) detuned pumping of the Doppler-broadened pump transition to achieve unidirectional operation (Weiss et al., 1985), (b) operation in the backward emission direction, and (c) involvement of high angular momentum states with the magnetic sublevels mixed by orthogonally polarized fields.

In a theoretical model with more than 1,000 equations (Laguarta et al., 1988) for single-mode FIR laser dynamics, factors (a) and (b) have been incorporated. That model finally recovered the experimental thresholds, detuning pump-bifurcation diagram, and intensity pulsation forms. More recent studies have separately considered the effects of factor (c) demonstrating that this contributes to Lorenz-like phenomena.

This experimental and theoretical situation motivated us to complement our first experimental results with new measurements (Hübner, Abraham, & Weiss, 1989, 1992; Hübner et al., 1989, 1990). A group of new data sets were taken which covered as many pulsation types as possible. Those data sets were then analyzed to obtain several of the important quantities pertinent in comparison to the parameters of numerical data sets which we obtained by the integration of the Lorenz-Haken equations.

This paper describes some of the theoretical background information (Lorenz model and its equivalence to laser physics in Section II), the experimental setup and the data sets collected (Section III), and their numerical analysis together with the comparison to the numerical data sets calculated by the integration of the Lorenz-Haken equations (Section IV). A final discussion (Section V) reveals small deviations from Lorenz-type dynamics in our data.

II. THE LORENZ-HAKEN MODEL

The laser experiment to be discussed here can be, as mentioned, directly related to the Rayleigh-Bénard convection experiment in fluid dynamics. In the convection experiment, two plates, separated vertically d units apart (Figure 1), have different temperatures. The lower plate is heated so that the temperature gradient points in the opposite direction of the gravitational gradient.

If the temperature lies below some critical value (i.e., the Rayleigh number R is below the critical Rayleigh number R_c), there will be only heat conduction. Above this value, convection starts forming rolls in the fluid which transport the heat. Furthermore, above a higher threshold there is chaotic motion of the fluid.

In the theoretical analysis of this experiment, Lorenz reduced the complexity of the nonlinear, partial differential equations of Navier and Stokes by expanding the velocity and the temperature fields into spatial Fourier series. The coefficients of the series were kept as time-dependent variables. Lorenz retained only the three main variables which we call X, Y, and Z. Thus, X, Y, and Z are some amplitudes. Their physical meaning does not matter in our present context. But the structure of the equations, the parameters, and their values are all directly related to our laser problem.

The Lorenz equations are:

$$\frac{dX}{dt} = \sigma Y - \sigma X\,,$$
$$\frac{dY}{dt} = -XZ + rX - Y\,, \tag{1}$$
$$\frac{dZ}{dt} = XY - bZ\,,$$

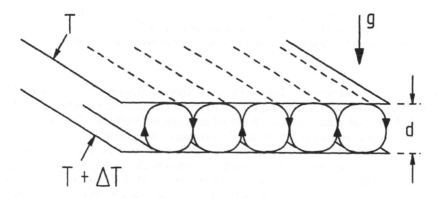

FIGURE 1 Convection rolls found in Rayleigh-Bénard experiments for $1 < r < 24.74$. The rolls are rectilinear and their axes are parallel. The gravitational (g) and the temperature (T) gradients point in opposite directions.

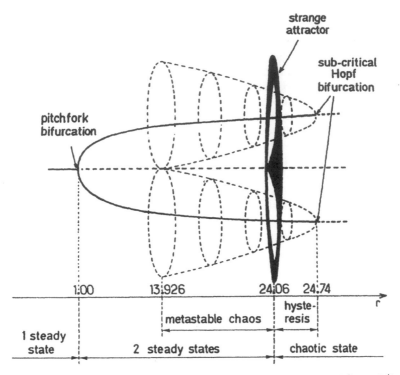

FIGURE 2 Bifurcation diagram of the Lorenz model for $\sigma = 10$ and $b = 8/3$. The ordinate can be taken as one of the amplitudes X or Y. Dashed lines represent unstable states and the solid lines stable states. (Taken from Tresser, 1981.)

with σ = Prandtl number (kinetic viscosity/thermal diffusivity); $r = R/R_c$ (R: Rayleigh number; R_c: critical Rayleigh number for the onset of convection); and $b = 4\pi^2/(\pi^2 + k^2)$ (k: dimensionless wave number connected with the rolls' periodicity).

The Lorenz equations have a simple structure and contain only two nonlinear terms, XZ and XY. Nevertheless, they can show quite irregular behavior. There is a vast amount of literature dealing with the Lorenz equations. An extensive analysis may be found in Sparrow (1982). Figure 2 from Tresser (1981) shows the bifurcation scheme for the case $\sigma = 10$ and $b = 8/3$ where r is the running control parameter.

There are two main bifurcations:

1. For $r = 1$, a supercritical pitchfork bifurcation (the first instability) happens which leads from the stationary trivial solution to two stable nontrivial stationary solutions which correspond to the two possible rotation directions of the rolls.

2. The next bifurcation (the second threshold) at

$$r = \frac{\sigma(\sigma + b + 3)}{\sigma - b - 1} \tag{2}$$

happens when $\sigma = 10$ and $b = 8/3$ for $r = 24.74$. It is a subcritical Hopf bifurcation which leads to chaotic pulsations connected with a strange attractor in phase space. It should be mentioned that chaos does not appear through the loss of stability of a periodic trajectory. This means that none of the traditional three routes to chaos (quasi-periodicity, period doubling, and intermittency) fit.

Because of the subcritical nature of the Hopf bifurcation at $r = 24.74$, there is hysteresis in the range $24.06 < r < 24.74$. For the values of r in this interval, three attractors coexist, the two nontrivial stationary solutions and the strange attractor.

For $30.1 < r < 214$, we find regimes alternating between chaos and windows of periodicity. As a prelude to a window of periodicity as r is increased, an inverse cascade of bifurcations is observed. We see a progressive decrease of the dynamical noise which ends with the basic limit cycle (stable pulsations). The end of a window can be identified by the appearance of type I intermittency. Above $r = 214$ the last window begins, and is then terminated above $r = 313$ by the basic limit cycle as a stable attractor.

We now show the connection between the Lorenz equations for fluid dynamics and the laser equations. The laser equations describe a travelling wave in a ring resonator which is excited by atoms or molecules with a homogeneously broadened line. Homogeneously means that, in the NH$_3$ laser case, we are only dealing with one velocity of the molecules.

Additional assumptions are used in the derivation of these equations: (1) Plane waves are used. (2) The atoms are supposed to have only two interacting levels. (3) We apply the so-called *rotating-wave approximation*, which means that the basic quantum mechanical equations were averaged over a time interval which is long compared to the atomic oscillation period, but short compared to times over which the amplitudes of the laser mode vary. This assumption can be used in most practical laser problems. (4) Further simplification of the basic quantum mechanical equations is achieved by the application of the *slowly varying amplitude approximation*, which means that, if we are taking the time derivative of the laser field, we then neglect the derivative of the slowly varying field amplitude while keeping the derivative of the rapidly oscillating field phase.

Thus, we start from the semiclassical laser equations (Haken, 1985):

$$\left(\frac{\partial}{\partial t} + \gamma_\perp\right)\hat{P} = \gamma_\perp \hat{E}\hat{D},$$

$$\left(\frac{\partial}{\partial t} + \gamma_\parallel\right)\hat{D} = \gamma_\parallel(\Lambda + 1) - \gamma_\parallel \Lambda \hat{E}\hat{P}, \tag{3}$$

$$\left(\frac{\partial}{\partial \kappa}\right)\hat{E} = \kappa\hat{P}.$$

Here, $\hat{E}$, $\hat{P}$, and $\hat{D}$ are real, normalized quantities and denote the field, polarization, and atomic inversion, respectively. Λ is the pump parameter, $\gamma_{\parallel}$ and $\gamma_{\perp}$ are relaxation constants of the inversion and polarization. The cavity decay rate of the field in the cavity is called κ. Of course, t is used for the time.

Using the transformation

$$t \rightarrow t' \frac{\sigma}{\kappa'}; \qquad \hat{E} \rightarrow \alpha X; \qquad \hat{P} \rightarrow \alpha Y; \qquad \hat{D} \rightarrow r - Z, \qquad (4)$$

and the substitutions

$$\alpha = \sqrt{b(r-1)}, \qquad r > 1; \qquad \gamma_{\parallel} = \frac{\kappa b}{\sigma}; \qquad \gamma_{\perp} = \frac{\kappa}{\sigma}; \qquad \Lambda = r - 1, \qquad (5)$$

we can transform Eq. (3) to the Lorenz equations (1). In the following we will call them the Lorenz-Haken equations.

We now state the correspondence:

Bénard convection	**Laser**	
σ = Prandtl number	$\sigma = \kappa/\gamma_{\perp}$	(6a)
$r = R/R_c$	$r = \Lambda + 1$	(6b)
$b = \dfrac{4\pi^2}{\pi^2 + k^2}$	$b = \gamma_{\parallel}/\gamma_{\perp}$	(6c)

Chaos for:

$\sigma > b + 1$	$\kappa > \gamma_{\parallel} + \gamma_{\perp}$	(6d)
$r > \dfrac{\sigma(\sigma + b + 3)}{\sigma - b - 1}$	$\Lambda > \dfrac{(\gamma_{\perp} + \kappa)(\gamma_{\parallel} + \gamma_{\perp} + \kappa)}{\gamma_{\perp}(\kappa - \gamma_{\parallel} - \gamma_{\perp})}$	(6e)

Now the variables in Eq. (1) can be interpreted as: X being the normalized electric field amplitude, Y, the normalized polarization, and Z, the normalized inversion.

Of course, the conditions for chaos in the laser are the same as described for the fluid, but σ, b, and r are replaced by their expressions of the relaxation constants and excitation rate. These formulae are more suitable when working with lasers.

The laser parameters σ, b, and r are given in the next section. For those values the subcritical Hopf bifurcation occurs for an r-value of 14 compared to 24.74 in the case of the fluids. No information was found for the hysteresis range in the laser.

For a further check of the robustness of the comparison between theoretical results and measured data, we include the detuning of the resonator which leads to the equations of Zeghlache and Mandel (1985):

$$\frac{dX_1}{dt} = -\sigma X_1 - \sigma d X_2 + \sigma Y_1\,,$$

$$\frac{dX_2}{dt} = -\sigma X_2 + \sigma d X_1 + \sigma Y_2\,,$$

$$\frac{dY_1}{dt} = -X_1 Z + r X_1 - Y_1 + d Y_2\,, \tag{7}$$

$$\frac{dY_2}{dt} = -X_2 Z + r X_2 - Y_2 - d Y_1\,,$$

$$\frac{dZ}{dt} = X_1 Y_1 + X_2 Y_2 - bZ\,,$$

where Z is the normalized inversion as stated before, X_1 and X_2 are the real and imaginary parts of the normalized complex amplitude of the electric field, and Y_1 and Y_2 are real and imaginary parts of the normalized complex amplitude of the polarization. d is the normalized steady-state laser frequency detuning from the molecular resonance frequency, $d = (\omega_A - \omega_L)/\gamma_\perp$, where ω_A denotes the molecular frequency and ω_L the laser frequency. Results from the numerical integration of Eq. (7) are given later.

III. EXPERIMENTAL SETUP AND DATA COLLECTION

Figure 3 shows the setup used in the study of chaotic emission of NH$_3$ lasers. It is essentially identical to that used in the experiments described in Hogenboom et al. (1985) and Weiss & Brock (1986).

The primary concern in the design of this system is to match, as closely as possible, the conditions of the Lorenz model. Thus, the laser must use a travelling-wave (ring) resonator. The laser transition must be homogeneously broadened, the emission must be unidirectional in a single mode, and the resonator must be dissipative enough to fulfill the bad-cavity condition (6d). Moreover, coherence effects—such as the AC Stark splitting of the pump transition—must be kept small enough even when pumping 10 to 20 times above the laser threshold.

The plane-wave condition, assumed in the derivation of the laser version of the Lorenz equations, cannot be realized in a laser. Instead, we used emission in a TEM$_{00}$ waveform, the closest single-mode approximation of plane-wave emission. On the other hand, this means that the laser resonator cannot utilize coupling holes for the admission of the pump radiation and for the outcoupling of the radiation

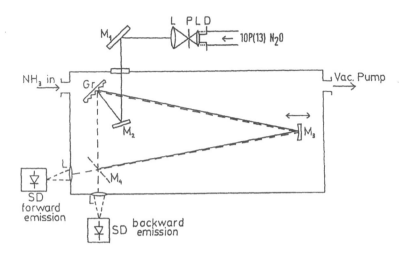

FIGURE 3   Ring laser for observation of chaotic laser dynamics. M: mirrors; Gr: 10 μm grating; SD: Schottky-barrier diodes used as detectors. The usual coupling holes which perturb modes are avoided by the incoupling of the pump radiation via a grating, and the outcoupling via a mesh reflector. Pumping takes place by a 10.7 μm N$_2$O laser. The wavelength generated is 81.5 μm. The pump beam is attenuated without change in geometry, direction, or frequency by a combination of a spatial filter (lens, pinhole diaphragm, lens (LPL)) and iris (diaphragm D).

generated—as is a common technique in FIR lasers. Instead, the pump radiation is introduced into the resonator by the first-order diffraction of a grating (Gr in Figure 3) whose specular reflection is then utilized for the generated far-infrared radiation.

The far-infrared radiation outcoupling is by means of a partially transmitting gold mesh (M$_4$ in Figure 3) of 37 μm (square grid) constant. The emission propagating in the opposite direction of the pump beam is utilized, since its gain line shape is less distorted by the AC-Stark effect than is the gain of the copropagating emission (Heppner et al., 1980).

The laser measurements to be discussed here are made on an 81.5 μm [14]NH$_3$ cw laser (rotational transition aR(7,7)), pumped optically by the P(13) line of an N$_2$O laser via the vibrational aQ(8,7) NH$_3$ transition. Since the pump laser frequency must be known and kept constant within $\approx$ 1 MHz, stable pump conditions were achieved by the control of pump laser frequency with respect to the lamb dip of the aQ(8,7) NH$_3$ absorption line center by the use of an NH$_3$ lamb dip cell.

The pump laser frequency detuning leads to different resonant frequencies for the forward and backward travelling waves (forward denotes FIR emission copropagating with the pump light) which are fully separated when the detuning exceeds the homogeneous linewidth. Laser emission in the opposite direction of the pump beam propagation can thus be achieved by the tuning of resonator length into resonance for the backward wave only. The backward emission line shows no AC Stark splitting (though there is AC Stark broadening). The emission in the direction of the pump beam propagation was nevertheless monitored to assure that there was no bidirectional emission while the data were taken. The backward emission of the FIR laser was detected by a micrometer-sized Schottky-barrier diode, since photon detectors of sufficient sensitivity and speed do not yet exist.

The ring laser parameters used in our studies of single-mode chaotic laser dynamics are approximately:

- resonator perimeter: 2 m,
- pressure of the active gas: 5 to 9 Pa,
- resonator loss: ≈ 20%, and
- pump strength: up to 20 times the laser threshold.

These values correspond to:

- homogeneous linewidth: ≈ 0.5 to 1 MHz, and
- resonator linewidth: ≈ 2 MHz,

so that the bad-cavity condition could be fulfilled. At high operating pressure the bad-cavity condition is no longer maintained because the pressure-broadened line width of the medium exceeds the resonator line width. As expected from the solutions of the Lorenz model, chaotic dynamics is then impossible. The resonator loss can, however, be increased by an iris inside the resonator, which then permits chaotic pulsations also at higher pressure.

At highest pump powers, one can estimate that the AC Stark splitting of the pump transition of the laser medium exceeds—particularly at lower homogeneous linewidths (i.e., at lower pressure)—the homogeneous linewidth, thus definitely violating the Lorenz model conditions. In fact, it is found that under these conditions the chaotic dynamics of the laser shows subtle differences from that of the Lorenz model.

The ratio of the relaxation rates $\gamma_\parallel/\gamma_\perp$ has been determined by experimentation (Leite et al., 1977) and by calculation (Heppner et al., 1980), both agreeing in a value equal to or slightly larger than 0.25.

From all the arguments cited, the typical set of parameters for the Lorenz-Haken equations to describe the NH$_3$ laser is: $r = 1$ to 20, $\sigma = 2$, and $b = 0.25$.

When intensity pulsations were observed on a spectrum analyzer, there was a spectral signal-to-noise ratio of as much as 70 dB in 100 kHz bandwidth limited by the detectors. This corresponds to a signal-to-noise ratio of about 300:1 in terms of peak pulse height to rms noise.

For numerical analysis the signals were digitized with:

- 8-bit resolution,
- an interval of 40 ns between samples[1],
- 25,000 samples per data set, and
- 26 data sets.

The different types of data sets taken included a wide range of pulsations:

- Periodic data sets (periods 2, 4, 8, 12 and periods 2 and 4),
- "period doubling" chaotic pulsations (2 data sets), and
- "Lorenz like" chaotic pulsations (15 data sets).

Short sections of different time series are shown in Figures 4 and 5. Figure 4 shows periodic pulsations and a closely related form of chaotic pulsation (in the following called "period doubling" chaos) which were observed by the detuning of the laser resonator around $d \approx 0.2$. The pressure was held constant at 9 Pa and the pump intensity was about 14 times above threshold. These include the onset of period-doubling chaos (traces (a) and (e)), stable period 8 (trace (b)), period 5 (see first third of trace (c)) which by inverse period doubling (perhaps because of a drifting parameter) switches to period 3, and stable period 12 (trace (d)).

Figure 5 shows a variety of measured "spiral" chaotic pulsations (in the following called "Lorenz-like chaos") found for near-resonant tuning of the FIR laser resonator. The envelopes of successive spirals differ notably in length and modulation depth. Differences in the spirals result primarily from changes in NH_3 pressure in the range of about 8 to 10 Pa. Unusual features are seen between the last large pulse and the very beginning of the next spiral, e.g., broad gaps, where oscillation is suppressed up to one pulse length (see trace (c)), small kinks before the start of the spirals (end of traces (d) and (e)), or overshooting compared to the next small pulses (midway in trace (d)). The pulsation frequencies increase from 1.05 MHz (about 25 samples per pulse) for trace (a) to 1.7 MHz (about 15 samples per pulse) for trace (e), primarily due to differences in gas pressure.

Periodic and period-doubling chaotic pulsations were found as a period-doubling cascade when the laser resonator was tuned towards the gain line center (Weiss et al., 1985). As the laser is tuned towards the line center, it starts with continuous emission which is then followed by periodic pulsation. This is followed by a period-doubling cascade ending in a chaotic range which, however, is not at line center and thus is not the "spiral" chaos of the (real) Lorenz model. This chaotic range has high periodic windows and then, closer to resonant tuning, has a relatively broad window of a stable or weakly chaotic period 3 attractor. Finally, the Lorenz-like chaotic pulsations arise at approximately line center corresponding to the (real) Lorenz attractor.

---

[1]The organizers of the Competition downsampled the data by a factor of two. The sampling time for Data Set A is the 80 ns.

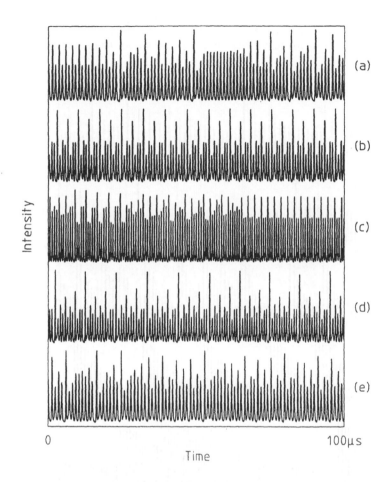

FIGURE 4 Periodic and period-doubling chaotic pulse trains for the detuned laser resonator with different detunings around $d \approx 0.2$. The pressure was held constant at 9 Pa and the pump intensity was about 14 times above threshold. 2,500 samples are shown per trace. Trace (a): Onset of period-doubling chaos; trace (b): stable period 8; trace (c): inverse period doubling to the stable period 3 window; trace (d): stable period 12; and trace (e): period-doubling chaos. Our internal number of the data sets plotted are: (a) → 26, (b) → 23, (c) → 9, (d) → 24, and (e) → 10. Successive traces are offset vertically for clarity.

Although the original (real) Lorenz equations do not account for laser detuning, the period-doubling cascade was found to be compatible with the complex Lorenz model. Zeghlache & Mandel (1985) have shown that the extension of the real Lorenz equations to the case of detuning, in which field and medium polarization become complex quantities, leads to the sequence of pulsation forms found experimentally.

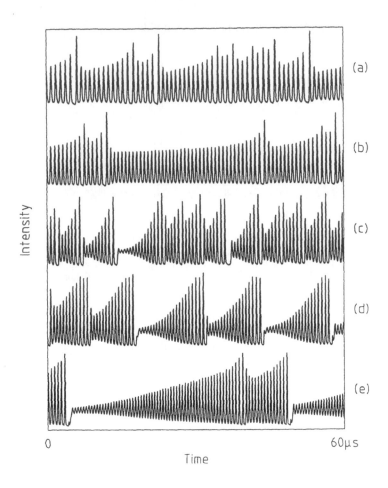

FIGURE 5   Lorenz-like pulsation of the laser intensity. The pressure varied for cases (a) to (e) between 8 and 10 Pa, and the pump intensity was about 14 times above threshold. Resonator tuning was kept as constantly at line center as experimentally possible. 1500 samples were plotted per trace. The average pulsation frequency increased from about 1.05 MHz (about 25 samples per pulse) for trace (a) to about 1.7 MHz (about 15 samples per pulse) for trace (e). Our internal number of the data sets plotted are: (a) → 3, (b) → 22, (c) → 1, (d) → 19, and (e) → 14. Successive traces are offset for clarity. Data Set no. 14 (trace (e)) was taken for the Santa Fe Time Series Prediction and Analysis Competition.

Furthermore, it was observed that the chaotic emission for central tuning sets in abruptly when pump strength is increased. This agrees with the onset of chaos in the real Lorenz equations by a subcritical Hopf bifurcation.

So far, we have not been able to measure precisely the locations of the bifurcation points or, more generally, to determine the two-dimensional (tuning, pump) laser phase diagram. Thus, a further quantitative comparison concerning the bifurcation parameter values has not been possible. It is not known if these parameter values correspond closely to those of the Lorenz model.

IV. NUMERICAL ANALYSIS OF THE EXPERIMENTAL DATA

We have used a variety of numerical methods to compare the physical properties of the experimental data sets with data the sets established by numerical integration of the Eq. (1) for the case of a tuned resonator, and Eq. (7) for the case of detuning of the resonator. The comparisons were done on

- pulse trains,
- phase portraits,
- autocorrelation functions,
- power spectra,
- correlation dimension, D_2,
- Renyi entropy of second order, K_2, and
- generalized dimensions, D_q (only 10 data sets).

Additionally, Section V adds analysis of some special properties of the data:

- the noise contained,
- small deviations from the dynamics of the real Lorenz model contained in the data (seen in pulse trains, phase portraits, and return maps), and
- evaluation of experimental field data.

As a first step, we explain the mathematical details of the methods cited above and apply those methods to data sets calculated by numerical integration of Eq. (1) in the tuned case and of Eq. (7) in the detuned case.

Figures 6 and 7 give some information about one data set calculated by the integration of the real Lorenz equations (Eq. (1)). The parameters used are $r = 15$, $b = 0.25$, and $\sigma = 2$. The Lorenz equations were integrated by a double-precision Runge-Kutta method of fourth order with self-adapting integration steps. The data of the X-variable (i.e., the electric field) was squared to correspond to an intensity quantity $I = X^2$. This intensity was then rounded to slightly less than 8 bits precision. Twenty-five thousand 8-bit data comprise one data set, as in the case of the experimental data.

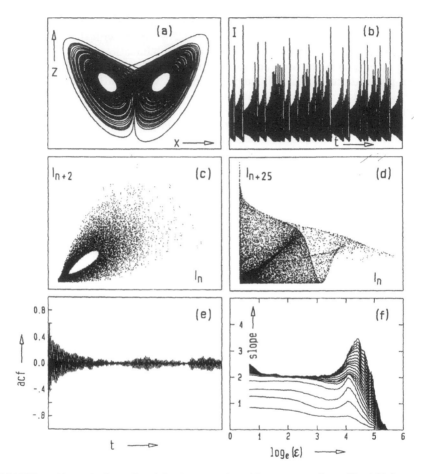

FIGURE 6 Numerical results of the integrated real Lorenz equations (Eq. (1)) for parameters $r = 15$, $b = 0.25$, and $\sigma = 2$. (a) Inversion Z vs. the field X (about 850 loops); (b) intensity pulsations of different spiralling lengths (5000 samples $\approx$ 175 pulses; average period $p \approx 28.6$ samples); (c) phase portrait I_n vs. I_{n+k}, $k = 2$; (d) phase portrait for $k = 25$; (e) autocorrelation function (maximum delay 4,000 $\approx 140n$, n = average pulse period in units of samples); and (f) slopes of the log-log plots of the correlation integral for embedding dimensions 1 (lowest curve) to 20 (uppermost curve) vs. $\ln(\varepsilon)$. For an explanation of the variable ε, see Eq. (10). Parts (a) and (b) may be found in the literature (Sparrow, 1982).

Part (a) of Figure 6 contains the inversion Z plotted versus the field X. A trace of about 850 loops on the symmetric attractor is shown. Typical Lorenz-like spirals of different lengths can be seen in part (b) where we plotted 5000 samples ($\approx$175 pulses) of the intensity I. The average number of samples per pulse is $n = 28.6$. Of course, both plots (a) and (b) are not new and may be found in the existing literature.

Two phase portraits were made, one with delay 2 (part (c)) and the other with delay 25 (part (d)). They show the hole of the attractor (part (c)) and the strange shape which projections of it can have (part (d)). The center of the hole locates the unstable steady-state laser solution $X = \pm(r - 1)^{1/2}$ and the hole itself indicates that the time-dependent solution in this case avoids that value.

The term *phase portrait* is used for a two-dimensional plot of the 8-bit data I_i ($i = 1, 2, \ldots, 25000$) in the form of I_{i+k} versus I_i. For different values of the delay k, one obtains some insight into the spatial structure of the attractor underlying the data set. Different delays provide different views of the attractor, by a process equivalent to looking at the attractor in phase space under different viewing angles. One can understand the structure of a spatial cluster of points by looking at various angles, which may reveal that a thickened line is the side view of a flat structure (see Figure 6(d)) or of nearly periodic and nearly coplanar structures (see the figures of the experimental data later). One can also judge how point density variations are responsible for the anomalous (nonfractal) structures seen in the slopes of the correlation integral C.

The *autocorrelation function* (ACF) also provides insight into the structure of the attractors. It is normalized to unity for zero delay τ using the definition

$$R(\tau) = \frac{\langle I(t)I(t+\tau)\rangle - \langle I(t)\rangle^2}{\langle I(t)^2\rangle - \langle I(t)\rangle^2} . \tag{8}$$

The maximum delay plotted (part (e)) is $\tau = 4000\,\Delta t$, where Δt is the time spacing between two adjacent samples. The ACF shows the typical small groups of peaks growing ("revivals") and decreasing nonperiodically, as found, in previous investigations of these Lorenz equations (Abraham et al., 1987).

Part (f) of Figure 6 shows the slopes of the *correlation integral C* versus the distance ε. C is calculated by using the method of Grassberger and Procaccia (1983a, 1983b). This method analyzes a single-variable time series by constructing representations of the attractor in E-dimensional phase spaces using the time-delay method as discussed, e.g., by Froehling et al. (1981). They define

$$\mathbf{X}(t) = (x(t), x(t+\tau), \ldots, x(t+(E-1)t)), \tag{9}$$

where $\mathbf{X}(t)$ is a vector built up out of E samples of the time series with successive time delays $\tau = n\Delta t$, n being an integer. The methods to optimize the choice of τ are described by Hübner et al. (1990). For discrete samples x_i the correlation integral C becomes

$$C(\varepsilon) = \lim_{N\to\infty}\left[\frac{1}{N(N-1)}(\mathcal{N}(i,k), \|\mathbf{X}_i - \mathbf{X}_k\| < \varepsilon)\right] . \tag{10}$$

$\mathbf{X}_i$ (similarly $\mathbf{X}_k, k \neq i$) denotes vectors $(x_i, x_{i+n}, x_{i+2n}, \ldots, x_{i+(E-1)n})$, $\mathcal{N}(i, k)$ is the number of pairs (i, k), and $\|\mathbf{X}_i - \mathbf{X}_k\|$ is the norm of the vector difference. E is called the "embedding dimension." The choice of the norm is of practical importance for data sets such as ours, although in Eq. (10) the Euclidean norm (length of the difference vector) and the maximum norm give the same correlation dimension D_2 and entropy K_2 (Takens, 1985). To reduce the time necessary for the computation of the *correlation integral* C for 25,000 data and 40 embedding dimensions $1 \leq E \leq 40$, we choose the maximum norm

$$\|\mathbf{X}_i - \mathbf{X}_k\| = \max_{0 \leq m \leq E-1} |x_{i+m} - x_{k+m}|. \tag{11}$$

The *correlation integral* C of the calculated data set is not shown. Instead, we plot its slope (part (f)) versus the natural logarithm of the distance ε, because the structure contained in C is enhanced and thus can be more easily identified. Also, it is easier to find the right plateau area, i.e., the scaling region described below. The ε-scale can be understood as follows: The 8-bit data can have integer values between 0 an 255, as do the absolute values of their differences. Thus, ε lies in the interval $0 \leq \varepsilon \leq 255$ which results in an interval of $0 \leq \ln(\varepsilon) < 6$ on the natural logarithmic scale.

In practice, one infers the existence of an attractor when one observes a plateau area in the slopes of C, where the flatness of the plateau is interpreted as evidence for a self-similar scaling region of some dimension and where the convergence of the slopes with increasing embedding dimension indicates the reconstruction of a topologically unique attractor. If the dimension is fractal and larger than two, it is taken as evidence of chaotic motion on a strange attractor.

The slopes in Figure 6(f) contain a special correction which we call the "shift correction." This means that the correlation integral $C(\varepsilon)$ is plotted versus $\varepsilon + p$, where p is half of the least significant bit. This is the most straightforward way to correct the errors in the calculation of C which result from the rounding of the data by the digitizer (Moeller et al., 1989). On an average, this correction procedure increases the slopes of $C(\varepsilon)$ in the plateau area (correcting for the erroneous bias resulting from using 8-bit data), and thus increases the calculated estimate of D_2 by about 5%, while not changing K_2.

Some more remarks are necessary about characteristic features of the plots of Figure 6(f). All slope values were calculated from the difference of two adjacent values of $\ln(C_E)$, divided by the difference of the corresponding values of $\ln(\varepsilon)$, and then centered onto the midpoint between those $\ln(\varepsilon + p)$ values. This is the reason that the slope curves do not start at the vertical axis. The calculated values of the slope were then connected by straight lines as a guide to the eye to discriminate between results for different values of E.

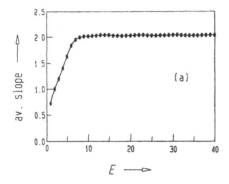
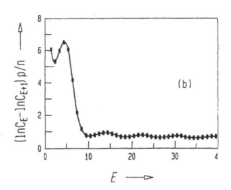

FIGURE 7 (a) Average slope in the "plateau region" versus embedding dimension E for the integrated real Lorenz equations (no detuning: $d = 0$). The correlation dimension D_2 is taken as the value for the embedding dimension $E = 15$. (b) $[\ln(C_E) - \ln(C_{E+1})]p/n$ versus embedding dimension E; p is the average number of data samples per average pulsation period and n (integer) means the delay between successive components of each vector $\mathbf{X}_i$. The entropy K_2 per average pulsation period T, K_2T, is taken as the estimated asymptotic from the highest values of E.

For small ε, as ε decreases towards zero, the slopes increase. An analysis shows (Moeller et al., 1989) that this increase is caused by noise and is proportional to E/ε^2 because noise blurs the attractor into the available E dimension, increasing to infinity for $E \to \infty$.

Besides the influence of noise, there is another structure on the slope curves which is connected with our choice of the maximum norm. It sharpens the frequency with which extreme values of the distance appear, in contrast to the Euclidean norm which gives somewhat smoother variations of the density of interpoint distances.

The average in the scaling region is plotted in Figure 7(a). This figure clearly exhibits the convergence (above $E_c = 7$) to a value of 2.03 which is expected for Lorenz-type chaos. Circles mark the calculated values, which are connected by a Lagrange interpolation curve of third order to make the shape of the curve more pronounced.

Figure 7 also contains the plot of the expression $[\ln(C_E) - \ln(C_{E+1})]p/n$ versus the embedding dimension E (Figure 7(b)), where p is the number of samples per average pulsation period T, and $n = \tau/\Delta t$ as stated together with Eq. (9). The above expression is calculated from the asymptotic formula (Caputo and Atten, 1987; Grassberger and Procaccia, 1983a, 1983b)

$$C(\varepsilon) \propto \varepsilon^{D_2} e^{-E\tau K_2} \qquad (12)$$

which holds for $r \to 0$ and $E \to \infty$. Figure 7(b) provides an estimate of K_2 which is taken as the average from the converging part for about $E > 10$. K_2 has the inverse dimension of a time variable. Therefore, we calculated a dimensionless K_2 in terms of the average pulse period T as $K_2 T$. Circles again mark the calculated values as in part (a).

The curve in Figure 7(b) contains an oscillating structure which can be interpreted as a residual influence of the average oscillation period of the intensity pulses. The E-difference of two successive maxima of the curve multiplied with τ is equal to the average pulsation period T. These oscillations and the fact that K_2 converges slower with E than D_2 with E are consistent with the observations of others (Albano et al., 1988; Caputo et al., 1987). We also find that the condition for the convergence of K_2 requires vector spans $(E-1)n\Delta t$ which are 5 to 10 times larger than E_c.

For a further check of the robustness of the comparison between theoretical results and the measured data, we included detuning by the use of Eq. (7). With the same values of the parameters ($r = 15$, $b = 0.25$, $\sigma = 2$) as before, we found a wide period 3 "window" near $d \approx 0.155$. Whereas for $d \lesssim 0.05$, we found Lorenz-type pulsations very similar to that in our Data Set 3 (see Figure 5(a)), but with a bigger hole in the attractor. Even with this rather large detuning, all other qualitative features differ only slightly from those of Figure 6. One is led to the conclusion that small detunings could have been present when the Lorenz-like chaos was measured, and that those detunings would not have led to noticeable differences from behavior in the perfectly tuned case.

Figure 8 presents three typical experimental data sets called Lorenz-like chaos (column 1 with Data Set 3; pressure 8 Pa, pump strength $r = 14$, detuning $d \approx 0$), period-doubling chaos (column 2 with Data Set 10; pressure 9 Pa, pump strength $r = 14$, detuning $d \approx 0.2$); and period 2 pulsation (column 3 with Data Set 7; pressure 9 Pa, pump strength $r = 14$, detuning $d \approx 0.2$).

Row 1 shows 1,500 data samples for each trace. Subtle differences in the no. 3 trace compared to that in the numerical integration (see Figure 6(b)) will be discussed in Section V. The phase portraits in row 2 are plots of 25,000 data samples for each data set.

The autocorrelation function (ACF) has been plotted in row 3 for 2,000 sample spacings Δt, i.e., the maximum delay plotted amounts to 80 μs. Data Set 3 shows the rapid decrease of the ACF envelope, as in the numerical data set (see Figure 6(e)); one also finds its characteristic revivals. Clearly length, amplitude, and separation of the revivals differ slightly among the 15 Lorenz-like data sets, but the overall structure is similar to the numerical data set. No such revivals can be found in the period-doubling chaos Data Set 10 (column 2). Despite a relatively small correlation time τ_c (see Table 1), a much longer second characteristic correlation is also found; it results from high periodicity (especially period 2) contained in the data. Of course, Data Set 7 (column 3) is highly correlated.

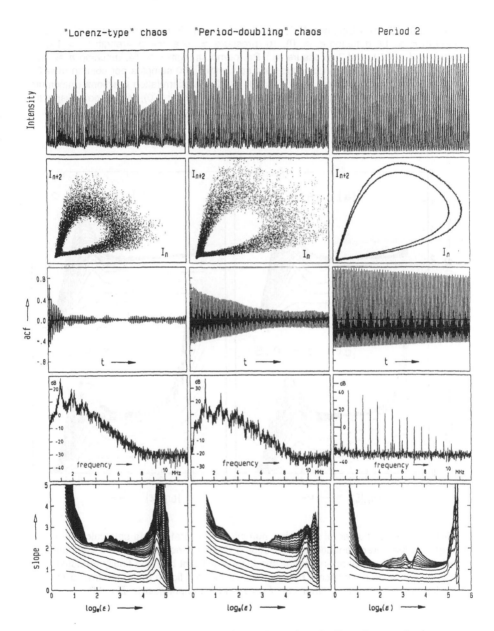

FIGURE 8 Examples of experimental data sets analyzed with various methods. (continued)

FIGURE 8 (Cont'd.) Columns 1, 2, and 3 are for Data Sets 3 (pressure ≈ 8 Pa, pump strength $r ≈ 14$, detuning $d ≈ 0$), 10 (pressure ≈ 9 Pa, pump strength $r ≈ 14$, detuning $d ≈ 0.2$), and 7 (pressure ≈ 9 Pa, pump strength $r ≈ 14$, detuning $d ≈ 0.2$). Row 1: Pulse trains of characteristic data sets with 1,500 samples per column; row 2: phase portraits (25,000 data; delay $k = 2$); row 3: autocorrelation functions (maximum delay: 2,000 sample spacings Δt; maximum vertical scale: 1.0); row 4: power spectra (maximum frequency: 12 MHz); row 5: slopes of the log-log plots of the correlation integral C vs. $\log(\varepsilon)$ for embedding dimensions 1 to 20 (lower to upper).

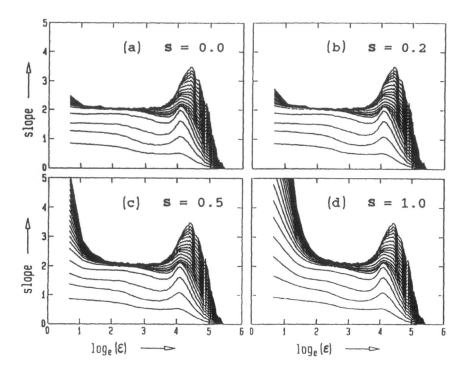

FIGURE 9 Slope curves of the noisy, numerically generated data sets calculated by integrating the Lorenz equations, then adding Gaussian noise of different standard deviations s (values taken in units of the least significant bit), and finally rounding the "noisy data" to 8 significant bits. In each of the four figure parts, the lowest curve belongs to the embedding dimension $E = 1$ and the uppermost to $E = 20$.

The power spectrum of each data set is an average over four spectra from consecutive groups of 5,000 data of the data set. The Fourier transform of each group is done with the Blackman-Harris window. For a sampling time of $\Delta t = 40\,\text{ns}$ the frequency resolution is $5\,\text{kHz}$, and we plotted the first 2,400 power data so that the whole frequency span covers $12\,\text{MHz}$. The chaotic data sets clearly exhibit the typical "noisy" background connected with chaos. On the other hand, there are relics of periodicity of the average pulsation period at $1.01\,\text{MHz}$ in the case of Data Set 3 (column 1) and at $1.32\,\text{MHz}$ for Data Set 10 (column 2). Of course, the periodic pulsation Data Set 7 shows the pure periodicity at $1.29\,\text{MHz}$. It is worth noting that there is typically no relic of this type of periodicity in the evolution of the X variable for chaos in the Lorenz model because of random changes of sign which are removed by squaring X to get I. This feature of Lorenz-like chaos was noted and discussed in Abraham et al. (1987).

Finally, row 5 shows the slopes of the correlation integral versus the natural logarithm of the distance ε with the embedding dimension E as the parameter ($1 \leq E \leq 20$, lowest to uppermost curve). Of course, all explanations and remarks, given in connection with the slope curves of the numerical data set (Figure 6(f)), are valid here. But some remarks need to be added:

1. While, in principle, the correlation integral C does not depend on the choice of the time delay τ, practical considerations of precision and noise limit satisfactory values of τ. As an estimate of the best value for time delay τ, denoted by n, we have found in our data sets that there is a relation $\tau E_c \approx T$ between τ, the average pulse period T, and the embedding dimension E_c. E_c marks the onset of the slope curves convergence, which, on the other hand, signals convergence of the method. It reflects the fact that the structural information used to reconstruct attractors which are nearly two-dimensional is connected with the time interval $E\tau$ of the data set covered by the vector, and not with the individual choices of E or τ (Albano et al., 1988; Caputo et al., 1987). Moreover, after one revolution on the attractor, the succeeding orbits come close to the previous orbit, so that the benefit for structural information is small when E or τ are further increased. After some empirical trials, we chose $\tau \approx T/7$, which leads to $n = 2$ or $n = 3$ for our data sets, as the best compromise.

2. The increase of the slope curves for decreasing to zero is slightly more pronounced, compared to the numerical data set. This indicates the enhanced influence of noise. To get a rough estimate of the noise influence in our experimental data, we added Gaussian noise to the exact numerical data before rounding them to 8 bits. Figure 9 shows four examples of slopes for different additive noise strengths s, which denotes the standard deviation in units of the least significant bit of the 8 bits used. One sees that the left part of the slope curves increases with increasing noise, whereas the rest of the curve remains unperturbed. The comparison with the experimental data (Figure 8) reveals that the noise contained in the experimental data is equivalent to $s \approx 0.5$, which is consistent with the signal-to-noise ratio of the measurements. An analytical discussion of the noise influence is published elsewhere (Moeller et al., 1989).

3. Compared to the slope curves of the numerical data set, there are additional structures (peaks) for intermediate values of ε in the measured data. These structures may be interpreted as follows: Density variations (e.g., lacunarity), especially found in attractors of periodic pulsation (e.g., Data Set 7) and easily seen in our phase portraits, cause sometimes abrupt variations in the slope increase of $C(\varepsilon)$ with ε and this, of course, is seen in the slopes. The consequence of the effect is that it is not smeared out by using more and more data. Thus, the determination of the so-called scaling region (plateau area) might be difficult or even impossible. The most reasonable location for such a plateau area is the interval, where ε is as small as possible, but where, on the other hand, the noise influence has been decreased sufficiently.

The aim of this work is to present an overview of the information available on Data Set 14 which was taken for the Santa Fe Time Series Prediction and Analysis Competition. Thus, Figures 10 and 11 are included to show some of the main results from our numerical evaluation of Data Set 14 that was selected (after down sampling by a factor of two, i.e., taking every other point) to be competition Data Set A. The intensity data were taken at a pressure of about 8 Pa and with a pump parameter $r \approx 14$. The detuning was held as close as possible to zero. Although these parameters are the same as for our Data Set 3 (Figure 8, column 1), the pulse train exhibits clear differences when compared to our Data Set 3. Such long spirals have not been seen in the data from the numerically integrated, real Lorenz equations ((Eq. 1)). The differences may especially indicate small detunings not seen when the data was measured.

Of course, those long spirals can also be seen in the power spectrum (part (b)). This shows the average pulsation frequency of 1.65 MHz (for the average pulsation period of 15.2 samples and the sampling time of 40 ns).

Relics of the long spiralling may also be found in the phase portrait for lag 2 (part (c)). The phase portrait part (d) shows the same linear point concentration as in the other Lorenz-like data sets. This seems to indicate a flat structure of the attractor reconstruction.

Part (e) shows the autocorrelation function with the same "revivals" as seen in other Lorenz-like data sets. It is calculated from 25,000 data with a maximum delay time of 2,000 sample spacings Δt, which means 0.08 ms.

The last part (f) of Figure 10 shows the slope curves of the correlation integral for the embedding dimensions E of 1 to 20 (lowest to uppermost). There is no special structure on these curves except the standard noise increase for decreasing to zero. These slopes even decrease steadily so that the determination of a plateau area is difficult. From our experience, an interval of $\ln(\varepsilon) \approx 2.7$ to $\ln(\varepsilon) \approx 3.2$ was chosen, where the noise influence seemed to be small enough.

Further information can be taken from Figure 11. When averaging the plateau areas of each of the twenty slope curves, we obtain the curve of part (a). Circles mark the calculated values and the continuous interpolation curve connecting them

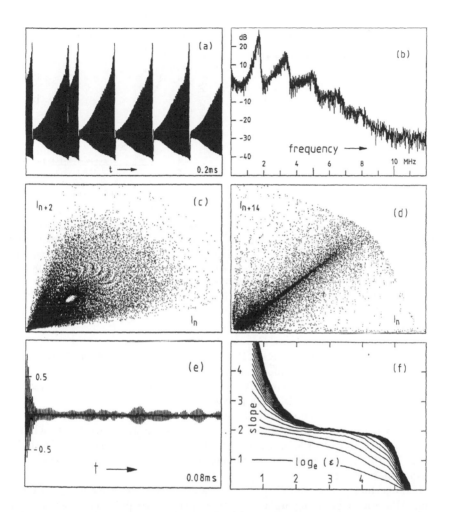

FIGURE 10 Evaluation of Data Set 14. Experimental parameters are: pressure 8 Pa, pump parameter $r \approx 14$, and detuning $d \approx 0$. Part (a): pulsation of the intensity; 5,000 points. Part (b): power spectrum; average of the power spectra from four consecutive groups of 5,000 points; 2,400 frequencies are plotted. Parts (c) and (d): phase portraits for indices 2 and 14; 25,000 points plotted. Part (e): autocorrelation function, calculated from 25,000 points with a maximum delay of 2,000 sample spacings. Part (f): slopes of the correlation integral $C(\varepsilon)$ for embedding dimensions E of 1 to 20 (lower to upper).

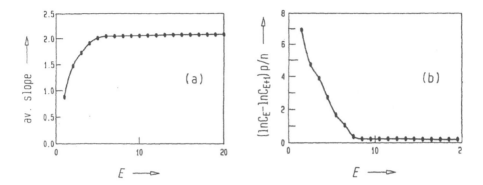

FIGURE 11   (a) Average slope in the "plateau area" versus embedding dimension E for Data Set 14. The correlation dimension D_2 is taken as the value for the embedding dimension $E = 15$. (b) $[\ln(C_E) - \ln(C_{E+1})]p/n$ versus embedding dimension E; p is the average number of data samples ($p = 15.2$) covering the average pulsation period, and n (integer) is the index delay between successive components of each vector $\mathbf{X}_i$. The entropy K_2 per average pulsation period T, K_2T, is taken as the estimated asymptotic from the highest values of E.

is plotted as a guide to the eye. From the converging part of the curve above $E \approx 5$, we estimate the correlation dimension to $D_2 = 2.06$. Figure 11(b) is calculated as is Figure 7(b). It shows a remarkably low value for the entropy K_2 which may be a result of the long spirals with high periodicity.

The dynamical characteristics of each of the data sets, as estimated by the various procedures, are collected in Table 1. Only two data sets were not included, 9 and 11. They consist of long parts alternating between periodic and chaotic behavior. We think that their average dimension, entropy, etc. would not yield any valuable information.

Periodic and period-doubling chaotic data sets were taken for a pressure of about 9 Pa and a detuning up to $d \approx 0.2$. Lorenz-like pulsation data sets are measured for a pressure of about 8 Pa, a pump intensity of about 14 times above threshold, and a detuning near zero. When these parameters were adjusted repeatedly, different pulsation types were seen so that 15 differing Lorenz-like data sets could be collected. Thus, the various Lorenz-like pulsation forms are a result of the above-mentioned uncertainty in the detuning (± 0.05), and other differences in the parameters difficult to control. Data Set 26 is the only one taken for a pressure of 6.5 Pa, with a detuning $d \approx 0$, and for a pump intensity 14 times above threshold as before. At this pressure, three-level coherence effects may be present and the dynamics of the laser may no longer be Lorenz-like.

Table 1 contains two columns related to K_2 which has the inverse dimension of a time variable. The sixth column uses the Kaplan-Yorke conjecture, a relation between K_2, the dimension, and the Lyapunov exponents for chaotic systems (Kaplan, 1979). If there is only one positive Lyapunov exponent (as is the case for chaotic behavior in the Lorenz model), $K_2 \approx \lambda_+$, where λ_+ is the positive Lyapunov exponent. The Kaplan-Yorke conjecture is then

$$D = 2 + \frac{\lambda_+}{|\lambda_-|} \tag{13}$$

and yields an estimate for λ_-. Because D_2 is a lower bound of D and is near 2, this estimation of the Lyapunov exponents of our experimental data can give only a very rough approximation. Nevertheless, it is rather robust as long as the estimate of D is not too close to 2.0.

It is worth noting the results of applying our algorithms to the periodic data sets. We found D_2 values close to 1 for the period 2 (P2) data, but values distinctly greater than 1 (in the range of 1–2) for more complex waveforms. This we attribute to the very nonuniform structures in complex periodic attractors on large and intermediate length scales. While on small scales a periodic attractor is one-dimensional, the difficulty of taking the limit of $\varepsilon \to 0$ before practical limits of noise and digitizing errors set in can lead to such spurious results. These kinds of distortions are the ones evident in row 5 of Figure 8 for the P2 data.

For K_2 for the periodic data sets, we see that, for all but the two most complex patterns, the values are less than 0.052, while for all the chaotic data sets the values fall in the range 0.18–0.74 with only two values less than 0.37. Periodic signals should give values of $K_2 = 0$, and while our values for periodic data sets are significantly lower than those for the chaotic data sets, we again attribute the nonzero numbers to the combined effects of noise and digitizing errors. As these values (0.02–0.05) are the errors of the algorithm in these circumstances, they may indicate that the values of K_2 for the chaotic data sets are systematically overestimated by a similar amount.

As a last step in the evaluation of the information contained in the measurements, we calculated the generalized dimensions of some data sets of the Lorenz-like pulsation type. As has been pointed out in Hentschel and Procaccia (1983), the geometric and probabilistic features of strange attractors may be characterized by the Renyi dimensions D_q with any q being not necessarily an integer.[2] Pawelzik and Schuster (1987) generalized the correlation integral method of Grassberger and Procaccia (1983a, 1983b) to calculate the D_qs via the *generalized correlation integral*

$$C^q(\varepsilon) = \left\{ \frac{1}{N} \sum_i \left[\frac{1}{N} \sum_j \Theta \left(\varepsilon - |\mathbf{X}_i - \mathbf{X}_j| \right) \right]^{q-1} \right\}^{1/(q-1)} \tag{14}$$

[2] A clear discussion of generalized dimensions is given by Pineda and Sommerer (this volume).

Udo Hübner et al.

TABLE 1 Collection of the most important results of the different experimental data sets with comparison to results of the numerically integrated Lorenz equations. The data sets are sorted so that the uppermost seven rows stand for data sets with periodic pulsation (P2, P4, . . .), the next two rows stand for period doubling (PDC), and the other rows for Lorenz-like (LC) chaotic data sets. Theoretical values are added in the last four rows for comparison. p is the average number of samples per pulse period, and τ_c is the decay constant of the autocorrelation function. C, WI, and I denote "constant," "weakly increasing," and "increasing"; this reflects the behavior of the last 10 points of the average slope curve versus the embedding dimension parameter E (see, e.g., part (a) of Figure 7). I means an increase of less than 5% over the full range $10 \leq E \leq 20$. Data Set 14 was taken for the Competition.

No.	p	τ_c/T	D_2	$K_2T \approx \lambda_+ T$	$\frac{K_2 T}{D_2 - 2} \approx \|\lambda_- T\|$	Remarks
7	19.4	525	1.03	0.018		P2; C
4	24.9	666	1.08	0.028		P2; C
8	20.0	531	1.23	0.046		P4; C
12	19.5	469	1.19	0.052		P4; C
23	23.1	1400	1.24	0.039		P8; C
24	21.4	24.3	1.79	0.33		P12; C
25	22.0	10.0	1.79	0.35		P2 + P4; C
26	28.6	7.5	2.30	0.54	1.8	PDC; C
10	19.0	2.7	2.29	0.61	2.1	PDC; I
1	18.7	1.9	2.16	0.74	4.6	LC; WI
2	19.0	2.8	2.09	0.73	9.1	LC; C
3	24.8	2.6	2.15	0.63	5.5	LC; C
5	17.3	2.5	2.20	0.46	2.3	LC; WI
6	16.9	3.2	2.18	0.45	2.5	LC; I
13	16.7	5.6	2.18	0.37	2.1	LC; WI
14	**15.2**	**4.9**	**2.06**	**0.18**	**2.7**	**LC; WI**
15	17.5	5.8	2.04	0.40	10.6	LC; I
16	17.6	2.8	2.23	0.50	2.2	LC; WI
17	17.5	3.4	2.29	0.59	2.1	LC; I
18	21.8	3.2	2.11	0.37	3.4	LC; WI
19	17.3	3.5	2.07	0.44	6.2	LC; WI
20	17.0	2.7	2.12	0.51	4.2	LC; WI
21	21.9	4.4	2.15	0.27	1.8	LC; WI
22	20.3	3.5	2.02	0.47	20.3	LC; WI
numer. integration			2.033	0.70		tuned, noise free
numer. integration			2.055	0.69		tuned + noise $(s = 0.5)$
numer. integration			2.066	0.71		detuned $(d = 0.05)$
Caputo et al. (1987)				0.69		numer. integration

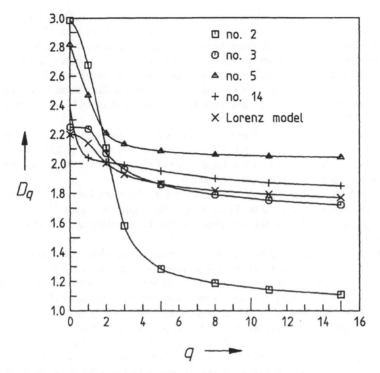

FIGURE 12 Plot of the calculated D_q values (marked points) vs. q for the numerically calculated Lorenz model data set, and for Data Sets 2, 3, 5, and 14. The smooth interpolation curves through the calculated points are plotted to better display the significant differences existing between the data sets.

where ε, N, and $\mathbf{X}$ are the same as defined for Eq. (10), and Θ denotes the Heaviside function. For $q = 2$ Eq. (14) becomes Eq. (10), and for $q \to 1$ Eq. (14) changes to

$$C^1(\varepsilon) = \exp\left(\frac{1}{N}\sum_i \ln\left[\frac{1}{N}\sum_j \Theta(\varepsilon - |\mathbf{X}_i - \mathbf{X}_j|)\right]\right) . \qquad (15)$$

The $D_q s$ are then calculated with

$$D_q = \lim_{\varepsilon \to 0}\left\{\frac{1}{\ln(\varepsilon)}\ln\left[C^q(\varepsilon)\right]\right\} . \qquad (16)$$

They reflect the density distribution of phase space points on the attractor. In particular, D_0 is the Hausdorff (or fractal) dimension, D_1 is called the information dimension, and D_2 is the correlation dimension with the relation $D_0 > D_1 > D_2 > \ldots$. This means that D_2, calculated so far, is only a lower bound for D_0.

The procedure used to calculate the correlation integrals D_q was similar to that used to calculate the D_2. Again we used the maximum norm in Eqs. (14) and (15), and the time delay $\tau = n \cdot \Delta t$ (n, integer; Δt, sampling time) was chosen to be about $1/7$ of the average pulsation period. The embedding dimension E was varied between 2 and 20 in steps of 2, and each D_q was then calculated as an average of the last four $D_q(E)$ values.

Figure 12 shows the D_q values of four measured Data Sets, 2, 3, 5, 14, and of the numerically calculated data set with detuning $d = 0$ versus q. The calculated D_q values are marked. To get a clear impression of the differences between the data sets, we also added their interpolation curves.

The similarity in the D_q values of Data Set 3 (the most Lorenz-like data set) and the Lorenz model is obvious. These values differ only by about 3%, indicating a good agreement between the structure of the laser attractor and the Lorenz attractor.

On the other hand, Data Sets 2 and 5 show significant differences compared to 3 and the Lorenz model. Both D_q curves increase to values near 3 for $q \rightarrow 0$. This shows that the commonly used correlation dimension D_2 appears to be quite insensitive to the attractor dimension, and is thus not very well suited for characterizing chaotic systems. The reason for this increase may be temporary emission of the ring laser in the opposite direction which was found recently (Tang et al., 1992). Consequently, the emission involves, at least at certain times, two laser modes so that this pulsation results from a system with a larger phase space than the Lorenz model. The reason for the strong decrease in the D_q curve of Data Set 2 for increasing q is not known.

V. DISCUSSION

For a variety of different comparison methods (pulse trains, phase portraits, autocorrelation functions, power spectra, and dimensions), we have found that chaotic pulsations of an NH_3-FIR ring laser can exhibit nearly the same features as pulsations generated by numerical integration of the Lorenz equations. Despite this similarity, we still find a few small but significant differences.

One of these differences was mentioned above in the discussion of the dimension spectra: Only a few of the 15 experimental Lorenz-like data sets prove to be quite similar to the Lorenz model when we compare the full dimension spectrum. Data Sets 2 and 5 are remarkably similar to the Lorenz model for all comparison methods except the dimension spectrum. Even though the correlation dimensions and the entropies are comparable, they still differ significantly in their overall dimension spectra.

Other differences were found recently (Hübner et al., 1992; Li et al., 1990; Tang et al., 1992). Some details of the dynamics of measured laser intensity spirals differ from those calculated by the integration of the Lorenz equations with parameters as close as possible to those in the experimental case. Measurements, in which weak counterpropagating emission (here: forward emission) occurs, were compared to simple model calculations allowing for such emission.

Figure 13 shows the experimental intensity pulsation (part (a)) under the influence of a counterpropagating emission (part (b)). It is strikingly similar to the pulsation of our Data Set 14. Contrary to this, the pulsation found in the Lorenz model shows "giant" pulses ending the sequence of increasing spirals after a rather large time delay. These "giant" pulses are missing in Figure 13(a) and in our Data Set 14. Figure 13(b) shows the reason for the absence of the "giant" pulses: They are suppressed at the expense of a smaller pulse built in the counterpropagating direction which reduces the gain for the "giant" pulse. Thus, it appears that the differences between the measured dynamics and the Lorenz dynamics can be attributed at least partly to the imperfect suppression of the counterpropagating emission.

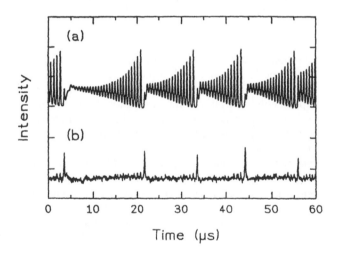

FIGURE 13 Intensity pulsation of a ring laser in the presence of a counterpropagating emission. NH₃ gas pressure is 9 Pa, the pump intensity is 6 W/cm², and the frequency offset amounts to 10 MHz. (a) Intensity pulsation of the backward emission vs. time; and (b) Intensity of the forward emission vs. time.

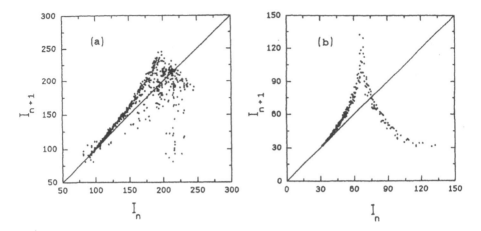

FIGURE 14 Intensity return maps of laser emission (peak intensity of pulse $n + 1$ versus peak intensity of pulse n). (a) Calculated from the experimental data shown in Figure 13(a). (b) Calculated from intensity data generated by the numerical integration of the real Lorenz equations with $r = 18$, $b = 0.3$, and $\sigma = 2$.

Further information about the deviations of our experimental Lorenz-like pulsations from the Lorenz model can be found in return maps where the peak intensity of pulse $n + 1$ is plotted versus the peak intensity of pulse n. Figure 14(a) shows such a return map, calculated from the intensity data shown in Figure 13(a). The difference from the cusp map generated by the real Lorenz equations (Figure 14(b)) is apparent, thus indicating substantial difference of the laser intensity spiral dynamics from that of the Lorenz model.

With a simple two-level model Tang et al. (1992) showed that the observed differences in the return maps between the laser intensity spiral dynamics and that of the Lorenz dynamics are caused by a counterpropagating emission of the ring laser.

However, the intensity pulsations of Figure 5(a) (our data set "nearest" to the Lorenz model) still show a difference with the Lorenz intensity spiral dynamics: the "giant" pulses are followed by a pulse of intermediate height before the next spiral begins with the smallest pulse in the spiral. Attempts to explain these intermediate pulses in terms of counterpropagating emission were unsuccessful. Instead, the full three-level laser model (Corbalan et al., 1989; Laguarta et al., 1988), without any counterpropagating emission, shows the same pulses of intermediate height (e.g., Data Set 3) if a higher loss of the resonator has been introduced in the numerical calculation of the chaotic pulsation.

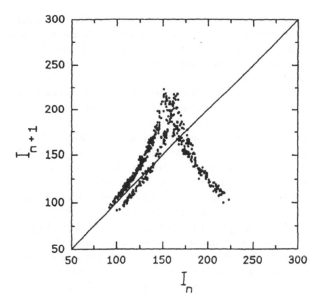

FIGURE 15   Intensity return map (peak intensity of pulse $n + 1$ versus peak intensity of pulse n) constructed from Data Set 3.

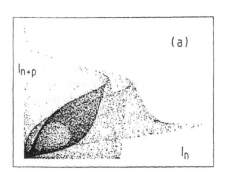 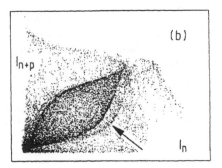

FIGURE 16   Phase portraits of the data set generated by numerical integration of the real Lorenz equations (part (a)) and of Data Set 3 (part (b)). The intensity data of index $n + p$ (p = average pulsation period) are plotted versus the intensity at index n. The arrow points to the shadow structure of the principal attractor which seems to indicate a second Lorenz attractor.

These three-level coherence effects are also evident in the return maps. Figure 15 shows the return map of Data Set 3 which contains the pulses of intermediate height. Two cusp maps can be seen, each of them resembling closely the Lorenz map of Figure 14(b). Obviously, the second cusp map is related to the pulses of intermediate height at the end of each spiral.

This finding is supported by Figure 16 which shows the phase portraits of Data Set 3 and of the data generated by numerical integration of the real Lorenz equations (see Figure 6). In both phase portraits the intensity data of index $n+p$ ($p =$ average pulsation period) is plotted versus the intensity data of index n. Figure 16(b) clearly repeats the structure of the Lorenz attractor, but additionally exhibits a second somewhat larger shadow structure of obviously the same form (arrow in Figure 16(b)). This seems to indicate the presence of a "doubling" of the attractor.

As mentioned, the "intermediate" pulses have been found in a rather complete model taking into account primarily three-level coherence effects and asymmetric gain line shapes which are typical for the experiment. Here they are caused by pump detuning necessary for achieving unidirectional ($\equiv$ single mode) laser emission (Tang et al., 1992). Recent numerical experiments of Tang (unpublished) have shown that asymmetric gain line profiles *alone* can generate these "intermediate" pulses. Whether coherence effects play an important role in the laser dynamics is thus at present not clear.

From the above, it was concluded that it should be possible to obtain dynamics corresponding precisely to the Lorenz model experimentally by avoiding all perturbing effects. Such "pure" Lorenz dynamics has then been experimentally demonstrated (Tang et al., 1992). However, since these results are too recent, the measurements could not be included in the Competition. Similarly, recent experiments (Tang et al., 1991; Tang et al., 1992) which have permitted measurement of high-quality data on the field dynamics of the laser could also not be included. The intensity measurements correspond to the square of the X variable of the Lorenz model, while the field corresponds to the X variable itself and reveals therefore the dynamics of the field amplitude and phase. Such vector component quantities might be even better suited to test mathematical techniques in the characterization of chaos than the scalar intensities reported by us up to now.

David R. Rigney, Ary L. Goldberger, Wendell C. Ocasio, Yuhei Ichimaru, George B. Moody, and Roger G. Mark
Harvard–MIT Division of Health Sciences and Technology, Harvard Medical School, MIT and Beth Israel Hospital, Department of Medicine (Cardiovascular Division), 330 Brookline Avenue, Boston MA 02215

Multi-Channel Physiological Data: Description and Analysis (Data Set B)

The concept of self-regulation occupies a central position in today's science of physiology. Though long in arriving, its principles are firmly established; yet it sadly lacks predictive potential.

—E. F. Adolph, 1961

INTRODUCTION

Data Set B of the 1991 Santa Fe Time Series Prediction and Analysis Competition is a multivariable physiological time series, consisting of 4 hours and 43 minutes of simultaneous heart rate, respiration, blood oxygen saturation, and sleep stage data. By way of general introduction to the data, we first address the reader who is not a physiologist, but who is looking for a case study with which to develop a particular forecasting method or who may wish to promote a larger agenda (Waldrop, 1992)

Times Series Prediction: Forecasting the Future and Understanding the Past, Eds. A. S. Weigend and N. A. Gershenfeld, SFI Studies in the Sciences of Complexity, Proc. Vol. XV, Addison-Wesley, 1993

in which physiological systems are governed by principles that might apply equally well to other types of series in the Competition.

Due in large measure to the influence of Walter B. Cannon (1929, 1932), physiologists spend much time investigating how organic systems regulate themselves under normal circumstances or fail to control themselves during disease. The canonical expectation is that under constant external conditions, normal physiological systems maintain a state of internal constancy, and when they are perturbed by changes in the external environment, by exercise or by trauma, these systems react in such a way as to return again to a constant and optimal state, a phenomenon known as *homeostasis*. Physiological systems are therefore considered to have machinelike properties, and a common objective in physiology is to discover how some system senses its current state and uses this information to respond automatically, e.g., to perform a visceral reflex with a sensory input to the nervous system (afferent signal) and an output from the nervous system to the controlled organ (efferent signal). For example, it has long been known that heart rate is regulated by the baro-reflex, in which specialized nerves in the aorta (and other blood vessels) sense the blood pressure and convey this information to the brainstem, which in turn sends signals to the pacemaker region of the heart that determines the heart rate (Kirchheim, 1976; Abboud & Thames, 1983). This example is representative and pertinent to Data Set B, but many other types of systems are amenable to analysis from the physiologist's perspective, as stated most clearly by Adolph (1961):

> "It is not supposed that all regulatory responses depend on transmission of information from one site to another in the organism. Rather, those responses that require transmission are the ones that have most often been studied, since physiologists can interrupt the transmission or can listen in during it. Numerous regulatory activities proceed within single cells; and if in them there be a transmission, it usually involves very short distances and unrecognized structures. Correspondingly, it is not supposed that all regulatory responses start from sensations aroused in organs of special sense. Many are the unknown ways in which information is impressed upon responding tissue, and many are the physical and chemical shifts that follow from influences that are often not even considered by physiologists to be stimuli. For, it is not always possible to distinguish those influences that directly distort the living unit from the influences that work only through specialized sensing elements in the unit."

This scheme of things has served physiologists well, but it is not without problems. In particular, neither the principle of homeostasis nor the metaphor of a machine apply unambiguously to Data Set B. We invite you to study the data and then decide the extent to which it is, in fact, meaningful to imagine that the data ever settle down to a state of constancy, even though they are from someone who is inactive (asleep) and who is subjected to minimal external influences. If this system were really machinelike, one would expect to have found that its provoked responses (e.g., the effect of respiration on heart rate) are perfectly reproducible, so that we could in principle forecast its state far into the future. This presumes that the system behaves in a purely automatic fashion. But if the control mechanism is in some sense anticipatory rather than reactive—as might conceivably exist for a system that never settles down to a constant state—then might the observed variability of responsiveness be due to conditioning by previous responses, so that the best prediction of the system's behavior would be based on a forecast of the system's own forecast of its own future state? If so, how does the system make its forecast? Is it capable of learning from its past experience, or are its predictive abilities more or less fixed during the course of embryonic and subsequent development? Alternatively, is the observed variability simply due to the presence of noise in an otherwise conventional machine?

Because the origin of a physiological control system is unlike that of any machine that we have constructed in our workshops, might it be necessary to understand how the system arose in the first place in order to answer the questions raised above? During embryonic development, does the anatomical basis of the system's control appear long before the system is called upon to exercise control, or does the control system arise only through the process of controlling? Is the construction of the control system a form of Darwinism (Edelman, 1987; Holland, 1992), in which many potential control circuits exist at the beginning, but only those that by chance succeed find their way into the mature system? If so, is functionality of the final system genetically inevitable, despite the multiplicity of possible outcomes; i.e., is the construction a form of orthogenesis (Mayr, 1982)? Or does construction of the control system proceed in a directed, self-organizing manner (Yates et al., 1987), through the unfolding of a more deterministic genetic process, with scaffolding, division of labor in the form of anatomical differentiation, and the like?

The need for a timely critique of the conceptual foundations of physiology is implicit in doubt that the notion of " machine with a mechanism," which emerged from nineteenth century habits of thought (Fleming, 1964), is always the best metaphor for a physiological control system. To the extent that time series analysts find that analytical ideas and methods apply equally well to Data Set B and to the other data sets of this Competition, they will have provided a service to physiologists in suggesting useful new metaphors. In that regard, we warn you that physiological time series like Data Set B exhibit peculiarities that make such comparisons difficult:

1. The data often consist of ten or more channels that may be recorded continuously for many hours.
2. A substantial amount of preprocessing is required to extract the variables of interest (e.g., the duration of the heart beat from an electrocardiogram);
3. Calibration of the signals is often difficult, indirect, or not assured and must therefore be interpreted with due skepticism;

4. The signals are ordinarily nonstationary due, for example, to the difficulty of controlling voluntary behavior (e.g., respiration or activity) or to circadian influences;
5. One is always aware of factors that influence the observed behavior of the system, but that were not measured;
6. Almost all physiological systems are controlled by the autonomic nervous system, for which there is no simple analog; and
7. A variety of nonlinearities abound in the interaction between the components of physiological systems.

Furthermore, because time series analysis as an autonomous discipline tends to model data without regard to their source, rather than attempting to model system-specific interactions, the nonphysiologists among you may be inclined to ignore the traditions of physiological modeling that take into account a great deal of experimental evidence concerning particular physiological systems (Grodins, 1963; Riggs, 1963, 1970; Schwan, 1969; Iberall & Guyton, 1973; Segel, 1984; MacGregor, 1987). Obviously, it is not necessarily undesirable to analyze a physiological system with general methods, but if you do so, the objectives of your modeling efforts may not be considered significant by practicing physiologists. They will not be impressed by models that give only a feeling for how things work—why not gain intuition by experimenting with the real thing? And if your objective is to find a dynamical principle that transcends particular physical implementations, physiologists may be less impressed by the transcendency than by whether the principle applies in detail to a particular physiological system. So, as matter of strategy, you may wish to use system-independent models primarily as parts of larger models, representing components for which no credible physiological model already exists. Accordingly, to aid you in the formulation phase of your modeling, we have included a bibliography on the physiology that is relevant to Data Set B, as well as references to the corresponding physiological models.

We hasten to point out, however, that it is by no means clear that existing physiological models are suitable for making forecasts. As described in the Discussion section, the validity of such models is ordinarily judged by criteria other than forecasting, so they have almost never been used to forecast the phenomena that they describe. In fact, with rare exceptions, the fitting of continuous physiological time series appears to have attempted only with linear models, and in those cases the authors do not even show goodness of fit or other calculations that a statistician would consider essential. Evidently, the fluctuations in physiological time series are being widely ignored, even though they may be used to test models and estimate parameters, in lieu of performing experiments that manipulate some of the parameter values. The most complete analysis, of course, would be one in which fluctuations are forecasted during such interventions.

Therefore, the state of modeling and simulation in physiology is currently very unsatisfactory from the point of view of time series prediction and analysis. This is quite remarkable because the ability to forecast life-threatening events such as sudden cardiac death (Amato, 1992), sudden infant death, stroke, diabetic and anaphylactic shock, and pileptic seizures, could have a public impact comparable to the ability to forecast earthquakes, the weather, or financial markets. We encourage you to analyze Data Set B if for no other reason than to help identify factors that would make such medically important forecasting possible.

SUMMARY OF THE CHAPTER

The remainder of the chapter is organized as follows. We describe how the heart rate, respiration, blood pressure, and sleep-stage data were collected and preprocessed to produce the Competition files (`B1.dat` and `B2.dat`). We then give examples of two signal patterns that are found throughout the data. One pattern corresponds to normal respiration, and the other exhibits repetitive episodes of apnea, i.e., periods during which the subject barely breathes for up to a minute. The latter pattern is one of several known as *periodic breathing* and is potentially life-threatening.

The physiological signals are not perfectly regular during either pattern of respiration, but instead exhibit erratic heartbeat-to-heartbeat, breath-to-breath, and apnea-to-apnea fluctuations. Analysts are asked to assess the extent to which these fluctuations are predictable manifestations of underlying deterministic, physiological control mechanisms, rather than being due to noise. The preferred method of analysis is forecasting, but conclusions based on the calculation of Lyapunov exponents, dimensions, and related dynamical attributes are also encouraged.

Forecasts might be made using well-established, general-purpose linear methods, e.g., autoregressive and moving average models, or by newer nonlinear methods, such as state space reconstruction and neural networks (see below). Forecasts may also be made using problem-specific models that explicitly represent known or postulated mechanisms of physiological control. It is an open question as to what improvement in forecasting may be obtained by making use of such prior physiological information.

As a benchmark, we summarize the results of a forecast we made using a nonlinear model that is based on the physiology of cardiovascular reflexes (Rigney et al., 1992, 1993). When the respiratory signal is treated as a given driving function, the model permits better-than-chance forecasts of up to five heartbeats into the future. If respiration were treated as a variable to be forecasted, the time that could be predicted would be on the order of several respiratory cycles, rather than five heartbeats. When it is exercised, the fitted model generates deterministic chaos for some breathing frequencies, as well as periodic dynamical patterns at other frequencies. Until someone makes equally good forecasts with a model that does not generate chaos, we conclude that the observed cardiovascular fluctuations may be due in part to **deterministic chaos**. Because we were able to forecast out to a shorter time would have been permitted by the estimated maximum Lyapunov exponent, we also conclude that noise contributes substantially to the system's unpredictability.

Data Set B raises many questions concerning the interrelation between the variables, which should be addressed by system modeling (see also Casdagli & Weigend, 1993). Specifically, we discuss forecasting problems concerning the effect of respiration on the fluctuating heart rate during normal breathing (respiratory sinus arrhythmia), the role of cardiovascular variables in producing fluctuations in normal involuntary breathing, and the interdependence of heart rate, respiration, and blood oxygen saturation during periodic breathing. We also pose the problem of determining whether periodic vs. normal breathing is a bistable phenomenon, or whether these breathing patterns always correspond to different sets of parameter values.

SOURCE OF DATA SET B

The physiological signals in Data Set B were recorded from a 49-year-old male in the sleep laboratory of Boston's Beth Israel Hospital. He had been tentatively diagnosed as suffering from **sleep apnea**, a potentially life-threatening disorder in which the subject stops breathing during sleep (Chase & Weitzman, 1983; Edelman & Santiago, 1986; Guilleminault & Partinen, 1990; Parkes, 1985). The primary symptom leading to his hospital admission was extreme daytime drowsiness, which is thought to result from sleep apnea for the following reason. When the subject starts to fall asleep, he stops breathing. This leads to a decrease in oxygen and an increase in CO_2 in the blood that perfuses certain centers in his brainstem, which in turn results in a reflex arousal and gasps of breath. The arousal also wakes the subject. When he starts to fall asleep again, the process repeats itself, so the subject is chronically deprived of sleep by this cycle of apnea and arousal.

As is customary in such cases, the subject went to the sleep laboratory shortly before midnight, where many of his vital signs were recorded as he tried to sleep for the next six hours. A large number of signals were measured in order to provide unambiguous documentation of the apnea and to diagnose its cause. They were recorded with a multichannel instrumentation recorder, several channels of which were subsequently played back and digitized at 250 Hz. The signals were:

1. a single electrocardiogram (ECG) lead, for measuring heart rate and detecting arrhythmia;
2. respiratory airflow, measured by changes of temperature in the nasal passage, using a thermistor;
3. lung volume, measured by the distension of wires strapped around his chest;
4. blood oxygen saturation, measured by ear oximetry (changes of hemoglobin color in a translucent region on his ear);

5. blood pressure, measured continuously from a fluid-filled catheter situated in his radial artery;
6. electroencephalographic (EEG) leads, for determining his state of arousal or sleep;
7. an experimental noninvasive device for measuring the volume of his heart chambers (Tamaki et al., 1988);
8. left and right eye muscle activity, for detecting the stage of sleep that is associated with rapid eye movement (REM);
9. muscle activity (electromyogram) from his chin, to measure mouth movement;
10. muscle activity from his chest, to identify breathing effort; and
11. a microphone, to identify airway obstruction associated with snoring. In addition, the entire session was videotaped to retrospectively identify artifacts associated with movement.

As described below, data derived from only four of these signals were distributed as part of the Competition. However, if additional data are needed to perform a more complete analysis, signals (1) to (7) are available on CD-ROM as part of the M.I.T./B.I.H. Polysomnographic Database (M.I.T. 20A-113, 77 Massachusetts Ave., Cambridge MA 02139). For example, one of our analyses (see below) makes use of the blood pressure signal, which was not distributed for the Competition. The CD-ROM contains night-time recordings from 15 subjects, including the one reported here (case slp60).

SELECTION AND PREPROCESSING OF DATA SET B

Data Set B was selected as a prototype of multichannel time series that are often encountered during physiological monitoring. In general, the physiological signals are correlated with one another due to reflex feedback. In this case, the blood oxygen saturation increases (after a delay) in response to respiration, but the respiration is itself partly controlled by the autonomic nervous system in response to the gas content of blood. The heart rate also participates in the feedback because it partly determines the rate at which oxygenated blood is transported from the lungs to sensors in the brain, and it is controlled by the autonomic nervous system, along with the respiration. Because of this interdependence, the ECG, respiration, and blood oxygen saturation were selected as the signals for the data set. As illustrated by the figures that are provided below, other variables—such as blood pressure and the brain arousal state (EEG state)—also participate in the feedback loop, but these variables provide information that is to some extent redundant, so the corresponding signals were not included in the basic data set.

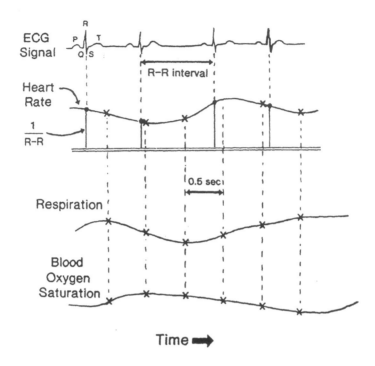

FIGURE 1   Preprocessing of the ECG, respiration, and blood oxygen saturation signals to produce the data in Set B. All three analog signals were digitized at 250 Hz. The R wave in each heart beat was detected automatically by a computer program. The reciprocal of successive R–R intervals was then interpolated to synthesize an "instantaneous heart rate" signal. It was sampled every 0.5 seconds along with the other two channels to generate the data files. See "Selection and Preprocessing of Data Set B" for details.

Because the subject was in bed for the entire recording, changes in the measured variables cannot be attributed to gross variation in the type and level of activity, or even posture. But it is not clear whether more subtle changes of activity nevertheless influence the stationarity of the variables under investigation. For example, digestion proceeds throughout the night, the blood volume changes due to urine formation, and the occurrences of small restless movements and hormone secretions (Yates, 1981) are to some extent a function of the varying stage of sleep. Consequently, we provided the sleep stage as part of the data set (Pack et al., 1989), obtained from a neurologist's interpretation of the EEG and eye movement.

HEART RATE DATA

The ECG signal has a complex waveform that requires sophisticated software to reliably detect features that are associated with normal heart beats. State-of-the-art software for the analysis of ECG waveforms corresponding to abnormal heartbeats is even more elaborate and cannot be trusted without being reviewed by a human expert (Mark & Moody, 1989). Because most investigators do not have access to such software, we preprocessed the ECG signal to determine the time at which a particular feature occurs in each heartbeat: the R wave, which corresponds to electrical depolarization of the heart's ventricles (see Figure 1). Some individuals have abnormal heartbeats that originate from regions of the heart other than the normal pacemaker region, in which case the identification of an R wave is problematic. So we deliberately selected a subject for which this complication never arose. By using only the R wave data, we are able to analyze the heart rate, but we ignore all other physiologically significant information in the ECG signal. This omission was considered appropriate because the additional information is most relevant to problems involving the mechanism of abnormal heart beats, which do not arise in these data.

The simplest condensed presentation of the data would have been to give the times of successive R waves in the ECG, along with the values of respiration and blood oxygen saturation at those times. However, the heart beat is not metronomically regular except under extraordinary circumstances, so such a presentation would have consisted of unevenly sampled data points. Because we expected most analysts of the data to have software that assumes the data are evenly spaced, we interpolated the R–R interval data as shown schematically in Figure 1, to produce the data that were provided for the Competition. The reciprocals of each R–R interval are assigned to the beginning of the corresponding intervals, which are then interpolated every 0.5 seconds to obtain an "instantaneous heart rate" signal. Details of the actual interpolation method are described by Berger et al. (1986) and Moody (1992). This widely practiced procedure filters the heart rate data and is therefore a potential source of artifact. However, it distorts the heart rate spectrum only at frequencies in the vicinity of the interpolation rate (DeBoer, Karemaker, & Strackee, 1985; Berger et al., 1986). Furthermore, some investigators prefer the interpolated signal to the original sequence of R–R intervals, on the grounds that it represents the continuously varying signal from the nervous system that actually regulates the heart rate. And when periodic external influences cause the heart rate to vary periodically, this procedure produces clear spikes in the corresponding heart rate spectrum, whereas the uninterpolated time series (indexed by beat number) has a spectrum that is more difficult to interpret.

The matter of how to preprocess the ECG signal is a good example of how conceptual issues arise in formulating physiological signal-processing problems. By transforming the continuous ECG signal into a series of R–R intervals (Figure 1), we make it convenient to view the system as a stochastic point process, about which there is a considerable analytical literature (Cox & Lewis, 1966; Lewis 1972). One may then analyze the multivariable problem in the standard terms of cross-spectra and coherence (Marmarelis, 1988), adapted as needed to accommodate the fact that these processes are not in general multivariate normal processes. However, physiologists may object to this formulation, saying that it is contrived because the actual process that underlies the synthesized point process is continuous and largely deterministic, and the only reason for converting the time series into a point process is to evade the biophysical basis of ECG waveforms. Similarly, by interpolating the point process to covert it back into a continuous one, new questions arise concerning the best way to perform spectral analysis on what was not actually a point process in the first place (Lewis, 1970; Press & Rybicki, 1989). But because the unprocessed data are available on CD-ROM, readers will be able to preprocess the data as they see fit. Our own preference is to model R–R interval data that have not been interpolated, but to calculate spectra and Lyapunov exponents using interpolated data.

RESPIRATION DATA

Two respiration signals were recorded and then digitized at 250 Hz. One is a function of the air flow (nose thermistor signal), and the other is a function of lung volume (signal from a wire around the chest). They provide redundant information, unless an airway obstruction prevents respiratory effort from generating an air flow. Therefore, only the flow signal was provided for the Competition.

Like the ECG signal, the respiratory signal is periodic, and for normal breathing, it is sometimes useful to use only the phase of respiration, ignoring information concerning lung volume. For example, one might use only the duration of inspiration and expiration (Carley et al., 1989). However, our subject exhibited periods of apnea, during which the depth of respiration is an important feature. Therefore, instead of preprocessing the respiratory signal to provide only the respiratory phase, we simply sampled it every 0.5 seconds, averaging its value over a window of 0.08 seconds to reduce noise (Figure 1). The nose thermistor signal is difficult to calibrate, and it tends to drift when the position of the head changes. Therefore, it was provided in uncalibrated digitization units.

BLOOD OXYGEN SATURATION DATA

The oxygen saturation signal was also digitized at 250 Hz. It is more easily calibrated than the respiration because it is based on the color of hemoglobin, which varies depending on its extent of oxygenation. However, this signal also drifts due to movement of the ear to which the color sensor is clipped, as well as to amplifier drift. Consequently, it too was provided in uncalibrated digitization units, after having been sampled every 0.5 seconds over an averaging window of 0.08 seconds (Figure 1).

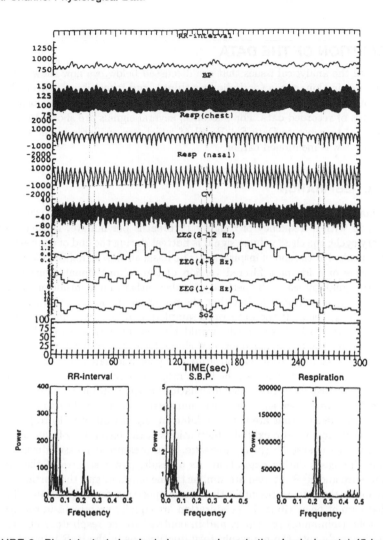

FIGURE 2 Physiological signals during normal respiration, beginning at 1:45 in Data Set B. The R–R interval was obtained as shown schematically in Figure 1. The remaining signals were measured as described under "Source of Data Set B," some of which were not included in the Competition data but are available on CD-ROM. BP is the blood pressure. Resp is a respiratory signal. CV is the cardiac volume. EEG is the electroencephalogram (filtered into frequency ranges). So2 is the blood oxygen saturation. S.B.P. is the systolic blood pressure (the peak pressure during each beat). The three insets at the bottom of the figure are power spectra of the corresponding signals. See "Description of the Data: Normal Respiration" for comments on these data.

DESCRIPTION OF THE DATA

To motivate the analytical issues that are discussed below, we now show examples of normal respiration, apnea (periodic breathing), and transitions between the two patterns of breathing. They are shown in Figures 2 to 4, each of which corresponds to 5 minutes of recorded data. For reference, several signals are shown in addition to those used to generate Data Set B, all of which are available in the CD-ROM database that was mentioned earlier.

NORMAL RESPIRATION

During normal respiration (Figure 2), both the chest volume and nasal air flow signals cycle with a frequency of roughly 12 breaths per minute. During a single respiratory cycle, the chest volume signal is flattest during the end of expiration and is sharply peaked at the end of inspiration. The nasal air flow signal is approximately the derivative of a low-pass filtered version of the volume signal. Neither of the respiratory signals appear to be perfectly regular, either in magnitude or frequency. This is also apparent from the fact that the power spectrum for respiration shows two neighboring peaks. If the data were sampled at a higher frequency, or if the respiration were much slower, peaks would be seen in harmonics of the average respiratory rate, due to the nonsinusoidal character of the respiratory cycle.

During normal respiration, cardiovascular signals fluctuate about their averages at the respiratory rate. In general, the heart rate accelerates during inspiration and deaccelerates during expiration. Such fluctuations in the R–R interval are prominent in Figure 2 and are known as respiratory sinus arrhythmia. The heart rate exhibits additional fluctuations that are not correlated in any obvious way to respiration.

The blood pressure also exhibits fluctuations at the respiratory frequency, most prominently in the systolic (peak) pressure. The pressure decreases during inspiration and increases during expiration. As a result, power spectra of the systolic blood pressure and R–R interval are similar to one another, and the portions above 0.15 Hz for both of them resemble the power spectrum for respiration. Fluctuations that cannot be attributed to respiration are equally prominent in the systolic and diastolic (minimum) pressures, and in contrast to the respiratory effect, they correlate inversely with the nonrespiratory fluctuations in the R–R interval signal. The cardiac volume, measured by the "CV" signal, exhibits the same respiratory and (to a lesser extent) nonrespiratory fluctuations as the blood pressure and heart rate. In contrast, the electroencephalogram filtered into different frequency ranges fluctuates with no apparent correlation with any of the other variables. The blood oxygen-saturation signal, SO_2, is almost constant, even though the respiration that drives it is periodic, but this is due in part to sluggishness of the oxygen sensor.

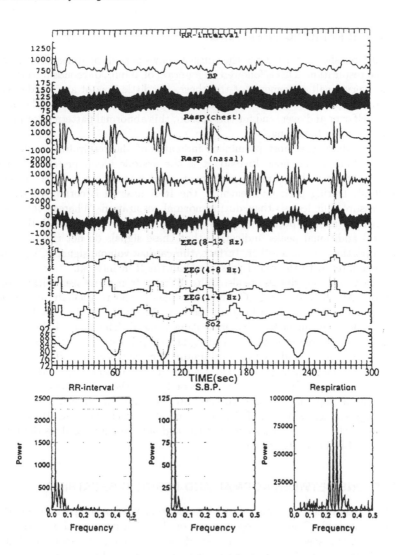

FIGURE 3 Physiological signals during peridic breathing, beginning at 0:25 in Data Set B. See the legend to Figure 2 for the meaning of acronyms. See "Description of the Data: Apnea and Periodic Breathing" for comments on these data.

APNEA AND PERIODIC BREATHING

In the example shown in Figure 3, the subject shows periods of apnea, during which there is no respiration. This is followed by a period of about 20 seconds during which there are very small movements of the chest, but little flow of air. Suddenly, the subject takes about four deep breaths and then stops breathing again. The process repeats itself over and over, and for this reason, this abnormal pattern of respiration is called "periodic breathing."

In these data, the most prominent indicator of the respiratory cycle is not the lung volume or air flow. Instead, it is the systolic blood pressure, which is cycling at the normal respiratory frequency even when there is little movement of the chest. During periodic breathing, the R–R interval is correlated with the blood pressure, but less so than during normal respiration. Whereas the systolic blood pressure shows spectral patterns at two low frequencies, the R–R interval has several additional peaks, indicating that these signals do not rise and fall in linear proportion to one another. In fact, the cardiac volume signal (CV) is more closely correlated with the blood pressure than the R–R interval.

The EEG and SO_2 signals are also quite different from those seen during normal respiration. All three of the EEG frequency bands tend to oscillate at the same frequency as the periodic breathing. In the 4- to 12-Hz bands, the signal increases abruptly before the deep breaths, indicating arousal. The 1- to 4-Hz band, on the other hand, decreases gradually during the deep breaths, and the cessation of breathing tends to coincide with its minimum. The SO_2 signal oscillates with the same period as the breathing pattern and is most nearly in phase with the slow EEG signal. It reaches a minimum roughly 10 seconds after the onset of deep breathing, an observation that may be used to estimate the time for blood in the lungs to pass through the left ventricle of the heart and then reach the head. However, the magnitude of this delay varies somewhat throughout the night, as does the relative phase of the EEG and SO_2 signal.

TRANSITIONS BETWEEN NORMAL AND PERIODIC BREATHING

Patterns of respiration that are intermediate between normal and periodic breathing seem to be unstable. They develop into one or the other patterns shown in Figures 2 and 3, but the speed with which they do so is variable. As seen in Figure 4, the transition from periodic to normal breathing may be abrupt and is associated with an arousal event that appears in all frequency bands of the EEG (85 seconds into Figure 4). Shortly thereafter, the R–R interval and blood pressure develop patterns that are difficult to distinguish from the normal patterns seen in Figure 2. The respiration pattern resembles the normal one, but is more erratic. There is a subsequent period during which the EEG indicates a prolonged arousal (100 to 160 seconds into Figure 4). This is followed by a quieter EEG pattern, the beginnings of periodic respiration, and the appearance of oscillations in the blood oxygen saturation signal (180 to 300 seconds into Figure 4).

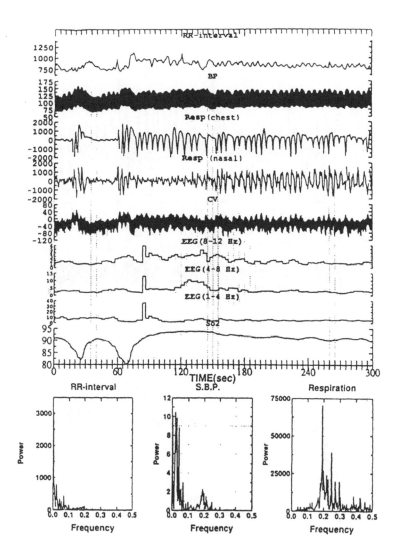

FIGURE 4 Physiological signals during transitions between normal and periodic breathing, beginning at 0:35 in Data Set B. See the legend to Figure 2 for the meaning of acronyms. See "Description of the Data: Transitions Between Normal and Periodic Breathing" for comments on these data.

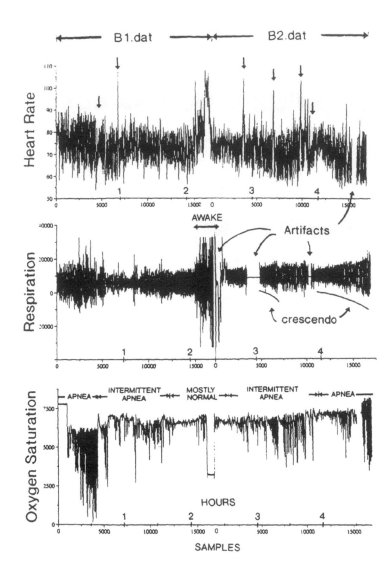

FIGURE 5 Overview of Data Set B, which consists of 4 hours and 43 minutes of
simultaneous heart rate data (beats per minute), respiration data (uncalibrated), and
blood oxygen saturation data (uncalibrated), divided equally into files B1.dat and B2.dat.
The arrows in the heart rate time series point to the most prominent arousal events.
See "Description of the Data: Overview of the Data Set" for details.

OVERVIEW OF THE ENTIRE DATA SET

The entire data set consists of 34,000 samples of the three signals, taken every 0.5 seconds (see Figure 1). For convenience of data storage and transmission, it was divided into two equal files, B1.dat and B2.dat. Instructions on how to obtain the data are provided in the Appendix of this book. The description file that accompanies the data contains information concerning the sleep stage as a function of time.

Figure 5 shows the entire data set, plotted as a function of clock time and sample number. Having described the signal patterns for normal and periodic breathing, we are now in a position to comment on their occurrence throughout the 4 hours and 43 minutes of data.

Periods of periodic breathing are most easily identified from the oxygen saturation signal. Periods of profoundly periodic breathing occurred at the beginning and end of the data set. They are characterized by wide excursions in the oxygen signal. In contrast, the periods of normal breathing, which appear in the middle of the data set, are characterized by relative constancy of the oxygen signal. The remaining periods of intermittent periodic breathing show limited excursions in the oxygen signal, primarily because arousals such as the one shown in Figure 4 limited their growth.

The arousals may also be identified from transient increases in the heart rate, the most prominent of which are indicated by arrows in the heart rate signal of Figure 5. When arousals do not interrupt the periodic breathing, it sometimes grows as a crescendo, as seen in the respiration signal. Finally, note that the data contain a number of artifacts that break the continuity of the recording. They are most frequent in the respiration signal, but they also occur at 4:30 in the heart rate signal and at 2:15 in the oxygen saturation signal. The latter break occurred during an intervention intended to test the subject's response to an intervention known as continuous positive airway pressure (CPAP), which is a countermeasure for apnea (Guilleminault & Partinen, 1990). As indicated in Figure 5, he was awake shortly before and after that intervention.

ANALYSIS AND FORECASTING OF FLUCTUATIONS IN PHYSIOLOGICAL TIME SERIES: THE GENERAL PROBLEM

Our description of Data Set B has until now been focused on the patterns that are readily apparent in the multichannel time series during normal and periodic breathing. We now emphasize the fact that these patterns are not perfectly regular. Although the successive respiratory cycles in Figure 2 resemble one another, none of the oscillating signals exhibit constant amplitudes, and they all fluctuate slowly about their averages. Similarly, although the successive episodes of apnea in Figure 3 have the same general morphology, they are not alike in every respect, and the

R–R interval time series is especially erratic. One's first impulse is to ignore these fluctuations as being distractions from the essence of the patterns. But if we cannot explain the fluctuations during periods of relative regularity, how can we hope to forecast the growth of the fluctuations during transitions from one pattern to another, as seen in Figure 4?

The extent to which we can in fact forecast the heart rate and other physiological variables depends on the unknown answer to the following question. To what extent are fluctuations in physiological time series intrinsic to their respective physiological control systems, and to what extent are they simply responses to unrecognized external perturbations? For example, it has long been known that a change in posture would cause a significant change in blood pressure, were it not for the fact that a certain reflex adjusts the heart rate rapidly in such a way as to maintain the blood pressure within safe limits (the baroreceptor reflex; see below). During daily activity, you are constantly changing posture, so some of your heart rate variability is simply a reflection of your variable posture.

But how do we explain the fluctutations that occur when activity is more or less constant? Are they simply transients due to influences—analogous to posture—that are unrecognized or that are difficult to control completely? Or are physiological control systems sufficiently nonlinear that they will fluctuate spontaneously, giving rise to complex sustained oscillations or to deterministic chaos (Degn, Holden, & Olsen, 1987)? *The answer may often be a combination of these two explanations, and the general problem is to distinguish the contribution of each.* Forecasting is a good method for doing so, because a good forecast is evidence for a deterministic mechanism, whereas a poor forecast is evidence for either unpredictable influences (noise or nonstationarity) or for chaos with a large Lyapunov exponent (see below).

However, a poor forecast may also merely demonstrate that our method is inadequate, so forecasts by all available methods are desirable. These methods include well-established, general-purpose linear methods such as autoregressive models (Abdel-Malek et al., 1988; Akaike, 1979; Bohlin, 1977; Box & Jenkins, 1976; Chatfield, 1989; Khoo & Marmarelis, 1989; Priestly, 1981; Rogowski, Gath, & Bental, 1981; Wada, Akaike, & Kato, 1986; Whittle, 1963; Wright, Kydd, & Sergejew, 1990), as well as recent general-purpose, nonlinear techniques (Crutchfield & McNamara, 1987; Korenberg & Hunter, 1990; Marmarelis, 1991; Mees, 1991; Packard, 1990; Sugihara & May, 1990; Tong, 1990; Tsonis & Elsner, 1992), including the use of neural networks (Weigend, Huberman, & Rumelhart, 1990), radial basis functions (Casdagli, 1989), and state-space reconstruction (Casdagli, 1991; Casdagli et al., 1992; Farmer and Sidorowich, 1987; Linsay, 1991). Such techniques have the virtue that they can be applied to any of the time series that are described in this volume. However, these techniques make no use of prior knowledge that we have about the physiology of heart rate, respiration, etc., which might constrain the forecasts or make possible better forecasts from a limited amount of data. For the benefit of investigators who wish to construct forecasting models specifically for Data Set B, the paragraphs that follow include a review of the pertinent physiology literature.

ANALYSIS AND FORECASTING WITH DATA SET B: SPECIFIC PROBLEMS

The analysis of Data Set B might best proceed in stages, each of which considers successively larger forecasting problems. We suggest that the initial forecasts be made using the examples shown in Figures 2, 3, and 4. Corroboration of the results may then be sought by analyzing similar examples that are found throughout the data set.

The first and simplest problem is to analyze heart rate fluctuations during normal breathing. From our description of the normal respiration seen in Figure 2, the heart rate is influenced by the respiration, which itself exhibits breath-to-breath fluctuations. For the initial problem, our suggestion is to treat the respiration as a given driving function, rather than as a time series to be forecasted. The advantage of doing so is that simulations may be subsequently undertaken to predict the heart rate under different patterns of respiration, which can be realized experimentally because breathing can be modulated voluntarily or forced by a mechanical respirator (Petrillo & Glass, 1984). In particular, it may be made metronomically periodic, to reveal frequency-dependent characteristics of respiratory sinus arrhythmia. It may also be driven simultaneously at two or more frequencies, to look for nonlinear effects on the heart rate, such as spectral sidebands. And it may be driven in complicated patterns, such as those observed during periodic breathing (Figure 3), in order to determine whether the physiology of heart rate control during normal respiration is the same as that during periodic breathing.

The second problem is to analyze and forecast the respiratory signal during normal involuntary breathing. The simplest approach would be to consider it independently of the heart rate. An improved forecast, however, might be obtained by treating the heart rate as a given driving function. Finally, the analysis of the first problem could be combined with that of the second to simultaneously forecast the cardiorespiratory fluctuations during normal involuntary breathing.

The third problem is to forecast respiration and blood oxygen during periodic breathing (Figure 3). Depending on the magnitude of the influence of heart rate on respiration, found during the analysis of the second problem, one may wish to initially ignore the heart rate as a model variable. However, in this case, the ability to model the blood-borne transport of oxygen may be an important consideration. One might also first consider the oxygen and respiration signals as driving functions of one another, then later treat them as simultaneous, interdependent variables.

A physiological model of periodic respiration might be applicable to normal respiration as well, provided that these patterns are bistable. That is to say, the prevailing pattern of respiration may be simply a matter of initial conditions, in which case forecasts with a single model would be equally applicable to the data shown in Figures 2 and 3. Such a model may also be applicable to the data of Figure 4 (which shows a transition from normal to periodic breathing), provided that the initial conditions that follow the arousal seen there result in a long transient, leading eventually to periodic breathing. The final problem is therefore to assess whether the physiology in those three figures is constant in the sense of parameter stationarity, or whether drift of physiological parameters is the best explanation for one or the other pattern of breathing. This type of question is ordinarily addressed experimentally by changing parameter values up and down in search of hysteresis (Labeyrie et al., 1987). At issue is whether it is possible to decide the matter on the basis of the time series alone, given that noise, as well as internal (brain arousal) and external (environmental) disturbances reset the initial conditions occasionally.

As background to these problems, we now summarize a benchmark forecast for the first problem and provide a literature review for the remaining problems.

HEART RATE FLUCTUATIONS DURING NORMAL RESPIRATION

There is a large literature to document the fact that the heart rate of normal individuals fluctuates erratically. Over long time periods, the heart rate has a $1/f$ Fourier spectrum (Kobayashi & Musha, 1982), and the spectrum over shorter time periods shows prominent peaks that have been attributed to physiological control systems involving respiration (see below), feedback from blood pressure sensors (see below), thermal regulation, and hormonal control of blood volume (Akselrod et al., 1981; DiRienzo et al., 1992; Goldberger et al., 1988; Kaplan & Talajic, 1991; Kitney & Rompelman, 1981). Although the existence of oscillatory patterns in the heart rate is clear, it is controversial as to whether the heart rate fluctuations may be chaotic. Tests for the presence of chaos in the heart rate have been affirmative, but the test results are equivocal for many technical reasons (Rigney et al., 1993).

Our approach to analyzing normal heart rate variability is to construct a physiological model for the system, fit the model to the data, forecast with the fitted model, and evaluate the role of noise by adding randomly sampled residuals of the fitted data to the model. A significant advantage of this approach is that questions concerning limited data and nonstationarity do not arise when analyzing simulations from the fitted model, rather than the original data. Our model is described in detail elsewhere (Rigney et al., 1992, 1993), including a forecast made with the data shown in Figure 2. Here, we will only sketch the physiological rationale for the model and summarize its properties, one of which is that the fitted model exhibits deterministic chaos for some patterns of respiration.

The primary mechanism of heart rate control is the baroreflex, which maintains arterial blood pressure within safe limits (Abboud & Thames, 1983; Kirchheim, 1976). This reflex operates by regulating the volume of blood in the arteries—as the volume increases, the blood vessel walls distend so that the pressure increases. The arterial blood volume is regulated by controlling the rate at which blood enters the aorta, by adjusting the heart rate and by adjusting the volume of blood ejected by the heart on each beat. Except during exercise, changes of heart rate are greater than changes of ejected volume per beat. Arterial blood volume is also controlled by adjusting the rate at which blood exits the small arteries to enter the capillaries.

This is accomplished by varying the diameter and stiffness of the arteries, through contraction of smooth muscle within the blood vessel walls. The stiffness also determines the blood pressure for a given arterial blood volume. At any instant, the heart rate and vessel stiffness are modulated by the secretion of hormones (neurotransmitters) from the endings of nerves that terminate in the heart and blood vessels. The magnitude of the secretion is determined by cardiovascular centers of the autonomic nervous system, located primarily in the brainstem (Calaresu et al., 1984; Ciriello, Rohlicek, & Polosa, 1983, 1985; Ciriello et al., 1982). These centers integrate sensory information from receptors throughout the body, the most important of which are pressure-sensitive nerves located in the walls of the aorta and carotid arteries—the baroreceptors (Brown, 1980).

Thus, a closed-loop model for the feedback control of heart rate should at least contain variables to represent the blood pressure, the change of heart rate as a function of changes in blood pressure, the change of blood vessel stiffness as a function of changes in blood pressure, and the change in arterial blood pressure (or volume) as functions of the heart rate and vessel stiffness. Our current model consists of five nonlinear equations for these variables, and it includes a subsidiary model for describing the mechanics of the blood pressure waveform (Ocasio et al., 1993). In its current form, the model assumes that the respiration is a given driving function, not a variable to be modeled. The model fits normal heart rate and blood pressure fluctuations, such as those shown in Figure 2, with a correlation coefficient on the order of 0.9. And it makes better-than-chance out-of-sample forecasts for up to five heartbeats, at which time the forecasts become dominated by the given respiratory signal.

We exercised the fitted model by driving it at different respiration frequencies. At many frequencies, the model generated chaos, as indicated by positive Lyapunov exponents, return maps that appear chaotic, and Fourier spectra with broad peaks. For other respiratory frequencies, the dynamics were periodic, with periods of up to 100 heart beats. Because normal respiration contains frequency components that correspond to chaotic dynamics, we conclude that the heart rate may also be chaotic. The maximum observed Lyapunov exponent was approximately 0.2/sec., which should permit forecasts of approximately 5 seconds (depending on the values of the entire spectrum of Lyapunov exponents). Because our actual forecasts were worse than this, we conclude that noise also contributes substantially to the system's unpredictability. Embellishing the model with more variables would not be likely to improve matters, because the residuals of the data fit (the noise) had auto- and cross correlations close to zero. To evaluate the effect of noise on the system, we randomly sample the residuals, adding the samples back to the corresponding model equations. The most striking effect of the noise is to produce a large increase in the low-frequency components of the heart rate spectrum, relative to the spectrum for the completely deterministic model.

RESPIRATORY SINUS ARRHYTHMIA

The variation in heart rate as a function of the respiratory phase might be due to at least four mechanisms. First, information from lung stretch receptors might be integrated directly by the cardiovascular centers of the brainstem. Second, information about respiratory status might radiate indirectly from the respiratory to the cardiovascular centers in the brainstem. Third, there may be a mechanical effect, for example, in which the thoracic pressure constrains diastolic filling of the heart's chambers with blood, the variable volume of which subsequently results in a variable systolic blood pressure and baroreflex, or which is sensed by atrial stretch receptors that convey this volume information directly to the cardiovascular centers. And fourth, respiration affects blood oxygen, carbon dioxide, and pH, which is sensed by chemoreceptors that convey this information to the cardiovascular centers of the brainstem. There may also be differential respiratory effects on signals from the two main branches of the autonomic nervous system (sympathetic and parasympathetic). Much of the literature on respiratory sinus arrhythmia is concerned with the relative importance of these mechanisms, as well as differences that occur between individuals due to aging and pathology (Anrep, Pascual, & Rossler, 1936; Bainbridge, 1920; Bernardi et al., 1989; Davies & Neilson, 1967; Eckberg, 1983; Eckberg, Kifle, & Roberts, 1980; Fouad et al., 1984; Freyschuss & Melcher, 1976; Haymet & McCloskey, 1975; Hellman & Stacy, 1976; Hirsch & Bishop, 1981; Hrushesky et al., 1984; Katona & Jih, 1975; Koepchen, Milton, & Trzebski, 1980; Levy, DeGeest, & Zieske, 1966; Lopes & Palmer, 1976; Melcher, 1976, 1980; Neil & Palmer, 1975). A few papers are concerned with the quantitative aspects of respiratory sinus arrhythmia, such as its dependence on the rate of breathing (Angelone & Coulter, 1964; Saul et al., 1989, 1991; Selman et al., 1982; Womack, 1971). The papers by Kitney and colleagues are most interesting because they describe nonlinear effects at abnormally low breathing rates, when respiratory and blood pressure oscillations apparently entrain one another (Giddens & Kitney, 1985; Kitney et al., 1985; Selman et al., 1982). Several mathematical models have also been proposed to account for such quantitative properties (Ahmed et al., 1986; Clynes, 1960; Giddens & Kitney, 1985; Kitney, 1981; Kitney et al., 1982; Saul et al., 1989, 1991).

NORMAL RESPIRATION

The most basic feature of normal respiration is that it is an oscillatory phenomenon. The respiratory rhythm is generated in the brainstem in neural networks known as "central pattern generators." Several mathematical models for these generators have been proposed, most of which produce limit cycle or "integrate-and-fire" oscillations (Benchetrit, Baconnier, & Demongeot, 1987; Feldman & Cowan, 1975; Geman & Miller, 1976; Glass, 1987; Glass & Mackey, 1988; Pham Dinh et al., 1983; Rubio, 1972). These networks drive the muscles that move the lungs during ventilation. However, the cycle of ventilation is not a correspondingly simple oscillation because it exhibits considerable breath-to-breath variation (Bendixen, Smith, &

Mead, 1964; Bertholon et al., 1987; Brusil et al., 1980; Goodman, 1964; Patil et al., 1987), as seen also in Figure 2. It has been suggested that the observed fluctuations in lung volume sometimes represent deterministic chaos (Bertholon et al, 1987; Donaldson, 1992; Yamashiro, 1989). If the fluctuations do represent chaos, they may result from the fact that the respiratory neural oscillator is part of several reflexes involving stretch receptors in the lung, blood gases (oxygen, carbon dioxide), blood pH, body heat, and cardiovascular status. Nonlinearity in the interactions between the neural oscillator and overall ventilation is demonstrated most convincingly by experiments in which phase locking occurs when ventilation is driven mechanically at different frequencies (Glass & Mackey, 1988; Petrillo & Glass, 1984). Models for analyzing control of the overall respiratory system are often constructed to simulate a pathological state, such as periodic breathing, as now described.

PERIODIC BREATHING

Apnea and periodic breathing may be induced by a variety of conditions (Chapman et al., 1988; Dowell et al., 1971; Khoo, 1991), such as heart failure, neurological disorders, respiratory obstruction during sleep, and hypoxia (which may occur, for example, when lowlanders are brought to high altitude). The general explanation for this breathing pattern is that it involves a physiological feedback for the control of blood gases and pH, plus a delay that is a consequence of the fact that the subject of control (oxygenated- and CO_2-depleted blood in the lungs) must be transported via the arteries to a distant site at which the information used for the feedback is actually measured (sensors in the brain). The combination of nonlinearities in the feedback and delay results in a situation in which periodic breathing either happens or does not, depending on the values of parameters that characterize the system, such as the duration of the transport delay. Several mathematical models of periodic breathing have been proposed, most of which are intended to predict the combination of parameters that stabilize the breathing pattern (Carley & Shannon, 1988; elHefnawy, Saidel, & Bruce, 1988; Grodins, Buell, & Bart, 1967; Horgan & Lange, 1962; Khoo, 1989; Khoo, Gottschalk, & Pack, 1991; Khoo et al., 1982; Longobardo, Cherniak, & Fishman, 1966; Mackey & Glass, 1977; Milhorn & Guyton, 1965; Nugent & Finley, 1987).

DISCUSSION

The difficulties mentioned in the Introduction may have deterred physiologists from attempting to make forecasts from their continuous time series data. Another deterrent may be the attitude that the fluctuations contain little information and are even a contaminating nuisance that obscures the signal one is trying to observe. Consequently, the modeling mentioned above is one of the first times that short-term forecasting has ever been undertaken for a continuous physiological time series (Rogowski, Gath, & Bental, 1981; Mees, 1991; Longtin, 1993), although the potential for forecasts is implicit in earlier work involving spectral analysis of the EEG, cardiovascular time series, and some discontinuous data, e.g., hormonal time series (Yates, 1981). In fact, several authors have used autoregressive models to fit beat-to-beat changes in cardiovascular time series (Bartoli, Baselli, & Certutti, 1985; Baselli et al., 1985; Giddens & Kitney, 1985; Kalli et al., 1988; Turjanmaa et al., 1990), but they do not report even a one-step prediction error. Some suggest that the main value of prediction lies in its ability to identify abnormal beats, such as premature ventricular contractions (Baselli et al., 1985). Similarly, linear models have been proposed for analyzing simultaneous heart rate, blood pressure, and respiratory data (Appel, 1992; Saul et al., 1989, 1991), but here again, forecasting was not attempted, and it is not clear from these publications how well the data actually fit individual time series. Linear systems analysis has also been used in the design of anesthetic delivery systems for the infusion of vasoactive substances (Sheppard, 1980, 1989), for the control of ventilation during inhalation anesthesia (Ritchie et al., 1990), and for the automatic administration of drugs to control ventricular arrhythmia (Jannet, Kay, & Sheppard, 1990). However, in these control applications, predictability occurs on a time scale of minutes, not the short-term time scale that is under consideration here.

If physiologists do not ordinarily fit time series data, much less make forecasts, then how do they judge the validity of their models? One method of evaluation is on the basis of verisimilitude, i.e., by deciding subjectively whether the overall features that are simulated by the model resemble the data. For example, one would have to agree that published models of nerve activity can generate bursting patterns that resemble those measured from nerve cells (Carpenter, 1981; Chay & Rinzel, 1985) or realistic-looking fluctuations in the EEG (Freeman, 1987). This is a significant accomplishment. But shouldn't the next step be taken, namely, to estimate parameters from the time series data, calculate goodness of fit, and perform all the other calculations that a statistician would consider essential? If this were done, the claim that different models can simulate data equally well might often be shown to be groundless. Furthermore, the recent reluctance of physiologists to construct models with more than a few parameters may be overcome by acknowledging that the large amounts of time series data that are not being used might greatly overdetermine the parameter values in even an elaborate model, like those that were popular two decades ago (Schwan, 1969; Iberall & Guyton, 1973).

A second method for verifying models is to vary parameters in search of bifurcation points. For example, the conditions for periodic breathing may be predicted by determining conditions for a model's instability (Mackey & Glass, 1977). This is certainly a useful approach and can be used to falsify models. But by itself, it greatly limits the number of parameters that can be estimated, and it is not well suited to situations involving continuous physiological monitoring, in which parameter values are usually not under experimental control.

So, in conclusion, we reiterate the need to perform detailed data fitting and forecasting of continuous time series, in order to validate physiologically realistic models. Potentially, much information may be extracted from fluctuations in the time series. A closer examination of the data can reveal the extent to which physiological systems are adaptive and predictable, which in turn may be of considerable medical significance.

ACKNOWLEDGMENTS

We thank A. Weigend and N. Gershenfeld for organizing a very stimulating Competition and workshop. We also thank F. Bennett and the reviewers for thoughtful comments on the manuscript. Support was provided by the National Institutes of Health (R01-HL42172, R01-AG11124, P30-AG08812), the National Institute on Drug Abuse (DA 06316), the National Aeronautics and Space Administration (NAG9-572), the G. Harold and Leila Y. Mathers Charitable Foundation, and the Colin Medical Instruments Corporation.

Jean Y. Lequarré
Union Bank of Switzerland, LHIS Lausanne Development Center, 6, Avenue de Provence, CH-1007 Lausanne, SWITZERLAND

Foreign Currency Dealing: A Brief Introduction (Data Set C)

1. INTRODUCTION

Series C of the competition is a series of currency exchange rate: the U.S. dollar/Swiss Franc exchange rate for the period 1990–1991. The seminar organizers selected ten segments of 3,000 points each of the series made available for this competition. For each segment they proposed to forecast the exchange rate 1 minute after the last tick, 15 minutes after the last tick, 60 minutes after the last tick, closing value of the day of the last tick, the opening value of the next trading day, and the prediction for the closing value of the fifth trading day after the day of the last tick. According to our experience, however, only the first four can be predicted with a reasonable accuracy.

Times Series Prediction: Forecasting the Future and
Understanding the Past, Eds. A. S. Weigend and N. A. Gershenfeld, SFI Studies
in the Sciences of Complexity, Proc. Vol. XV, Addison-Wesley, 1993

2. THE FOREIGN EXCHANGE MARKET

2.1 SIZE

The foreign exchange market is very large. In 1989, the volume exchanged was estimated at about $650 billion dollars per day. Of that amount, a very small portion (less than 3%) covers actual commercial transactions. The rest is made of speculative transactions between market markers.

2.2 CURRENCIES

The most important currencies in the market are the U.S. dollar (which acts as a reference currency), the Japanese yen, the British pound, the German mark, and the Swiss franc. Close followers in terms of importance are the ECU, the French franc, the Italian lira, and the Dutch gulden.

Deals made to exchange a currency against U.S. dollars are called direct deals. Deals made to exchange any currency pair not including the dollar are called cross deals. Currencies are normally quoted against the U.S. dollar.

The large part (60%) of foreign exchange trading concerns spot deals, i.e., deals which must be settled one or two days (depending on the currency) after the deal is done. There exist however an active market in forward trading (deals which will only be settled at a given time in the future) and in currency swaps. There exists also foreign currency futures which are traded in a centralized exchange. There is also a very active market in derivative instruments (options).

2.3 DEALING

There is no centralized exchange for normal spot deals. Deals are made over the telephone directly between major players or via a broker. Direct exchanges between financial institutions make up about two-thirds of the market in volume with the brokers being involved in the rest.

The foreign exchange dealer who wants to trade, calls a market maker with the currency he is interested in and requests a quote. Market makers are obliged to quote both sides (bid and ask). Exchange rate are normally expressed in terms of unit of foreign currency needed to buy one U.S. dollar, but there are exceptions.[1]

The spread between the bid and ask prices is generally constant and depends on the currency. Prices are generally quoted to the fourth decimal place.[2] Unless specified otherwise the price quoted is good for amounts of 5 to 10 million dollars.

[1] The most notable exception is the pound which is quoted in terms of numbers of U.S. dollars needed to buy one British pound.

[2] When the exchange rate is of the order of 100 to the dollar, like the Japanese yen, then the price is quotes only to two decimal places.

If the amount to be traded is significantly larger or smaller, it must be specified when the quote is requested. The price quoted is only good for immediate trading and the decision to trade or not must be taken within five seconds. If the requesting dealer decides to trade, he then just specifies the exact amount and the settlement instructions.

2.4 INFORMATION DISTRIBUTION

Since no central trading location exists for spot currency trading, this market could be quite slow and inefficient. That would be the case if a trader had to call continuously all the participants in the market to find out the current price and therefrom decide if he should keep, liquidate, or turn his position.

To fill the information need, specialized companies publish electronically the most recent market prices. Those prices are contributed by the various market makers and immediately distributed to all service subscribers.

It is, however, noteworthy that the prices published in this service are not binding. So, a dealer willing to deal can select the counterparty to call based on the indication on the monitor but has to request a new quote from this counterparty before he can actually deal. Reputable institutions will always quote on the phone a price that does not differ much from their published price.

2.5 BROKERS

Brokers act as a central market place. They offer a possibility for one market player to quote anonymously. The parties matched by a broker must only identify themselves once the deal is done. Also by combining several quotes, the brokers can often offer a better price, but they charge a fee which can destroy that advantage.

2.6 ELECTRONIC DEALING

Recently several organizations have tried to set up electronic brokeraging. The foreign currency dealers would enter, on a CRT, the prices at which they are willing to deal and the corresponding currency amounts and the system would try to match automatically bids and offers. At the time of this writing, those systems still amount for a very small part of the transaction volume.

2.7 DAILY VOLUME

Since there is no central exchange, daily exchange volume is impossible to measure with accuracy. A market maker can estimate the volume in his market by assuming that his share of the market remains constant, but no reliable source exists about other currencies.

3. THE DATA SERIES

Data Set C is extracted from the Union Bank of Switzerland's foreign exchange department internal information system at the main office in Zurich, Switzerland.

A dealer is responsible for U.S. dollar/Swiss franc trading. Each time that it is required, this dealer enters a new exchange rate for this currency pair in the information system. This rate is then forwarded to Reuters, Telerate, and other information vendors to be displayed as the current price at which the bank is willing to deal. This rate is also sent to various internal pricing systems which will use it to price other instruments, the price of which depends on the U.S. dollar/Swiss franc rate.

There is no prescribed time at which the dealer must enter a new rate. However, if there is a lot of action in the market, the dealer will enter a price more frequently than during a quiet period. The price will be marked as invalid in the internal display system if it is older than 5 minutes.

Prices are updated during business hours in Zurich; that is, from 7:00 in the morning until 18:00 at night. Discontinuities are often observed around 14:00–14:30 when the market opens in New York.

4. EXCHANGE RATE FORECAST

4.1 FACTOR INFLUENCING THE RATES

Generally speaking the economic indicators, monetary parameters, and interest rates are the most important factors influencing exchange rates.

Given the large amount of speculative dealing in the market, psychological factors also play a very large role in determining the exchange rate of a currency.

As an example, one could recall the coup against Mikahaïl Gorbatchov in August 1991 which triggered a sudden and massive increase in the dollar value when the general trend was downward. A few days later, when it became obvious that the coup had failed, the dollar went back to its "normal" rate and the downward trend resumed (see Figure 1).

The details appear better on Figure 2 which shows the hourly evolution of the price. The news broke over the weekend and the price had already skyrocketed when the market opened in Zurich on Monday, August 19, 1992. But the return towards the downward trend started already well before Gorbatchov's return, showing that the market was already "digesting" the news. It is likely that even if Gorbatchov had not returned, the downward trend would have resumed albeit more slowly.

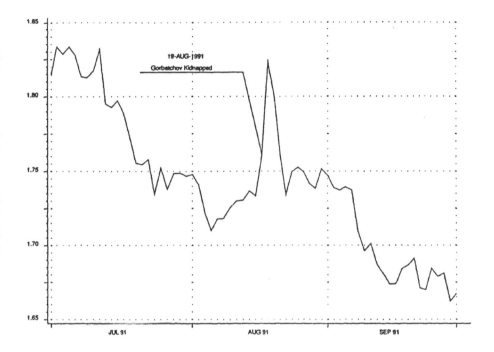

FIGURE 1   Daily U.S. dollar/Deutsche mark exchange rate. (Note that this is not the Competition data set; the organizers of the Competition selected the U.S. dollar/Swiss franc exchange rate.)

Each currency has a known image. The Deutsche mark, for example, is tied very closely to the political stability and the economical success of eastern European countries. That explains why it went down so much when Gorbatchov was kidnapped. The U.S. dollar is very sensitive to political unrest and the Swiss franc plays the role of safe haven. Those reputations, much more than actual economic factors, regulate the evolution of the currencies in time of turmoil.

4.2 FUNDAMENTAL VERSUS TECHNICAL ANALYSIS

Two schools of thought compete in the forecasting field. The fundamentalists believe that the forecasting process should model at least approximately the mechanisms that underlie the exchange rate determination. They use models that account for the interest rate differential, the balance of payments, and many econometric quantities.

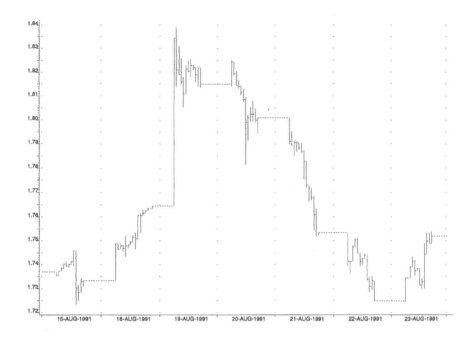

FIGURE 2   Hourly U.S. dollar/deutsche mark exchange rate.

One of the problems with that thinking is that many of the quantities involved in the model are only known much later. They are thus useful to explain *a posteriori* why things happen but could not be use in a forecast.

In contrast, the technical analysts believe that all the information available has already been discounted by the market and that the prices themselves already include all the possible information. Therefore, they will look at nothing else in trying to predict their future.

4.3 FORECAST ACCURACY

Although it would be most profitable to know the exact rate of exchange in the future, a rough estimation is often enough to make a reasonable profit on dealing operation. For speculative purposes, only trend turning points are actually important.

Therefore, many of the financial institutions that are trying to predict foreign exchange rates generally rate algorithms by evaluating the percentage of time when the algorithm gives the right trend from today until some time in the future.

In the spot market, the figures quoted are often related to the forecast one, two, and three business days ahead. For this type of duration, the forecast in the forex market accuracy can reach 60% which is sufficient for a market maker with low transaction costs to run a profitable desk (Orlin Grabbe, 1986).

5. CONCLUSIONS

Financial forecasting will remain a frustrating field. Indeed, for the market to remain efficient and even alive, it must remain essentially unpredictable or the amount of predictability must be low enough so that only a few of the market participants can take advantage of it.

Therefore, the scientists wishing to earn money with technology have to keep their secret as long as possible and keep forging ahead with always more potent algorithms which alone would allow them to remain ahead of the pack. This inability to discuss their findings in the open is often frustrating for many of those involved in this activity and specially the ones who come from academia.

J. Christopher Clemens
Astronomy Department, RLM 15.308, University of Texas, Austin, TX 78756

Whole Earth Telescope Observations of the White Dwarf Star PG1159–035 (Data Set E)

INTRODUCTION

In March 1989 an international collaboration called the Whole Earth Telescope (WET) obtained a unique series of measurements of the brightness changes of the pre-white dwarf star PG1159–035. Astronomical measurements at visual wavelengths are ordinarily interrupted by the inconvenient intrusion of daylight at the observing site. By using a cooperative network of observers at telescopes well dispersed in longitude, we were able to minimize these interruptions. We acquired a nearly continuous data set which spans 231 hours, sampled at 10-second intervals. We submitted these data as a candidate for the Santa Fe Time Series Prediction and Analysis Competition where it was labelled "Data Set E."

Our primary purpose for acquiring the observations was not an exercise in abstract analysis, but an attempt to measure fundamental physical properties of the star; therefore, I will begin with a brief description of the scientific importance of white dwarf stars. Subsequently, I will discuss how we acquired the data set and processed it to produce the reduced set we submitted. This will include a description of the sources of noise we have identified and their effects on the data. Finally, I will present a summary of our analysis and the conclusions which result.

Times Series Prediction: Forecasting the Future and
Understanding the Past, Eds. A. S. Weigend and N. A. Gershenfeld, SFI Studies
in the Sciences of Complexity, Proc. Vol. XV, Addison-Wesley, 1993   **139**

Since the techniques we applied differ substantially from most of those discussed at this workshop, I will discuss whether we might benefit by applying the analysis tools used in the Competition to our data.

ASTEROSEISMOLOGY OF WHITE DWARF STARS

The variations in the brightness of PG1159 are caused by global oscillations of the star similar to oscillations of the earth which occur during earthquakes or seismic tests. Like seismic waves in the earth, stellar oscillations can reveal the structure of a star's interior, since their distribution in frequency depends upon the resonant mechanical properties of the star.

The multiperiodic pulsations we observe in white dwarfs are nonradial normal modes of oscillation of the star. Since stars are (roughly) spherical, the mass motions at the surface are best described by spherical harmonics Y_{lm}, where l is the number of nodes along a meridian of longitude and m is the number around a line of latitude. Another index, k, denotes the number of nodes in the radial direction.

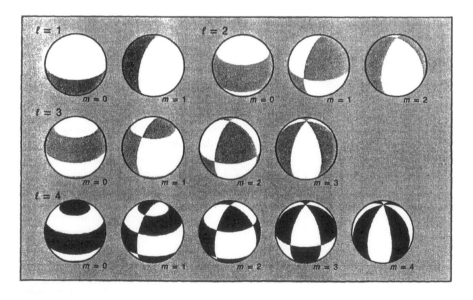

FIGURE 1   A plot of the spherical harmonics for $l = 1 - 4$. The light and dark regions crudely represent areas of higher and lower brightness on the surface of a pulsating star. (Reprinted from *Sky & Telescope*, September 1982.)

Figure 1 shows the spherical harmonics for a variety of l and m. For a star, much as in quantum mechanics, the different m modes for a particular l are degenerate in frequency unless the star is rotating. Rotation raises the degenerate modes into $2l + 1$ different frequencies. Since PG1159 is only about the size of the earth and is about 1000 light-years away, we see only the luminosity integrated over the whole surface, so geometric cancellation effects make modes of high l difficult to detect. For detailed descriptions of the properties of nonradial pulsations of stars, see Cox (1980) or Unno et al. (1979).

Typically, the periods of the pulsations we observe in white dwarfs range from 100 to 2000 s. Using the observed pulsation frequencies to probe stellar interiors is referred to as "asteroseismology." Asteroseismology of white dwarfs is potentially very rewarding since white dwarfs contain within them the answer to many fundamental astrophysical questions.

White dwarf stars are the final phase in the life of most stars, including our own sun. Once a star like our sun consumes all the fuel available for thermonuclear fusion in its core, the core contracts and heats up while the outer envelope expands. Soon this outer envelope is cast off via some mysterious process, exposing the very dense, hot core. This core, a pre-white dwarf, quickly cools to a surface temperature of around 100,000 K, contracting further in the process. After 100,000 K the white dwarf cools at essentially constant radius.

Since the white dwarf has exhausted its available nuclear fuel, the rest of its life is spent radiating away its remaining store of energy. If this were the entire story, white dwarfs would not be very interesting or illuminating. However, as a white dwarf cools, it passes through at least three temperature regions, called instability strips, where the flow of energy through its outer layers is modulated, exciting pulsations such as those we observe in PG1159.

The details of the mass loss which occurs in white dwarf formation are currently completely obscure. Using pulsations to learn about the interior structure and composition of white dwarfs will tell us how this and other processes occur in the formation of white dwarfs from their progenitor stars. The range of internal structures exhibited by white dwarfs will also tell us about the evolution of stars of various masses before they reach the white dwarf stage.

Perhaps more interesting are studies of the detailed temperature distribution of white dwarfs. We find few white dwarfs at high temperature since these cool very quickly. As we move to lower temperatures, we find progressively more white dwarfs since they cool slower as the temperature decreases. Finally, we find the largest number of white dwarfs with surface temperatures around 2500 K; at temperatures cooler than that, we find effectively none. This temperature cut-off reflects the finite age of the galaxy. There are no white dwarfs cooler than 2500 K because the cooling rate is so slow that the first white dwarfs to form in our galaxy have not had time to cool beyond this temperature. Hence, if we can measure the cooling rate of white dwarfs, we will obtain an independent measure of the age of the galaxy. Furthermore, the white dwarfs in each temperature domain were formed at different times in the life of the galaxy, so the detailed temperature distribution contains a record of star formation rates for the entire history of our galaxy. To decipher this record requires only that we know how white dwarfs cool. But the details of white dwarf cooling are regulated by internal structure, which can only be probed using the tools of asteroseismology. For a review of the advances in seismology of white dwarfs, see Winget (1988).

DATA ACQUISITION

To acquire a record of luminosity variations in PG1159, we measured its brightness using astronomical photometers. These photometers incorporate photomultiplier tubes (PMTs) wired for "photon counting." In this configuration, a digital counter counts each pulse from the PMT if the pulse is larger than a fixed threshold. Except for thermal emissions, which are infrequent for the PMT we use, each pulse represents the arrival of a photon at the photocathode of the PMT. By reading and resetting the counter at equal intervals, 10 seconds for the PG1159 data, we obtain a record of the intensity of starlight falling on the PMT. We refer to this record as a "light curve." A typical light curve for PG1159 is plotted along the bottom of Figure 3. Calculating the Fourier transform (FT) of this light curve permits identification of any periodic signals it may contain.

The key to asteroseismology is the accurate identification of the various pulsation frequencies present in a star. Figure 2 illustrates a problem with this identification. On the left is the FT of a single sine wave sampled as indicated by the darkened regions of the timeline displayed above it. The sidelobes or "aliases" in the transform do not confuse identification of the correct frequency here because there is no noise and only one periodic signal is present. In real data there is always noise and often many frequencies. This can confound attempts to separate real frequencies from their aliases

The solution to this problem is depicted on the right in Figure 2. This time the sine wave has been sampled continuously, resulting in a reduction of the alias sidelobes. Continuous sampling is difficult in astronomy due to the rotation of the earth, but we have confronted this difficulty by assembling interested observers at all the sites in Figure 3. At least one, and usually more, of these sites can observe our target at any given time. The redundancy at many longitudes is necessary to insure adequate coverage in cases of cloudy weather. The PG1159 data was acquired using this network.

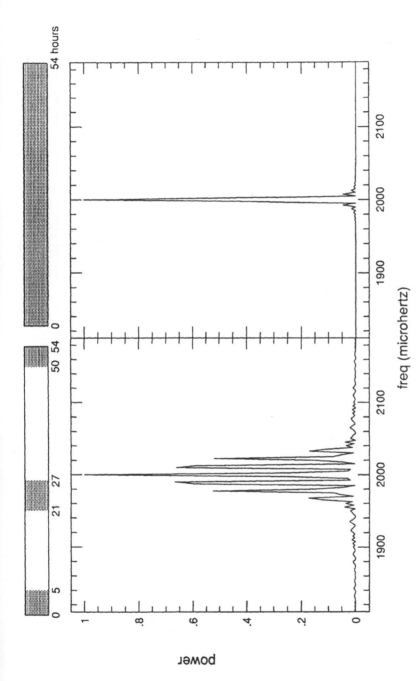

FIGURE 5 (Left) The Fourier transform of a single sine wave sampled as in the shaded box at the top. This sampling is typical of data acquired from a single site. (Right) The Fourier transform of the same sine wave sampled continuously. (Reprinted from the *Astrophysical Journal*, September 1990.)

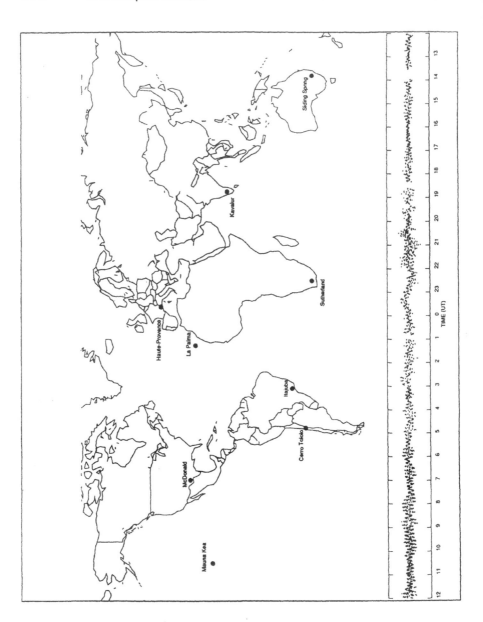

FIGURE 3   (Cont'd.) A plot of the sites which participated in Whole Earth Telescope observations of PG1159–035. The box along the bottom contains a sample of the light curve. (Reprinted from the *Astrophysical Journal,* September 1990.)

DATA REDUCTION AND NOISE

After we acquire data, we must process it to remove the effects of background sky brightness before we can calculate the FT. The PMT views the target star through a small aperture to eliminate the light from nearby stars, but the aperture cannot exclude all of the background sky. We sample the background sky brightness by moving the telescope to a location where no star is in the aperture from time to time to measure the sky alone, then we subtract that value from each measurement of the star. Since the sky brightness is not sampled continuously, we must interpolate between samples before subtracting. In one of our photometers, a separate PMT provides a continuous measure of the sky brightness, permitting much better sky removal. Eventually all photometers in the network will have this capability.

After removing sky brightness we must correct the light curve for atmospheric extinction. As a star rises, the amount of atmosphere through which the starlight must pass to reach the telescope decreases. This changes the apparent brightness of the star. All of our photometers have a second PMT which records the brightness of a constant comparison star to monitor changes in atmospheric transparency. We can use information from this channel to remove extinction effects or we can fit and remove a polynomial from the light curve of the target star directly. Usually the extinction changes have much lower frequency than the star's pulsation frequencies.

Errors in removal of sky and extinction are by far the dominant source of noise in the region of the FT including periods of 2,000 seconds and longer. Outside this region there are two other major sources of noise: photon shot noise and scintillation or "twinkling." Photon counting noise is proportional to the square root of the number of photons detected; therefore, it increases as the count rate goes up. But while the signal is increasing as the number of counts, n, the noise due to photon counting increases only as $\sqrt{n}$. Thus the signal-to-noise ratio improves as $\sqrt{n}$. Scintillation noise is caused by turbulent cells in the earth's atmosphere which act as lenses to disperse or condense the starlight, causing apparent changes in brightness. Scintillation can occur at all frequencies of interest to us. Its amplitude is proportional to the inverse of the diameter of the telescope; for larger telescopes scintillation noise is smaller. Scintillation noise does not depend on the brightness of the object. Thus, for faint objects, photon-counting noise usually dominates; for bright objects, scintillation noise is more important. The exact crossover between the two depends on the telescope size.

The final step in our reduction is to correct for changes in light travel time between the source and our telescope. We apply corrections to the time of each measurement to account for the earth's motion around the barycenter (center of gravity) of the solar system. We do not calculate this correction for each data point, but for each observatory's start time every night. Since the correction changes during the night, but these changes are not applied to the times, we introduce a one or two second jitter in the timings of our measurements.

For a more thorough description of the WET observing and reduction procedures, see Nather et al. (1990).

ANALYSIS OF THE PG1159 PHOTOMETRY

Figure 4 shows the FT of the entire PG1159 data set. If we investigate the large amplitude peaks in this transform, we see that they are arranged in groups of 3 equally spaced in frequency (triplets). For example, there is a large triplet at 1940 μHz. We also see groups of 4 or 5 equally spaced in frequency but with a spacing different than the groups of 3. This suggests that the groups of 3 represent $l = 1$ modes split into $m = -1, 0, 1$ by rotation of the star, and that the groups of 5 are $l = 2$ modes split into $m = -2, -1, 0, 1, 2$. We cannot firmly conclude that this is correct based on the number of modes alone, since some m modes might have undetectable amplitude; e.g., the groups of 3 might really be $l = 2$ with $m = +2, -2$ too small to detect. To decide on the l value of the multiplets we have measured, we must appeal to the theory of nonradial pulsations.

The pulsations we observe in white dwarfs are called g-mode pulsations because the restoring force is gravity, or more properly, buoyancy. In this respect they are more like water waves than sound waves. From a theoretical perspective, the properties of nonradial g-mode oscillations are governed by the conservation equations for mass, energy, and momentum as they apply to stars. Perturbed versions of these equations, subjected to boundary conditions and various simplifying assumptions yield oscillatory solutions with periods P_{kl}. Unno et al. (1979) derives the following g-mode dispersion relation for large values of k:

$$P_{kl} = \frac{k \, \Delta\Pi}{\sqrt{[l(l + 1)]}} + \text{constant},$$

where k is the number of radial nodes for a particular mode and l is its spherical harmonic index. The value of $\Delta\Pi$ depends on the mass and structure of the star. For a star of homogeneous composition, i.e., with no discontinuities in the density, $\Delta\Pi$ is a constant; thus, modes of a given l will be exactly uniformly split in period. In real white dwarfs, we expect gravitational settling to separate the heavy elements from the light creating density discontinuities between the layers. This will cause slight deviations in period spacings.

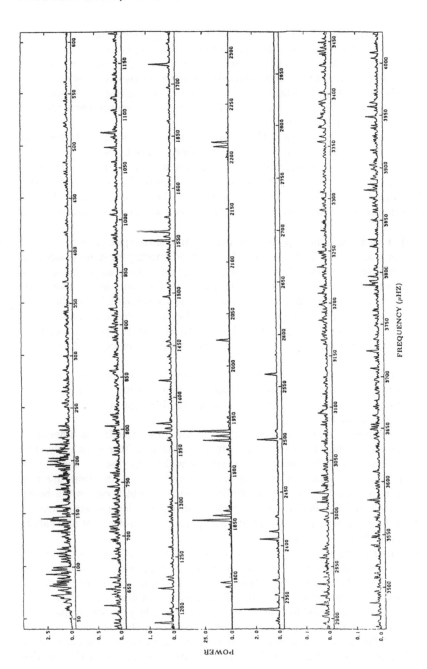

FIGURE 6 A plot of the power spectrum of PG1159–035 from 50 to 4000μHz. The vertical scale varies to accommodate the large dynamic range of the pulsations. (Reprinted from the *Astrophysical Journal*, September 1991.)

If we now return to the FT and ask whether the triplets are uniformly spaced in period, we find that they are. The mean period spacing is 21.5 seconds, with slight deviations. Furthermore, the groups of 5 are also roughly uniformly spaced in period, with a mean spacing of 12.5 seconds. From the dispersion relation above, we expect the ratio of the spacing for $l = 1$ to that for $l = 2$ to be $\sqrt{3}$, since $\Delta\Pi$ does not depend on l. Since $21.5/12.5 = 1.72$, we are now more confident that the triplets are $l = 1$ and the quintuplets $l = 2$.

There is another test we can apply to certify these identifications. In the limit of slow rotation and large k, we expect the frequency splitting between modes of different m for $l = 1$ and the splitting between different m, $l = 2$ to obey the ratio 0.6. The frequency splitting for the modes we have identified as $l = 1$ is 4.22 μHz while that for $l = 2$ is 6.92 μHz. The ratio of these is 0.61, which confirms beyond reasonable doubt our identification of the triplets as $l = 1$ and the quintuplets as $l = 2$.

Now that we know a mean period spacing and frequency splitting for $l = 1$ and $l = 2$, we can search the FT at intervals of the period spacing for more peaks with the proper frequency splitting. Note that this greatly relaxes our amplitude criteria for deciding which peaks are statistically significant. Before, we had to choose only peaks which were substantially larger in amplitude than could be produced by chance, now we may apply our expectation that they fit a predetermined pattern of period and frequency spacings. This allows us to identify peaks of significantly lower amplitude. This advantage increases the total number of peaks identified as pulsation modes of specific k, l, and m to 101.

Having identified 101 quantized pulsation modes, we may use their exact periods to derive physical values for some of the stellar parameters. Acquiring these values requires comparing the measured periods to those generated by detailed numerical stellar models.

First, we calculate equilibrium models by solving numerically the differential equations which describe the basic equilibrium conditions a star must obey. The solution requires introducing additional equations which describe the behavior of the gases in the star (equation of state and opacity). For white dwarf models, the solution must also incorporate gravitational settling, which causes the heavy elements to sink and the lighter ones to rise, resulting in a layered structure. At the interface between layers of different atomic species, the model must also satisfy diffusion equilibrium conditions.

After constructing an equilibrium model, we can perturb it to find its spectrum of pulsation frequencies. By comparing these to the frequencies measured from observations of a real star, and then adjusting the model until its frequencies match those of the star, we can derive the value of fundamental stellar parameters.

The 101 modes identified in PG1159 are enough to tightly constrain the model parameters for this star. Comparison to models calculated thus far show that the mass of PG1159 is 0.586 ± 0.003 times the solar mass. This is the most accurate mass ever determined for any star other than the sun. We also know that the rotation rate of the star is 1.38 ± 0.01 days. We have confirmed theoretical expectations that the star is compositionally stratified, most likely into three layers with carbon and oxygen in the center, hydrogen at the surface, and a layer of helium in between. The mass of this helium layer is about 10^{-3} of the total stellar mass.

These results are from a preliminary comparison to theoretical models. Future model calculations will yield even better knowledge of PG1159's structural details. This preliminary analysis is described in greater detail in Winget et al. (1991).

TIME SERIES PREDICTION AND ANALYSIS COMPETITION TECHNIQUES

The time series analysis we conducted for PG1159 relied solely on a very simple tool—the Fourier transform. Many more sophisticated techniques were discussed at the NATO ARW workshop in May 1992 that followed the Competition. In this final section, I will discuss which of these techniques may be useful for analyzing our astronomical data. These are my subjective impressions as a nonexpert, so I invite dissent.

Much of the emphasis of the was on prediction. Serre et al. (1991) have actually applied the techniques of state space reconstruction to the light curve of another pulsating white dwarf (PG1351+489), for the purpose of predicting its light curve. The most fundamental question anyone who applies this analysis to the light curves of pulsators must answer is: "Why do it?"

The justification set forth by Serre et al. (1991) is that the predicted light curves could be used for "filling gaps in the observational data" and "replacing or cleaning up noisy sections." As an observational astronomer I will not address this suggestion, although I think the time series workshop participants will agree that my job is safe.

The fundamental problem with reconstructions is that they do not currently move us any closer to the goal of determining the structure of a pulsating star from its light curve. The information that the reconstruction contains about the dynamical system which produced a time series is not easily accessible. Perhaps future advances will change this situation, but at the moment the usefulness of prediction techniques as applied to pulsating white dwarfs remains limited.

Much more useful are tests for determining whether a time series was generated by a linear dynamical system. Implicit in our analysis of the data on PG1159 is the assumption of linearity. If there are nonlinear interactions between modes, resonant mode coupling for instance, it will affect the results of our analysis. We are fairly confident that our assumption of linearity is correct since the usual signatures of nonlinear behavior—sum and difference frequencies—are not large in the FT. Unfortunately, the FT is not the best diagnostic for detecting nonlinear behavior. Sum and difference frequencies can be caused by pulse shape distortions which have nothing to do with nonlinear coupling of modes. Since the FT cannot discriminate between these effects, we need better methods.

ACKNOWLEDGMENTS

I would like to thank the Santa Fe Institute and the organizers of the time series workshop for encouraging interaction between the experts who develop new time series analysis techniques and those who can benefit by applying them in their various disciplines.

Matthew Dirst† and Andreas S. Weigend‡
†Department of Music, Stanford University, Stanford, CA 94305;
E-mail: mdirst@leland.stanford.edu
‡Xerox PARC, 3333 Coyote Hill Road, Palo Alto, CA 94304;
Address for correspondence: Andreas Weigend, Department of Computer Science
and Institute of Cognitive Science, University of Colorado, Boulder, CO 80309-0430;
E-mail: weigend@cs.colorado.edu

Baroque Forecasting:
On Completing J. S. Bach's Last Fugue

An unfinished work of art is eternally provocative. Unfinished musical works, particularly those by important composers, are rarely left incomplete: midwife musicians learn to recreate the style of a composer in order to finish them. The recent availability of musical texts in machine-readable form allows us to apply methods of statistics, machine learning, and artificial intelligence to the formerly exclusive domain of historically minded composers and musicologists. The scientific approach, while it may begin with the same goal as traditional inquiry—i.e., How do we finish the piece?—leads to new questions and points in new directions.

To provide a testing ground for these new questions and methods, the organizers of the Santa Fe Time Series Analysis and Prediction Competition selected one of the most enigmatic unfinished works in music history: J. S. Bach's last fugue, Contrapunctus XIV from *Die Kunst der Fuge*.

We address three different tasks: analysis, continuation, and completion. While we make no attempt to actually complete the fugue, we apply statistical methods in order to characterize the data set, and we relate the

Times Series Prediction: Forecasting the Future and
Understanding the Past, Eds. A. S. Weigend and N. A. Gershenfeld, SFI Studies
in the Sciences of Complexity, Proc. Vol. XV, Addison-Wesley, 1993   **151**

musical text to perception and cognition. We emphasize the importance of hierarchical structure and discuss the effects of different representations.

We contrast this "data-driven" approach (in which features are learned from the data) to rule-based expert systems, and to various completions by composers and musicologists. Finally, by reexamining Bach's manuscript, we add a new twist to the detective story of Contrapunctus XIV.

1. INTRODUCTION

In modern time series analysis and prediction, one of the recurring themes is the tension between randomness and order, between stochastic and deterministic models. The realization that simple functions can produce sequences that look complex and even pass standard tests for randomness has had an impact on time series prediction. Unfortunately, it is sometimes difficult to separate the hope from the hype. Our hope in this paper is to see whether this fact—that simple nonlinear models can generate apparently complex behavior—has any relevance for music. To clarify our scope, we are not addressing issues of sound (such as timbre), performance (such as articulation), or instrumentation: we focus on the notes.

Musical examples of this tension between randomness and order range from stochastic instruments—e.g., wind-driven chimes and Aeolian harps (the use of which nearly cost Saint Dunstan (d. 988) his life for suspected sorcery)—to deterministic compositional techniques. Some examples of the latter include: Johannes Kepler's (1619) calculation of melodies based on the orbits of the planets, the use of Bach's name as a melodic figure (the notes B.A.C.H. in German notation correspond to B♭.A.C.B♮. in English notation), or the composition of a melody inspired by images, such as the skyline of San Francisco or Hong Kong, or a bunch of bent nails strewn on the ground.[1] Mixing randomness and order, Samuel Pepys (1639–1703) used decks of cards to "draw" melodic tunes, and W. A. Mozart (1787) constructed algorithms for the random combination of subsequences and called them *Musikalisches Würfelspiel*, a musical game of dice.[2]

The tension between randomness and order is also important in music perception. Our perception of music is controlled by expectations, which are generated by music's deterministic structure. In principle, deterministic structure—the regularities—can be extracted by using artificial intelligence techniques. The more

---

[1] Loy (1991) mentions that this technique of generating melodies by casting bent nails on the ground (suggested by Vogt in Prague around 1719) served mainly "to prime the pump, so to speak, of a composer's imagination," and was not intended to be completely deterministic.

[2] Two examples are KV Anh. 294d and KV 516f, reprinted in Cope (1991) and in Schwanauer and Levitt (1993), respectively.

TABLE 1 Transformations of the fugue subject.

musical term	operation		
transposition[1]	$x \leftarrow x + c$	(translation)	move to a different pitch level
retrograde	$t \leftarrow -t$	(time reversal)	play backward
inversion	$d \leftarrow -d$	(pitch reflection)	play mirror image
diminution	$t \leftarrow 2t$		play twice as fast
augmentation	$t \leftarrow 0.5t$		play twice as slow

[1] The operation of *transposition* sometimes requires slight changes in the intervallic structure of the theme. For example, in the opening measures of a fugue, where each voice enters in turn, the theme is sometimes adjusted in order to remain within the key (the tonality) of the fugue. This type of alteration produces a *tonal* answer; a *real* answer replicates exactly the intervallic structure of the subject.

structure in a piece of music, the higher the chance that a machine learning approach will succeed. The organizers of the Competition selected a fugue, because it has a high amount of structure, certainly more than the foreign exchange rate data set of the Competition.

A **fugue** typically has one primary *theme* (a fugue *subject*) and may or may not have secondary themes (*countersubjects*). These themes are processed by a number of symmetry transformations, shown in Table 1.

A good fugue uses the theme(s) in all the voices, combining the theme(s) with transformations as often, and in as many artful ways, as possible. But a random combination of the theme and its modified versions in different keys is hardly the essence of a fugue. There are many constraints: thematic and nonthematic material must fit together for a musical work to make harmonic and rhythmic sense. Composing a fugue can be viewed as an optimization problem.

The music of Johann Sebastian Bach (1685–1750) contains an embarrassment of riches. His *Kunst der Fuge*, BWV 1080 (hereafter KdF), a multimovement summation of the fugal art, includes 14 *Contrapuncti* (fugues), 2 inversions of these fugues, and 4 canons—all of which are related by the use of a single theme, the *KdF theme*, which during the course of the work undergoes subtle variations in rhythm and melody that serve to distinguish the individual pieces.

Contrapunctus XIV from Bach's *Art of Fugue* (hereafter abbreviated Cp. XIV)[3] is incomplete, making it a prime candidate for analysis and attempted continuation and completion.[4] Although Cp. XIV, as it stands in Bach's manuscript, does not contain the KdF theme, it has long been acknowledged as part of KdF.[5] The issue of *why* Cp. XIV is incomplete will be addressed from a musicological perspective in Section 4. (The first and last pages of Bach's manuscript of Cp. XIV are reproduced in Figure 1.)

Before turning to technical issues, we offer a historical perspective. In 1961 John Pierce commented on the *Illiac Suite for String Quartet*, composed by Lejaren Hiller, Leonard Isaacson, and their computer (1957):

> The work of Hiller and Isaacson does demonstrate conclusively that a computer can take over many musical chores which only human beings had been able to do before. A composer...might very well rely on a computer for much routine musical drudgery.... [T]he computer could be used to try out proposed new rules of composition....
>
> In these days we hear that cybernetics will soon give us machines which learn.... Why couldn't they learn what we like, even when we don't know ourselves? Thus, by rewarding or punishing a computer for the success or failure of its efforts, we might so condition the computer that when we pressed a button marked Spanish, classical, rock-and-roll, sweet, etc., it would produce just what we wanted in connection with the terms. (Pierce, 1961 [2nd ed., 1980, p. 260f.]).

Have the last thirty years brought us closer to this vision? The impressive collection on music and connectionism edited by Todd and Loy (1991) and the recent volume by Schwanauer and Levitt (1993) contain ambitious ideas for automatic composition and computer music. The purpose of the present article is more modest: to show how both standard and modern time series techniques can be applied to music. Our focus is more on ideas and methodology than on specific results.

---

[3] There is some confusion in the literature (and in the editions) about Bach's intended order for the individual pieces of *Die Kunst der Fuge*. We will follow Butler's (1983) numbering scheme, in which the unfinished fugue is Cp. XIV.

[4] The organizers of the Competition selected Cp. XIV as Data Set F and posted it after the official close of the competition in January 1992 because of requests for more data. Since the hope was to inspire creative responses, no specific goal was set for this time series. Although the origin of the "mystery data set" was not revealed until the NATO workshop in May 1992, several participants discovered its source in their explorations. Terry Sanger's prediction went far into the future: he replaced Bach's theme with the theme song from *Gilligan's Island.*

[5] In the late nineteenth century, three separate (and nearly simultaneous) claims were made for the discovery that all three themes of Cp. XIV can be combined with the KdF theme: Higgs (1877), Nottebohm (1881), and Ziehn (1894). See Kolneder (1977), pp. 280ff.

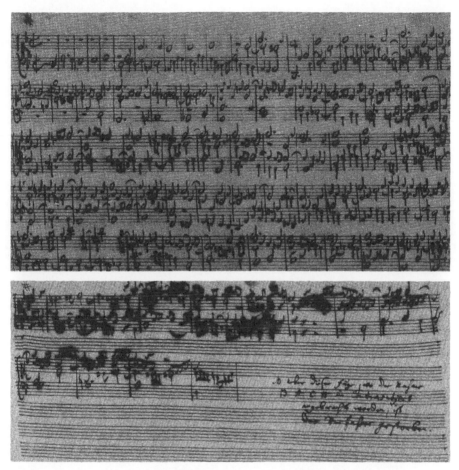

FIGURE 1   First and last page of Bach's manuscript of Cp.XIV. Reprinted with permission.

2. A PHYSICIST'S PERSPECTIVE

What can musicians expect from physicists, data analysts, statisticians, computer scientists, or artificial intelligence workers?

- Computer-assisted *analysis*, ranging from the extraction of the main theme(s) or the structure (e.g., the locations and types of thematic transformations), to the discovery of rules used in composing.

- Computer-assisted *composition*, both for low-level tasks (transitions, episodes, etc.) and high-level structure (such as the possible combinations of themes). In the case of Cp. XIV, we distinguish between two different goals: mere continuation and full-scale completion.

For any of these enterprises, the first step is to choose a representation for the data.

2.1 REPRESENTATION

Musical characteristics vary over seven orders of magnitude in time, extending from 10^{-4} to 10^3 seconds, as shown in Table 2.[6]

Because of this remarkably large range of time scales, the choice of representation is important. If, for example, we were to choose a representation based on the waveform of a recording, the natural focus of analysis would be on rhythmic or timbral aspects. Since our focus is on the notes themselves, we use three (related) representations that closely resemble the musical text:

TABLE 2 Music spans seven orders of magnitude in time.

characterization	time scales from. . .	. . .to
timbre pitch (3 orders of magnitude)	0.0001 sec $\left(\sim \frac{1}{10\text{kHz}}\right)$	0.1 sec $\left(\sim \frac{1}{10\text{Hz}}\right)$
rhythm (1 order of magnitude)	0.1 sec ($\sim$sixteenth note in Cp. XIV)	1 sec
melody (1 order of magnitude)	1 sec	10 sec
large-scale form (2 orders of magnitude)	10 sec	$1{,}000$ sec ($\approx$15 min)

[6] Seven orders of magnitude in music are large compared to less than half an order of magnitude (one octave) in color perception. However, compared to eleven orders of magnitude present in a river basin with drainage area (Montgomery & Dietrich, 1992), they seem modest.

TABLE 3 The last two measures of F.dat in the x-representation. The four voices are indicated by S (Soprano), A (Alto), T (Tenor) and B (Bass). The numbers indicate pitch: a value of 60 corresponds to "middle C." NA denotes a rest. Time t is given in units of sixteenth notes from the beginning of Cp. XIV. The vertical lines (|) indicate bar lines.

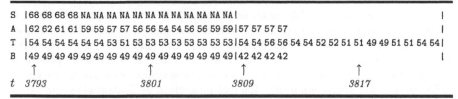

```
S  |68 68 68 68 NA NA NA NA NA NA NA NA NA NA NA NA|                                          |
A  |62 62 61 61 59 59 57 57 56 56 54 54 56 56 59 59|57 57 57 57                               |
T  |54 54 54 54 54 54 53 51 53 53 53 53 53 53 53 53|54 54 56 56 54 54 52 52 51 51 49 49 51 51 54 54|
B  |49 49 49 49 49 49 49 49 49 49 49 49 49 49 49 49|42 42 42 42                               |
      ↑                          ↑                    ↑                        ↑
t   3793                       3801                 3809                     3817
```

- The x-**representation** gives the pitch values for the four voices ($i = 1, \cdots, 4$) as a function of time. $\mathbf{x}_t$ is the four-dimensional vector at time t, and x_t^i denotes its components. t indicates the time and is given in units of sixteenth notes (semiquavers).[6] Table 3 shows the last two measures of Cp. XIV as it stands in Bach's manuscript.

- For certain tasks, alternative representations can be more appropriate. The **difference representation** is given by $\mathbf{d}_t := \mathbf{x}_t - \mathbf{x}_{t-1}$. The d-series gives the number of semitones of each interval between successive notes. The major advantage of the d-representation over the x-representation is that (exact) transpositions of the theme are identical. The disadvantage of the d-representation is that it ignores absolute pitch—the control that keeps continuations from walking off randomly in harmonic space.

- Alternatively, the **run length representation** gives each note as a pair (p, l)— its pitch number and length. This representation is convenient if rhythm and pitch (or intervals, when the d-series is run length encoded) are to be studied, but it is less suited for analyzing polyphony, since vertical alignment is lost.

All these representations can be augmented with additional explicit information about the placement of each note within the measure. In a connectionist implementation, a sensible metric of similarity is induced if this "phase" is represented by four binary units (the unit corresponding to the most significant bit encodes the location of the note within the first or second half of the measure, etc.).

We first analyze horizontal structure, ignoring the relation of each voice to the others. We then analyze vertical structure, i.e., the interaction of different voices

---

[6]The fact that this representation does not distinguish between tied and repeated notes is not a serious shortcoming, since the Cp. XIV themes do not contain repeated notes.

at each time step. We then consider polyphonic structure (the unfolding of vertical structure in time) and suggest an automated way of extracting higher level structure. In Section 3, we turn to the more difficult task of continuation.

2.2 HORIZONTAL ANALYSIS (MELODY)

A first step in exploratory data analysis (Tukey, 1977) is to histogram the data. A **histogram** counts the number of occurrences of a specific event.[8] For example, the number of occurrences of each interval in one voice throughout the entire piece can be plotted against interval size (in semitones). Most of the statistics presented in this section can be applied to all three representations given above. However, different representations emphasize different properties, as illustrated by this list of histograms:

1. Distribution of *pitch, allowing for note length*: histogram the x values. This statistic takes the length of each note into acount, e.g., the pitch level of a quarter note is counted four times.
2. Distribution of *intervals* between consecutive notes: histogram d, the number of semitones, from the difference representation. This histogram contains information about relative pitch only.[8]
3. Distribution of *pitch, irrespective of length*: histogram p from the run length encoding. Unlike the first histogram, this method counts the number of occurrences of each pitch, without taking note length into account.
4. Distribution of note *lengths*: histogram l from the run length encoding.

Histograms 1 and 3 can be collapsed over octaves, focusing on pitch class.

In the description of dynamical systems, one-dimensional histograms are used when only minimal information about the system is available. They approximate the probability of states without taking into account any knowledge of the previous state of the system. In literature, histograms have been used in authorship disputes: Thisted and Efron (1987) attribute a poem (discovered in 1985) to Shakespeare by comparing the words in the poem with the entire Shakespeare corpus.

From a horizontal analysis of music, we want more than static statistics. In particular, we want information about temporal progression. The simplest way to collect this information is with a *first-order* **Markov model**.

[8] For a discrete alphabet, the cells of the histogram simply correspond to the characters of the alphabet. Note that there is no "natural" metric between the characters. This is different from histograms of continuous-valued data that are quantized (or binned into the histograms cells).

[8] Fucks (1962) gives the histograms for the x- and the d-representations, as well as the autocorrelation functions. Hsü and Hsü (1990) find that the Fourier transform of the d-histogram obeys a power law (i.e., the log-log plot of the Fourier transform of the intervals resembles a straight line).

Markov (1913) studied the patterns of individual letters in written Russian. He used Alexander Pushkin's *Eugene Onegin* to fill a two-dimensional histogram, along whose sides were the letters of the Cyrillic alphabet. Each time letter i is followed by letter j, the counter in the corresponding (i, j)-cell is incremented. This allowed Markov to extract fuzzy rules for the spelling of Russian.

First-order Markov models, described by two-dimensional histograms or tables, are similar to phase portraits used in the study of dynamical systems. A phase portrait is a plot of x_t against x_{t-1}.

Brown et al. (1992) use 583 million words of written English to build a *second-order* Markov model, which predicts an ASCII character as a function of the two preceding characters. By computing the cross-entropy between their model and a balanced sample of English, they obtain an upper bound for the average amount of information in a printed English character: 1.75 bits. In comparison, the standard Lempel-Ziv algorithm (see Cover and Thomas, 1991) reduces the amount of information from 8 bits per ASCII character to 4.43 bits. The additional compression ratio obtained by Brown et. al. shows that a second-order Markov model, despite its simplicity, captures a significant amount of information about the sequence of letters.

One way to characterize a simple deterministic time series is by the order m of the Markov model, where the next value becomes a single-valued function of the previous m values (i.e., each column in the table of transition probabilities has only one nonzero cell).

Although music is certainly not an entirely deterministic process, histograms can quantify the similarity between pieces, composers, and styles. Suitable histograms can be constructed for pitch (x or p), intervals (d), or length (l). The histograms can be compared in raw form or through summary measures, such as moments, or *entropy* (for one-dimensional histograms), *mutual information* (for two-dimensional histograms), or *redundancy* (for higher orders).[10]

So far, we have dealt with structure that is local in time. A complementary approach is to find structure that is global in time, such as a description obtained by a **Fourier transform**, which yields the spectral coefficients corresponding to the average amount of energy for each segment of the spectrum. Voss and Clarke (1978) take the waveform of a recording (the amplitude as a function of time) and study both its audio power and the rate of zero-crossings ("instantaneous frequency"). They find that the Fourier transforms of both time series are inversely proportional

---

[10]The redundancy measure describes the information gained by increasing the order of the Markov model. It is based on incremental mutual information as a function of the order of the model (the number of past time steps that are taken into account). Redundancy is a nonlinear generalization of partial autocorrelation, just as mutual information is a generalization of (ordinary auto-) correlation. See Gershenfeld and Weigend (1993).

to the frequency f over several orders of magnitude.[11] An important feature of $1/f$ spectra is that correlations in time are important for the entire range of the spectrum. The Fourier spectra obtained from the audio representation can be compared and contrasted to the spectra obtained from pitch-based representations.

2.3 VERTICAL ANALYSIS (HARMONY)

The previous subsection on horizontal analysis focused on the progression of each individual voice. (The individual voices of a fugue have similar statistics.) We now look at the vertical dimension, ignoring the progression in time. In this section, we use only the x-representation. The term *chord* denotes simultaneously sounding pitches, i.e., the (up to four) pitch numbers in a vertical slice.

The "weakest" model for vertical analysis (the model with the fewest assumptions) simply counts the number of occurrences of each chord. The most simplistic approach is to provide a large array and to increment for each time step the contents of the corresponding cell. The resulting numbers in the array characterize the piece as a whole. They can also be used to assign a "surprise" value to each chord, defined by the negative logarithm of its probability. The number of different chords (i.e., cells that have one or more entries) can be plotted against the total number of chords (i.e., the total number of entries in the array). Gabura (1970) analyzed this average occupation number of the nonempty cells and found the differences between composers to be significant.

Gabura also tried to classify different composers on the basis of pitch structure in their music by using a **neural network** with a binary output unit. His network had no hidden units: backpropagation had not yet been invented. We now show how hidden units allow the extraction of structure from music, and we relate this structure to a number of fields, including cognitive psychology.

An auto-associator neural network is a simple connectionist architecture with hidden units. As a method of encoding the chords, the network is trained to reproduce the input pattern at the output, after piping it through a bottleneck of hidden

---

[11] Voss and Clarke (1978) analyze the low frequency variations of the audio power by taking the Fourier transform (below 20 Hz) of the squared waveform, after the waveform was bandpassed (100 Hz to 10 kHz). The Fourier coefficient at f (say, 0.1 Hz, to fix an idea) measures the degree of loudness variation at that time scale. The variable in Fourier space, f, is usually called frequency. We have to be careful not to confuse f with pitch: in this context, the Fourier spectrum characterizes temporal periodicities of variations in both loudness and "instantaneous frequency." The spectral information can be presented "back" in the time domain as the autocorrelation function (the inverse Fourier transform of the power spectrum).

A more general, yet still linear technique is **quefrency alanysis** (Bogert, Healy, and Tukey, 1963). It can be applied to each voice in order to extract information such as echo-like repetitions of the themes. It is also interesting to analyze the cross-spectra between voices, i.e., the (complex) covariance of the complex Fourier coefficients between two voices as a function of the relative time delay.

units.[12] Once the network has learned to encode the patterns (i.e., to reproduce them as faithfully as possible, given the limiting number of hidden units), we can try to learn what it has learned. We consider two network responses to a chord at the input: the reconstruction error of the output and the activation values of the hidden units. Standard connectionist analyses of these responses include:

- A scatter plot of *error vs. surprise value* (as defined above). This plot relates the features extracted by the network to the number of occurrences of each chord.
- A plot of the *principal components of the hidden unit activations* (the eigenvalues of the covariance matrix). This plot estimates the effective dimension of pitch space (see Weigend and Rumelhart, 1991).

Since we are dealing with music, we can relate network features—reflecting only the statistics of the musical text—to theories from other fields, including physics, composition theory, and cognitive psychology.

PHYSICS/ACOUSTICS. We live in a world in which most sound generators (vocal chords, string instruments, wind instruments, etc.) are one-dimensional objects, which implies that their spectra contain only integer multiples of the fundamental frequency.[13] To what degree are such physical contingencies reflected in music? Is there structure in a scatter plot of the *network error vs. the spectral overlap* (Kameoka & Kuriyagawa, 1969)?

COMPOSITION THEORY. Eighteenth-century composers followed a general set of rules governing the use of consonances and dissonances (e.g., Fux, 1725). Such "common-practice" rules are implemented in an expert system by Maxwell (1992). He assigns a "dissonance level" to chords (e.g., consonant intervals are assigned dissonance level 1, augmented fifth dissonance level 3, etc.). Is there structure in a scatter plot of the *network error vs. the level of dissonance*? To what degree does the network error "explain" the concept of dissonance, and where are the discrepancies?

---

[12] There is one input unit for each pitch value in the piece. Each input pattern corresponds to one vertical slice in the x-representation. For each note in the chord, the corresponding input is set to 1; all the other inputs are set to 0. (An alternative representation is suggested by Forte, 1964.) For each chord, the reconstruction error is given by the distance between the target (the given chord) and the prediction (the network output). A general discussion of auto-associator networks can be found in Weigend (1993).

[13] This is different for higher dimensional objects. A circular drum, for example, has noninteger harmonics and subharmonics, located at the zeros of the Bessel functions.

COGNITIVE PSYCHOLOGY. Although pitch numbers are indeed numbers, the proximity of those numbers does not imply perceptual similarity. For example, substituting a C♯ for a C in a C major chord is usually less acceptable than replacing the C with a G, or with a C from another octave. If we want to compute distances between notes with the simple Euclidean metric, the notes must be embedded in a higher dimensional space.

Shepard (1982a,b) constructs a geometrical model that reflects perceived similarity between musical tones. The pitch number x is enhanced by the location on two circles: the circle of chroma describing the sequence C, C♯, D, . . . , B, C, and the circle of fifths C, G, D, . . . , F, C. Pitch and one of these two "wrap-around" variables can be visualized as a helix in three-dimensional space. By adding the other cyclical variable, we can construct a helix of a helix in five-dimensional Euclidean space. Although all five coordinate values are functions of only one pitch number, this embedding allows the Euclidean norm to do justice to cognitive-structural constraints.

Shepard shows that listeners' judgments induce a metric in pitch space. The activation values of the auto-associator's hidden units also induce a metric. Are the *similarities in network response* related to Shepard's?[14]

The network encoding/decoding error can also be used to characterize the temporal evolution of the fugue. We plot a smoothed version of the *error vs. time*. (Short-term fluctuations are removed by averaging over a bar of music or by applying a standard smoothing convolution filter or some denoising by wavelets.) Furthermore, we plot the *volatility* of the error as a function of time. (The volatility is the running standard deviation of the errors computed in a sliding window in time; see Weigend et al., 1992, p. 419.) These ideas can be traced back to Jackson (1970), who plots a "rate of dissonance" against the measure number.

2.4 POLYPHONIC ANALYSIS

We now come to the most important part of fugue analysis: polyphonic structure—the unfolding in time of vertical elements (i.e., how the individual voices fit together and relate to one another). Music theorists of the Renaissance and Baroque wrote innumerable treatises on counterpoint, each elaborating the various techniques and rules that govern the combination of two or more melodic lines (the essence of counterpoint). Many of these rules are formulated as prohibitions: "Do not use parallel fifths or octaves." This approach can be contrasted to a data-driven analysis, which finds (often implicit) generative rather than restrictive rules.[15]

[14]Useful methods that can be used to relate these two representations include clustering and multidimensional scaling, as well as visualization with the help of self-organizing feature maps (Kohonen, 1990).

[15] We are aware of only one computer program for automatic counterpoint (Schottstaedt, 1989).

The following three examples of four-part chorale harmonization in the style of J. S. Bach illustrate the transition from artificial intelligence (AI) without learning, through traditional AI with learning, to connectionism. Ebcioglu (1988) uses an expert system with some 350 rules for a "generate-and-test method." These rules were hand-coded in a form of first-order predicate calculus—not learned. Schwanauer (1993) describes a rule-based system with a chunker that combines successful sequences of rules to new rules. Hild et al. (1992) take a connectionist approach. Their neural network has 70 hidden units. By learning to extract regularities from the training examples, it produces convincing harmonizations for new melodies.

The task of chorale harmonization differs from the goal of automatic continuation in one important respect: in chorale harmonization the melody is always given, whereas in automatic continuation there are no pre-existent parts. The key to automatic analysis and continuation is the recognition of not only small-scale patterns, but also higher order structure. In the next section, we suggest a method of extracting higher order structure.

2.5 HIGHER ORDER STRUCTURE

In a traditional fugue analysis, one of the first steps is to look at thematic structure—that is, to identify the themes, their recurrences, and transformations. Let us pretend we do not know the themes: can we extract them from the data alone? We suggest an automated analysis based on **clustering**:

1. Use the difference representation. Define a window of length w. (For example, in the Cp. XIV data set, setting w to 32 or 48 corresponds to two or three measures, respectively.) Start from the first note of the first voice. Each time the window is "sat down" on the data, it produces a point in w-dimensional space. Record the first point. Advance the window by half a bar or a full bar (8 or 16 time steps), generating the next point in w-dimensional space, and continue through the entire piece with one voice after another. This will produce several hundred points in the w-dimensional space.
2. Cluster these points. The clusters with low variance and a relatively large number of points correspond to parts of the fugue themes. (There is some variation within these clusters because tonal answers require slight modifications.)
3. Treat each cluster center as a symbol. Since each point in the w-dimensional space is assigned to one of the clusters, the original series is now transformed into a sequence of symbols. Although the number of significant clusters is comparable to the number of pitch values, this procedure captures structure at a time scale an order of magnitude slower than the original representation, because the window is moved by 8 or 16 steps every time.
4. Analyze the transitions between these symbols. Some of the symbols are followed consistently by a single symbol. These pairs correspond to adjacent parts of one of the themes. They can be combined into *compound symbols* (Simon &

Sumner, 1968; Redlich, 1993). Symbols which have a large number of successors (stochastic transitions) signal the end or the absence of a theme.

In order to visualize the results of this automatic analysis, we can construct matched filters (corresponding to the cluster centers), convolve them with the data, and plot the results. If we are lucky, the thematic structure of the fugue will appear.

3. CONTINUATION AND EXPECTATIONS

Problems in scientific analysis and inference have two dimensions: theory and data. Traditionally, music analysis has been theory-rich and data-poor; music theorists seldom construct their theories from the data ("bottom up"). Because of the widening availability of musical texts in computer readable form,[16] musical analysis can now incorporate data-rich modeling. Data-rich/theory-poor modeling starts from the data, not from first principles or theories. We use the data to construct a model that makes predictions. We then analyze where the predictions went wrong, modify the model, predict, analyze, modify, etc. This method is sometimes called **analysis by synthesis.**

In the case of Cp. XIV, the idea is to build a model from Bach's fragment, generate continuations, and analyze the shortcomings. This approach will suggest improvements for the model. It may also broaden our understanding of human cognition and musical creativity. In Section 3.1 we list some approaches to continuation; in Section 3.2 we address the question of how expectations, central in music cognition, might be modeled from the data. In Section 4 we contrast this inductive approach to the deductive, approach of traditional musicology.

3.1 CONTINUATION

A straightforward implementation of an inductive approach is to find a part of the past that resembles the present, and to predict the same continuation. This **nearest-neighbor** approach taken by Zhang and Hutchinson (this volume).[17]

[16] The complete works of J. S. Bach (and works by other composers) will soon be available from the *Center for Computer-Assisted Research in the Humanities* (Hewlett and Selfridge-Field, 1989; Hewlett, in preparation).

[17] Zhang and Hutchinson (this volume) use run length encoding. They consider the last two notes in all four voices (pitch values and length × 2 previous values × 4 voices = 16 numbers). In this 16-dimensional space they find the 10 nearest neighbors in the training set to the present point. For a continuation (of $4 + 4$ dimensions, i.e., the next values for pitch and length for the four voices), there are two possibilities: (1) to stochastically pick one of the past continuations (with equal probability or with a probability that reflects the distance), or (2) to use an average of the ten neighbors, rounded to integers. One problem is the choice of metric in this space of pitch × length. Zhang and Hutchinson use the sum of the component-wise differences. Unfortunately,

Another approach is to express the next value in the series as a function of previous values. This is called an **autoregressive model** (AR model). Assumptions have to be made about the interpolating function: in the simplest case, it is linear.[18]

Feedforward **neural networks** with sigmoid hidden units are a nonlinear generalization of linear AR models. The input into each hidden unit can be viewed as a linear filter, the activation of the hidden unit as a squashed version of the filter, and the output of the entire network (its prediction) as a weighted superposition of the squashed filters. The filter coefficients and the weights to the output are adjusted to fit the training data.

In the last four years, connectionist networks have successfully emulated dynamical systems and predicted their time series. The application of connectionist networks to music, however, differs in two important respects:

1. In dynamical systems, the squared difference between the predicted and the target value is a reasonable error measure. Music is different: a semitone error is usually worse than an octave error.

2. For systems governed by differential equations, a sufficiently large number of past values provides all information necessary for prediction. Music is different: it has structure on a hierarchy of time scales.

The first point can be addressed in two ways: by using an appropriate metric that reflects cognitive-structural constraints, or by avoiding a metric altogether. Shepard's representation (see Section 2.3) can serve as a "good" metric; it has been used by Mozer (1991) in a network to generate Bach-like tunes. There are two ways of avoiding a metric: Markov models and a local representation.

Markov models need not assume any metric or distance function between the symbols.[19] Having weak assumptions requires large amounts of training data: the training set size increases exponentially with the order of the model ("curse of dimensionality"). In order to learn from data without having an intractable number of cells to fill, Kohonen et al. (1991) use a *dynamically expanding context*. The size of the context starts at zero and is expanded until either all ambiguity has been

this mixes pitch with duration. Even in pitch space by itself, this metric is inappropriate. Better representations are discussed in the main text. Furthermore, run length encoding destroys any vertical structure.

[18] Musha and Goto (1989) use a linear autoregressive model for Schubert's *Gute Nacht*. Their filter takes the past 64 pitch values in an x-representation into account; each time step corresponds to an eighth note. Creating different pieces with the same filter is equivalent to using surrogate data with a spectrum smoothed by a fit with 64 parameters. In surrogate data, the idea is to Fourier transform a time series into frequency space, to randomize the phases (keeping the amplitudes), and to inverse Fourier transform back to the time domain. The power spectrum (the squares of the amplitudes) of the surrogate series is, by definition, identical to that of the original series. See Theiler et al. (this volume) and Kaplan (this volume).

[19] A review of Markov models for composition (from a nonlearning perspective) is given by Ames (1989).

resolved (deterministic continuation) or until a maximal context length (8 in their run length encoding) is reached—whichever comes first. Todd (1988, 1989), who also tries to avoid an inappropriate metric, uses a *local representation*: each unit corresponds to a single pitch value, similar to the representation discussed above.[20]

We turn to the second point—the hierarchy of time scales. If we want to use an ordinary Markov model or a feedforward network, we can explicitly incorporate knowledge about the hierarchical structure (obtained, for example, from the clustering method discussed in Section 2.4).[21] **Hidden Markov models** and **recurrent networks** have more computational power than ordinary Markov models and feedforward networks because the former can learn to represent the past internally. Mozer (1993) gives an overview of different recurrent network architectures and discusses their advantages and disadvantages.

3.2 EXPECTATIONS

In the Introduction we mentioned the importance of the tension between randomness and order. For music to "work," some balance has to be struck between the realization and the violation of deterministic predictions (expectations).

Meyer (1956) is the foremost exponent of the metaphor of expectations (or expectancies) in music criticism. His method of musical analysis is based on music's tendency to arouse expectations on both large and small time scales. Unfortunately, his ideas remain peripheral to most music theorists and critics: it may be that expectations are too subjective or too imprecise for traditional music analysis.

Nevertheless, the fact is that musical expectations do exist. However, the question of how they arise—as a function of both past musical experience and present input—is not easily answered. Expectations generated by a musical phrase may be due to a number of historical factors, ranging from common musical practices to peculiarities of specific traditions, schools, or composers. Other influences on musical expectations include the myriad musics (and musaks) of modernity.

Leonard Bernstein, in a discussion of Beethoven's Sixth Symphony, articulates the "formalist" method of looking at musical expectations:

> We are concerned not with the birds and bees, but with the F's and G's, the notes themselves which form the intrinsic metaphors of music, metaphors that evolve out of syntactic and phonological transformations. (Bernstein, 1976, p. 154)

---

[20] Todd's goal is to generate tunes. His network differs from the chord encoder discussed in Section 2.3. in three ways: his network predicts the next note (rather than the same chord), it learns to represent the past internally through recurrent connections, and it has some extra inputs that represent musical style.

[21] Cope (1991) approaches this problem with a hybrid system that does not fit our classification. His system composes "recombinant music" by chopping up a piece of music and recombining the parts in new ways.

By relating music to Chomsky's ideas on linguistics, Bernstein suggests that music ("the F's and G's") is more crucial than an extraneous program ("the birds and bees"—the representational reading of Beethoven's *Pastoral Symphony*). Like Bernstein, we are more interested in the notes and their resulting structures than in extra-musical information. But unlike Bernstein, our interest in the notes is purely statistical. We want to see whether musical expectations can be derived solely from the musical text.

There have been some recent experiments that model the formation of expectations by using synthetic (computer-generated) data. In music, Bharucha and Todd (1989) model tonal expectancy with connectionist techniques. They use series of isolated chords in succession, and sequences of seven successive chords each.

In linguistics, Elman (1990) generates sentences with a simple grammar, using a set of one thousand words. His network predicts the next letter, differing from the auto-associator presented above in Section 2.3 in two respects: Elman's network is trained to predict the next step ("hetero-associator"), and it contains recurrent connections that encode relevant features of the past. The network generates an expectation that reflects the probability of the next letter. An error is obtained by comparing the prediction with the actual letter. A large error indicates a violation of the expectation and often signals a boundary between words.

With a large amount of music available in computer-readable form, real data (rather than computer-generated data) can be used to build models for expectations in music. A data-driven approach may help separate the truly creative from the merely mechanical, and thus distinguish "Bachian Creativity" from "Bachian Noise."

We have not yet addressed the issue of how to complete Bach's unfinished fugue. But a fugue has a beginning and an end—unlike dynamical systems theory, where initial transients are usually considered to have decayed and the system is assumed to be in a stationary state. If the computer is to complete Bach's last fugue, it needs to know how Bach completed other fugues. With all of Bach's fugues (and more) in the computer, will modern learning algorithms on powerful machines be able to generate satisfactory completions for Cp. XIV?

The challenge posed by this question is hardly new; many musicians have struggled with Bach's most ambitious fugue and admitted defeat. Others have chosen to play the Baroque forecasting game, and have left their mark upon Cp. XIV with published completions and elaborate detective stories.

4. A MUSICOLOGIST'S PERSPECTIVE
4.1 PUBLISHED COMPLETIONS TO CP. XIV

Composers, musicologists, and musicians of all stripes have taken turns completing Cp. XIV: there are numerous published endings by an array of Bach "wannabes." Most of the completions share a common thematic content—the three fugue themes from Cp. XIV plus the KdF theme. There is no such agreement on the scope of the missing portion, although Butler (1983) has determined how many pages were allotted for Cp. XIV in the plan for the original edition.[22]

The issue of overall length must be decided before completion can be attempted. Cp. XIV, as it stands in Bach's manuscript, is a fugue in three large sections, each of which has a different theme. The manuscript breaks off in measure 239, on the heels of the first combination of all three themes, a maneuver which may be the second part of section three (the B.A.C.H. section)—or, the beginning of an entirely new closing section.

Some latter-day Bachs bring in the KdF theme right away, making the third section the final section. Others extend the third section beyond measure 239 (where Bach stopped) and include an entirely new closing section in which all four themes are combined. The partisans of the three-section approach include a number of first-rate scholars and musicians: Tovey (1931), Walcha (1967), Wolff (1975), Butler (1983), Moroney (1989), and Schulenberg (1992). Wolff and Butler present solid arguments for a fairly brief concluding section, while Tovey, Walcha, Moroney, and Schulenberg each try their hand at composition. The four-section sympathizers— Busoni (1912), Husmann (1938), and Bergel (1985)—are less numerous, but considerably more adventurous. Using Bach's other multisection fugues as models,[23] they prefer to bring the third section to a close with a full cadence before bringing in the KdF theme in a new fourth section.

The various completions range from the perfunctory to the outrageous: Schulenberg provides a four-bar solution (for the faint of heart), while Bergel proposes a monumental four-section fugue of some 381 measures. There are, predictably, a few iconoclasts in the crowd: Busoni embeds Bach's fugue within a gigantic improvisatory fantasy for piano (44 pages worth!), while Martin (1948) offers two different completions, each of which merely completes section three without introducing the KdF theme.

[22]This is not to say that Cp. XIV was conceived to fit into the allotted space (six pages by Butler's calculation) but rather, that six pages were allotted for it. This fugue was probably not complete at the time of the original pagination scheme, and the six-page requirement may have been one of the reasons why Bach never finished (or gave up copying) the piece. (This issue is discussed later in the main text.)

[23]Among others, Cp. VIII and XI from the KdF and the E♭ organ fugue, BWV 552b.

4.2 THE MUSICOLOGICAL "DETECTIVE STORY" OF CP. XIV

Most of the published completions to Cp. XIV follow traditional methods of composition.[24] But musicologists are seldom composers; they tend toward primary source study. This type of inquiry has created a never-ending, ever-changing detective story that seeks to "explain" Bach's unfinished fugue. In keeping with this tradition, we suggest yet another answer to the question, "Why did Bach stop?" We return to Bach's manuscript of the unfinished fugue with the following questions in mind:

1. What kind of a score is Bach's manuscript of Cp. XIV? (Is it a composing score, a printer's final copy, or something in between?)
2. Why is Cp. XIV in two-stave format? (All the other contrapuncti are in four-stave format.)
3. How can we explain the infamous final bar? (Why did Bach stop writing in the middle of the fifth page without any remark?)

The first question invites a comparison with another movement from KdF. The final version of the *Canone per augmentationem in motu contrario* is the only other piece in Bach's personal copy of KdF written on the same type of paper as Cp. XIV. Both pieces are on loose sheets; they were either revisions or additions to the collection of stitched folios known as the *P200* manuscript. The two separate movements are on oblong paper with five systems of two staves each. Bach had a good reason for making a second version of the canon: the final version is a rearrangement (in larger note values and in ₵ time) of the earlier *Canone per augment. in motu contr.* in P200 (which was in C time).

Bach's ₵ time version of the canon is clearly the final version: the cleanness of the copy and its layout—a page turn falls conveniently when one hand is resting—are proof that this was the version Bach wished to print. Since this loose-leaf copy of the canon is a revision, we might assume that the existing fragment of Cp. XIV is also a revision—they are both laid out in the same two-stave format on the same type of paper, after all. But if the Cp. XIV manuscript is a revision copy, what did Bach revise?

A change in meter (the reason for the revision of the canon) does not seem to be the right answer. A reduction of note values in the first section of Cp. XIV would put the fugue subject and the answer into different metric positions—not a likely possibility. Nor will the first exposition of Cp. XIV work in ₵ time (with two whole notes per bar, as in Cp. I, II, and III); there is the same problem of shifting metric position for each entry of the theme.

There is, however, one clue in the manuscript of Cp. XIV that supports the hypothesis that there was some sort of metric change between versions. On the last two systems of the fourth page (in the B.A.C.H. section), Bach writes partial

[24] Weigend and Dirst (in preparation) address the question of whether the computer can statistically "authenticate" the completions listed in the Appendix.

bar lines every two bars: precisely the sort of error one would expect if Bach was copying from an earlier version in a different meter—particularly if that meter was ¢ time with four half notes per bar. In the midst of copying the B.A.C.H. portion into his manuscript, Bach must have momentarily forgotten that he was (now) in ¢ time; he inadvertently reproduced the barring from the previous version of the (lost?) B.A.C.H. fugue.[25]

How can we explain Bach's mistake? Perhaps the B.A.C.H. section was written first. This would fit in nicely with Wolff's idea concerning "Fragment X" (the lost completion), which Wolff supposes Bach must have written (or at least sketched out) first, *before* composing sections one through three. Perhaps Bach's procedure for this fugue was totally backwards: he may have worked out the concluding section first (as Wolff suggests), and only then figured out how he was going to get there. If B.A.C.H. had to appear in third place, why not use an already existing fugue on this theme?

Our answer to the first question—What kind of score is this?—thus posits the following scenario: Bach began work on Cp. XIV by inventing three new themes that could be combined with the KdF theme. He then resurrected an earlier fugue exposition on B.A.C.H.[26] Finally, after planning the general shape of the entire fugue, he composed the rest. Bach's manuscript of Cp. XIV is, then, a combination of revision and composing score.

Our answer to the second question—Why the two-stave format?—begins with Butler's conclusion that Cp. XIV was supposed to fill six pages in the original publication. Butler's reconstruction of the (lost) original pagination scheme is quite clever; it neatly solves some lingering problems associated with the whole work.[27] But Butler does not answer all the questions. If Cp. XIV was supposed to serve as the final contrapunctus (Butler's idea, after all), then why is it not in four-stave format like all the other contrapuncti? Even if Bach was revising, why would he have condensed the score at the same time?

---

[25] There are no other fully authenticated examples of B.A.C.H. fugues by J. S. Bach: BWV 898 is spurious; BWV Anh. 45 has been ascribed to both Justin Heinrich Knecht (Schmieder, 1990, p. 899) and Johann Christian Bach (Kobayashi, 1973, p. 391); BWV Anh. 107, 108, and 110 are probably the work of Georg Andreas Sorge (Schmieder, 1990, p. 920); BWV Anh. 109, the only other B.A.C.H. fugue in Schmieder, is stylistically so anomalous as to be irrelevant for serious comparison with the B.A.C.H. portion of Cp. XIV. Although Bach may have never written a complete fugue on B.A.C.H., he used this chromatic motive occasionally in larger works. For a listing of B.A.C.H. motives in works of J. S. Bach, see the preface to *B.A.C.H. Fugen der Familie Bach*, edited by Fedtke (1984).

[26] Schulenberg (1992, p. 368) also wonders whether the various sections of this fugue were composed separately. He ventures that "Bach composed [Cp. XIV] in sections, linking them by bridges that perhaps were worked out only during the writing of the surviving autograph."

[27] Bach died before the first printed edition of the work was finished, and the executors of his estate (C. P. E. Bach and Agricola) misunderstood his intentions for the projected KdF publication. The confused state of the original (1751–2) publication explains the radical differences between (even recent) editions of KdF.

The most likely reason is that Bach was behind schedule and needed to save some space. We have no way of knowing whether Bach finished Cp. XIV before the pagination scheme was drawn up. But if we suppose that he had not, the two-stave format can be easily explained. Bach's printer, having already engraved Cp. I–X, needed to know the order of the remaining pieces in the volume. Cp. XIV, for which Bach or his printer (or both) allotted six pages, was to appear after Cp. XIII. When Bach finally began composing (or perhaps just revising and copying) Cp. XIV, he knew he had only six pages for the job. This may have been why he changed the format: in two staves, the work occupies less space.

The two-stave arrangement evidently worked fine until page 5, where Bach ran out of properly ruled two-stave paper (in the middle of the B.A.C.H. section) and was forced to continue on a badly ruled, smaller piece of paper. (Page 5 of the manuscript, reprinted in Figure 1, is about 1 centimeter less in width and length than the previous four pages.) Midway through the second system on this page, after the (first and only) combination of all three themes, the music stops. The appearance of the final measure is puzzling: Bach wrote the tenor part through to the next bar line, while the other voices simply stop on the downbeat of measure 239.

Without a *nota bene* indication or Wolff's "Fragment X" (which would have to begin with the missing voices in measure 239), we may never be sure what Bach's intentions were. But we can make an educated guess: Bach was copying from an earlier version of this fugue (or at least a sketch of the fugue's completion), and he probably did his copying much like we do—one voice at a time, bar by bar. The last measure (the subject of our final question) thus can be understood if we realize that Bach was *copying, not composing*. In measure 239 Bach copied only one voice and stopped (for whatever reason) without copying the other parts through to the end of the measure.

Now for the big one: Why did he stop? Perhaps he was unhappy with the lousy paper (Bergel). Perhaps he did not need to recopy a completion that already existed (Wolff). Or perhaps Bach realized that to continue would be foolhardy, because the fugue was already too long and would never fit within the allocated number of pages. He copied as far as his previous copy went and was unable (or unwilling) to complete this tour-de-force fugue, choosing instead to exclude the five separate sheets of Cp. XIV from the P200 manuscript. Who knows: he may have deliberately destroyed Wolff's "Fragment X" in order to keep anyone from trying to publish the unfinished final contrapunctus of his *Kunst der Fuge.*

Bach's failure to complete Cp. XIV need not doom our efforts; it may, in fact, be an instructive lesson in humility. Although we have no solution yet, the desire to complete this (in)famous work remains strong, particularly since its completion can now be considered from more than just a purely musical perspective.

ACKNOWLEDGMENTS

We thank our friends and colleagues for all the discussions related to this enterprise, in particular Neil Gershenfeld (MIT Media Lab) and Eckhard Kahle (IRCAM).

APPENDIX

Contrapunctus XIV and some completions are available at `ftp.santafe.edu` in the directory `pub/Time-Series/Bach`. Table 4 lists the file names, composers, year, and number of measures for each completion. Bach's unfinished measure 239 is counted as measure 1 of each completion. We invite further "predictions." Please contact one of the authors with your continuations, suggestions, questions, or comments. General directions for accessing the data of the Competition are given in the Appendix of the volume.

Note. A version of this paper has been submitted to *Music Theory Online* (MTO) at `mto-serv@husc.harvard.edu` .

TABLE 4 Cp. XIV in different representations and some completions.

file name	composer	year	length (in measures)	comments
README				description of files
F.dat	J. S. Bach	c. 1750	239	x-representation
F.dif				difference series
F.rl1				run length (soprano)
F.rl2				run length (alto)
F.rl3				run length (tenor)
F.rl4				run length (bass)
Tovey	D. F. Tovey	1931	79	
Martin1	B. Martin	1948	52	
Martin2	B. Martin	1948	41	
Walcha	H. Walcha	1967	72	
Bergel	E. Bergel	1985	143	
Moroney	D. Moroney	1989	31	
Schulenbg	D. Schulenberg	1992	42	

II. Time Series Prediction

Tim Sauer
Department of Mathematical Sciences, George Mason University, Fairfax, VA 22030

Time Series Prediction by Using Delay Coordinate Embedding

We present a numerical algorithm for short-term prediction of a time series based on delay coordinate embedding. Filtered delay coordinates are used to reconstruct the time series in a space large enough to unfold the dynamical attractor. Within this space, local linear models are built of dimension at or less than the dimension of the attractor, by projecting the embedded data down to the top few singular value decomposition modes. These low-dimensional linear models are used for prediction ahead in time.

The algorithm is used to make short-term predictions for Data Set A of the Santa Fe Time Series Prediction and Analysis Competition, as well as for artificial data generated by the Lorenz attractor.

1. INTRODUCTION

The analysis of time series produced by linear processes is based on the fact that a finite-dimensional linear system produces a signal characterized by a finite number of frequencies. There exist highly successful methods of time series prediction based on the exploitation of this fact in the frequency domain or the time domain (such

Times Series Prediction: Forecasting the Future and
Understanding the Past, Eds. A. S. Weigend and N. A. Gershenfeld, SFI Studies
in the Sciences of Complexity, Proc. Vol. XV, Addison-Wesley, 1993   **175**

as ARMA models). The key considerations in this pursuit are the effect of noise on the model development and use of the model for prediction and other analysis. Brockwell and Davis (1987) is an example of this direction in time series analysis.

These conventional methods are generally ineffective for solving problems such as prediction and filtering for time series produced by nonlinear processes. In particular, a time series measured from a chaotic process typically has a continuous Fourier spectrum instead of a discrete set of frequencies. As a result of this fact, frequency domain methods lack applicability. Time domain methods such as ARMA models become inappropriate because a single global model no longer applies to the entire state space underlying the signal; see Gershenfeld and Weigend (this volume). Some beginning work on state-dependent models for nonlinear signals can be found in Priestley (1988) and Tong (1990).

Along with the discovery and recognition of chaotic systems during the past quarter-century, researchers have looked for an effective means of analyzing the time series produced by these systems. The major goals have been to use the measured time series for system identification and reconstruction, and for prediction and control of future behavior of the system.

One line of investigation that has become very successful is the reconstruction of the state space of a dynamical system using delay coordinates. Packard et al. (1980), in a letter on the reconstruction of chaotic attractors using independent coordinates from a time series, attribute the suggestion of using delay coordinates to D. Ruelle. Mathematical results on delay coordinates for nonlinear systems were first published by Takens (1981).

Eckmann and Ruelle (1985) took the idea one step further and suggested examining not only the delay coordinates of a point, but also the relation between the delay coordinates of a point and the point which occurs a number of time units later. In principle, one can then approximate not only the attractor, but also its dynamics. Their discussion centers on the computation of Lyapunov exponents from experimental data, but there are many other applications, including time series prediction. Implementations of prediction algorithms which followed include those of Farmer and Sidorowich (1985) and Casdagli (1989).

In this paper we describe an implementation which builds on the Eckmann-Ruelle suggestion by making use of conventional filters, interpolation, and the singular value decomposition to optimize prediction accuracy. We demonstrate the use of this algorithm on a time series produced by the Lorenz attractor, and on Data Set A of the Santa Fe Time Series Prediction and Analysis Competition, which consists of laser intensity data collected from a laboratory experiment as described by Hübner et al. (this volume). This prediction algorithm follows closely along the lines of the noise reduction algorithm in Sauer (1992).

Our implementation uses the information from a length w window of the time series to compute an m-dimensional vector with which to represent the current state of the system. This process is called a *low-pass embedding*. The dynamics in the m-dimensional reconstruction space is then modeled locally by a l-dimensional linear map, where the best l-dimensional subspace is chosen for the local map domain. The numbers w, m, and l, respectively, should be chosen according to the sampling rate of the time series, the approximate dimension needed to embed (unfold) the dynamical attractor in the reconstructed state space, and the approximate dimension of the attractor, respectively.

In most cases, we expect $w \geq m \geq l$. For best results, we recommend l significantly smaller than m (by at least a factor of two; see the theory in Section 2). We also recommend m smaller than w except in the case of a poorly sampled flow. In particular, the equality $w = m = l$ is not required. This "decoupling" of the three parameters w, m, and l allows more accurate predictions than the case $w = m = l$ implicit in previous implementations (see, for example, Farmer & Sidorowich, 1988).

Section 2 of the paper provides background on delay coordinate embedding of attractors, including the low-pass embedding used in the algorithm. In Section 3, the details of the prediction algorithm are described, and the choices that were made to arrive at our implementation are discussed informally. Section 4 contains the computational results of the algorithm.

2. EMBEDOLOGY

The *state* of a deterministic dynamical system is the information necessary to determine the entire future evolution of the system. A useful paradigm to keep in mind is a system of n ordinary differential equations in n variables. Under minimal requirements on the smoothness of the equations, the existence and uniqueness properties hold. That means that through any point $\mathbf{a} = (x_1, \ldots, x_n)$ in R^n, there exists a unique trajectory with $\mathbf{a}$ as the initial condition. Therefore, the future of the system that unfolds from state $\mathbf{a}$ is uniquely determined by $\mathbf{a}$. In this case the set of all possible states, or *state space*, is the Euclidean space R^n.

A time series is a list of numbers which are assumed to be measurements of an observable quantity over time. The system on which the observable is being measured is evolving with time: that is, it is a dynamical system. An observable quantity, such as the one generating the time series, is a function only of the state of the underlying system. If the system returns to the same state later, the observable will yield the same value for the time series.

This is the foundation of our approach to prediction. We identify the present state of the system which is producing the time series, and search the past history for similar states. By studying the evolution of the observable following the similar states, information about the future can be inferred.

To be more precise about our approach to prediction, we introduce some notation. Denote the present state of the system by $\mathbf{a}$, and call the measured quantity at the present time $h(\mathbf{a})$. By assumption, the present state $\mathbf{a}$ contains all information needed to produce the state t time units into the future, which we will call

$F_t(\mathbf{a})$. In these terms, the prediction problem is to calculate the observed quantity t time units in the future knowing only the present state. That is, given $\mathbf{a}$, find $P_t(\mathbf{a}) = h(F_t(\mathbf{a}))$. We might call P_t the *prediction function*.

Viewed in this way, the prediction problem breaks down into two subproblems: the *representation problem* and the *function approximation problem*. The representation problem consists of transforming the present state $\mathbf{a}$, which is an abstraction, into something concrete enough to be manipulated in a computer program. The function approximation problem consists of doing the same for the prediction function P_t.

In the representation problem, one attempts to reconstruct the present state from the time series or, in other words, to use the available data to find a vector $\mathbf{b}$ with which to replace the theoretical state $\mathbf{a}$. A particularly elegant solution to this problem is *delay coordinate embedding* (see Takens, 1981, and Sauer, Yorke, & Casdagli, 1991). Each state $\mathbf{a}$ is identified with a unique vector $\mathbf{b}$ in a Euclidean space R^m. A prediction of the time series t units into the future can be made by computing $P_t(\mathbf{b})$. We describe this approach to the representation problem in the remainder of Section 2.

Solving the second problem, the function approximation problem, means finding an efficient approximation of the function $P_t : R^m \to R$ using available data. If a good approximation can be found, then prediction involves locating the present state vector $\mathbf{b}$, and evaluating P_t there. We treat the function approximation problem in Section 3.

2.1 DELAY COORDINATES

The technique of delay coordinate reconstruction is to reproduce the set of dynamical states of a system using vectors derived from a time series measured from the system. Let A denote a compact finite-dimensional set of states of a system. For example, this set may consist of an equilibrium, periodic orbit, or chaotic attractor. Let $g : A \to R$ be an observation function which is a measurement of some quantity of the system, and let τ be a real number greater than zero. For each state $\mathbf{a} \in A$, one can define the m-dimensional vector

$$\mathbf{b} = [g(\mathbf{a}), g(F_{-\tau}(\mathbf{a})), \ldots, g(F_{-(m-1)\tau}(\mathbf{a}))]. \tag{1}$$

This vector is called a *delay coordinate vector* because its components consist of time-delayed versions of the observable of the system. It is an interesting fact that under quite reasonable conditions on the dynamics F_t of the system, this correspondence $D(\mathbf{a}) = \mathbf{b}$ is a one-to-one correspondence, as long as m is greater than twice the box-counting dimension of A, and the observation function g is chosen generically. This fact is called the Fractal Delay Coordinate Embedding Prevalence Theorem in Sauer, Yorke, and Casdagli (1991).

The vector **b** above is a segment of a time series with equally spaced data produced by the measurement function g. That is,

$$\mathbf{b} = [x_t, x_{t-\tau}, \ldots, x_{t-(m-1)\tau}] \tag{2}$$

where $x_t = g(\mathbf{a})$ is the value of the time series at time t and $\mathbf{a}$ is the state. Thus **b** is readily available.

Any tangent manifold structure that exists on A is also carried over by the correspondence D. In the original formulation of the theorem by Takens (1981), A is assumed to be a smooth manifold, and the conclusions are that the image $D(A)$ is also a smooth manifold and that D is a diffeomorphism. It is shown in Sauer, Yorke, and Casdagli (1991) that more generally, if A is a fractal attractor, then not only does the one-to-one correspondence hold, but also any manifold structure that exists, such as an unstable manifold, is faithfully represented on $D(A)$. This motivates the belief, for example, that the positive Lyapunov exponents of A can be measured on the reconstructed set $D(A)$, even when A is fractal and not a smooth manifold.

The one-to-one correspondence is useful because the state $\mathbf{a}$ of a deterministic dynamical system, and thus its future evolution, is completely specified by the corresponding vector **b**. Suppose, at a given time, one observes the vector **b** in reconstruction space R^m, and that this is followed one second later by a particular event. If the correspondence D of vectors with states is one-to-one, then each appearance of the measurements represented by **b** will be followed one second later by the same event. There is predictive power in finding a one-to-one map.

In practice, the precise measurements represented by **b** may not be repeated exactly. Moreover, noise in the measuring apparatus may prevent the exact reconstruction vector **b** that corresponds to $\mathbf{a}$ from being known. Yet if the correspondence is reasonably smooth (the correspondence is as smooth as the dynamics), similar measurements will predict similar events. This solution to the representation problem, giving concrete proxies for the invisible states, provides a foundation for attempting prediction.

2.2 FILTERED DELAY COORDINATES

Correspondences can be devised that are more sophisticated than Eq. (1). These are more adept at handling noise in practical situations, yet can be mathematically proved to yield a one-to-one correspondence. A long-term goal is to merge, as much as possible, the useful filtering techniques used in conventional signal processing with embedding ideas. In this work, we demonstrate a first step, using very simple filters based on the Fourier transform.

Defining a *filtered delay coordinate embedding* consists of replacing Eq. (1) with the more general

$$\mathbf{b} = M[g(\mathbf{a}), g(F_{-\tau}(\mathbf{a})), \ldots, g(F_{-(w-1)\tau}(\mathbf{a}))]^T. \tag{3}$$

In this formulation, M is an $m \times w$ matrix of rank m (so $w \geq m$). As before, $\mathbf{b}$ is an m-dimensional vector. Translating back to the time series, which is defined by $x_t = g(\mathbf{a})$, we are setting

$$\mathbf{b} = M[x_t, x_{t-\tau}, \ldots, x_{t-(w-1)\tau}]^T. \tag{4}$$

The vector $\mathbf{b}$ that corresponds to the state $\mathbf{a}$ has entries which are linear combinations of the entries of Eq. (2).

It can be shown that this filtered delay coordinate map, like the original delay coordinate map, is virtually assured to give a one-to-one correspondence of state $\mathbf{a}$ with vector $\mathbf{b}$. Specifically, if m is greater than twice the box-counting dimension of the attractor A, then almost every observation function g will lead to a one-to-one correspondence, except for possible collapsing of periodic points by M (see details in Sauer, Yorke, & Casdagli, 1991).

The theoretical results state that given a series of data that is noise-free, the hidden attractor A and the reconstruction $D(A)$ will have identical topological and dynamical properties. This fact motivates using the same approach in practice, where the data are collected with noise.

In the present case, we recommend defining $M = M_3 M_2 M_1$ to be the composition of the following three linear operations:

1. $M_1 =$ FFT (discrete Fourier transform) of order w;
2. M_2 sets to zero all but the lowest $m/2$ frequency contributions; and
3. $M_3 =$ inverse FFT of order m, using the remaining $m/2$ frequencies.

Multiplication by the matrix M has the effect of a low-pass filter on the length w window of the time series. This type of filtered delay coordinate map could be called a *low-pass embedding*. One starts with a length w section of the signal, and represents it in the reconstruction by a real vector of length m. For example, if $w = 32$ and $m = 16$, the information that is contained in Eq. (2) but missing from Eq. (4) is essentially the upper half of the Fourier spectrum. If the sampling rate is not too low, it is likely that the lower half of the spectrum will be sufficient to give a good approximation to the length w section of the clean signal underlying the noise.

The basic philosophy of the low-pass embedding is to use all available data for the purpose of attractor representation. Typically, a window containing a number of oscillations of the time series is used to develop a representation of the current state of the system. The window length w, in terms of samples of the time series, is dependent on the sampling rate and should be treated independently of the dimension m of the representation space.

The low-pass embedding is a way to intelligently "downsample" the data by a factor of w/m. In the absence of noise, there is no advantage of this approach over simply decimating the series by a factor of w/m (choosing every w/m point). In the presence of high-frequency noise (through digitization error, for example), however, increased representation accuracy is achieved by this type of downsampling.

2.3 INTERPOLATION

In the next section, we describe the use of local linear models of the dynamics in the reconstructed state space for prediction. The local model of the dynamics near the present state vector is built using the nearest neighbors of the present state. However, if the available data set is sparse, there may not be enough neighbors within the "radius of validity" of the linear model with which to fit the parameters of the model, and large errors will result. A further difficulty is that near neighbors that do exist will be translated in time an amount that is less than one sampling step.

A solution to this problem, illustrated by Figure 1, is upsampling: to artificially increase the sampling rate by interpolating the time series. This augments the existing reconstructed states by filling in the sampling interval. There are many types of interpolation that could be used. We chose to fit a section of the time series with a Fourier polynomial, and then sample the polynomial s equally spaced times each sampling period. A choice of a power of 2 for s makes the interpolation particularly convenient using the Fast Fourier Transform; we used $s = 16$ in all results reported in this paper.

Although in principle this interpolation is done prior to the filtering step of the previous section, in practice it may be inconvenient in terms of storage to explicitly interpolate and store the entire expanded series. As a matter of implementation, the interpolation can be done on an as-needed basis. Moreover, since Fourier interpolation uses the same frequency component information as the filtering step of the previous subsection, some efficiency can be gained by combining the two steps.

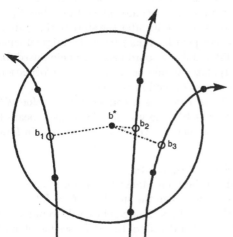

FIGURE 1　The reference point (b^*) is shown together with its nearest neighbors (black dots) in reconstructed state space. By interpolating the time series, the more appropriate neighbors b_1, b_2, b_3 (open dots) can be found. Our algorithm uses the nearest interpolated neighbor from each pass of the trajectory.

3. PREDICTION

In the last section we presented a solution to the first half of the prediction problem, that of the representation of the present state by a computable entity. Next, we proceed to the remaining problem of estimating the prediction function, which has as input the present state and as output a future value of the time series.

We refer to the known data series as the *training set*. Through filtered delay coordinate embedding, we can reconstruct the underlying dynamics. In fact, for each reconstructed state **b** in the training set, we can look up the value of the series t time units later, and call it $X = P_t(\mathbf{b})$.

To continue a time series, we find the reconstruction vector $\mathbf{b}^*$ corresponding to the end of the series, and use our knowledge of the training set to estimate $P_t(\mathbf{b}^*)$. There are many possible ways in which to do this estimation. We discuss one approach in the next section.

3.1 EVALUATION OF THE PREDICTION FUNCTION

Assume that $\mathbf{b}^*$ is a vector in R^m representing a state of the system. For example, $\mathbf{b}^*$ could be a filtered delay coordinate vector. We want to evaluate $P_t(\mathbf{b}^*)$, where P_t is the function that gives the value of the time series t time units into the future. To do this, search the training set for the nearest k neighbors $\mathbf{b}_1, \ldots, \mathbf{b}_k$ of $\mathbf{b}^*$ in R^m, subject to the condition that only one neighbor, the one nearest to $\mathbf{b}^*$, is chosen from each nearby trajectory segment. The interpolation described in the previous section should be fine enough that the direction from $\mathbf{b}^*$ to the nearest point on the trajectory should be virtually perpendicular to the trajectory, as in Figure 1.

Since each $\mathbf{b}_i$ was found from the training set, we also know the value $X_i = P_t(\mathbf{b}_i)$, which is the value of the observable t time units after the state of the system was $\mathbf{b}_i$. Note that X_i is a value (or an interpolated value) from the original time series, and not a filtered value. Even if filtered delay coordinates are being used for $\mathbf{b}_i$, the raw time series value (or its interpolate) should be used for X_i. Information will be lost if the entire time series is low-pass filtered in advance.

The remaining task is to use the X_i to calculate a corresponding value X^* for the observable t units after the state was $\mathbf{b}^*$. The way in which the k nearest neighbors of $\mathbf{b}^*$ are used to estimate the prediction $P_t(\mathbf{b}^*)$ is summarized in four steps:

1. Find the center of mass **c** of the neighbors $\mathbf{b}_1, \ldots, \mathbf{b}_k$.
2. For a fixed dimension $l \leq m$, find the best low-dimensional linear space R^l passing through the point **c**. By "best," we mean the subplane R^l of R^m which minimizes the squared distances to the neighbors. This can be easily found using the singular value decomposition (SVD). For details of the SVD, see Golub and Van Loan (1989). To be precise, define the matrix A whose rows consist of the vectors $\mathbf{b}_1 - \mathbf{c}, \ldots, \mathbf{b}_k - \mathbf{c}$. The SVD of A will have form $A = U \Sigma V^T$, where U and

V are orthogonal matrices and Σ is diagonal with nonincreasing nonnegative entries. The l dominant right singular vectors (the first l columns of V) span the desired space R^l.

3. Project the data points $\mathbf{b}_1 - \mathbf{c}, \dots, \mathbf{b}_k - \mathbf{c}$ down to this R^l, and form the affine (constant + linear) model $L : R^l \to R$ which best fits the data points

$$(\Pi(\mathbf{b}_1 - \mathbf{c}), X_1), \dots, (\Pi(\mathbf{b}_k - \mathbf{c}), X_k)$$

where $\Pi : R^m \to R^l$ is the projection onto the best R^l. The model L has form

$$L(\mathbf{x}) = \mathbf{a} \cdot \mathbf{x} + d$$

where $\mathbf{a}$ is an l-vector and d is a constant scalar.

4. Project $\mathbf{b}^* - \mathbf{c}$ onto R^l and evaluate the model there. The result will be $L(\Pi(\mathbf{b}^* - \mathbf{c})) = X^*$, an estimate for $P_t(\mathbf{b}^*)$.

3.2 DISCUSSION

There are many unresolved issues relating to the way in which P_t is estimated. To the extent that the training set is large and noise is low, these issues become of lesser importance. But, in realistic situations, these optimal conditions are often absent, and predictions will be sensitive to the estimation method for P_t.

THE BIAS/VARIANCE DILEMMA AND THE CHOICE OF LOCAL MODEL. The challenge of continuing Data Set A of the Competition brings these issues into the foreground. The given time series has *length* 1,000 during which the amplitude undergoes periods of growing oscillations followed by collapses and reinsertion into a growth period. There are only four such collapses in the available training set. In this sense, there are only four examples with which to predict the next collapse and reinsertion level. By this measure, the training set is rather short.

Noise is also a major issue with Data Set A, which consists of 8-bit data. The known series consists of 1,000 integers between 3 and 255. At best, we can assume there is digitization noise of 0.5 units, which corresponds to 10–20% noise during the low-intensity period just after the collapse. Assuming that this part of the signal is important for determination of the reinsertion level at which the next growth period begins, this noise level is significant.

These twin problems, lack of data and presence of noise, have a large impact on the method used for approximating the prediction function P_t. Once it has been decided to use local models, there is a choice between linear and nonlinear local models. Even if linear models are chosen, the question remains "what should be the dimension l of the model?"

Severe constraints are put on this choice if very few independent neighbors (four, for example, in the case of Data Set A) of the reference point $\mathbf{b}^*$ representing the present state are available in the training set. A linear model $L : R^l \rightarrow R$ is of form

$$L(\mathbf{x}) = \mathbf{a} \cdot \mathbf{x} + d$$

and as such has $l + 1$ free parameters. If a nonlinear model were used in place of L, presumably far more parameters would be needed. Since each independent neighbor in the training set can contribute one constraint on the parameters, there are strict upper bounds on the sophistication of the model that can be used.

Of course, one can widen the search for neighbors, and accept training points that are farther away from the reference point for the purposes of determining the local model $L : R^l \rightarrow R$. In fact, this is where our method meets the bias/variance dilemma that is inherent in the study of nonparametric functional inference. (See Casdagli & Weigend, this volume, and Geman, Bienenstock, & Doursat, 1992, for a discussion.) Using a small number of neighbors increases the variance of the model's estimation, because of the presence of noise. If more far-flung neighbors are accepted in order to decrease the variance, the assumption that the model is locally valid is compromised (for example, the approximation of a curved manifold by the linear tangent plane) and the bias is increased. To summarize, noise causes high variance, but because of the restricted amount of data, any attempt to decrease the variance will tend to increase the bias.

These considerations motivated our choice of embedding dimension m and model dimension l. It is important for m to be high enough so that the delay coordinate vector $\mathbf{b}$ locates the true state $\mathbf{a}$ as closely as possible. It is also important, in the presence of noise and small data sets, to have the freedom to choose l relatively low.

The use of the singular value decomposition helps to concentrate the usefulness of the available data. The idea of sorting out principal directions on the reconstructed attractor using the SVD appears in the papers of Broomhead and King (1986) and Broomhead, Jones, and King (1987). By finding the dominant directions of the data and projecting onto them, the regression step is restrained from trying to fit parameters in the less relevant directions. This decoupling of the reconstruction dimension, represented by m, and the model dimension l, allows prediction to proceed in the presence of small data sets without the ill-conditioning problems associated with requiring $m = l$.

In this report we emphasize the importance of allowing independent choice of the signal window length w, the dimension m of the representation space of the reconstructed attractor, and the model dimension l. Roughly speaking, the choice of w depends on the sampling rate of the time series, the choice of m depends on the degrees of freedom explored by the dynamical attractor, and the choice of l is constrained, because of the bias/variance dilemma, by the amount of available data. This is one of the major differences between the present method and previous implementations of nonlinear forecasting using delay coordinates, for which the choice $w = m = l$ is commonly made.

WEIGHTED REGRESSION. Once a linear map $L : R^l \to R$ is chosen for the local model and the number of neighbors that will determine L are fixed, there are further choices of how to use the data to fit the model. Our experiments showed that weighted regression improved prediction performance. We weighted the contribution of neighbor $\mathbf{b}_i$ to the regression by the factor $(1 - (1/2)d_i^2)^3$, where

$$d_i = \frac{\text{dist } (\mathbf{b}_i, \mathbf{b}^*)}{\text{dist } (\mathbf{b}_k, \mathbf{b}^*)},$$

and where $\mathbf{b}_k$ is the furthest of the k neighbors from $\mathbf{b}^*$. This weighting strategy was ad hoc and is not vital to the success of the prediction method. We did not do extensive testing of different weighting strategies, but we can say that this was slightly superior to flat weights $d_i = 1$. No claim is made that this weighting scheme is optimal.

There are two approaches to the weighting issue that we have to suggest, and in fact, these approaches are also relevant to the other choices referred to in the preceding discussion. One approach is to use a combination of theory, heuristic, and exhaustive search to find the optimal weighting strategy. A second is to take an adaptation approach at this stage of the process, and to use the training set to optimize a simple model whose few weights vary locally through reconstruction space. Neither of these approaches have been systematically explored by this author. The same general procedure could also be applied to the model selection discussion above.

DIRECT VERSUS ITERATED PREDICTION. Assume we are given a time series $x_1, \ldots, x_N$ and asked to provide a continuation. We apply our method to predict one time unit ahead and get an estimate $\hat{x}_{N+1}$. To get an estimate for x_{N+2}, there are two obvious choices. The *direct prediction* method means that the original method is applied to $x_0, \ldots, x_N$ to predict two time units ahead. In contrast, *iterated prediction* means applying the method to $x_0, \ldots, x_N, \hat{x}_{N+1}$ to predict one unit ahead.

Much discussion has ensued over which choice is superior. The reliability of direct prediction is suspect because it is forced to predict farther ahead. On the other hand, iterated prediction uses $\hat{x}_{N+1}$, which is possibly corrupted data. Farmer and Sidorowich (1988), for example, argue that iterated prediction is superior, although under ideal conditions that may not be realized in practice.

We would suggest that the discussion should be refocused. The question should not be which is better, but how both approaches can be used to optimize prediction. One suggestion would be to average the two approximations for x_{N+2}; that is essentially what is done in our algorithm. This is a clear opportunity to exploit the availability of two approximations to minimize the variance of the estimate.

3.3 ALGORITHM

In this section we put together the pieces from the previous sections and describe the prediction algorithm. There are five parameters to be chosen ahead of time. They are:

1. s = interpolation steps per sample period;
2. w = window length (length of input window for filtered embedding);
3. m = low-pass embedding dimension;
4. l = model dimension (number of dominant SVD modes projected onto); and
5. k = number of neighbors used to fit a one-dimensional linear model.

We begin with a scalar time series $\{x_1, \ldots, x_N\}$. By the interpolation step discussed above, we can instead consider the time series to have length sN, where s denotes the number of interpolation steps per sample period. Then each window of length w, in the original time units, can be made into a filtered delay coordinate vector, of dimension m, as discussed in Section 2.2. There will be approximately sN such vectors.

We will call the time interval over which the prediction is made the *prediction horizon*. For each prediction horizon t and each filtered delay vector $\mathbf{b}$, one can look up the value $P_t(\mathbf{b})$ of the interpolated, unfiltered time series corresponding to the time t units after the state vector was $\mathbf{b}$. These are used as in Section 3.1 to evaluate the prediction function P_t when predictions are needed.

Continuing the time series $\{x_1, \ldots, x_N\}$ means producing $x_{N+1}, x_{N+2}, \ldots$ which are extrapolations of the given series. The method for producing a value x_i, for $i > N$, of the series continuation takes advantage of the fact that there are many possible estimators for x_i. In fact, for some previous time $j < i$, let $\mathbf{b}_j$ be the filtered delay coordinate vector that represents the state at time j. Then $P_{i-j}(\mathbf{b}_j)$ is an estimator for x_i. Our suggestion is to average many of these available estimators to continue the series.

We calculate each series continuation value x_i for $i > N$ as

$$x_i = \frac{1}{w} \sum_{j=i-w}^{i-1} P_{i-j}(\mathbf{b}_j)$$

where $\mathbf{b}_j$ is the filtered delay coordinate vector at time j. In other words, the prediction for time i is the average of the predictions for time i from the previous w time steps. This is a mixture of direct and iterated prediction.

4. EXAMPLES

In this section we exhibit the result of applying the algorithm outlined above to two different time series. The first is computer-generated data from the Lorenz attractor. The second is laboratory data measured from a pumped far-infrared laser in a chaotic state, provided as Data Set A of the Competition. (The relation between the Lorenz equations and Data Set A is explored by Hübner et al., this volume.)

We report on several runs of the algorithm in the following. For comparison purposes, the parameters of the algorithm were set identically on all runs:

- s = interpolation steps per sample period = 16;
- w = window length = 32;
- m = low-pass embedding dimension = 16;
- l = model dimension = 1; and
- k = number of neighbors used to fit a one-dimensional linear model = 4.

4.1 LORENZ EQUATIONS

The Lorenz attractor time series was generated by solving the Lorenz equations (Lorenz, 1963):

$$\dot{x} = \sigma(y - x)$$
$$\dot{y} = \rho x - y - xz \qquad (6)$$
$$\dot{z} = -\beta z + xy$$

where the parameters are set at the standard values $\sigma = 10$, $\rho = 28$, $\beta = 8/3$. Solutions to this system of three differential equations exhibit the sensitive dependence on initial conditions which is characteristic of chaotic dynamics. In realistic situations, knowledge of the true state of a system can be done only in finite precision. In such cases, sensitivity to initial conditions rules out long-term prediction. On the other hand, short-term prediction is possible to the extent that the current position can be estimated and that the dynamics can be approximated.

A long trajectory of the Lorenz attractor was generated using a differential equation solver, and the x-coordinate of the trajectory was sampled (with a sampling period of $\Delta t = 0.05$) to create a univariate time series. At this sampling rate, the x-coordinate completes an oscillation every 15 to 20 samples.

For the purposes of this exercise, no dynamical information about the Lorenz equations was used by the algorithm. The time series up to a certain point was input to the algorithm, and the continuation was produced.

In Figure 2, we show an example of using 10,000 training points to predict the continuation of the clean Lorenz time series. The dashed curve is the true continuation of the Lorenz data for the next 200 time steps (.05 per time step), and the solid curve is the series predicted by the algorithm of this paper.

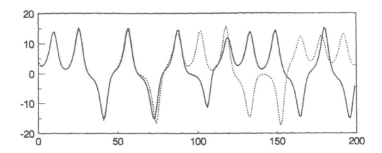

FIGURE 2   Time series prediction for signal from Lorenz attractor. Solid curve is predicted time series, and dashed curve is true time series. Predictions start at time 0 and stay near correct signal for around 90 time units.

Knowledge of the dynamics of the Lorenz system shows us that each time the signal passes through zero, the corresponding state on the attractor passes through a region of high sensitivity, and the predicted signal becomes susceptible to error. Most of the time, we see the divergence between predicted and true at these points. A positive aspect of the predictions in Figure 2, which is also true for the predicted continuations of the experimental data in the next section, is that even after the predictions diverge from the true trajectory determined by the given time series, the predictions seem to follow *some* true trajectory of the system. Thus, this algorithm may be useful not only for short-term prediction, but for long-term simulation.

Fifty examples of continuations as in Figure 2 were run, each using 10,000 training points, which corresponds to approximately 600 oscillations of the Lorenz signal. The results are summarized in Figure 3, where the root mean square error of the prediction is graphed against the prediction horizon. For the purposes of this graph, the average errors over ten neighboring predictions have been gathered together. Each data point plotted at t on the prediction horizon is the RMS of the errors between $t - 4$ and $t + 5$ over all 50 runs.

The graph shows a monotonic increase in error up to around 8, the (root mean square) size of the Lorenz attractor, after which the prediction is uncorrelated with the true time series. The error bars represent the sample standard deviations, and are quite large, considering that they represent an aggregate of 50 repetitions. Their size points to the large amount of variance in the accuracy of the predicted continuations.

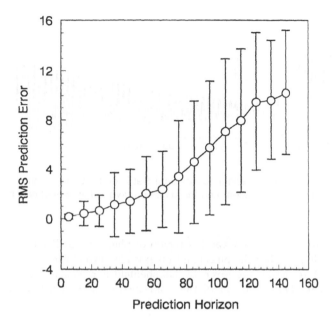

FIGURE 3   Root mean square prediction error is plotted as a function of the time interval of the prediction. The errors are an average over 50 runs. A training set of 10,000 points, sampled at $\Delta t = 0.05$, was used.

4.2 EXPERIMENTAL DATA

Our algorithm was also applied to the laser data analyzed in Hübner et al. (1989). Data Set A provided by the organizers of the Competition consisted of a time series of this data of length 1,000. The goal of the competition was to produce a continuation of the provided data of length 100.

One of the motivations of the laser experiment described in Hübner et al. (1989) was to develop a physical system that was closely modeled by the Lorenz equations. The parameter values of the Lorenz equations needed for the correspondence are different than those used in the previous subsection, so no direct comparison can be made with those results. (See Hübner et al., this volume.)

We show the continuation of the length 1,000 series produced by our algorithm in Figure 4. As before, the dashed curve is the true continuation, and the solid curve is given by the algorithm. A continuation of 400 points is given. The oscillatory signal grows slowly in intensity until it reaches a transitional state, followed

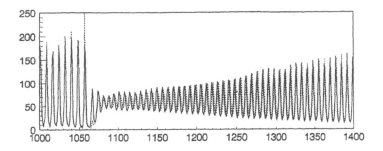

FIGURE 4   A continuation of length 400 of Data Set A of the Competition. The solid curve is the predicted continuation, and the dashed curve is the correct continuation.

by a reinsertion into the slowly varying region of phase space. The transition time can be well estimated, but the phase and magnitude beyond the transition can be problematic. More details are given by Gershenfeld and Weigend (this volume), comparing our prediction with predictions obtained with other algorithms such as neural networks.

After the close of the competition, an expanded version of the laser data was released. The longer data set contained 10,000 points, including the original set of 1,000 points and its continuation. Using this extra data, more continuations could be attempted, to test algorithm performance in a more statistically significant way. Figure 5 shows the results of four further attempts at continuing the time series, using the same training set, but with continuations beginning at other points. We emphasize that the training set for the four runs in Figure 5 is the same as for the first (Figure 4), namely, the original 1,000 points provided for the competition. Predicted continuations of 200 points are given by the solid curve, compared with the true continuation in the dashed curve.

A measure of prediction accuracy is given by the *normalized mean-squared error* (Gershenfeld & Weigend, this volume)

$$\text{NMSE} = \frac{1}{\sigma^2 N} \sum_{i=1}^{N} (x_i - \hat{x}_i)^2, \tag{7}$$

where x_i is the true value of the ith point of the series of length N, $\hat{x}_i$ is the predicted value, and σ is the standard deviation of the true time series during the prediction interval. In other words, NMSE is the ratio of mean squared errors of the prediction method in question and the method which predicts the mean at every step.

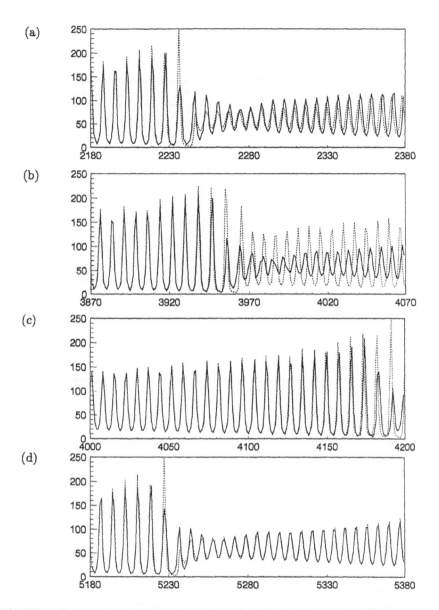

FIGURE 5   Four continuations of length 200 of Data Set A, each with a different starting point. The training set is the same 1,000 points provided as part of the competition. The solid curve is the predicted continuation, and the dashed curve is the true continuation.

TABLE 1 Performance Statistics for Predictions of Set A, Using Original 1,000 Points as the Training Set[1]

Starting point	Duration	NMSE	*Wan, this volume*[2]
1000	100	0.077	(0.027)
	200	0.199	
2180	100	0.174	(0.065)
	200	0.366	
3870	100	0.183	(0.487)
	200	0.617	
4000	100	0.006	(0.023)
	200	0.254	
5180	100	0.111	(0.160)
	200	0.093	

[1] Normalized mean-squared error (NMSE) is given for ten different prediction runs with five different starting points. The first two runs correspond to the first half of Figure 4. The remaining runs are displayed in Figure 5.

[2] The corresponding values of the neural network by Wan (this volume) are provided for comparison by the editors.

Table 1 summarizes the prediction error measured by NMSE of the algorithm on the laser data. Prediction errors are given for continuations of length 100 and of length 200. The first two runs of Table 1, starting at 1,000, refer to the first quarter and first half of Figure 4, respectively. The remainder of Table 1 refers to the continuations of Figure 5, which are of length 200. Therefore, the NMSE of the continuations from point 2180 of the laser data set are given in Table 1, for the 100-point and 200-point continuation, which correspond to the first half and full length of Figure 5(a), respectively. The other three starting points for continuation are 3870, 4000, and 5180, and similarly, the errors in Table 1 refer to the respective continuations pictured in Figure 5.

As in the case of prediction of the Lorenz data, there is a large variance in the results, depending on the success with which the continuation handles the reinsertion (or, more accurately, mode switching, in the physical system). An example with particularly large error for prediction horizon over 100 is the predicted continuation beginning at 3870 in Figure 5(b). The probable cause of the large error in these predictions presumably is due to the fact that the behavior of the time series in the range 3960–3980 appears atypical, compared to the behavior seen in the training set composed of the range 1–1000.

5. SUMMARY

Eckmann and Ruelle (1985) explained how the dynamics of a process can be reconstructed from an experimental time series. It follows that, in principle, short-term prediction of a deterministic time series is possible, even in the case that the system producing the time series is chaotic. The implementation described in this paper follows their suggestion of using delay coordinate embedding to reconstruct the attractor, and local models to fit the dynamics of the system.

In contrast to earlier implementations of their idea, we allow the parameters $w = $ window length, $m = $ embedding dimension, and $l = $ model dimension to be chosen independently. This is essential for successful prediction because the three parameters govern quite separate aspects of state space representation and prediction.

The prediction algorithm is applied to artificial data generated from the Lorenz system of differential equations and to laboratory data from a laser experiment. The summary data collected in Figure 3 and Table 1 document the prediction accuracy attained.

Eric A. Wan
Department of Electrical Engineering, Stanford University, Stanford, CA 94305-4055
e-mail: wan@isl.stanford.edu

Time Series Prediction by Using a Connectionist Network with Internal Delay Lines

A neural network architecture, which models synapses as Finite Impulse Response (FIR) linear filters, is discussed for use in time series prediction. Analysis and methodology are detailed in the context of the Santa Fe Time Series Prediction and Analysis Competition. Results of the Competition show that the FIR network performed remarkably well on Data Set A.

INTRODUCTION

The goal of time series prediction or forecasting can be stated succinctly as follows: given a sequence $y(1), y(2), \ldots, y(N)$ up to time N, find the continuation $y(N+1)$, $y(N+2), \ldots$. The series may arise from the sampling of a continuous time system, and be either stochastic or deterministic in origin. The standard prediction approach involves constructing an underlying model which gives rise to the observed sequence. In the oldest and most studied method, which dates back to Yule (1927), a linear autoregression (AR) is fit to the data:

$$y(k) = \sum_{n=1}^{T} a(n)y(k-n) + e(k) = \hat{y}(k) + e(k).$$

Times Series Prediction: Forecasting the Future and
Understanding the Past, Eds. A. S. Weigend and N. A. Gershenfeld, SFI Studies
in the Sciences of Complexity, Proc. Vol. XV, Addison-Wesley, 1993   **195**

This AR model forms $y(k)$ as a weighted sum of past values of the sequence. The single-step prediction for $y(k)$ is given by $\hat{y}(k)$. The error term $e(k) = y(k) - \hat{y}(k)$ is often assumed to be a white noise process for analysis in a stochastic framework.

More modern techniques employ *nonlinear* prediction schemes. In this paper, neural networks are used to extend the linear model. The basic form $y(k) = \hat{y}(k) + e(k)$ is retained; however, the estimate $\hat{y}(k)$ is taken as the output $\mathcal{N}$ of a neural network driven by past values of the sequence. This is written as:

$$y(k) = \hat{y}(k) + e(k) = \mathcal{N}[y(k-1), y(k-2), \ldots, y(k-T)] + e(k). \qquad (2)$$

Note this model is equally applicable for both scalar and vector sequences.

The use of this nonlinear autoregression can be motivated as follows. First, *Takens Theorem* (Takens, 1981; Packard et al., 1980) implies that for a wide class of deterministic systems, there exists a *diffeomorphism* (one-to-one differential mapping) between a finite window of the time series

$$[y(k-1), y(k-2), \ldots, y(k-T)]$$

and the underlying *state* of the dynamics system which gives rise to the time series. This implies that there exists, in theory, a nonlinear autoregression of the form

$$y(k) = g[y(k-1), y(k-2), \ldots, y(k-T)],$$

which models the series exactly (assuming no noise). The neural network thus forms an approximation to the ideal function $g(\cdot)$. Furthermore, it has been shown (Irie & Miyake, 1988; Honnik, Stinchcombe, & White, 1989; Cybenko, 1989) that a feedforward neural network $\mathcal{N}$ with an arbitrary number of neurons is capable of approximating any *uniformly* continuous function. These arguments provide the basic motivation for the use of neural networks in time series prediction.

The application of neural networks to time series prediction is not new. Previous work includes that of Werbos (1974, 1988), Lapedes (1987), and Weigend et al. (1990), to cite just a few. The connectionist entries in the Competition attest to the success and significance of networks in the field. In this paper, we focus on a method for achieving the nonlinear autoregression by use of a Finite Impulse Response (FIR) network (Wan, 1990a, 1990b). We start by reviewing the FIR network structure and presenting its adaptation algorithm called *temporal backpropagation*. We then discuss the use of the network in a prediction configuration. The results of the Competition are then presented with step-by-step explanations on how the specific predictions were accomplished. We conclude by reevaluating our original prediction model of Eq. (2) and propose various classes of network schemes for both autoregressive and state space models.

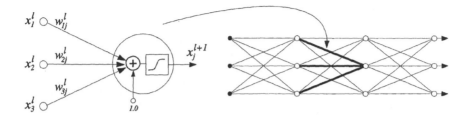

FIGURE 1   Static neuron model and feedforward network: each neuron passes the weighted sum of its inputs through a sigmoid function. The output of a neuron in a given layer acts as an input to neurons in the next layer. In the network illustration, each line represents a synaptic connection. No feedback connections exist.

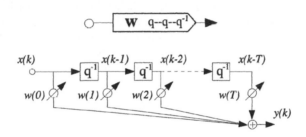

FIGURE 2   FIR filter model: A tapped delay line shows the functional model of the FIR "synapse" (q^{-1} represents a unit delay operator, i.e., $x(k-1) = q^{-1}x(k)$).

FIR NETWORK MODEL

The traditional model of the multilayer neural network is shown in Figure 1. It is composed of a layered arrangement of artificial neurons in which each neuron of a given layer feeds all neurons of the next layer. A single neuron extracted from the lth layer of an L-layer network is also represented in the figure. The inputs x_i^l to the neuron are multiplied by variable coefficients $w_{i,j}^l$ called *weights*, which represent the synaptic connectivity between neuron i in the previous layer and neuron j in layer l. The output of a neuron, x_j^{l+1}, is simplistically taken to be a sigmoid function[1] of the weighted sum of its inputs:

$$x_j^{l+1} = f\left(\sum_i w_{i,j}^l x_i^l\right). \tag{3}$$

A *bias* input to the neuron is achieved by fixing x_0^l to 1. The network structure is completely defined by taking x_i^0 to be the external inputs, and x_i^L to be the final outputs of the network. Training of the network can be accomplished using the familiar *backpropagation* algorithm (Rumelhart, Hinton, & Williams, 1956). (In difference to Weigend, 1991, and Gershenfeld & Weigend, this volume, we here do not allow for direct connections between input and output.)

The model of the feedforward network described above forms a complex mapping from the input of the first layer to the output of the last layer. Nevertheless, for a fixed set of weights, it is a *static* mapping; there are no internal dynamics. A modification of the basic neuron is accomplished by replacing each static synaptic weight by an FIR linear filter.[2] By FIR we mean that for an input excitation of finite duration, the output of the filter will also be of finite duration. The most basic FIR filter can be modeled with a tapped delay line as illustrated in Figure 2. For this filter, the output $y(k)$ corresponds to a weighted sum of past delayed values of the input:

$$y(k) = \sum_{n=0}^{T} w(n)x(k-n). \tag{4}$$

Note that this corresponds to the *moving average* component of a simple Autoregressive Moving Average (ARMA) model (Ljung, 1987; Wei & William, 1985). The FIR filter, in fact, was one of the first basic adaptive elements ever studied (Widrow & Stearns, 1985). From a biological perspective, the synaptic filter represents a Markov model of signal transmission corresponding to the processes of axonal transport, synaptic modulation, and charge dissipation in the cell membrane.

---

[1] The sigmoid is a continuous *S*-shaped function chosen to roughly model the thresholding properties of a real neuron. Mathematically, it is common to use the hyperbolic tangent function, tanh.

[2] The term "finite impulse response" filter and "infinite impulse response" (IIR) filter are explained in Gershenfeld and Weigend (this volume).

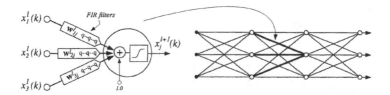

FIGURE 3 FIR neuron and network: In the temporal model of the neuron, input signals are passed through synaptic filters. The sum of the filtered inputs is passed through a sigmoid function to form the output of the neuron. In the feedforward network all connections are modeled as FIR filters.

Continuing with notation, the coefficients for the synaptic filter connecting neuron i to neuron j in layer l is specified by the vector $\mathbf{w}_{ij}^l = [w_{i,j}^l(0), w_{i,j}^l(1), \dots, w_{i,j}^l(T^l)]$. Similarly $\mathbf{x}_i^l(k) = \left[x_i^l(k), x_i^l(k-1), \dots, x_i^l(k-T^l)\right]$ denotes the vector of delayed states along the synaptic filter. This allows us to express the operation of a filter by a vector dot product $\mathbf{w}_{i,j}^l \cdot \mathbf{x}_i^l(k)$, where time relations are now implicit in the notation. The output $x_j^{l+1}(k)$ of a neuron in layer l at time k is now taken as the sigmoid function of the sum of all filter outputs which feed the neuron (Figure 3):

$$x_j^{l+1}(k) = f\left(\sum_i \mathbf{w}_{i,j}^l \cdot \mathbf{x}_i^l(k)\right).$$

Note the striking similarities in appearance between these equations and those of the static model (Eq. (3)) along with their associated figures. Notationally, scalars are replaced by vectors and multiplications by vector products. The convolution operation of the synapse is implicit in the definition. As we will see, these simple analogies carry through when comparing standard backpropagation for static networks to *temporal backpropagation* for FIR networks.

ALTERNATIVE REPRESENTATIONS OF THE FIR TOPOLOGY

A similar network structure incorporating embedded time delays is the *Time-Delay Neural Network* (TDNN) (Lang & Hinton, 1988; Waibel, 1989; Waibel et al., 1989). TDNNs have recently become popular for use in phoneme classification. A TDNN is typically described as a layered network in which the outputs of a layer are buffered several time steps and then fed fully connected to the next layer (see Figure 4). The TDNN and the FIR network can, in fact, be shown to be functionally equivalent. Differences between the topologies are a matter of their pictorial representations and

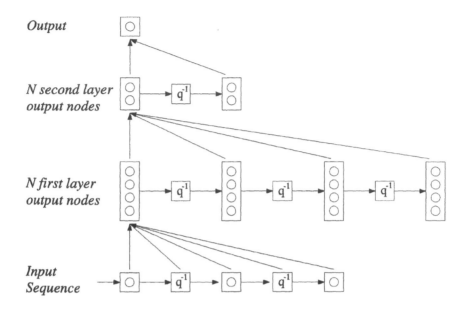

FIGURE 4 Time-delay neural network: All node outputs in a given layer are buffered over several time steps. The outputs and the buffered states are then fed fully connected to the next layer. This structure is *functionally* equivalent to an FIR network.

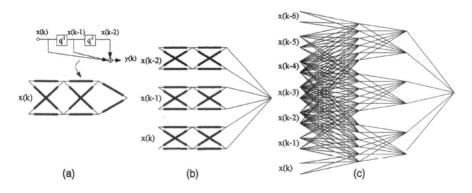

FIGURE 5 An FIR network with second-order taps for all connections is unfolded into a *constrained* static network. The original structure has 30 variable filter coefficients while the resulting network has 150 static synapses.

TABLE 1 FIR Network vs. Static Equivalent

Network Dimension		Variable Parameters	Static Equivalent
Nodes[1]	Order[2]		
2×2×2×1	2:2:2	30	150
5×5×5×5	10:10:10	605	36,355
3×3×3	9:9	180	990
3×3×3×3	9:9:9	270	9,990
3^n	9^{n-1}	$(n-1)90$	$10^n - 10$

[1] Number of Inputs × Hidden Neurons × Outputs.

[2] Order of FIR synapses in each layer.

an added formalism in notation for the FIR network. In addition, the FIR network is more easily related to a standard multilayer network as a simple temporal or vector extension. As we will see, the FIR representation also leads to a more desirable adaptation scheme for on-line learning.

An alternative representation of the FIR network (and TDNN) can be found by using a technique referred to as *unfolding-in-time*. The general strategy is to remove all time delays by expanding the network into a larger equivalent static structure. As an example, consider the very simple network shown in Figure 5(a). The network consists of three layers with a single output neuron and two neurons at each hidden layer. All connections are made by second-order (two-tap) synapses. Thus, while there are only 10 synapses in the network, there are actually a total of 30 variable filter coefficients. Starting at the last layer, each tap delay is interpreted as a "virtual neuron" whose input is delayed the appropriate number of time steps. A tap delay is then "removed" by replicating the previous layers of the network and delaying the input to the network accordingly (Figure. 5(b)). The process is then continued backward through each layer until all delays have been removed. The final unfolded network is shown in Figure 5(c).

This method produces an equivalent *constrained* static structure where the time dependencies have been made external to the network itself. Notice that whereas there were initially 30 filter coefficients, the equivalent unfolded structure now has 150 static synapses. This can be seen as a result of redundancies in the static weights. In fact, the size of the equivalent static network grows *geometrically* with the number of layers and tap delays (see Table 1). In light of this, one can view an FIR network as a compact representation of a larger static network with imposed symmetries. These symmetries force the network to subdivide the input pattern into local overlapping regions. Each region is identically processed with the results being successively combined through subsequent layers in the network. This is in contrast to a fully connected network which may attempt to analyze the scene all at once. Similar locally symmetric constraints have been motivated for use in pattern classification using "shared weight" networks (LeCun, 1989; LeCun et al., 1989).

ADAPTATION: TEMPORAL BACKPROPAGATION

Given an input sequence $x(k)$, the network produces the output sequence $y(k) = \mathcal{N}[W, x(k)]$, where W represents the set of all filter coefficients in the network. For now, assume that at each instant in time, a desired output $d(k)$ is provided to the network (we will reformulate this in terms of time series prediction in the next section). Define the instantaneous error $e^2(k) = \|d(k) - \mathcal{N}[W, x(k)]\|^2$ as the squared Euclidean distance between the network output and the desired output. The objective of training corresponds to minimizing over W the cost function:

$$C = \sum_{k=1}^{K} e^2(k)$$

where the sum is taken over all K points in the training sequence. Regularization terms (e.g., constraints on W) are not considered in this paper. The most straight forward method for minimizing C is stochastic gradient descent. Synaptic filters are updated at each increment of time according to:

$$\mathbf{w}_{ij}^l(k+1) = \mathbf{w}_{ij}^l(k) - \eta \frac{\partial e^2(k)}{\partial \mathbf{w}_{ij}^l(k)} \tag{7}$$

where η controls the learning rate.

The most obvious way to obtain the gradient terms involves first unfolding the structure into its static equivalent, and then applying standard backpropagation. Unfortunately, this leads to an overall algorithm with very undesirable characteristics. Backpropagation applied to a static network finds the gradient terms associated with each weights in the network. Since the constrained network contains "duplicated" weights, individual gradient terms must later be carefully recombined to find the *total* gradient for each unique filter coefficient. Locally distributed processing is lost as global bookkeeping becomes necessary to keep track of all terms; no simple recurrent formula is possible. These drawbacks are identical for the TDNN structure.[3]

A more attractive algorithm can be derived if we approach the problem from a slightly different perspective. The gradient of the cost function with respect to a synaptic filter is expanded as follows:

$$\frac{\partial C}{\partial \mathbf{w}_{ij}^l} = \sum_k \frac{\partial C}{\partial s_j^{l+1}(k)} \cdot \frac{\partial s_j^{l+1}(k)}{\partial \mathbf{w}_{ij}^l}, \tag{8}$$

where $s_j^{l+1}(k) = \sum_i \mathbf{w}_{i,j}^l \cdot \mathbf{x}_i^l(k)$ is the input to neuron j prior to the sigmoid. We may interpret $\partial C/\partial s_j^{l+1}(k)$ as the change in the total squared error over all time, due to a change at the input of a neuron at a single instant in time. Note that we are not expressing the total gradient in the traditional way as the sum of instantaneous gradients. Using this new expansion, the following stochastic algorithm is formed:

$$\mathbf{w}_{ij}^l(k+1) = \mathbf{w}_{ij}^l(k) - \eta \frac{\partial C}{\partial s_j^{l+1}(k)} \cdot \frac{\partial s_j^{l+1}(k)}{\partial \mathbf{w}_{ij}^l}. \tag{9}$$

A complete derivation of the individual terms in this equation is provided in Appendix A. The final algorithm, called *temporal backpropagation*, can be summarized as follows:

$$\mathbf{w}_{ij}^l(k+1) = \mathbf{w}_{ij}^l(k) - \eta \delta_j^{l+1}(k) \cdot \mathbf{x}_i^l(k) \tag{10}$$

$$\delta_j^l(k) = \begin{cases} -2e_j(k)f'(s_j^L(k)) & l = L \\ f'(s_j^l(k)) \cdot \sum_{m=1}^{N_{l+1}} \bar{\delta}_m^{l+1}(k) \cdot \mathbf{w}_{jm}^l & 1 \le l \le L-1, \end{cases}$$

where $e_j(k)$ is the error at an output node, $f'()$ is the derivative of the sigmoid function, and $\bar{\delta}_m^l(k) \equiv [\delta_m^l(k)\delta_m^l(k+1)\ldots\delta_m^l(k+T^{l-1})]$ is a vector of propagated gradient terms. We immediately observe that these equations are seen as the vector generalization of the familiar backpropagation algorithm. In fact, by replacing the vectors $\mathbf{x}$, $\mathbf{w}$, and $\bar{\delta}$ by scalars, the above equations reduce to precisely the standard backpropagation algorithm for static networks. Differences in the *temporal* version are a matter of implicit time relations and filtering operations. To calculate $\delta_j^l(k)$ for a given neuron, we *filter* the δ's from the next layer backwards through the FIR synapses for which the given neuron feeds (see Figure 6). Thus δ's are formed not by simply taking weighted sums, but by backward filtering. For each new input and

---

[3]TDNN's are typically used for classification in a batch mode adaptation. Training consists of fully buffering the states until the entire pattern of interest is captured and then using backpropagation though multiple "time-shifted" versions of the network. This can be shown to be equivalent to using a similar unfolded network as above.

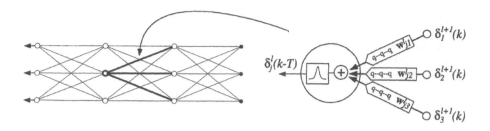

FIGURE 6 In temporal backpropagation, delta terms are *filtered* through synaptic connections to form the deltas for the previous layer. The process is applied layer by layer, working backward through the network.

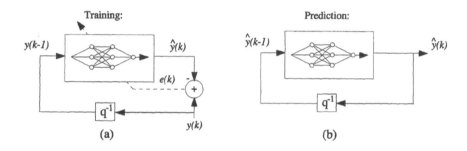

FIGURE 7 Network Prediction Configuration: The single-step prediction $\hat{y}(k)$ is taken as the output of the network driven by previous sample of the sequence $y(k)$. During training the single-step squared prediction error, $e^2(k) = (y(k) - \hat{y}(k))^2$, is minimized using temporal backpropagation. Feeding the estimate $\hat{y}(k)$ back forms a closed loop process used for long-term iterated predictions.

desired response vector, the forward filters are incremented one time step and the backward filters one time step. The weights are then adapted on-line at each time increment.

Temporal backpropagation preserves the symmetry between the forward propagation of states and the backward propagation of error terms. Parallel distributed processing is maintained. Furthermore, the number of operations per iteration now only grow *linearly* with the number of layers and synapses in the network. This savings comes as a consequence of the efficient recursive formulation. Each unique coefficient enters into the calculation only once in contrast to the redundant use of terms when applying standard backpropagation to the unfolded network.[4]

EQUATION-ERROR ADAPTATION AND PREDICTION

We are now in a position to discuss the use of the FIR network in the context of time series prediction. Recall that we wish to model the series $y(k)$ (for simplicity, we assume a scalar series). Figure 7(a) illustrates the basic predictor training configuration. At each time step, the input to the FIR network is the known value $y(k-1)$, and the output $\hat{y}(k) = \mathcal{N}_q[y(k-1)]$ is the single-step estimate of the true series value $y(k)$. Our model construct is thus:

$$y(k) = \mathcal{N}_q[y(k-1)] + e(k). \tag{11}$$

As explained earlier, the FIR network $\mathcal{N}_q$ may be represented as a constrained network acting on a finite window of the input (i.e., $\mathcal{N}_q[y(k-1)] \equiv \mathcal{N}_c[y(k-1), y(k-2) \ldots y(k-T)]$). This allows us to rewrite Eq. (11) as:

$$y(k) = \mathcal{N}_c[y(k-1), y(k-2), \ldots, y(k-T)] + e(k), \tag{12}$$

which emphasizes the nonlinear autoregression.

During training, the squared error $e(k)^2 = (y(k) - \hat{y}(k))^2$ is minimized by using the temporal backpropagation algorithm to adapt the network ($y(k)$ acts as the desired response). Note we are performing *open-loop* adaptation; both the input and desired response are provided from the known training series. The actual output of the network is not fed back as input during training. Such a scheme is referred to as *equation-error* adaptation (Mendel, 1973; Gooch, 1983). The neural network community has more recently adopted the term *teacher-forcing* (Williams & Zipser, 1989).

A simple argument for adapting in this fashion is as follows: in a stationary stochastic environment, minimizing the sum of the squared errors $e(k)^2$ corresponds to minimizing the *expectation* of the squared error:

$$E[e^2(k)] = E[y(k) - \hat{y}(k)]^2 = E[y(k) - \mathcal{N}_c[\mathbf{y}_1^T(k)]]^2, \tag{13}$$

where $\mathbf{y}_1^T(k) = [y(k-1), y(k-2), \ldots, y(k-T)]$ specifies the regressor.[5] The expectation is taken with respect to the joint distribution on $y(k)$ through $y(k-T)$. Now consider the identity:

$$E[y(k) - \mathcal{N}_c[\mathbf{y}_1^T(k)]]^2 = E[y(k) - E[y(k)|\mathbf{y}_1^T(k)]]^2 + E[\mathcal{N}_c - E[y(k)|\mathbf{y}_1^T(k)]]^2. \tag{14}$$

[4] Temporal backpropagation may also be applied to TDNNs to for an efficient on-line algorithm in contrast to the typical batch mode training method.

[5] In Eq. (13), $y(k)$ correspond to *random variables* in a stationary stochastic process, and not specific points in a given series.

The first term on the right-hand side of this equation is independent of the estimator, and hence the optimal network map $\mathcal{N}_c^*$ is immediately seen to be

$$\mathcal{N}_c^* = E[y(k)|\mathbf{y}_1^T(k)]\,, \tag{15}$$

i.e., the conditional mean of $y(k)$ given $y(k-1)$ through $y(k-T)$, which is what we would have expected. Again we should emphasize that this only motivates the use of training a network predictor in this fashion. We cannot conclude that adaptation will necessarily achieve the optimum for a give structure and training sequence. Issues concerning *biased* estimators in the context of network learning are presented in Geman et al. (1992).

Once the network is trained, long-term *iterated* prediction is achieved by taking the estimate $\hat{y}(k)$ and feeding it back as input to the network:

$$\hat{y}(k) = \mathcal{N}_q[\hat{y}(k-1)]\,. \tag{16}$$

This closed-loop system is illustrated in Figure 7(b). Equation (16) can be iterated forward in time to achieve predictions as far into the future as desired. Note, for a linear system, the roots of the regression coefficients must be monitored to insure that the closed-loop system remains stable. For the neural network, however, the closed-loop response will always have *bounded output* stability due to the sigmoids which limit the dynamic range of the network output.

Since training was based on only single-step predictions, the accuracy of the long-term iterated predictions cannot be guaranteed in advance. One might even question the soundness of training open-loop when the final system is to be run closed-loop. In fact, *equation-error* adaptation for even a linear autoregression suffers from convergence to a *biased* closed-loop solution (i.e., $\theta = \theta^* + bias$, where θ^* corresponds to the optimal set of closed loop autoregression parameters.) (Gooch, 1983; Shynk, 1989). An alternative configuration which adapts the closed loop system directly might seem more prudent. Such a setup is referred to as *output-error* adaptation. For the linear case, the method results in a estimator that is not biased. Paradoxically, however, the linear predictor may converge to a local minimum (Landau, 1979; Johnson, 1984; Stearns, 1981). Furthermore, the adaptation algorithms themselves becomes more complicated and less reliable due to the feedback. As a consequence, we will not consider the *output-error* approach with neural networks in this paper.

RESULTS OF THE SFI COMPETITION

The plot in Figure 8 shows the chaotic intensity pulsations of an NH_3 laser (Hübner et al., this volume). distributed as part of the Competition. For the laser data, only 1,000 samples of the sequence were provided. The goal was to predict the next 100

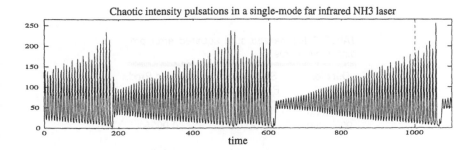

FIGURE 8   1100 time points of chaotic laser data.

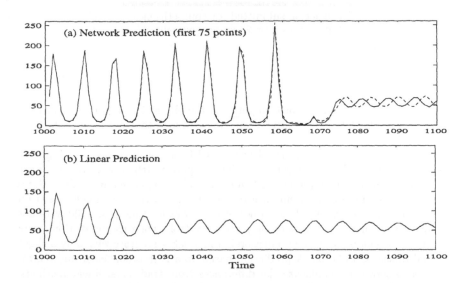

FIGURE 9   Time series predictions: (a) Iterated neural network prediction (first 75 points). The remaining 25 points were selected by adjoining a similar sequence taken from the training set. The prediction is based only on the supplied 1,000 points. Dashed line corresponds to actual series continuation. (b) 100-point iterated prediction based on a 25th-order linear autoregression. Regression coefficients were solved using a standard least squares method.

TABLE 2 Normalized sum squared error prediction measures

Duration	Single-Step Pred. NMSE	Iterated Pred. NMSE
100-1000	0.00044	-
900-1000	0.00070	0.0026
1001-1050	0.00061	0.0061
1001-1100	0.02300	0.0551
1001-1100[1]	-	0.0273

[1] Prediction submitted for competition (75 iterations plus 25 smoothed values).

samples. During the course of the Competition, the physical background of the data set, as well as the 100-point continuation, was withheld to avoid biasing the final prediction results.

The 100-step prediction achieved using an FIR network is shown in Figure 9 along with the actual series continuation for comparison (the last 25 points of the prediction necessitated additional smoothing as will be explained under *Long-Term Behavior*). It is important to emphasize that this prediction was made based on only the past 1,000 samples. True values of the series for time past 1000 were not provided to the network nor were they even available when the predictions were submitted. As can be seen, the prediction is remarkably accurate with only a slight eventual phase degradation. A prediction based on a 25th-order linear autoregression is also shown to emphasize the differences from traditional linear methods. Other submissions to the Competition included methods of $k - d$ trees, piecewise linear interpolation, low-pass embedding, SVD, nearest neighbors, Wiener filters, as well as standard recurrent and feedforward neural networks as summarized by Gershenfeld and Weigend (this volume), who show that the FIR network outperformed all other methods on this data set. While this is clearly just one example, these initial results are extremely encouraging. In the next sections we report on details concerning performance measures, selection of network parameters, training, testing, and postcompetition analysis.

PERFORMANCE MEASURE. A measure of fit is given by the *normalized mean-squared error* (Gershenfeld & Weigend, this volume):

$$\text{NMSE} = \frac{1}{\sigma^2 N} \sum_{k=1}^{N} (y(k) - \hat{y}(k))^2, \tag{17}$$

where $y(k)$ is the true value of the sequence, $\hat{y}(k)$ is the prediction, and σ^2 is the variance of the true sequence over the prediction duration, N. A value of NMSE = 1 thus corresponds to predicting the unconditional mean. Table 2 summarizes various NMSE values for both single-step and iterated predictions within the training set and for the continuation.

SELECTION OF NETWORK DIMENSIONS. The FIR network used in the Competition was a three-layer network with $1 \times 12 \times 12 \times 1$ nodes and $25 : 5 : 5$ taps per layer. Selection of these dimensions were based mostly on trial and error along with various heuristics. Since the first layer in the network acts as a bank of linear filters, selection of the filter order was motivated from linear techniques. As seen in Figure 10, the single-step error residuals using linear AR predictors show negligible improvement for order greater than 15, while the autocorrelation indicates substantial correlation out to roughly a delay of 60. Candidate networks evaluated included 10th-, 50th-, and 100th-order filters in the first layer with a varying number of units in the hidden layers. Attempts to perform analysis on the individual filters for the final converged network did not prove illuminating. In general, selection of dimensions for neural networks remains a difficult problem in need of further research.

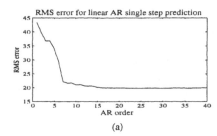
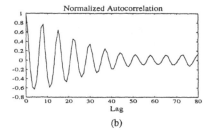

(a)   (b)

FIGURE 10   (a) Single-step prediction errors using linear AR(N) filters.
(b) Autocorrelation of laser data.

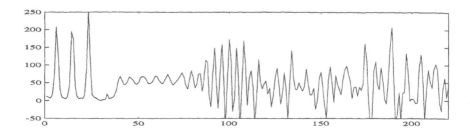

FIGURE 11   Extended iterated prediction.

TRAINING AND CROSS VALIDATION. For training purposes, the data was scaled to zero mean and unit variance. Initial weight values of the network were chosen randomly and then scaled to keep equal variance at each neuron output. The learning rate η was nominally set at 0.01 (this was selected heuristically and then varied during the course of training). Actual training occurred on the first 900 points of the series with the remaining 100 used for cross validation. A training run typically took over night on a Sun SPARC2 system and required several thousand passes through the training set (attempts to optimize training time were not actively pursued). Both the single-step prediction error and the closed-loop iterated prediction were monitored for the withheld 100 points. No overfitting was observed for the single-step error; the larger the network, the better the response. However, a larger network with a lower single-step error often had a worse iterated prediction. Since the true task involved long-term prediction, the iterated performance measure was ultimately used to evaluate the candidate networks. It was clear from the nature of the data that predicting downward intensity collapses would be the most important and difficult aspect of the series to learn. Since the withheld data contained no such features, iterated predictions starting near the point 550 were run to determine how well the network predicted the known collapse at point 600.

After the Competition, the same network structure was retrained with different starting weights to access the sensitivity to initial conditions. Three out of four trials resulted in convergence to equivalent predictions.

PARAMETER FITTING AND LONG-TERM BEHAVIOR. The network used to learn the series had a total of 1,105 variable parameter and was fit with only 1,000 training points. (A statistician would probably have a heart attack over this.) We speculate that this was necessary to accurately model the occurrence of a signal collapse from only two such examples in the training set. In a sense, we were forced to deal with learning what appears statistically to be an outlier. One must also realize that for this nonlinear system, the degrees of freedom do not correspond directly to the number of free parameters. We are really trying to fit a function which is constrained by the network topology. Issues concerning *effective* degrees of freedom are discussed in Weigend and Rumelhart (1991b) and J. Moody (1992).

One consequence of the large number of parameters can be seen in the extended long-term iterated prediction (Figure 11). Due to the excessive number of degrees of freedom, the signal eventually becomes corrupted and displays noisy behavior. While a smaller network (or the addition of weight regularization) may prevent this behavior, such networks were unable to accurately predict the location of the intensity collapse. Note that the prediction still has *bounded output* stability due to the limiting sigmoids within the network.

Because of the eventual signal *corruption*, only the first 75 points of the actual iterated network prediction were submitted to the Competition. This location corresponds to a few time steps after the detected intensity collapse and was chosen based on visual inspection of where the iterated prediction deteriorated. Since after an intensity collapse, the true series tends to display a simple slow-growing oscillation, the remaining 25 points were selected by adjoining a similar sequence taken from the training set.

ERROR PREDICTIONS. For the laser data, an estimate of the uncertainty of the prediction was also submitted. It was assumed that the true observed sequence, $y(k)$, was derived from an independent Gaussian process with a mean corresponding to the prediction $\hat{y}(k)$ and variance $\hat{\sigma}(k)^2$. Thus the standard deviations $\hat{\sigma}(k)$ determine *error bars* for each prediction value. A measure of the probability that the observed sequence was generated by the Gaussian model is given as the *negative average log likelihood*:

$$\text{nalL} = -\frac{1}{N} \sum_{k=1}^{N} \log \left(\frac{1}{\sqrt{2\pi\hat{\sigma}(k)^2}} \int_{y(k)-0.5}^{y(k)+0.5} \exp -\frac{(\tau - \hat{y}(k))^2}{2\hat{\sigma}(k)^2} d\tau \right) . \qquad (18)$$

(See the Appendix of Gershenfeld & Weigend, this volume for full explanation of this likelihood measure.) In order to estimate $\hat{\sigma}(k)^2$, we averaged the known iterated squared prediction errors starting at time count 400 through 550 (i.e., 150 separate iterated predictions were used). It clearly would have been more desirable to base these estimates on segments of data outside the actual training set; however, due to the limited data supplied, this was not possible. The error bars found using the above approach are shown with the prediction in Figure 12, and correspond to nalL $= 1.51$. (The actual error bars submitted were scaled to much smaller values due to a missinterpretation of the performance measure, nalL $= 3.5$.) Alternative Bayesian methods for estimating uncertainties have been suggested by Skilling, Robinson, and Gull (1990) and used in the context of neural networks by MacKay (1992a, 1992b).

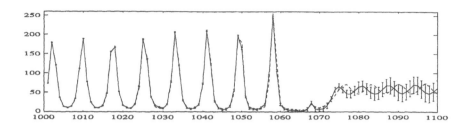

FIGURE 12 Estimated standard deviation error bars are centered at each prediction point. The dashed curve corresponds to the actual series.

ADDITIONAL PREDICTIONS. The complete 10,000-point series continuation was provided after the Competition. Figure 13 shows various iterated predictions starting at different locations within the series. The original network (trained on only the first 1,000 points) is used. While there is often noticeable performance degradation, the network is still surprisingly accurate at predicting the occurrence of intensity collapses.

COMMENTS ON OTHER DATA SETS IN COMPETITION. The organizers of the Competition also distributed a financial series and 100,000 points of a high-dimensional computer-generated chaotic series. Most of the effort, however, was made on the laser data. The financial series was not evaluated using an FIR network. Insufficient time and resources were the major stumbling blocks with the synthetic series. Cursory results with the FIR network were not encouraging; however, additional testing is necessary to draw solid conclusions.

NEW DIRECTIONS AND CONCLUSIONS

In this paper we have focused on the basic autoregressive model; however, it should be clear that the general methodology presented may be easily extended to other configurations. The most trivial extension is to the nonlinear *Autoregressive Moving "Average,"* ARMA, as suggested in Weigend (1991):

$$y(k) = \mathcal{N}[y(k-1), y(k-2), \ldots, y(k-T), e(k-1), e(k-2), \ldots, e(k-T2)] + e(k), \quad (19)$$

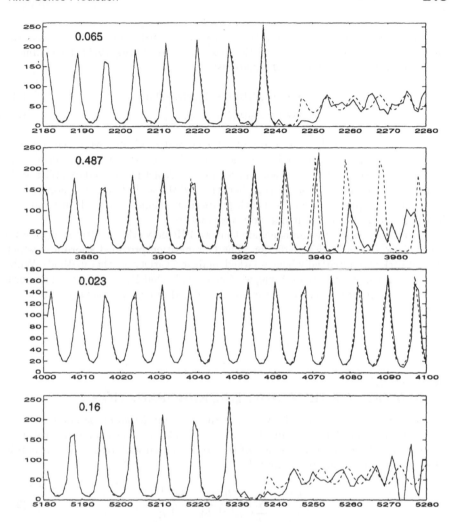

FIGURE 13 The original network, trained on the first 1000 points, is used to make
iterated predictions at various starting locations. The dashed curves correspond to the
true time series. The numbers in the figure give the normalized mean-squared errors.

where $\mathcal{N}$ may be an FIR network, a standard feedforward network, or any variety of
network topologies. The *ARMA* model is most applicable for single-step prediction
where we can additionally regress the network equations on past known prediction
error residuals. Both the neural network *AR* and *ARMA* models, however, are
extrapolated from rather old methods of linear difference equations. A more modern

approach draws from *state space* theory (Kailath, 1980). In the linear case the predictor corresponds to a Kalman Estimator (Bryson & Ho, 1975). Extending to neural networks yields the set of equations:

$$\mathbf{x}(k) = \mathcal{N}_1[\mathbf{x}(k-1), e(k-1)] \tag{20}$$

$$y(k) = \mathcal{N}_2[\mathbf{x}(k)] + e(k), \tag{21}$$

where $\mathbf{x}(k)$ corresponds to a vector of internal states which govern the dynamic system and must be learned by the network. This construct may form a more compact representation than that capable in an *ARMA* model and exhibit significantly different characteristics. Together, the *ARMA* and *state space* models form a taxonomy of possible neural network prediction schemes. Such schemes must be investigated as the field of neural networks matures.

From this paper, we can conclude that an FIR network constitutes a powerful tool for use in time series prediction. The Competition provided a forum in which the network was impartially benchmarked against a variety of other methods. While this was only one concrete example, we feel strongly that FIR networks, and neural networks in general, must be seriously considered when approaching time series problems.

APPENDIX A:
DERIVATION OF TEMPORAL BACKPROPAGATION

Provided here is a complete derivation of the *temporal backpropagation* algorithm. We wish to minimize the cost function $C = \sum_k e^2(k)$ (i.e., the sum of the instantaneous squared errors). The gradient of the cost function with respect to a synaptic filter is expanded using the chain rule:

$$\frac{\partial C}{\partial \mathbf{w}_{ij}^l} = \sum_k \frac{\partial C}{\partial s_j^{l+1}(k)} \cdot \frac{\partial s_j^{l+1}(k)}{\partial \mathbf{w}_{ij}^l}, \tag{22}$$

where

$$s_j^{l+1}(k) = \sum_i s_{i,j}^{l+1}(k) = \sum_i \mathbf{w}_{i,j}^l \cdot \mathbf{x}_i^l(k) \tag{23}$$

specifies the input to neuron j in layer l at time k. Note this expansion differs from the traditional approach of writing the total gradient as the sum of instantaneous gradients: $\partial C/\partial s_j^{l+1}(k) \cdot \partial s_j^{l+1}(k)/\partial \mathbf{w}_{ij}^l \neq \partial e^2(k)/\partial \mathbf{w}_{ij}^l$. Only the sums over all k are equivalent.

From Eq. (22) a stochastic algorithm is formed:

$$w^l_{ij}(k+1) = w^l_{ij}(k) - \eta \frac{\partial C}{\partial s^{l+1}_j(k)} \cdot \frac{\partial s^{l+1}_j(k)}{\partial \mathbf{W}^l_{ij}} \, . \tag{24}$$

From Eq. (23) it follows immediately that $\partial s^{l+1}_j(k)/\partial w^l_{ij} = x^l_i(k)$ for all layers in the network. Defining $\partial C/\partial s^l_j(k) \equiv \delta^l_j(k)$ allows us to rewrite Eq. (24) in the more familiar notational form

$$\mathbf{w}^l_{ij}(k+1) = \mathbf{w}^l_{ij}(k) - \eta \delta^{l+1}_j(k) \cdot \mathbf{x}^l_i(k) \, . \tag{25}$$

To show this holds for all layers in the network, an explicit formula for $\delta^l_j(k)$ must be found. Starting with the output layer, we have simply

$$\delta^L_j(k) \equiv \frac{\partial C}{\partial s^L_j(k)} = \frac{\partial e^2(k)}{\partial s^L_j(k)} = -2e_j(k)f'(s^L_j(k)) \, , \tag{26}$$

where $e_j(k)$ is the error at an output node. For a hidden layer, we again use the chain rule, expanding over all time and all N_{l+1} inputs $s^{l+1}(k)$ in the next layer:

$$\begin{aligned}
\delta^l_j(k) &\equiv \frac{\partial C}{\partial s^l_j(k)} \\
&= \sum_{m=1}^{N_{l+1}} \sum_t \frac{\partial C}{\partial s^{l+1}_m(t)} \frac{\partial s^{l+1}_m(t)}{\partial s^l_j(k)} \\
&= \sum_{m=1}^{N_{l+1}} \sum_t \delta^{l+1}_m(t) \frac{\partial s^{l+1}_m(t)}{\partial s^l_j(k)} \\
&= f'(s^l_j(k)) \sum_{m=1}^{N_{l+1}} \sum_t \delta^{l+1}_m(t) \frac{\partial s^{l+1}_{jm}(t)}{\partial x^l_j(k)} \, .
\end{aligned} \tag{27}$$

But recall

$$s^{l+1}_{jm}(t) = \sum_{k'=0}^{T^l} w^l_{jm}(k')x^l_j(t - k') \, . \tag{28}$$

Thus

$$\frac{\partial s^{l+1}_{jm}(t)}{\partial x^l_j(k)} = \begin{cases} w^l_{jm}(t - k) & \text{for } 0 \le t - k \le T^l; \\ 0 & \text{otherwise,} \end{cases} \tag{29}$$

which now yields

$$
\begin{aligned}
\delta_j^l(k) &= f'(_j^l(k)) \sum_{m=1}^{N_{l+1}} \sum_{t=k}^{T^l+k} \delta_m^{l+1}(t) w_{jm}^l(t-k) \\
&= f'(s_j^l(k)) \sum_{m=1}^{N_{l+1}} \sum_{n=0}^{T^l} \delta_m^{l+1}(k+n) w_{jm}^l(n) \qquad (30) \\
&= f'(s_j^l(k)) \cdot \sum_{m=1}^{N_{l+1}} \bar{\delta}_m^{l+1}(k) \cdot \mathbf{w}_{jm}^l ,
\end{aligned}
$$

where we have defined

$$
\bar{\delta}_m^l(k) = [\delta_m^l(k), \delta_m^l(k+1), \ldots, \delta_m^l(k+T^{l-1})] . \qquad (31)
$$

Summarizing, the complete adaptation algorithm can be expressed as follows:

$$
\mathbf{w}_{ij}^l(k+1) = \mathbf{w}_{ij}^l(k) - \eta \delta_j^{l+1}(k) \cdot \mathbf{x}_i^l(k) \qquad (32)
$$

$$
\delta_j^l(k) =
\begin{cases}
-2e_j(k) f'(s_j^L(k)) & l = L \\
f'(s_j^l(k)) \cdot \displaystyle\sum_{m=1}^{N_{l+1}} \bar{\delta}_m^{l+1}(k) \cdot \mathbf{w}_{jm}^l & 1 \le l \le L-1.
\end{cases} \qquad (33)
$$

A CAUSALITY CONDITION. Careful inspection of the above equations reveals that the calculations for the $\delta_j^l(k)$'s are, in fact, noncausal. The source of this noncausal filtering can be seen by considering the definition of $\delta_j^l(k) = \partial C / \partial s_j^l(k)$. Since it takes time for the output of any internal neuron to completely propagate through the network, the change in the *total* error due to a change in an internal state is a function of future values within the network. Since the network is FIR, only a finite number of future values must be considered, and a simple reindexing allows us to rewrite the algorithm is a causal form:

$$
\mathbf{w}_{ij}^{L-1-n}(k+1) = \mathbf{w}_{ij}^{L-1-n}(k) - \eta \delta_j^{L-n}(k-nT) \cdot \mathbf{x}_i^{L-1-n}(k-nT) \qquad (34)
$$

$$
\delta_j^{L-n}(k-nT) =
\begin{cases}
-2e_j(k) f'(s_j^L(k)) & n = 0 \\
f'(s_j^{L-n}(k-nT)) \cdot \displaystyle\sum_{m=1}^{N_{l+1}} \bar{\delta}_m^{L+1-n}(k-nT) \cdot \mathbf{w}_{jm}^{L-n} & 1 \le n \le L-1.
\end{cases}
$$

$$(35)$$

While less aesthetically pleasing than the earlier equations, they differ only in terms of a change of indices. These equations are implemented by propagating the delta terms backward continuously *without* delay. However, by definition this forces the internal values of deltas to be shifted in time. Thus one must buffer the states $\mathbf{x}(k)$ appropriately to form the proper terms for adaptation. Added storage delays are necessary only for the states $\mathbf{x}(k)$. The backward propagation of the delta terms require no additional delay and is still symmetric to the forward propagation. The net effect of this is to delay the actual gradient update by a few time steps. This may result in a slightly different convergence rate and misadjustment as in the analogous linear *Delayed LMS* algorithm (Kabal, 1990; Long et al., 1989).

For simplicity we have assumed that the order of each synaptic filter, T, was the same in each layer. This is clearly not necessary. For the general case, let T_{ij}^l be the order of the synaptic filter connecting neuron i in layer l to neuron j in the next layer. Then in Eqs. (34) and (35), we simply replace nT by $\sum_{l=L-n}^{L-1} \max_{ij}\{T_{ij}^l\}$. The basic rule is that the time shift for the delta associated with a given neuron must be made equal to the total number of tap delays along the longest path to the output of the network.

ACKNOWLEDGMENTS

I would like to thank Andreas Weigend and Neil Gershenfeld for organizing the Santa Fe Workshop, with special thanks to Andreas for his enthusiastic feedback and ever-stimulating conversations. This work was sponsored in part by NASA under contract NCA2-680 and the Department of the Navy under contract N00014-86-K-0718.

Xiru Zhang† and Jim Hutchinson‡
†Thinking Machines Corporation, 245 First Street, Cambridge, MA 02142;
and ‡MIT Artificial Intelligence Laboratory, 545 Technology Square, Cambridge, MA 02139

Simple Architectures on Fast Machines: Practical Issues in Nonlinear Time Series Prediction

This paper describes our work on predicting the Data Set C (financial), Data Set D (computer generated), and Data Set F (music) in the 1991–92 Santa Fe Time Series Prediction and Analysis Competition. Due to the nature of the tasks, we adopted *state-space* embedding models, and our approach focused on finding a good data *representation* for each prediction task. For predictive techniques, we investigated multilayer perceptrons, radial basis functions, and nearest-neighbor techniques, though the first technique was most heavily used in our analyses. Encouragingly, we were able to do significantly better than guessing either the mean or the last value of the series for up to 25-step-ahead predictions on Data Set D, which was generated from a complex nine-dimensional chaotic dynamical system. Heavy use of a Connection Machine supercomputer greatly reduced the effort and time required to estimate and test complex nonlinear models and handle large data sets. Although the hope for this work was to assess the applicability of a variety of nonlinear techniques to prediction, our main conclusion is that the most important ingredient for good performance is *knowing the data and your modeling technique*.

Times Series Prediction: Forecasting the Future and
Understanding the Past, Eds. A. S. Weigend and N. A. Gershenfeld, SFI Studies
in the Sciences of Complexity, Proc. Vol. XV, Addison-Wesley, 1993   **219**

1. INTRODUCTION

In time series analysis, linear models have been most frequently used (Chatfield, 1989; Shumway, 1988), though often there is no principle reason to restrict consideration to such models. There is a great appeal in employing more general nonlinear methods, especially for building empirically motivated models. Even when a model is theoretically motivated, linearity is often thought of more as a convenient starting point rather than as an absolute truth. New methods arising from advances in chaos theory (Farmer & Sidorowich, 1988; Casdagli, 1989; He & Lapedes, 1991), neural networks (Broomhead & Lowe, 1988; Moody & Darken, 1989; Weigend et al., 1990; Poggio & Girosi, 1990), and statistics (LeBaron, 1989; Granger & Terasvirta, 1992; Brock, 1991) offer hope of providing practitioners with the tools necessary to handle various types of nonlinearity.

Unfortunately, despite the reassuring scientific nomenclature, time series analysis is still very much an art. Even in the relatively well understood linear world of ARIMA models and correlation-based measures, there is still considerable room for human interpretation, such as in model specification, in diagnostics evaluation, and (perhaps most importantly) in weighing the overall merit of a resulting model. This requirement for expert care is, if anything, amplified with the introduction of new nonlinear tools, especially in the area of model specification. None of the new tools are comprehensive enough to warn their users of all possible pitfalls, and thus they cannot be applied indiscriminately.

This paper describes the approach, analysis, and results of our entry on Data Set C (financial), Data Set D (computer generated), and Data Set F (music) of the Competition. Our main attention is on how to develop good representations for nonlinear models through detailed, careful analysis of the given data sets. The paper is organized as follows: Section 2 presents our guiding principles for approaching the competition problems; Sections 3, 4, and 5 outline our analysis and results for each part of the competition; and finally Section 6 gives some discussion and comments.

2. APPROACH

2.1 A GENERAL STATE SPACE MODEL

We can classify the techniques for time series analysis and prediction into roughly two general categories. (1) In the case that a lot of information about the underlying model of a time series is known (such as whether it is linear, quadratic, periodic, etc.), the main task left is then to estimate a few parameters of the model to fit the observation data. Sufficient observations can make this kind of model quite accurate and powerful. Unfortunately, for many real-world problems, the underlying models are often unknown. (2) At the other extreme, the only thing available is a set of observations. The Competition problems belong to this class. For such problems, people often assume that the underlying model has some "state variables," which determine what the values of the time series should be (Chatfield, 1989).

Formally, we define a general state-space model as follows: Let $\ldots x_{t-j}, \ldots, x_{t-1}, x_t, x_{t+1}, \ldots$ be a time series, we assume:

$$x_{t+1} = F(y^1_{t+1}, \ldots, y^i_{t+1}, \ldots, y^m_{t+1}) + N_{t+1}$$

where N_{t+1} represents random noise at time $t+1$ and $y^1_{t+1}, \ldots, y^m_{t+1}$ are *state variables*, and

$$y^i_{t+1} = G_i(y^1_t, \ldots, y^k_t, \ldots, y^m_t, x_t, x_{t-1}, \ldots), \qquad i = 1, 2, \ldots, m.$$

F and G_i's are some functions. This formulation is a generalization of Chatfield (1989) to approximate nonlinear models. The motivation of using state variables is that they often correspond to certain features or properties of the time series and can help us understand and characterize the series. They can also help to simplify the computations for analysis and prediction. In our work, F is some general tool for function approximation (*model fitting*), and the G_i's transform the original "raw data" into a new representation (*model building*).

2.2 REPRESENTATION

A major lesson from Artificial Intelligence (AI) research is that *representation* plays an important role in problem solving (Winston, 1984; Brachman, 1985). A good representation can make useful information explicit and strip away obscuring clutter. Two different representations can be equivalent in terms of expressive power (e.g., the class of functions expressible in both representations is the same), but differ dramatically in the efficiency or ease with which to solve problems. In the context of state space models for time series prediction, representation amounts to two things. First, we need to decide on the input/output variables for such models. For example, what data points should be used as inputs? Should the series be preprocessed or normalized somehow? Second, there is the practical question of encoding the input/output values for a particular nonlinear prediction algorithm; see Feldman and Ballard (1982) for discussion of different encoding schemes.

Once the representation issue is settled, we can generate a set of examples of input/output mappings from a given time series, which can then be used to build a model for predicting the future values of the series. Several nonlinear algorithms are available for such purposes, that is, to approximate a function based on a set of "training examples." To be sure, our emphasis on representation does not suggest that the task left for the function approximation is necessarily easy. Our contention is simply that in terms of getting good performance on real-world problems, finding the right representation for the problem can be significantly more important than deciding, for instance, whether to use sigmoidal multilayer perceptrons or radial basis functions as function approximator.

For the time series in the Competition in particular, our design of representation relied heavily on exploratory data analysis. In addition to the traditional tools such as autocorrelation and linear models, we did extensive nonlinear model estimation and testing using a Connection Machine supercomputer system, which helped to reduce the cycle time needed for model building and model fitting and, enabled us to painlessly attack the problems with large amounts of data. We chose to use multilayer perceptrons trained by backpropagation as our primary function approximation method, in large part because of the quality of a backpropagation implementation on the Connection Machine, and the familiarity of one of the authors with that code.[1] For details of the backpropagation algorithm and its implementation on the Connection Machine system, see Feldman and Ballard (1982); Rumelhart, Hinton, and Williams (1986); Zhang et al. (1990); and Singer (1990). Some experimentation was also done with the method of radial basis functions, a method equivalent to generalized splines. See Poggio and Girosi (1990) for a complete description of this method.

3. PARTICLE MOTION DATA

This section describes our analysis of Data Set D, which was generated by applying a nonlinear observation function to the position of a particle moving through a time varying potential field. The resulting univariate time series is 100,000 points long, has an underlying dimensionality of nine, and was designed to be weakly nonstationary and chaotic.

3.1 EXPLORATORY DATA ANALYSIS

Stationarity is an important property of time series and was one of the first things we checked. Plots of Data Set D, running means, and running standard deviations reveal that, if the series is nonstationary, it is only mildly so (see Figure 1). The persistence of the sample autocorrelation function (ACF) at long lags (see Figure 2), however, indicates that some periodicity or nonstationarity is likely to be present in the data (nonperiodic and stationary time series tend to have an exponential decay of ACF; see Chatfield (1989) for a detailed discussion on this subject). The irregularity and strength of the ACF (and partial ACF) also advocates the use of a reasonably large window of data to be used in modeling, although there is no principled way of chosing the optimal window size from the ACF without knowing the correct functional form of the model.

---

[1] This backpropagation code is publicly available—contact Zhang for details.

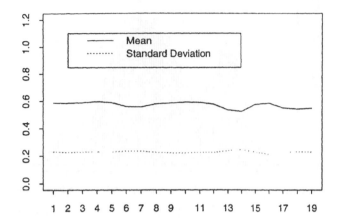

FIGURE 1   Running average and standard deviation for Data Set D time series. Window size is 10,000 points, and the window is moved 5,000 points each step.

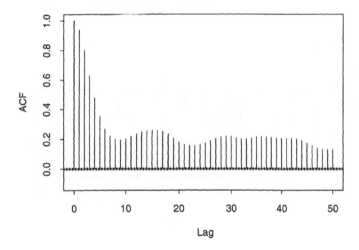

FIGURE 2   Autocorrelation function of Data Set D.

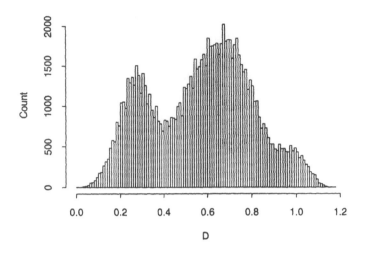

FIGURE 3   Histogram of data points in Data Set D.

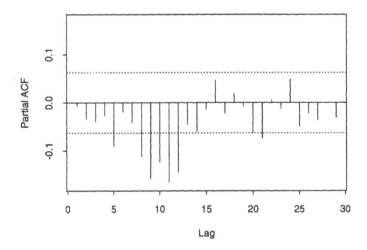

FIGURE 4   Autocorrelation function of the residuals from a linear ARIMA(5,1,0) model of Data Set D.

Histograms of the data show two significant peaks and one minor peak, suggesting that the underlying dynamical system has at least three attractors[2] (see Figure 3).

To test the linearity of the time series, we first fit a set of simple ARIMA models to the series with lag length up to 20, and then computed the autocorrelation of the residuals (i.e., the differences between the original time series and the one-step-ahead prediction of the ARIMA model). Figure 4 shows an example of the result for an ARIMA(5,1,0) model. There are a fair number of statistically significant correlations within the residuals for these linear models, indicating that they were inadequate.

3.2 REPRESENTATION

This problem asks us to predict 500 steps into the future, the furthest ahead of any of the problems in the Competition. Some comments are in order about how we might have attained this goal. There are a number of ways to predict multiple steps into the future. Consider the univariate problem of predicting x_{t+L} from the available data at time t (i.e., $x_t, x_{t-1}, \ldots$). One possibility is for us to construct a single function f which predicts one point into the future, and simply iterate this function on its own outputs to predict further into the future:

$$\hat{x}_{t+1} = f(x_t, x_{t-1}, \ldots)$$
$$\hat{x}_{t+2} = f(\hat{x}_{t+1}, x_t, x_{t-1}, \ldots)$$
$$\vdots$$
$$\hat{x}_{t+L-1} = f(\hat{x}_{t+L-2}, \hat{x}_{t+L-3}, \ldots, x_t, x_{t-1}, \ldots)$$
$$\hat{x}_{t+L} = f(\hat{x}_{t+L-1}, \hat{x}_{t+L-2}, \ldots, x_t, x_{t-1}, \ldots)$$

Alternatively, we could construct one function that uses only past data as inputs to directly predict one desired future point, i.e.,

$$\hat{x}_{t+L} = f_L(x_t, x_{t-1}, \ldots).$$

Finally, we proposed a mixed strategy where the constructed functions take both previous predictions and past data as inputs, and have the direct prediction of the future point as an output, i.e.,

$$\hat{x}_{t+1} = f_1(x_t, x_{t-1}, \ldots)$$
$$\hat{x}_{t+2} = f_2(\hat{x}_{t+1}, x_t, x_{t-1}, \ldots)$$
$$\vdots$$
$$\hat{x}_{t+L-1} = f_{L-1}(\hat{x}_{t+L-2}, \hat{x}_{t+L-3}, \ldots, x_t, x_{t-1}, \ldots)$$
$$\hat{x}_{t+L} = f_L(\hat{x}_{t+L-1}, \hat{x}_{t+L-2} \ldots, x_t, x_{t-1}, \ldots)$$

[2] Actually the system has four attractors, but two were indistinguishable due to the observation function.

Note that the direct and the mixed approaches both require constructing multiple functions to span the specified range of steps ahead, and that the functions in the mixed approach depend on each other. Furthermore, the number of input variables in the mixed approach grows with each time step.

After experimenting with each of the above strategies using backpropagation as the function constructor, we used a combination of the direct and mixed approaches. The mixed approach seemed the most powerful, but it was prohibitively complex to adopt for predictions too far out into the future (both in terms of the computation needed to estimate the models and in terms of the excessive number of parameters involved). Thus we used it only to make the first ten predictions ("short-term predictions"). The direct approach was then used to make the remaining 490 predictions ("long-term predictions") as described below. Altogether we used a total of 108 different networks of two types.

3.2.1 SHORT-TERM PREDICTION.

The first type of network predicts a certain number of values at future temporal positions $t + 1, \ldots, t + L$. The inputs of such a network contain the last 20 data points from the data set (i.e., data points at $t-19, t-18, \ldots, t-1, t$) plus the predictions from the network before it (i.e., the network whose outputs were the predictions for temporal positions $t+1, \ldots, t+L-1$). Thus network 1 had as inputs a window of 20 points from the data set and its output was a prediction for $t + 1$; network 2 had as inputs the 20 data points plus the output of network 1 and its outputs were predictions for $t+1$ and $t+2$; network 3 had as inputs the 20 data points plus the outputs of network 2 and its outputs were predictions for $t+1$, $t+2$, and $t+3$; etc, up to network 10, which had as inputs the 20 data points plus the outputs of network 9, and its outputs were predictions for $t + 1$ through $t + 10$. However, for the competition we only needed one value for each temporal position. To choose between the different predictions for a given temporal position $t + 1$ to $t + 10$, we picked the network output whose difference from the true values had the lowest variance across the entire training set. Note that each of these networks has a different number of input units (network i has $20 + i - 1$ inputs) and output units (network i has i outputs), but all used two hidden layers of 30 and 30 units. Figure 5 shows the input-output relations of these networks.

The number of past values used in the above networks, 20, was chosen from a set of values we tested because it gave the smallest in-sample prediction error and it matched nicely with the drop-off in the sample autocorrelation function. In hindsight, we can show further evidence of the sufficiency of this number by appealing to a theorem due to Takens (1981), which says that an embedding dimension (i.e., number of past values) of $2 \times N + 1$ is sufficient to reconstruct a noise-free system of dimension N. In our case, this is $2 \times 9 + 1 = 19$, which is a surprisingly close match.

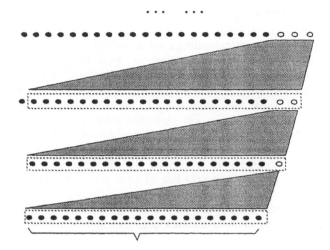

FIGURE 5   Input/output relations among different neural networks for the short-term prediction for Data Set D.

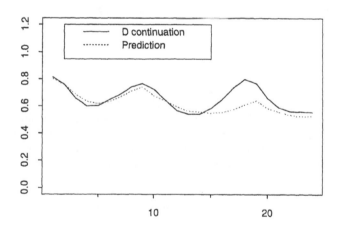

FIGURE 6   True vs. predicted values for the first 25 steps ahead in the "D" series.

3.2.2 LONG-TERM PREDICTION. The second type of network predicts five temporal positions per net, using the last 20 data points from the data set plus ten more sparsely sampled data points extending further back in history. The sampling rate for the 10 sparse data points was matched to the distance into the future that the network was supposed to predict. Thus the first network of this type, network 11, had as inputs a window of 20 points from the past data (i.e., data points at $t, t - 1, t - 2, \ldots, t - 19$) plus some sparsely sampled data (i.e., data points at $t - 20, t - 30, t - 40, \ldots, t - 110$). The network's outputs were the predictions for $t+11, t+12, \ldots, t+15$. Network 12 used as inputs data points at $t, t-1, t-2, \ldots, t-19$, and $t - 20, t - 35, t - 50, \ldots, t - 155$, and its outputs were the predictions for temporal positions $t + 16, t + 17, \ldots, t + 20$. This pattern repeats up to network 108. Note that each of these networks has 30 input units, 100 hidden units, and 5 output units.

In the above networks for both short- and long-term prediction, we did not use the standard sigmoid function as the unit activation function. Rather, we used

$$f(x) = \frac{e^x - e^{-x}}{e^x + e^{-x}}.$$

$f(x)$ has the same shape as the sigmoid function ("continuous threshold"), except that its range is $[-1, 1]$ rather than $[0, 1]$. For Data Set D, the data points were rescaled to fall into $[-1, 1]$.

3.3 RESULTS

The above networks were trained on a 8192 processor Connection Machine CM-2, for roughly one hour per network (i.e., roughly 100 hours total). Experiments with randomly selected sets of 10,000 data points gave similar in-sample and out-of-sample errors (i.e., there was no evidence of overfitting), and thus we felt safe in using the entire data set for training these networks for our final predictions. The predictions for the Competition entry are shown in Figure 6. Taking RMS error[3] as our performance measure, these predictions are significantly better than predicting either the mean of the data or the last known value for the first 25 future values. The predictions after 25 steps degenerate to the mean of the series, which shows that the networks could not find anything better than predicting the mean.

---

[3] Given a time series $x_1, x_2, \ldots, x_n$, the Root-Mean-Squared (RMS) error for prediction values $\hat{x}_1, \hat{x}_2, \ldots, \hat{x}_n$ is defined as:

$$\sqrt{\frac{1}{N} \sum_{i=1}^{n} (x_i - \hat{x}_i)^2}.$$

TABLE 1 Accuracies for the first 25 step predictions on the chaotic time series (Data Set D), as well as the errors by predicting either the mean or the last value of series. Set 0 is the competition entries. Set 100 to set 400 are our additional test sets, which start at position 100, 200, 300, and 400 of the continuation of the data set which we obtained after the competition. "Last value" prediction uses the value right before each test set. "Mean value" prediction uses the mean value of the entire series (100,000 points).

Test Data	RMS Error (networks)	RMS error by predicting the mean	RMS error by predicting the last value
Set 0	0.0665	0.1219	0.1472
Set 100	0.0616	0.3552	0.1257
Set 200	0.1475	0.1973	0.3608
Set 300	0.0541	0.3371	0.0679
Set 400	0.0720	0.1058	0.0830
Average	0.0801	0.2235	0.1569
Error ratio	1.0	0.3584	0.5105

After the Competition, when it was evident that none of the algorithms submitted did better than predicting the mean for more than 25 steps ahead, we decided to do some further out-of-sample tests of the networks' ability for 1- through 25-step-ahead predictions. To do this, we made four new test sets by dividing the competition test set (i.e., the 500 point continuation of the 100,000 data points) into 5 segments and testing 1- through 25-step-ahead predictions at the start of each these segments. Thus the new test sets started at 100, 200, 300, and 400 points from the end of the in-sample series. These new tests indicate that the results for the competition (i.e., the first 25 out of the 500 points) were representative, and thus add confidence in our belief that the networks have predictive ability in this time range (see Table 1).

To test the importance of using multilayer perceptrons in the above results, we also attempted the 1-step-ahead prediction task using the *radial basis function* (RBF) approximation technique (Poggio & Girosi, 1990). We used a RBF network with twenty inputs and one output (as above), 100 radial (Gaussian) basis functions as centers (i.e., hidden units), and a diagonal 20×20 weight matrix for computing distances to centers. Selection of centers was done using K-means clustering, similar to the work by Moody and Darken (1989), and were not moved subsequently. The RBF coefficients were estimated using least-squares regression, and the weights and widths of the Gaussians were estimated using the stochastic function minimization method as in Caprile & Girosi (1990). The network's resulting in-sample RMS error had the same order of magnitude, although not quite as good, as the multilayer perceptron error. Note, however, that much less experimentation was done on tuning the RBF network.

4. FINANCIAL DATA

This section describes our analysis of Data Set C, which was tick-by-tick Swiss Franc to U.S. dollar exchange rates. The data and Competition tasks are described by Lequarré (this volume) and LeBaron (this volume).

4.1 EXPLORATORY DATA ANALYSIS

Plots of Data Set C data reveal obvious nonstationarity (see Figure 7). A simple transformation to make it more stationary is to take the first-order differences (Chatfield, 1989). Histograms of the differenced values indicate that three distinct values are much more likely than all others: -0.0005, 0, and 0.0005 (i.e., a "downtick," no change, and an "uptick"; see Table 2). The differences of trading prices that are 15-minutes and 60 minute apart shows a Gaussian-like distribution (see Figure 8).

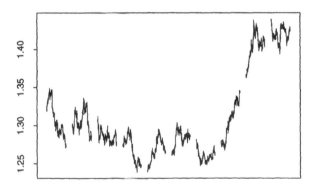

FIGURE 7   Simple time series plot of Data Set C (i.e., financial data). There are 10 sets of 3,000 points, with unknown time periods between them. Note that this plot assumes points are equally spaced in time, and does not make use of the time information given.

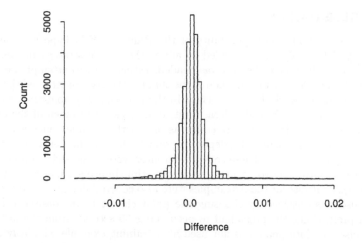

FIGURE 8   Distribution of the first-order differences for points that are 15 minutes apart in Data Set C.

TABLE 2   The histogram of the first-order differences of Data Set C.

First-Order Difference (d)	Occurrences
$d < -0.001$	290
$d = -0.001$	1067
$-0.001 < d < -0.0005$	9
$d = -0.0005$	**10781**
$-0.0005 < d < 0$	45
$d = 0$	**5531**
$0 < d < 0.0005$	60
$d = 0.0005$	**10655**
$0.0005 < d < 0.001$	14
$d = 0.001$	1116
$d > 0.001$	318

4.2 REPRESENTATION

Given concerns about the applicability of the Random Walk hypothesis and the exploratory data analysis, we decided to adopt the simple strategy of predicting which *direction* the financial series was headed, rather than to attempt the bolder task of predicting the exact magnitude of the change. We also constructed the inputs to our predictive networks by coding them in the following way, to extract the most significant information from the data. First, differences were taken of the original series,[4] and these differences were coded as one of three values: increasing (+1), decreasing (−1), or no change (0), depending on whether the difference was positive, negative, or (close to) zero. These coded differences were then used as inputs to the multilayer perceptron networks, along with some time information described below. The "correct" outputs for these examples were also coded differences, and the exact point to use was obtained by choosing the point closest to the desired time. For 1-minute predictions, the point had to occur 60 ± 30 seconds from the last input point or else the data point was not used as a training example. Similarly for 15-minute and 60-minute predictions, the time ranges used were 900 ± 90 seconds and 3600 ± 360 seconds, respectively. We considered this to be appropriate sampling strategy for the Competition because it was implied that there would be a value near the specified time interval, although this does make the prediction conditional on this fact. A final detail of our coding scheme was to compute predictions of the real (uncoded) differences by multiplying the outputs of the networks (which were real numbers in $[-1, 1]$) by the mean absolute value of the differences across all of the data.

For the predictions of the end of the last day, the beginning of the next day and a week after, we did not feel that there were enough data points in the training set for us to make reliable statistics. Thus we followed the Random Walk model ourselves and used the last value of each set as our predictions, except for the one-week prediction, where we used the average of the last value of the current set and the first value of the next set.[5] This actually produced better results than the Random Walk model for the one-week predictions because *future* information was used. This was only possible due to the peculiarities of the Competition setup, and thus we will not discuss them further. In the following, we will concentrate on the 1-minute, 15-minute, and 60-minute predictions.

Separate multilayer perceptron networks were trained for each of the different prediction intervals of 1 minute, 15 minutes, and 60 minutes. All networks had two hidden layers of 20 and 20 units and one output unit. The number and content of input units varied for each network as follows: (a) The 1-minute network had 8 inputs. Seven of the inputs were the last seven coded price differences as described above. The eighth input was the change in time from the first to the last of

---

[4] Without resampling to a uniform time grid.

[5] For the tenth set we used the last observed value.

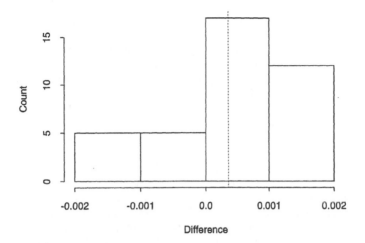

FIGURE 9 Output distribution for input pattern $[0, -1, +1, 0, 0, 0, -1]$ on Data Set C 15-minute prediction task. The vertical line shows the network's output for this input pattern.

the seven price differences, coded into a $\{-1, 0, +1\}$ value depending on which third of the distribution (i.e., across all of the data) it fell in. The breakpoints were 180 and 720 seconds. (b) The 15-minute network had seven inputs. The first six inputs were the coded price differences for every 15 minutes for the last 1.5 hours. The seventh input was the frequency of transactions in that 1.5-hour window, coded into $\{-1, 0, +1\}$ depending on which third of the distribution (i.e., across all of the data) it fell in. The two breakpoints were 6 and 11 transactions, respectively. (c) The 1-hour network had six inputs. The first four inputs were the coded price differences every 30 minutes for the last 2 hours. The fifth input was the frequency of transactions in that 2-hour window, coded into $\{-1, 0, +1\}$ depending on which third of the distribution (i.e., across all of the data) it fell in. The sixth input was the hour of the day of the end of the 2-hour window, coded into the three values $\{-1, 0, +1\}$ as above. The breakpoints were third and seventh hours for the 10-hour trading day. Support for our choice of input variables can be found in LeBaron (this volume), which shows interesting correlations between these variables.

4.3 RESULTS

The 1-, 15-, and 60-minute networks were trained with very small learning rate for a few thousand epoch cycles each until the error leveled off, which took a 8192 processor Connection Machine CM-2 roughly 2 hours per network. Our first question is: what do the trained networks actually compute? Due to the simple representations described in the previous section, the number of distinct input patterns[6] is much smaller than the number of actual training patterns. Therefore, many input patterns must have occurred more than once in the training set, possibly with different outputs each time. Figure 9 shows the distribution of the real outputs for a particular input pattern in the 15-minute training set ($[0, -1, +1, 0, 0, 0, -1]$) and the output of the network[7] for it. We see that the real output distribution has a positive mean. The network output for this input is roughly at the center of this distribution. In other words, the input-output relation is one-to-many, which is not a function, and the network is not approximating a function, but rather it tries to estimate the center of a distribution.

One obvious question is then: why do we need to use the neural networks in the first place? Why can't we simply do a table look-up? The problem is that similar input patterns have similar output distributions; and they influence each other. To estimate a pattern's output, we need to look at not only the outputs of the identical input patterns, but also those that differ by one bit, two bits, etc. The neural network gives a smooth estimate of the posterior probability for this type of classification task (Hinton & Nowlan, 1990). Figure 10 shows the output distribution of an input pattern in 1-minute training set ($[+1, -1, -1, -1, +1, +1, -1, 0]$) together with the patterns that differ from it by one bit. Also shown in the figure is the output of the network for this pattern.[8] Again, the network's output is in the center of the distribution.

Not all output distributions are centered at a positive or negative number. Quite a few are centered at zero, in which case the network's prediction is no change (zero), which is the same as that of the Random Walk model. However, the network can still beat the Random Walk model if it can do better for *some* patterns (i.e., the ones having positively or negatively centered output distributions), and do the same for the rest. Motivated by this observation, we actually filled in all the input patterns that have not appeared in the training set and simply assumed their outputs to be zero. This "space filling" technique ensures that the network has seen all the possible input patterns during training, and thus should do no worse than the Random Walk model for all inputs.

[6] There are three possible values ($\{-1, 0, +1\}$) for each input variable. If we have n input variables, there are $m = 3^n$ distinct input patterns. $3^6 = 729$; $3^7 = 2187$; and $3^8 = 6561$.
[7] This is one of the Competition entries.
[8] This is also an entry for the Competition.

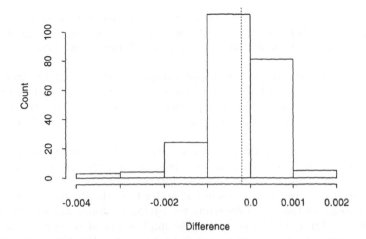

FIGURE 10 Output distribution for input pattern $[+1, -1, -1, -1, +1, +1, -1, 0]$ and all other input patterns that differ by one position with it, for Data Set C 1-minute prediction task. The vertical line shows the network's output for this input pattern.

TABLE 3 Prediction errors for both the training and competition data for Data Set C of financial data.

	Training data			Competition data		
	Neural Network	Random Walk	Ratio	Neural Network	Random Walk	Ratio
1 minute	0.000662	0.000702	0.943	0.000404	0.000484	0.835
15 minutes	0.00151	0.00160	0.944	0.00170	0.00167	1.018
60 minutes	0.00270	0.00287	0.941	0.00134	0.00135	0.994
average			0.943			0.949

In what situations will the general approach discussed above work? It should work if the output distribution for an input pattern does not change over time. Take the case of Figure 9, for example. If that distribution persists into the future, then the network's prediction will statistically have smaller errors than predicting no change (zero) over the long run. Table 3 shows the networks' prediction errors on the training set and the Competition entries. On average for 1-minute, 15-minute, and 60-minute predictions, the networks' predictions seemed to have about 5% improvements in RMS errors over the Random Walk model for both the training and the competition data, although the error distribution is quite different in the two cases.

Unfortunately, the Competition as planned had a worrisome aspect: there were only ten predictions to test for each of the six different prediction intervals requested. Given the large amount of unpredictable noise in the data, this is not a sufficient number of points to reach significant conclusions about small differences in performance.

To gain more insight on the problem, we performed more tests after the Competition. First, we used the data from the gaps between the ten competition sample sets as test data and applied our networks without retraining. The results for all networks were negative—none of them was better than the Random Walk predictions. The ratios of the two RMS errors for 1-minute, 15-minute, and 60-minute predictions are 1.044, 1.050, and 1.048. The implications of these results are not clear. They may indicate nonstationarity of the data during that time period (8/90–4/91), such as due to the interest rate changes (see LeBaron, this volume); or it may be that the network has overfitted the training data.

To test our models for different time periods, we retrained a network using the same representation and network architecture as before on one year's data of the same exchange rate (6/85–5/86) and tested the network on the next year's data (6/86–5/87) for 1-minute prediction. There are 17,096 test cases (out-of-sample tests) and the network outperformed Random Walk model slightly—the RMS error ratio was 0.982, which is statistically significant considering the large number of test cases. We also checked the percentage of correctly predicted classes (increasing, decreasing, no change) by the network; it was 52.4% correct, whereas the best one can do by random guessing is to always predict the largest class, which is only 41.4% correct. Thus the network's performance was more impressive in this measure. This brings up the question of what is a good measure of performance for predicting financial time series, or exchange rate in particular. It seems that practitioners in the financial industry rarely use RMS errors; "change of direction" may be a more useful measure to them. We also repeated the above experiment for another period, using the data of 8/87–7/88 as the training set and the data in the following year (8/88–7/89) as the test set. Again the network outperformed the Random Walk model—the RMS error ratio is 0.987; the correct class prediction rate was 51.2% (versus 44.5% by random). In conclusion, we found that the financial time series used in the Competition is highly nonstationary in some fundamental ways, such as: (1) a nonlinear network trained with data in one time period may not work for another period and (2) a general algorithm may work for some periods but not others.

As in our Data Set D analysis, we again wanted to determine how important the use of multilayer perceptrons was in obtaining the above performance. The comparison we made was for 1-minute-ahead predictions on the competition training set, again using a radial basis function network. The network used the same inputs and outputs as the 1-minute multilayer perceptron network above, had 100 Gaussian functions as centers, and used a full 8×8 weight matrix for computing distances to the centers (otherwise, the training procedure was the same as for the Data Set D RBF network). The in-sample RMS error on the competition training set for the resulting network was 6.842×10^{-4}, versus a variance of 7.023×10^{-4} for the desired training outputs, for a ratio of roughly 0.97. Again, the similarity of the in-sample performance with multilayer perceptrons encouraged us to think that a more thorough comparison of these methods would find relatively minor differences in performance.

As a final comment, note that we did *not* resample the Data Set C onto a fixed time grid to make the (resampled) series have one point per minute, for instance. A resampled series might be a better choice to use as inputs to the predictive system, so that the inputs span a fixed amount of time and all possible windows of data can be used, as in the approach of Mozer (this volume). Unfortunately, we did not have time to adequately explore this interesting issue.

---

5. MUSIC DATA

This section describes our (post-competition) analysis of the mysterious multivariate Data Set F, which in fact was Bach's Last Fugue, a famous unfinished piece of music. It is discussed in Dirst and Weigend (this volume). The values in the series were encodings of the musical notes,[9] with zeroes inserted both to represent rests and boundaries between notes. This is a multivariate time series containing four values at each time point, which corresponded to the four parts of the music. Although some participants realized the series was music (and who the composer was!), we did not. Nevertheless, we found our analysis of the data was an interesting exercise in what one can discover from data that one knows nothing about.

5.1 EXPLORATORY DATA ANALYSIS

Inspection of the data revealed that values were integers in the range of 25 to 75, and that apart from having distinct mean values, the four series had similar statistical properties, such as autocorrelation and variance. Another glaring property of the data was the stretches of repeated values.[10] Because of this, we decided to look at a *run-length encoding* of the data, which is formed by replacing runs of repeated values by the value and a count of how many times it was repeated. Thus, we can consider two separate series, the series of nonrepeated values and the series of counts

---

[9] Using the music industry standard MIDI encoding convention.

[10] Since each value represented a fixed-duration musical note, repeated values then corresponded to notes of longer duration.

(i.e., durations). Sample autocorrelations of both of these series were significant for a number of lags, indicating structure in both the progression of values and the durations of values (see Figures 11 and 12).

5.2 REPRESENTATION

Because of the limited time available for this analysis, we used a simple nearest-neighbor scheme to produce predictions, rather than the multilayer perceptrons used with the other problems. The inputs to the system were the run-length encoded value and duration series mentioned above, taking the past two values and two durations from each of the four series, for a total of 16 inputs. The outputs of the system were the predictions for the next four values, each for one series. Our nearest-neighbor scheme first computed the "distances" between the inputs and every training example to find the nearest ten neighbors, and then combined these ten nearest neighbors to produce the predictions. The distance metric used was the sum of the absolute values of the componentwise differences between the inputs and the training examples; and the predictions were the average of the 10 nearest neighbors, weighted inversely by their distance to the inputs. Multiple-step-ahead predictions were obtained by iterating this procedure, using the predicted values at step i as known values when predicting step $i + 1$.

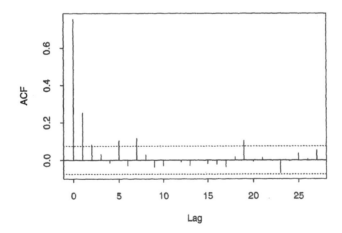

FIGURE 11   Partial autocorrelation function of the "value component" of the music data.

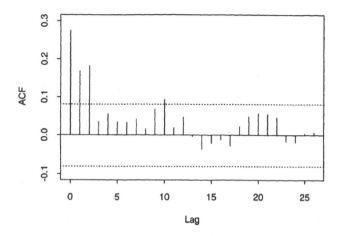

FIGURE 12   Partial autocorrelation function of the "frequency component" of the music data.

TABLE 4 Prediction errors by nearest-neighbor method for Data Set F of music data on the cross-validation tests.

Dimension	1	2	3	4
Nearest neighbor	20.4	19.3	16.5	14.2
"Last value"	37.0	35.9	28.6	24.1
"Mean value"	30.1	26.0	24.2	20.1

5.3 RESULTS

Cross-validation experiments on the given data show that the system does better than predicting either the mean or the last observed value (see Table 4). However, since there is no "true" continuation for this time series, it is impossible to make uniquely "correct" measurements of the resulting system's performance. Indeed, the best evaluation of the system may be listening to it!

6. DISCUSSION

Our main conclusion from this work is the often-repeated message of statistical data analysis: *get to know the data and your methods.* Take the time to look at the data (from many different "angles"), and let the data guide what representation is appropriate for subsequent analysis. Understand what is happening at each step of the modeling process, why it is happening, and whether assumptions you (or the prediction method) are making are satisfied. Only after this kind of careful and thorough analysis is it meaningful to compare methods; blind application of the best methods is likely to end in disappointment.

That being said, we feel that there is reason for optimism about the use of nonlinear methods in time series prediction. In particular, we found the performance of multilayer perceptrons and radial basis functions to be surprisingly good on Data Set D of high-dimensional chaos—much better, for instance, than our attempts with linear ARIMA models. Of course, it shouldn't be too surprising since nonlinear models embody a much broader class of functions than linear models. Nevertheless, the point is that it is possible to profitably explore the rather unconstrained world of nonlinear models.

A danger of all modeling is that of *overfitting* the known data, i.e., memorizing the noise rather than finding the true regularities in the data.[11] The overfitting problem is often accentuated in nonlinear models due to the large number of free parameters used. There are two issues concerning overfitting: (a) how to make an objective estimate of a system's performance and avoid the false optimism due to the seemingly very accurate prediction on the in-sample tests, and (b) how to *prevent* the system from overfitting so that the performance on the in-sample tests reflects the true performance.

For issue (a), a general approach is to use part of the known data as training data to fit the model and the rest as test data to estimate the performance. There are different ways to divide the training/test sets in multiple experiments, resulting in different procedures such as jackknife, cross validation and bootstrapping (Bradley & Gong, 1983). Issue (b) is more complicated. To prevent overfitting, a general guideline is either to increase the amount of training data or reduce the number of free parameters in the system. For some classes of systems and problems, people have come up with certain "rules of thumb" on the ratio of these two factors based on their experience. However, there are fundamental limitations with this general approach when nonstationarity is suspected. For example, it is often felt that we should not look too far back into the past for many financial time series because of their changing nature, but there are painfully few data points available in shorter periods of time! Therefore, short-lived nonlinear correlations (if they exist) require complex models, but we may not have enough data points to train them!

---

[11] Actually, underfitting is in general a concern as well, especially for linear models. But it is not usually as much of a concern for the more powerful nonlinear models.

In this work, we approached the overfitting problem in several different ways depending on the problem at hand. Where there was a lot of training data (such as Data Set D), we simply chose the number of parameters to be relatively small compared with the number of training examples (about 1:50 in Data Set D). For Data Set F, the nearest-neighbor technique only needed a few parameters. Our work on the financial (Data Set C) data was motivated by the idea of designing a suitable representation to avoid overfitting. There seems to be no *automatic* procedure to prevent overfitting: each particular problem has to be treated differently.

A goal of the Competition was to compare different modeling and analysis methods on a standard set of problems in an attempt to assess the strengths and weaknesses of these methods. We feel, however, that this comparison is susceptible to confounding two features of an approach: the problem representation adopted and the core modeling technique used. Choosing a good representation for a problem is a significant part of the whole solution; the right factors, lags, and normalization must be selected in order to get good performance. For example, although our system gave the best performance on Data Set D of high-dimensional chaos using backpropagation as our function constructor, we think it is too simplistic and, indeed, not accurate to claim that "neural networks are the best" for such tasks, without noting the importance of other factors. Also, to get good performance on a real-world problem, we feel that the "best" algorithm is not necessarily the algorithm with best theoretical properties, but the algorithm that one really understands and knows how to "tweak."

ACKNOWLEDGMENTS

We would like to thank Neil Gershenfeld and Andreas Weigend for organizing an informative and enjoyable competition and workshop. Tomaso Poggio first suggested this project to us, and this work benefited greatly from discussions with him. Woody Lichtenstein and Bruce Boghosian at Thinking Machines Corporation gave helpful comments and suggestions. Geraldine Caulfield at CSVI London provided helpful proofreading assistance.

Michael C. Mozer
Department of Computer Science and Institute of Cognitive Science, University of Colorado, Boulder, CO 80309–0430

Neural Net Architectures for Temporal Sequence Processing

I present a general taxonomy of neural net architectures for processing time-varying patterns. This taxonomy subsumes many existing architectures in the literature and points to several promising architectures that have yet to be examined. Any architecture that processes time-varying patterns requires two conceptually distinct components: a short-term memory that holds on to relevant past events and an associator that uses the short-term memory to classify or predict. My taxonomy is based on a characterization of short-term memory models along the dimensions of form, content, and adaptability. Experiments on predicting future values of a financial time series (U.S. dollar–Swiss franc exchange rates) are presented using several alternative memory models. The results of these experiments serve as a baseline against which more sophisticated architectures can be compared.

Times Series Prediction: Forecasting the Future and
Understanding the Past, Eds. A. S. Weigend and N. A. Gershenfeld, SFI Studies
in the Sciences of Complexity, Proc. Vol. XV, Addison-Wesley, 1993   **243**

INTRODUCTION

Neural networks have proven to be a promising alternative to traditional techniques for nonlinear temporal prediction tasks (e.g., Curtiss, Brandemuehl, & Kreider, 1992; Lapedes & Farber, 1987; and Weigend, Huberman, & Rumelhart, 1992). However, temporal prediction is a particularly challenging problem because conventional neural net architectures and algorithms are not well suited for patterns that vary over time. The prototypical use of neural nets is in *structural pattern recognition*. In such a task, a collection of features—visual, semantic, or otherwise—is presented to a network and the network must categorize the input feature pattern as belonging to one or more classes. For example, a network might be trained to classify animal species based on a set of attributes describing living creatures such as "has tail," "lives in water," or "is carnivorous"; or a network could be trained to recognize visual patterns over a two-dimensional pixel array as a letter in $\{A, B, \ldots, Z\}$. In such tasks, the network is presented with all relevant information simultaneously.

In contrast, *temporal pattern recognition* involves processing of patterns that evolve over time. The appropriate response at a particular point in time depends not only on the current input, but potentially on all previous inputs. This is illustrated in Figure 1, which shows the basic framework for a temporal prediction problem. I assume that time is quantized into discrete steps, a sensible assumption because many time series of interest are intrinsically discrete, and continuous series can be sampled at a fixed interval. The input at time t is denoted $\mathbf{x}(t)$. For univariate series, this input is a scalar but, to allow for a multivariate series, $\mathbf{x}$ should be considered as vector valued. The output of the predictor at time t, $\mathbf{y}(t)$ is based on the input sequence $\mathbf{x}(1) \ldots \mathbf{x}(t)$ and represents some future value of the input or possibly some function of a future value; $\mathbf{y}$ can also be vector valued. A prediction $\mathbf{y}$ can be made at every time step or only at selected times. Prediction involves two conceptually distinct components. The first is to construct a short-term memory that retains aspects of the input sequence relevant to making predictions. The second makes a prediction based on the short-term memory. In a neural net framework, the predictor will always be a feedforward component of the net, while the short-term memory will often have internal recurrent connections. ZhangZhangX. and Hutchinson (this volume) and de Vries and Principe (1992) also make the distinction between these two components.

In designing a neural net model, one must consider the following issues:

- **Architecture:** What is the internal structure of the short-term memory and predictor networks? Answering this question involves specifying the number of layers and units, the pattern of connectivity among units, and the activation dynamics of the units.

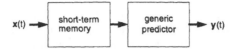

FIGURE 1   Abstract formulation of the temporal prediction task. $\mathbf{x}(t)$ denotes a vectorial representation of the input at time t; $\mathbf{y}(t)$ denotes a prediction at time t based on the input sequence $\mathbf{x}(1)\ldots\mathbf{x}(t)$. The short-term memory retains aspects of the input sequence that are relevant to the prediction task.

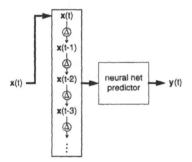

FIGURE 2   Tapped delay-line memory model. Each "Δ" operator introduces a one-time step delay.

- **Training**: Given a set of training examples, how should the internal parameters in the model—the connection *weights*—be adjusted? Each training example i consists of an input series, $\{\mathbf{x}_i(1),\ \mathbf{x}_i(2),\ \mathbf{x}_i(3),\ \ldots,\ \mathbf{x}_i(t_i)\}$, and an associated series of predictions, $\{\mathbf{p}_i(1),\ \mathbf{p}_i(2),\ \mathbf{p}_i(3),\ \ldots,\ \mathbf{p}_i(t_i)\}$, where t_i is the number of steps in the example and $\mathbf{p}_i(\tau)$ is a target prediction at time τ. Training a neural net model involves setting its weights such that its predictions, $\mathbf{y}_i(\tau)$, come as close as possible to the target predictions, $\mathbf{p}_i(\tau)$, usually in a least-squares sense.
- **Representation**: How should the time-varying input sequence be represented in the short-term memory? The nature and quantity of information that must go into the memory is domain dependent.

In this chapter, I focus on the representation issue. However, the issues of architecture, network dynamics, training procedure, and representation are intricately related and, in fact, can be viewed as different perspectives on the same underlying problems. A given choice of representation may demand a certain neural net architecture or a particular type of learning algorithm to compute the representation. Conversely, a given architecture or learning algorithm may restrict the class of representations that can be accommodated.

I address the representation issue by characterizing the space of neural network short-term memory models. In the following sections, I present three dimensions along which the memory models vary: *memory form*, *content*, and *adaptability*. This framework points to interesting memory models that have not yet been explored in the neural net literature.

FORMS OF SHORT-TERM MEMORY
TAPPED DELAY-LINE MEMORY

The simplest form of memory is a buffer containing the n most recent inputs. Such a memory is often called a *tapped delay-line* model because the buffer can be formed by a series of delay lines (Figure 2). It is also called a *delay space embedding* and forms the basis of traditional statistical autoregressive (AR) models. Tapped delay-line memories are common in neural net models (e.g., Elman & McClelland, 1986; Elman & Zipser, 1988; Lapedes & Farber, 1987; Plaut & Hinton, 1987; Waibel et al., 1989; Wan, this volume; and Zhang & Hutchinson, this volume). These memories amount to selecting certain elements of a sequence $\mathbf{x}(1) \ldots \mathbf{x}(t)$, say, a total of Ω elements, and forming a state representation $(\bar{\mathbf{x}}_1(t), \bar{\mathbf{x}}_2(t), \bar{\mathbf{x}}_3(t), \ldots \bar{\mathbf{x}}_\Omega(t))$, where $\bar{\mathbf{x}}_i(t) = \mathbf{x}(t - i + 1)$.

A minor extension of this formulation to permit nonuniform sampling of past values involves specifying varying delays such that $\bar{\mathbf{x}}_i(t) = \mathbf{x}(t - \omega_i)$, where ω_i is the (integer) delay associated with component i.

Let me present another formalism for characterizing the delay-line memory which is broad enough to encompass the other forms of memories to be discussed. One can treat each $\bar{\mathbf{x}}_i(t)$ as a convolution of the input sequence with a kernel function, c_i:

$$\bar{\mathbf{x}}_i(t) = \sum_{\tau=1}^{t} c_i(t - \tau)\mathbf{x}(\tau),$$

where, for delay-line memories,

$$c_i(t) = \begin{cases} 1, & \text{if } t = \omega_i \ ; \\ 0, & \text{otherwise.} \end{cases}$$

Figure 3(a) shows the kernel function. By substituting a different kernel function, one obtains the forms of memories discussed in the next three sections.

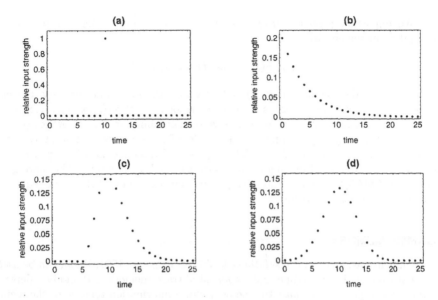

FIGURE 3 The kernel functions for (a) a delay-line memory, $\omega = 10$; (b) an expontial trace memory, $\mu = .8$; (c) a gamma memory, $\omega = 6$; and $\mu = .4$, and (d) a Gaussian memory.

EXPONENTIAL TRACE MEMORY

An exponential trace memory is formed using the kernel function

$$c_i(t) = (1 - \mu_i)\mu_i^t,$$

where μ_i lies in the interval $[-1,1]$ (Figure 3(b)). This form of memory has been studied by Jordan (1987), Mozer (1989), and Stornetta, Hogg, and Huberman (1988). Unlike the delay-line memory, the exponential trace memory does not sharply drop off at a fixed point in time; rather, the strength of an input decays exponentially. This means that more recent inputs will always have greater strength than more distant inputs. Nonetheless, with no noise, an exponential memory can preserve all information in a sequence. For instance, consider a sequence of binary digits, i.e., $x_i(t) \in \{0,1\}$ and a memory with $\mu = .5$. The memory becomes a bit string in which the kernel function assigns the input at time $t - \tau$ to bit position $2^{-(\tau+1)}$ of the string. If the memory has finite resolution or is noisy, the least significant bits—the most distant inputs—will be lost. Even with no information loss, however, it is difficult for a neural network to extract and use all the information contained in the memory.

An important property of exponential trace memories is that the $\bar{x}_i$ can be computed incrementally:

$$\bar{x}_i(t) = (1 - \mu_i)x_i(t) + \mu_i\bar{x}_i(t - 1)$$

with the boundary condition $\bar{x}_i(0) = 0$. As with the delay-line memory, the exponential trace memory consists of Ω state vectors, $(\bar{x}_1(t), \bar{x}_2(t), \bar{x}_3(t), \ldots, \bar{x}_\Omega(t))$. One can view exponential trace memories as maintaining moving averages (exponentially weighted) of past inputs. The μ_i allow for the representation of averages spanning various intervals of time. The exponential trace memory is a special case of traditional statistical moving-average models, the MA(1) model. The more general case of MA(q) cannot readily be represented using a trace of the input sequence alone and is dealt with below.

GAMMA MEMORY

Two dimensions, *depth* and *resolution* (de Vries & Principe, 1991, 1992), can be used to characterize the delay-line and exponential-trace memories. Roughly, "depth" refers to how far into the past the memory stores information relative to the memory size, quantified perhaps by the ratio of the first moment of the kernel function to the number of memory state variables. A low-depth memory only holds recent information; a high-depth memory holds information distant in the past. "Resolution" refers to the degree to which information concerning the individual elements of the input sequence is preserved. A high-resolution memory can reconstruct the actual elements of the input sequence; a low-resolution memory holds coarser information about the sequence. In terms of these dimensions, the delay-line memory is low depth but high resolution, whereas the exponential-trace memory is high depth but low resolution.

de Vries and Principe (1991, 1992) have proposed a memory model that generalizes across delay lines and exponential traces, allowing a continuum of memory forms ranging from high resolution and low depth to low resolution and high depth. In continuous time, the kernel for this memory is a gamma density function, hence the name *gamma memory*. Although de Vries and Principe deal primarily with continuous time dynamics, there is a discrete equivalent of the gamma density function, the negative binomial, which can serve as the corresponding kernel for discrete-time gamma memories:

$$c_i(t) = \begin{cases} \begin{pmatrix} t \\ \omega_i \end{pmatrix} (1 - \mu_i)^{\omega_i+1}\mu_i^{t-\omega_i} & \text{if } t \geq \omega_i; \\ 0 & \text{if } t < \omega_i . \end{cases}$$

As before, ω_i is an integer delay and μ_i lies in the interval $[0, 1]$. Figure 3(c) shows a representative gamma kernel. With $\omega_i = 0$, the gamma kernel reduces to an exponential trace kernel. In the limit as $\mu_i \to 0$, it reduces to a delay-line kernel.

As with the exponential trace kernel, the convolution of the gamma kernel with the input sequence can be computed incrementally (de Vries & Principe, 1991, 1992). However, to compute $\bar{\mathbf{x}}$ for a given μ and ω, denoted here as $\bar{\mathbf{x}}_{\mu,\omega}$, it is also necessary to compute $\bar{\mathbf{x}}_{\mu,j}$ for $j = 0 \ldots \omega - 1$. The recursive update equation is

$$\bar{\mathbf{x}}_{\mu,j}(t) = (1 - \mu)\bar{\mathbf{x}}_{\mu,j-1}(t - 1) + \mu\bar{\mathbf{x}}_{\mu,j}(t - 1)$$

with boundary conditions

$$\bar{\mathbf{x}}_{\mu,-1}(t) = \mathbf{x}(t + 1) \text{ for all } t \geq 0, \text{ and}$$
$$\bar{\mathbf{x}}_{\mu,j}(0) = 0 \text{ for all } j \geq 0.$$

In constructing a gamma memory, the designer must specify the largest ω, denoted Ω. The total number of state vectors is then $\Omega + 1$: $\bar{\mathbf{x}}_{\mu,0} \ldots \bar{\mathbf{x}}_{\mu,\Omega}$. Given Ω, the tradeoff between resolution and depth is achieved by varying μ. A large μ yields a high-depth, low-resolution memory; a small μ yields low-depth, high-resolution memory.

Each set of $\Omega+1$ state vectors with a given μ forms a *gamma family*. The gamma memory can consist of many such families, with family i characterized by μ_i and Ω_i. Of course, this leads to a large and potentially unwieldy state representation. Based on assumptions about the domain, one can trim down the state representation, selecting only certain vectors within a family for input to the neural net predictor, e.g., the subset $\{\bar{\mathbf{x}}_{\mu_i,\Omega_i}\}$. One can view this as a direct extension of the exponential trace memory, which is of this form with $\Omega_i = 0$ for all i. The nonselected vectors must be maintained as part of the recursive update algorithm but need not be passed to the neural net predictor.

OTHER FORMS OF MEMORY

The three memories just described are not the only possibilities. Any kernel function results in a distinct memory form. For instance, a Gaussian kernel (Figure 3(d)) can be used to obtain a symmetric memory around a given point in time. What makes the gamma memory and its special cases particularly useful is that they can be computed by an incremental update procedure, whereas forms such as the Gaussian require evaluating the convolution of the kernel with the input sequence at each time step. Such convolutions, while occasionally used (Bodenhausen & Waibel, 1991; Tank & Hopfield, 1987; and Unnikrishnan, Hopfield, & Tank, 1991), are not terribly practical because of the computational and storage requirements.

Radford Neal (personal communication, 1992) has suggested a class of kernels that are polynomial functions over a fixed interval of time beginning at a fixed point in the past; i.e.,

$$c_i(t) = \begin{cases} \sum_{j=0}^{\Omega} \mu_{ij} t^j, & \text{if } t_i^- \leq t \leq t_i^+; \\ 0, & \text{otherwise.} \end{cases}$$

He has proposed an incremental update procedure for memories based on this kernel. The number of memory state vectors, Ω, is the order of the polynomial; the update time is independent of the interval width $t_i^+ - t_i^-$. This appears to be a promising but as yet unexplored alternative to the gamma memory.

For any kernel function, c, which can be expressed as a weighted sum of gamma kernels,

$$c(t) = \sum_{j=0}^{\Omega} q_j c_{\mu,j}(t). \tag{1}$$

Where q_j are the weighting coefficients, the resulting memory can be expressed in terms of the gamma memory state vectors:

$$\bar{\mathbf{x}}(t) = \sum_{j=0}^{\Omega} q_j \bar{\mathbf{x}}_{\mu,j}.$$

de Vries and Principe (1991) show that the gamma kernels form a basis set, meaning that, for an arbitrary kernel function c that decays exponentially to 0 as $\tau \to \infty$, there exists a value of Ω and a set of coefficients $\{q_j\}$ such that Eq. (1) holds. This result is easy to intuit for the case of delay kernels, but one can see from this simple case that the required computation amounts essentially to convolving a complex kernel with the input sequence at each time step and, consequently, is not of great practical utility.

CONTENTS OF SHORT-TERM MEMORY

Having described the *form* of the memory, I now turn to the *content*. Although the memory must hold information pertaining to the input sequence, it does not have to be a memory of the raw input sequence, as was assumed previously. The memory encoding process can include an additional step in which the input sequence, $\mathbf{x}(1)\ldots\mathbf{x}(t)$, is transformed into a new representation $\mathbf{x}'(1)\ldots\mathbf{x}'(t)$, and it is this transformed representation that is encoded in the memory. Thus, $\bar{\mathbf{x}}_i$ are defined in terms of $\mathbf{x}'$, not $\mathbf{x}$.

The case considered in previous sections is an identity transformation,

$$\mathbf{x}'(\tau) = \mathbf{x}(\tau).$$

Memories utilizing this transformation will be called *input* or *I* memories. Models in the neural net literature make use of three other classes of transformation. First, there is a transformation by a nonlinear vector function f,

$$\mathbf{x}'(\tau) = f(\mathbf{x}(\tau)).$$

This transformation results in what I will call a *transformed input* or *TI* memory. Generally f is the standard neural net activation function, which computes a weighted sum of the input elements and passes it through a sigmoidal nonlinearity:

$$f_{\mathbf{w}}(\mathbf{v}) = \frac{1}{1 + e^{-\mathbf{w} \cdot \mathbf{v}}}. \tag{2}$$

Second, the nonlinear transformation can be performed over not only the current input but also the current internal memory state:

$$\mathbf{x}'(\tau) = f(\mathbf{x}(\tau), \bar{\mathbf{x}}_1(\tau), \ldots, \bar{\mathbf{x}}_\Omega(\tau)). \tag{3}$$

This leads to a *transformed input and state* or *TIS* memory. Such memories can be implemented in a recurrent neural net architecture in which $\bar{\mathbf{x}}_i$ and $\mathbf{x}'$ correspond to activities in two layers of hidden units (Figure 4(a)). Note that this architecture is quite similar to a fairly common, recurrent, neural net sequence-processing architecture (Figure 4(b); e.g., Elman, 1990; Mozer, 1989), which I'll refer to as the *standard* architecture.

Third, for *autopredictive* tasks in which the target output, $\mathbf{p}(\tau)$, is a one-step prediction of the input; i.e.,

$$\mathbf{p}(\tau) = \mathbf{x}(\tau + 1),$$

one can consider an alternative content to the memory. Rather than holding the actual sequence value, the preceding prediction can be used instead:

$$\mathbf{x}'(\tau) = \mathbf{p}(\tau - 1).$$

This transformation will be called an *output* or *O* memory. Of course, *TO* and *TOS* memories can be constructed by analogy to the TI and TIS memories, suggesting a characterization of memory content along two subdimensions, one being the transformation applied and the other being whether the transformation is applied to the input or previous output. Since it does not make a great deal of sense to ignore input sequence information when it is available, I will include in the category of output memories those that combine input and output information, e.g., by taking a difference between input and output, $\mathbf{x}(\tau + 1) - \mathbf{p}(\tau)$, or by using the input information when it is available, such as during training, and the output information otherwise.

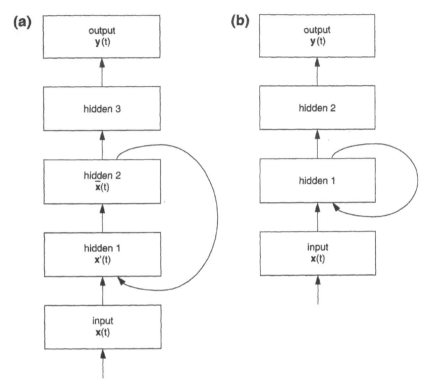

FIGURE 4   (a) A TIS memory in a neural net architecture. Each rectangle represents a layer of processing units, and each arrow represents complete connectivity from one layer to another. The input layer corresponds to the current element of the time series, $x(t)$, the first hidden layer to the TIS representation, $x'(t)$, and the second hidden layer to the memory state, $\{\bar{x}_i(t)\}$. (b) A standard recurrent neural net architecture, which is like the TIS architecture except that the first two hidden layers are collapsed together; i.e., $x'(t) = [\bar{x}_1(t) \dots \bar{x}_\Omega(t)]$.

MEMORY ADAPTABILITY

The final dimension that characterizes short-term memories is the *adaptability* of the memory. A memory has various parameters—$\{\mu_i\}$, $\{\Omega_i\}$, and the w of Eqs. (2) and (3)—that must be specified. If the parameters are fixed in advance, one can call the memory *static* because the memory state, $\{\bar{x}_i(t)\}$, is a predetermined function of the input sequence, $x(1) \dots x(t)$. The task of the neural net is make the best predictions possible given the fixed representation of the input history. In contrast, if the neural net can adjust memory parameters, the memory representation is *adaptive*. Essentially, by adjusting memory parameters, the neural net selects, within limits on the capacity of the memory, what aspects of the input sequence are available for making predictions. In addition to learning to make predictions, the neural net must also learn the characteristics of the memory—as specified by its parameters—that best facilitate the prediction task.

The memory model in Figure 2 is an example of a static memory. The model sketched in Figure 4 is intrinsically an adaptive memory because without training the hidden unit responses, the network seems unlikely to preserve sufficient relevant information.

Interesting cases of adaptive memory models in the neural net literature include: the learning of delays (Bodenhausen & Waibel, 1991; and Unnikrishnan, Hopfield, & Tank, 1991), the learning of exponential decay rates (Bachrach, 1988; Frasconi, Gori, & Soda, 1992; and Mozer, 1989), and the learning of gamma memory parameters (de Vries & Principe, 1992). Every sequence-processing, recurrent, neural net architecture trained with *backpropagation through time* (or *BPTT*; Rumelhart, Hinton, & Williams, 1986) or *real-time recurrent learning* (or *RTRL*; Robinson & Fallside, 1987; Schmidhuber, 1992; and Williams & Zipser, 1989) is also an adaptive memory model in that the training adjusts the **w** of Eq. (2). The Elman SRN algorithmSRN (Elman, 1990) lies somewhere between an adaptive and static memory because the training procedure—which amounts to backpropagation one step in time—is not as powerful as full-blown backpropagation through time or RTRL.

Static memory models can be a reasonable approach if there is adequate domain knowledge to constrain the type of information that should be preserved in the memory. For example, Tank and Hopfield (1987) argue for a memory that has high resolution for recent events and decreasing resolution for more distant events. The argument is based on a statistical model of temporal distortions in the input. If there is local temporal uncertainty in the occurrence of an input event, then the uncertainty relative to the present time increases with the expected temporal lag of the event. The decreasing-resolution-with-increasing-depth memory suggested by this argument can be achieved by a gamma family, $\{\bar{\mathbf{x}}_{\mu,\omega} : 0 \leq \omega \leq \Omega\}$. As ω increases, the state vector $\bar{\mathbf{x}}_{\mu,\omega}$ represents an increasing-depth, decreasing-resolution trace of the input.

MEMORY TAXONOMY

I have described three dimensions along which neural net memory models vary: memory form (delay line, exponential trace, gamma trace), content (I, TI, TIS, O, TO, TOS), and adaptability (static, adaptive). The Cartesian product of these dimensions yields 36 distinct classes.

Table 1 presents some combinations of memory form and content, along with existing models in the literature of each class.[1] The I-delay memory is the simplest class and corresponds to a feedforward network with a delay space embedding of the input sequence. TI-delay memories are the basis of the TDNN (Waibel et al., 1989) and the FIR neural net (Wan, this volume). In these models, each hidden unit— which is a nonlinear transformation of the input—maintains a history of its n most recent values, and all these values are available to the next layer. TI-delay models have also been studied extensively within the more physically oriented neural net community (Kleinfeld, 1986; Sompolinksy & Kanter, 1986; and for an overview, see Kühn & van Hemmen, 1991). Herz (1991) proposes a TIS-delay memory in a network whose dynamics are governed by a Lyapunov function under certain symmetry conditions on the time-delayed weights. Connor, Atlas, and Martin (1992) study a nonlinear, neural network ARMA model, whose MA component is constructed from an O-delay memory that retains the difference between the predicted and actual outputs (the latter being identical to the next input). In general, nonlinear $MA(q)$ models can be based on O-delay memories.

Moving to the second column in Table 1, Bachrach (1988), Frasconi, Gori, and Soda (1992), and Mozer (1989) have proposed an TI-exponential memory model and described a computationally efficient algorithm for adapting the μ_i parameters. Mozer's (1992) multiscale integration model is a TIS-exponential memory. The motivation underlying this work was to design recurrent hidden units that had different time constants of integration, the slow integrators forming a coarse but global sequence memory and the fast integrators forming a fine-grained but local memory. This memory was adaptive in that the hidden-unit response functions were learned but static in that the μ_i were fixed. Jordan (1987) has incorporated an O-exponential memory in a sequence production network. During training, the network is given target output values, and these are fed back into the output memory; during testing, however, the actual outputs are used instead.

As the third column of the table suggests, the gamma memory is the least studied form. de Vries and Principe's initial simulations have been confined to an I-gamma memory, although they acknowledge other possibilities.

Beyond the simple memory classes, hybrid schemes also can be considered. The TDNN, for example, is an I-delay memory attached to a series of TI-delay memory modules. I have proposed architectures combining an I-delay memory with a TIS-exponential memory (Mozer & Soukup, 1991) and an I-delay memory with a TI-exponential memory (Mozer, 1989). The possibilities are, of course, diverse, but the certain conclusion is that the best class of memory depends on the domain and the task.

---

[1] The TO and TOS varieties have been omitted because they have received little study.

TABLE 1 Taxonomy of neural net architectures for temporal processing

memory contents	memory form		
	delay	exponential	gamma
I	Elman & Zipser, 1988; Lapedes & Farber, 1987; Zhang & Hutchinson, this volume	?	de Vries & Principe, 1991, 1992
TI	Waibel et al., 1989; Wan, this volume; Kleinfeld, 1986; Sompolinsky & Kanter, 1986	Bachrach, 1988; Frasconi, Gori, & Soda, 1992; Mozer, 1989	?
TIS	Herz, 1991	Mozer, 1992	?
O	Connor, Atlas, & Martin, 1992	Jordan, 1987	?

While this taxonomy represents a reasonable first cut at analyzing the space of memory models, it is not comprehensive and does not address all issues in the design of a memory model. Although some models in the literature can be pigeonholed into a certain cell in the taxonomy, they have critical properties that are ignored by the taxonomy. For example, Schmidhuber (1992; Schmidhuber, Prelinger, & Mozer, in preparation) has proposed a type of I-delay memory in which the memory size and the ω_i delay parameters are dynamically determined based on the input sequence as it is processed. The idea underlying this approach is to discard redundant input elements in a sequence. (See Myers, 1990, and Ring, 1991, for variants of this idea.)

For the most part, I also have sidestepped the issue of network dynamics, assuming instead the typical backpropagation-style activation dynamics. There is an interesting literature on temporal associative networks that seek energy minima (e.g., Herz, 1991; Kleinfeld, 1986; and Sompolinsky & Kanter, 1986) and networks that operate with continuous time dynamics (Pearlmutter, 1989). However, this work more directly addresses the issue of network architecture, not representation, and as I indicated at the outset, I have focused on representational issues.

THE INADEQUACY OF STANDARD, RECURRENT, NEURAL NET ARCHITECTURES

The standard, recurrent, neural net architecture for sequence-processing tasks in Figure 4(b) amounts to a trivial case of a TIS-exponential memory with $\mu_i = 0$ or, equivalently, a TIS-delay memory with $\Omega = 0$. I will refer to this limiting case as a *TIS-0* memory. Although I and others have invested many years exploring TIS-0 architectures, the growing consensus seems to be that the architecture is inadequate for difficult temporal-processing and prediction tasks.

The TIS-0 architecture seems sufficiently powerful *in principle* to handle arbitrary tasks. One is tempted to argue that, with enough hidden units and training, the architecture should be capable of forming a memory that retains the necessary task-relevant information. This is achieved by adjusting the **w** parameters in Eqs. (2) and (3) using a gradient-descent procedure. *In practice*, however, many have found that gradient descent is not sufficiently powerful to discover the sort of relationships that exist in temporal sequences, especially those that span long temporal intervals and that involve extremely high order statistics. Bengio, Frasconi, and Simard (1993) present theoretical arguments for inherent limitations of learning in recurrent networks. Mozer (1989, 1992) illustrates with examples of relatively simple tasks that cannot be solved by TIS-0 memories. Although TIS-0 memories were inadequate for these tasks, TI-exponential or TIS-exponential memories yielded significantly improved performance. In both cases, the exponential trace components allowed the networks to bridge larger intervals of time. As further evidence, consider the active neural net speech-recognition literature. I know of no serious attempts at speech recognition using TIS memories. However, the TDNN architecture—a TI-delay memory—is viewed as quite successful. There have even been interesting explorations using TI-exponential memories (Bengio, De Mori, & Cardin, 1990; and Watrous & Shastri, 1987).

With TIS-0 architectures, gradient descent often cannot find meaningful values for the recurrent-connection strengths. Gradients computed by backpropagation, using either BPTT or RTRL, tend to be quite small.[2] For an intuition as to why this is so, consider the network "unfolded" in time, which is necessary for computing the gradients with BPTT. In the unfolded network, there is a copy of the original network (units and connections) for each time step of the temporal sequence. This turns the recurrent network into a deeply layered feedforward network. Now consider the consequence of changing a weight embedded deep in the unfolded network, say, a weight connecting a hidden unit at time 1 to a hidden unit to time 2. Although it will have some consequence at a later time t, the effect is likely to be small, hence so will the gradient. Further, the impact of a connection weight—even

[2] With careful selection of the weights, one can prevent the gradients from shrinking through time (Hochreiter, 1991); however, this imposes serious restrictions on the sorts of memories the network can form.

if appropriate—will be masked by other weights if their values are inappropriate. This situation is true for feedforward nets as well as recurrent nets, but the problem is exacerbated by the deeply layered structure of the unfolded net. The result is an error surface fraught with local optima.

While this situation may hold for TIS-0 memories, it may not be the case for other recurrent architectures that have a more specialized connectivity, e.g., TI-exponential memories. Further, used *in conjunction with* other types of memory that may help bootstrap learning, TIS-0 memories are a promising possibility. It also seems well worth investigating classes of recurrent network memory in the taxonomy that have yet to be explored. To the best of my knowledge, these classes include TIS-delay, TI-gamma, TIS-gamma, nonlinear MA models of the O-delay variety, as well as MA models based on O-gamma and O-exponential and the TO and TOS varieties.

EXPERIMENTS WITH A FINANCIAL DATA SERIES

In the remainder of this chapter, I present experiments examining a prediction task from financial data series using three memory types: I-delay, TIS-0, and a hybrid approach combining I-delay and TIS-0. Although these experiments do not explore the more sophisticated memory classes described earlier, they provide a baseline against which future experiments can be compared.

The data series that I studied was Data Set C of the Santa Fe Time Series Prediction and Analysis Competition, the tickwise bids for the exchange rate from Swiss francs to U.S. dollars, recorded by a currency trading group from August 1990 to April 1991. "Tickwise" means that the samples come at irregular intervals of time, as demanded by the traders. Each sample is thus indexed by a time of day. There were six prediction tasks in the Competition, but I focused on three: predicting the value of the series 1, 15, and 60 minutes in the future.

The data set appears challenging because the average sampling rate is high relative to the rate of trends in the data. Although the nature of the data is that it is obviously noisy, smoothing is inappropriate because there may be information in the high frequency changes. Consequently, a high-depth memory would seem necessary. This data set seems ideal for a TIS architecture which, in principle, does not have a fixed memory depth.

The training data consisted of ten contiguous spans of time with unspecified gaps between them, in total 114 days of data. I treated each day's data as an independent sequence, i.e., I assumed no useful information was carried over from one day to the next. For each day, I massaged the data to transform tickwise samples to fixed interval samples, one minute apart, operating on the assumption that the value of the series remained constant from the time of a sample until the time of the next sample. This sequence was subsampled further, at an interval of Δ_s

minutes. For the 1-, 15-, and 60-minute prediction tasks, I used $\Delta_s = 2$, 5, and 10, respectively. The average trading day spanned 575 minutes, resulting in daily series of lengths 57.5 to 287.5. Of the 114 days of training data, 6 complete days—chosen at random—were set aside to be used for validation testing. It was on the basis of this validation data that many parameter values, such as the Δ_s, were selected.

The input to the neural net predictor consisted of three values: the day of the week, the time of day, and the change in the series value from one sample to the next, $s(\Delta_s t) - s(\Delta_s(t-1))$, where $t = 1...T$ is an index over the input sequence and $s(\tau)$ denotes the sample at minute τ from the first sample of a day. The target output of the neural net is a prediction of the change in the series value Δ_p minutes in the future, i.e., $s(\Delta_s t + \Delta_p) - s(\Delta_s t)$.

Each of the input and output quantities is represented using a *variable encoding* (Ballard, 1986), which means that the input and output activities are monotonic in the represented quantity. This is the obvious encoding; I mention it simply because connectionist approaches to representation (e.g., Ballard, 1986; and Smolensky, 1990) offer a range of alternatives that could be fruitfully explored, but I have not done so in the present work.

The activity of each input unit was normalized to have a mean of 0.0 and variance 1.0 over the training set. This leads to a better-conditioned search space for learning (le Cun, Kanter, & Solla, 1991; and Widrow & Stearns, 1985). The target output activity was normalized to have mean 0.0 and variance 100.0 over the training set. Despite this large variance, the observed variance in the unit was much closer to 1.0 because strong predictions could not be made due to the uncertainty in the data.

The general architecture studied, shown in Figure 5, includes two memories—types TIS-0 and I-delay—which feed, along with the inputs directly, into a layer of hidden units, and then map to an output prediction. A symmetric activation function (activation in range -1 to $+1$) was used for the hidden units, including the hidden units of the TIS-0 memory. (Recall that the TIS-0 memory is implemented as a recurrent network with internal hidden units; see Figure 4(b).) The output unit had a linear activation function.

The actual architectures tested were three special cases of this general architecture. The first case eliminated the TIS-0 memory and used a six-step I-delay memory. The number of hidden units was varied. When the number of hidden units is 0, this architecture reduces to a linear autoregressive model. The second case eliminated the I-delay memory in favor of the TIS-0 memory. The TIS-0 memory consisted of 15 internal hidden units, plus 5 hidden units in the penultimate layer. The third case—a hybrid architecture—included both the TIS-0 and I-delay memories, with parameters as above. The choice of number of hidden units and number of delays was based on validation testing, as described below.

The various networks were trained using stochastic gradient descent, whereby the weights were updated following presentation of an entire day's data. Backpropagation through time (Rumelhart, Hinton, & Williams, 1986) was used to train

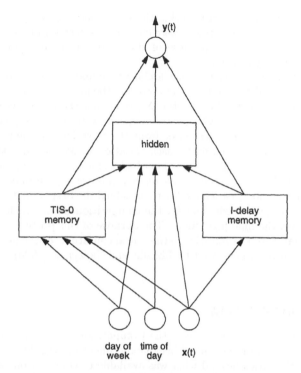

FIGURE 5   General architecture used for time series experiments. The figure depicts three inputs feeding into two types of memory which then map, through a layer of hidden units, to a single output—the prediction unit. Arrows represent connections between units or sets of units. The architecture includes direct connections from "day of week" and "time of day" to the output, which are not depicted.

the recurrent connections in the TIS-0 memory. Sufficiently small learning rates were chosen such that the training error decreased reliably over time; generally, I set the learning rate to .0002 initially and decreased it gradually to .000005 in the first 500 passes through the training data. The networks were initialized with weights chosen randomly from a normal distribution, mean zero, and were normalized such that the sum of the absolute values of the weights feeding into a unit totaled 2.0.

Following each pass through the training data, the errors in the training and validation sets were computed. Rather than training the network until the training set error converged, as is commonly done, training was stopped when the validation error began to rise, a method suggested by Weigend, Rumelhart, and Huberman (1990) and Geman, Bienenstock, and Doursat (1992). In actuality, this involved training the network until convergence, observing the point at which the validation error began to rise, and then restoring the saved network weights at this critical point. At this point, ten final passes were made through the training data, annealing the learning rate to 0. The purpose of this step was to fine tune the weights, removing the last bit of jitter.

I explored many variations of the architecture and training procedure, including: varying the number of hidden units, varying the number of delays in the I-delay memory, trying different values for the Δ_s, removing the day of week and time of day inputs, and generating a larger training set by transforming each day's data into Δ_s different streams, each offset from the next by one minute. Generalization performance was estimated for each variation using the validation set.[3] The parameters and choices as previously described yielded performance as good as any.

For each architecture studied, 25 replications were performed with different random initial weights. The ten models that yielded the best validation performance were selected to make predictions for the competition data, which were averaged together to form the final predictions. The purpose of this procedure was to obtain predictions that were less sensitive to the initial conditions of the networks. Similar procedures have been suggested by Lincoln and Skrzypek (1990) and Rosenblatt (1962).

COMPETITION TEST DATA

I did not submit an official entry to the Competition. However, shortly after the close of the competition deadline, I began experimenting with Data Set C, with no knowledge of the series beyond what was available to competition entrants. I used the TIS-0 version of the architecture to obtain 15- and 60-minute predictions for the competition data. Performance on this series is measured in units of *normalized mean squared error* or NMSE. The normalization involved is to divide the mean squared error by the error that would have been obtained assuming that the future value of the series is the same as the last observed value. This normalization term is the least-squares prediction assuming a random walk model for the series. NMSE values less than 1.0 thus indicate that structure has been extracted from the series. After submitting my predictions to Andreas Weigend, he reported back that I'd done a bit better than entrants to the Competition: the 15-minute prediction yielded an NMSE of .859 and the 60-minute prediction an NMSE of .964.

At the Time Series Workshop, Xiru Zhang quite validly argued that the competition data set simply did not contain sufficient test points to ascertain reliability of the predictions. We agreed to perform additional experiments using larger test sets.

---

[3] Because the validation set was also used to determine when to stop training, a bias is introduced in using the validation set for selecting the architecture. In retrospect, two distinct validation sets should have been used.

EXTENDED COMPETITION TEST SET

The first set of experiments utilized the competition training set and, hence, the models developed for the Competition, but test data came from the gaps in the training set. Recall that the training set consisted of ten contiguous time spans with gaps in between. The gaps covered 69 days, more than sufficient for an extended test set. All data in the gaps was used for testing, with the exception of the first hour of each day, in order to provide an initialization period for the recurrent TIS-0 architecture.

The results for 1-, 15-, and 60-minute predictions are shown in Table 2. The I-delay architecture was tested with different numbers of hidden units and the TIS-0 model in the configuration described earlier. The I-delay architecture with zero hidden units is equivalent to a linear autoregressive model, AR(6). To a first order, all architectures performed the same, slightly better than a random walk model and also better than Zhang and Hutchinson (this volume). I attempted to determine the reliability of differences in the NMSEs but did not get very far. A straight t-test on the squared errors for the individual predictions yielded no significant differences. Noting that the data distribution strongly violated the assumption of normality, I applied a log transform to the squared errors, which still did not achieve a normal distribution but came a bit closer. After the log transform, reliable differences between NMSEs were obtained for pairs of conditions whose NMSEs differed by roughly .001 or more. No matter, performance was discouragingly poor in all conditions.

Finnoff, Hergert, and Zimmermann (1992) report performance comparisons across a broad spectrum of tasks for a variety of different training techniques. While their explorations included variations on the techniques I used, they found that a relatively simple technique—stopping training based on a validation set and then training further using weight decay—appeared to work very well across tasks.

TABLE 2 NMSE for extended competition test set

architecture	1-minute prediction (18,465 data points)	15-minute prediction (7,246 data points)	60-minute prediction (3,334 data points)
I-delay, 0 hidden	.998	1.001	.999
I-delay, 1 hidden	.998	1.001	.996
I-delay, 3 hidden	.998	1.000	.998
I-delay, 5 hidden	.998	1.000	.998
I-delay, 10 hidden	.998	.999	.999
TIS-0	.998	.999	.997

262 Michael C. Mozer

TABLE 3 NMSE for 1985–87 test data

architecture	1-minute prediction (57,773 data points)
I-delay, 0 hidden	.999
I-delay, 5 hidden	.985
I-delay, 10 hidden	.985
I-delay, 20 hidden	.985
TIS-0	.986
hybrid TIS-0 and I-delay	.986

I experimented with this method, but obtained results essentially the same as those reported in Table 2. They also used a much larger validation set. Being concerned that my small validation set was responsible for performance, I ran further experiments in which the validation corpus was increased to 25% of the training set, but this manipulation also had little effect on performance.

THE 1985–1987 CORPUS

The extended competition test set was drawn from the same time period as the training data. Xiru Zhang suggested an additional data set in which the testing corpus came from a different time period than the testing corpus. He proposed training data from the same exchange rate series spanning the time period 6/85–5/86 and testing data spanning the time period 6/86–5/87. This encompassed 248 days of training data and a comparable amount of testing data. I selected 37 days of the training data—roughly 15%—for validation testing. Following Zhang and Hutchinson, I explored only the one-minute prediction task. The first hour of each day's data was excluded from testing, for reasons indicated earlier.

The results are summarized in Table 3 for a variety of architectures. Here, the I-delay architecture with 0 hidden units—equivalent to a linear AR(6) model—performed significantly worse than the other architectures, but there was no reliable difference among the other architectures.

Zhang and Hutchinson report an NMSE of .964 on the same data set, better than the .985 I obtained. However, it is important to emphasize that the two performance scores are *not* comparable. Zhang and Hutchinson threw out cases—from both training and test sets—for which a tickwise series value was not available 60 ± 30 seconds from the time at which a prediction was made. Hence, their predictions were *conditional on assumptions about the time of occurrence of future ticks*. At the point at which a prediction is made, they cannot possibly know whether the assumptions are valid. Therefore, their approach cannot be used as a predictive model without a secondary component that predicts whether their prediction criteria will be satisfied. Alternatively, they could retrain and retest their network on all data, not just the data satisfying their criterion.[4]

Making such a conditional prediction is likely a simpler task than making an unconditional prediction, because it ensures some homogeneity among the cases. In effect, it throws out a large number of the "no change" predictions; in my test set, the fraction of "no change" future values was 71% versus only 41% in Zhang and Hutchinson's set.

A perhaps more interesting measure of performance is whether the network can correctly predict the change in direction of the series: down, no change, or up. To explore this measure, it is necessary to quantize the network outputs, i.e., to specify a criterion, call it c^+, such that all predictions greater than c^+ are classified as "up," and a second criterion, c^-, such that all predictions less than c^- are classified as "down." Predictions between c^- and c^+ are classified as "no change." The principled approach to selecting c^- and c^+ involves selecting the values that yield the best classification performance on the training set; this involves an exhaustive search through all possible values. Instead, I used a simple technique to determine c^+: c^+ was chosen such that the fraction of training set predictions greater than c^+ equaled the fraction of training cases whose target prediction was "up," and likewise for c^-. Using these values of c^- and c^+ on the test set, the 10-hidden I-delay network yielded a performance of 63.2% correct predictions. This sounds pretty good, but in reality is quite poor because 71% of all cases were "no change" and the net could have obtained this performance level simply by always responding "no change." The network has a much greater ability to predict the direction of change *conditional upon a change having occurred*: It correctly predicts the direction on 58.5% of test cases, whereas always predicting the most frequent category in the training set obtains a performance of only 49.9%. If one is seriously interested in predicting direction of change, it seems sensible to train a network with three output units, one for each mutually exclusive prediction, perhaps using a normalized exponential output function (Bridle, 1990; Rumelhart et al., 1993) to obtain a probability distribution over the alternative responses.

CONCLUSIONS

In this work, I have begun an exploration of alternative architectures for temporal sequence processing. In the difficult time series prediction problem studied, a non-linear extension of autoregressive models—the basic I-delay architecture—fared no worse than a more complex architecture, the hybrid I-delay and TIS-0. However, it is impossible to determine at present whether the prediction limitations I found

[4] In defense of Zhang and Hutchinson's decision, they believed that the competition set-up implied there would be a value near the prediction time, and thus it was to their advantage to make use of this information.

are due to not having explored the right architecture, the right preprocessing or encoding of the input sequence, the right training method, or are simply due to the absence of further structure in the data that can be extracted from the series alone without the aid of other financial indicators. The studies of Zhang and Hutchinson (this volume) of the same data series using an I-delay memory but with a different input/output encoding and training method yielded comparable results, supporting the hypothesis that the encoding and method are not responsible for the poor predictability. At very least, my results can serve as a baseline against which researchers can compare their more sophisticated architectures, representations, and training method. I look forward to reports showing improvements over my work.

My aim in presenting a taxonomy of architectures is to suggest other possibilities in the space to be explored. I don't claim that the taxonomy is all encompassing or that it is necessarily the best way of dividing up the space of possibilities. However, it seems like a reasonable first attempt in characterizing a set of models for temporal processing commonly used in the field.

ACKNOWLEDGMENTS

My thanks to Andreas Weigend and Neil Gershenfeld for organizing the Santa Fe workshop, and in particular to Andreas for his abundant feedback at all stages of the work and suggestions for extensions. Jim Hutchinson, Radford Neal, Jürgen Schmidhuber, and Fu-Sheng Tsung, and an anonymous reviewer provided very helpful comments on an earlier draft of the chapter. This research was supported by NSF Presidential Young Investigator award IRI–9058450, grant 90–21 from the James S. McDonnell Foundation, and DEC external research grant 1250.

Andrew M. Fraser and Alexis Dimitriadis
Systems Science Ph.D. Program, Portland State University, Portland, Oregon 97207-0751;
email: andy@ee.pdx.edu

Forecasting Probability Densities by Using Hidden Markov Models with Mixed States

We note similarities of the state-space reconstruction ("embedology") practiced in numerical work on chaos, state-space methods of stochastic systems theory, and the hidden Markov models (HMMs) used in speech research. We review Baum's EM algorithm in general and the specific forward-backward algorithm that optimizes a class of HMM that has a mixed state space consisting of continuous and discrete parts. We then describe forecasts based on models fit to Data Set D.

1. INTRODUCTION

In the first part of this paper, we hope to provide an intuitive explanation of hidden Markov model (HMM) methods that builds on the notion of state-space reconstruction. Later, we provide enough details about the approach to enable a careful reader to develop new variants and to write his own programs. We begin with some thoughts on forecasting and state spaces. We were drawn to work on varieties of hidden Markov models for scalar time series from chaotic dynamics, by the similarity of the notion of *reconstructed state space* in the chaos literature to the notion of *hidden state* in the HMM literature. The approach provides forecasts that consist

Times Series Prediction: Forecasting the Future and
Understanding the Past, Eds. A. S. Weigend and N. A. Gershenfeld, SFI Studies
in the Sciences of Complexity, Proc. Vol. XV, Addison-Wesley, 1993   **265**

of probability densities instead of single guesses of future values. In Section 6, such a forecast of Data Set D suggests that the approach is quite powerful.

The most direct method of forecasting is to search the past for times when conditions matched the patterns of recent observations and then guess that what happened before will happen again. While forecasts for discrete-valued periodic sequences like $(0, 1, 2, 3, 0, 1, 2, 3, 0, 1, \ldots)$ are trivial, forecasting sequences like Data Set D, $(0.643, 0.558, 0.484, 0.434, 0.422, \ldots)$, is difficult because there is no segment in the recorded history that exactly matches recent observations. In such circumstances one may proceed by *guessing* how *close* conditions at various times in the past are to present conditions and then appropriately averaging near matches to make forecasts; here *closeness* corresponds to distance in *state space*, and the guesses are implemented by conditional probability density functions for location in state space, given observations.

The notion of state space has also been essential in the efforts over the past dozen years in which researchers have claimed that various experimental time series arise from chaotic dynamics. Characteristically, scalar time series are converted to vector time series in a procedure called *state-space reconstruction*. The simplest procedure is the use of *delay vectors*. Using a notation in which a sequence of scalar observations is denoted by $y_1^T \equiv (y(1), y(2), \ldots, y(T))$, we write delay vectors as $\mathbf{x}(t) \equiv y_{t-m}^{t-1}$, and observe that $\mathbf{x}_{m+1}^T$ can be obtained from y_1^T. Having reconstructed a vector time series, investigators generally argue that the dynamics are deterministic, i.e., $\exists \; \mathcal{F} \colon \mathbf{R}^m \to \mathbf{R}^m$ such that $\mathbf{x}(t+1) = \mathcal{F}(\mathbf{x}(t)) \; \forall t$, and then estimate invariants such as dimensions, entropies, or Lyapunov exponents from experimental measurements.

While the practice of explaining scalar time series in terms of vector dynamics and using observed time series segments to specify locations in state space has a long history,[1] the more recent literature in chaos usually cites Packard et al. (1980) or Takens (1981). The idea is that an "original" state-space variable $\mathbf{z}$ evolves via a diffeomorphism[2] F of some low-dimensional manifold $\mathbf{Z}$ and gives rise to a scalar observable $y \in \mathbf{R}$

$$\mathbf{z}(t+1) = F(\mathbf{z}(t)) \tag{1a}$$
$$y(t) = g(\mathbf{z}(t)), \tag{1b}$$

and a reconstruction function $\phi : \mathbf{R}^w \to \mathbf{R}^m$ is used to map windows of observations y_{t-w}^{t-1} of size w to reconstructed vectors $\mathbf{x}(t) \in \mathbf{X} = \mathbf{R}^m$. Combining ϕ, F^{-1}, and g one can write $\Phi : \mathbf{Z} \to \mathbf{X}$ with

$$\mathbf{x}(t) = \Phi(\mathbf{z}(t)) = \phi\left(g(F^{-1}(z(t))), g(F^{-2}(z(t))), \ldots, g(F^{-w}(z(t)))\right) .$$

[1]Most control theory text books have a section titled *Observability* which addresses the question of whether or not sequences of observations uniquely determine locations in state space.
[2]A one-to-one differentiable function with a one-to-one differentiable inverse.

Takens showed that if g and ϕ are differentiable and m is large enough, then it is a generic property that Φ is a diffeomorphism, and thus one expects coordinate invariant properties of trajectories and limit sets in $\mathbf{Z}$ to be the same as the properties of their images in $\mathbf{X}$. Although Takens' result is insensitive to the details of ϕ and g, when experimenters implemented the procedure, they found that their estimates of invariants varied with changes in ϕ and g.

This variability of invariants lead to a literature (recently called "embedology" (Sauer et al., 1991)) concerned with defining and finding *good reconstructions* (Fraser, 1989a). The variability is usually explained by observing that the procedures used to estimate invariants converge in the limit of infinite amounts of noise-free data, but that experimenters work with short noisy data sets instead. While it may be true that different applications lead to different optimal reconstructions, we suspect[3] that practical embedology will best be developed as an aspect of modeling techniques for optimizing likelihood or a variant such as MDL or AIC.

For noisy or stochastic dynamics and observations, Eq. (1) is not appropriate. Instead, the y's are functions of a Markov process and are characterized by the conditional densities[4] $P_{\mathbf{z}(t+1)|\mathbf{z}(t)}$ and $P_{y(t)|\mathbf{z}(t)}$ with

$$P\left(\mathbf{z}(t+1)|\mathbf{z}_{-\infty}^t, y_{-\infty}^t\right) = P\left(\mathbf{z}(t+1)|\mathbf{z}(t)\right) \tag{2a}$$

$$P\left(y(t)|\mathbf{z}_{-\infty}^t, y_{-\infty}^\infty\right) = P\left(y(t)|\mathbf{z}(t)\right). \tag{2b}$$

Given these conditional density functions and a natural measure or stationary density μ with $\mu(\mathbf{z}') = \int P_{\mathbf{z}(t+1)|\mathbf{z}(t)}(\mathbf{z}'|\tilde{\mathbf{z}})\mu(\tilde{\mathbf{z}})d\tilde{\mathbf{z}}$, forecasting formally reduces to iterating a recursion

$$P\left(y(T+1)|y_1^T\right) = \int P\left(y(T+1)|\mathbf{z}(T+1)\right) P\left(\mathbf{z}(T+1)|y_1^T\right) d\mathbf{z}(T+1) \tag{3a}$$

$$P\left(\mathbf{z}(T+1)|y_1^T\right) = \int P\left(\mathbf{z}(T+1)|\mathbf{z}(T)\right) P\left(\mathbf{z}(T)|y_1^T\right) d\mathbf{z}(T) \tag{3b}$$

$$P\left(\mathbf{z}(T)|y_1^T\right) = \frac{P\left(y(T)|\mathbf{z}(T)\right) P\left(\mathbf{z}(T)|y_1^{T-1}\right)}{P\left(y(T)|y_1^{T-1}\right)} \tag{3c}$$

starting with $P(\mathbf{z}(1)) = \mu$. The key idea is that in state space there is an evolving cloud of locations that are consistent with observations up to the time t; i.e.,

---

[3] This was suggested to us by Henry Abarbanel (Abarbanel and Kadtke, 1990) and is similar to the notion of Casdagli et al. (1991) that reconstruction and prediction are related.

[4] Our notation blurs the distinction between probability distributions of discrete variables and probability densities of continuous variables. For the probability at x, we use the notation $P(x)$, for the function, we use P_x; i.e., we use subscripts to specify a function and parentheses to denote the value of a function at a point. We resolve ambiguities such as $P(5.23)$ by using subscripts or $=$, e.g., $P_x(5.23)$ or $P(x{=}5.23)$. We attempt to balance opacity and imprecision in our notation.

$P(\mathbf{z}(t)|y_1^t)$. If the conditional densities of Eq. (2) correspond to adding Gaussian noise to Eq. (1), they can be written as

$$P(\mathbf{z}(t{+}1)|\mathbf{z}(t)) = \left[\frac{|\Sigma^{-1}|}{\sqrt{2\pi}}\right]^n \exp\left(\frac{-\left[\mathbf{Z}(t{+}1) - F(\mathbf{Z}(t))\right]\Sigma^{-1}\left[\mathbf{Z}(t{+}1) - F(\mathbf{Z}(t))\right]}{2}\right) \quad (4a)$$

$$P(y(t)|\mathbf{z}(t)) = \frac{1}{\sqrt{2\pi\sigma_y^2}} \exp\left(\frac{[y(t) - G(\mathbf{Z}(t))]^2}{2\sigma^2}\right) \quad (4b)$$

where Σ^{-1} is an inverse covariance matrix. Further, if the functions F and G are linear, the recursion of Eqs. (3a–c) constitutes Kalman filtering.[5]

The simplicity of the Eqs. (3a–c) obscures the following difficulties:

- The conditional densities that specify a stochastic process may be complex, requiring descriptions with infinite numbers of parameters.
- The derived intermediate terms like $P(\mathbf{z}(t)|y_1^t)$ are not simply values, but functions.
- And most importantly, the conditional densities that specify a stochastic process must be estimated on the basis of observations alone.

Rather than attempting to estimate the "true" stochastic process on the basis of observations, in view of these difficulties, we reconsider our pragmatic goals. For forecasting, we are interested only in the y's, the hidden $\mathbf{z}$'s are just computational intermediates. Consequently, we consider a *model* of the process to be a sequence of probability density functions $P_{y_1^t} : \mathbf{R}^t \to \mathbf{R}, \{t = 1, 2, 3, \ldots\}$. In selecting model classes and fitting their parameters, our goal is to obtain a set $\left\{P_{y_1^t} : \forall t \in \mathbf{Z}^+\right\}$ or equivalently $\left\{P_{y(t{+}1)|y_1^t} : \forall t \in \mathbf{Z}^+\right\}$ that performs well in our application (forecasting) rather than discovering a "true" generating mechanism.[6] We are still exploring several model types and, not having carefully accounted for free parameters in our comparisons, we simply use maximum likelihood methods.

2. MODEL CLASSES

In this section we begin by introducing basic discrete-state discrete-output HMMs. Then we turn to mixed state models, of which the hidden-filter hidden Markov models HFHMMs[7] described in Sections 4 and 5 are a special case.

[5] Sorenson (1970) observes that the basic ideas go back to Gauss.

[6] Our view of forecasting has been influenced by Williams' book on data compression Williams 1991) and indirectly by the work of Rissanen (1989).

[7] We called these models autoregressive hidden Markov models (ARHMMs) until we found that Poritz (1988) had already described them and called them HFHMMs.

Equation 2 describes a process in which the observable is a probabilistic function of an underlying Markov process. If the state variables z and the observations y are drawn from discrete finite sets, the process is what is called a hidden Markov model (HMM). Much HMM development work is motivated by applications in natural language. The original work was done in the Communications Research Division of the Institute for Defense Analysis in Princeton and the methods were reviewed at an open symposium in 1980. In the proceedings Ferguson (1980) described HMMs:

> ...a Markov chain with state space $\mathcal{S}$, having S states..., a finite output alphabet, $\mathcal{K}$, which we may take to be the integers $1, 2, \ldots, K$, and a collection of probability distributions. Explicitly, we need a transition matrix (a_{ij}), $i, j \in \mathcal{S}$, where
>
> $$a_{ij} = \text{Prob}\{\text{next state} = j \text{ given current state} = i\}$$
>
> and we need an output probability matrix $(b_j(k))$, $j \in \mathcal{S}, k \in \mathcal{K}$, where
>
> $$b_j(k) = \text{Prob}\{\text{observation} = k \text{ given current state} = j\}$$
>
> For completeness, we need an initial distribution on states, to get us started. Let $(a(i)), i \in \mathcal{S}$ be this distribution.

HMMs are useful because the probability distributions can be adjusted by the Baum-Welch, or forward-backward, algorithm to maximize the likelihood of a given set of training data. We describe the version of the algorithm needed for HFHMMs in Section 5.

The following points about discrete HMMs merit emphasis:

1. Although the hidden process is first-order Markov, the output process may not be Markov of any order.
2. Even if the dynamics and observations (the functions F and G in Eq. (4)) are nonlinear, a discrete HMM can approximate the continuous case arbitrarily well by using large numbers of states S and possible output values K.
3. Larger numbers of training data are required as S and K are increased.

As an illustration of point 1, consider the process described by

$$a = \begin{bmatrix} .9 & .1 & 0 & 0 \\ 0 & 0 & 1 & 0 \\ 0 & 0 & .9 & .1 \\ 1 & 0 & 0 & 0 \end{bmatrix} \qquad b = \begin{bmatrix} 1 & 0 & 0 \\ 0 & 1 & 0 \\ 1 & 0 & 0 \\ 0 & 0 & 1 \end{bmatrix}$$

which produces output strings with runs of about seven 1s interspersed with occasional 2s and 3s. In the output stream, the 2s and 3s alternate no matter how many 1s fall in between. Such behavior cannot be captured by a simple Markov process of any order.

A discrete HMM fit to continuous observations of continuous dynamics (e.g., Eq. (4)) disregards useful properties of the data. The large number of training data in point 3 above can be reduced by preserving a measure of nearness; parameters for situations that do not occur in the training data can be fit on the basis of *interpolations* of nearby situations that do. To build HMM-like models that interpolate in this sense, we introduce the notion of a *mixed state* $\psi(t) = (s(t), \mathbf{x}(t))$ which consists of a discrete part $s \in \{1, 2, \ldots, n_{\text{states}}\}$ and a continuous part $\mathbf{x} \in \mathbf{R}^n$.

As a simplifying assumption, we let the continuous part be a deterministic function of past observations, i.e., $\mathbf{x}(t) = \text{Funct.}(y_1^{t-1})$. The mixed states are meant to summarize histories; thus we assume that $P(y(t)|\psi(t), y_1^{t-1}) = P(y(t)|\psi(t))$ and

$$P(y(t)|y_1^{t-1}) = \sum_{s(t)} P(y(t)|\psi(t)) \, P(\psi(t)|y_1^{t-1}). \qquad (5)$$

By putting all of the uncertainty about location into the discrete part of a mixed state, the integral in Eq. (3a) has been simplified to the sum in Eq. (5). While we doubt natural time series are actually generated by mixed-state processes, we use them as models because they achieve such operational simplifications and their observable aspects, i.e., $\left\{ P_{y(t+1)|y_1^t} : \forall t \in \mathbf{Z}^+ \right\}$, provide high likelihood fits to complex behavior using relatively few free parameters.

In our discussions as we develop models and write programs to implement them, we have found sketches like those in Figure 1 helpful.

3. INCOMPLETE DATA: THE EM ALGORITHM

In this section we review the EM algorithm[8] which adjusts model parameters θ to maximize the likelihood of observations $\mathbf{y}$. It operates on models[9] which include unobserved data $\mathbf{s}$, $P_{\mathbf{y},\mathbf{s},\theta}$. In Section 4 we describe a specific model for time series, and in Section 5 a version of the EM algorithm tailored for that model is presented. The steps in any EM algorithm are:

1. Guess a starting value of θ.
2. Choose $\hat{\theta}$ to maximize[10] $Q(\theta, \hat{\theta}) \equiv \left\langle \log p_{\mathbf{y},\mathbf{s},\hat{\theta}}(\mathbf{y}, \mathbf{s}) \right\rangle_{\mathbf{s}|\mathbf{y},\theta}$

---

[8] Our development follows the 1970 paper by Baum et al. (1970) In a 1977 paper Dempster, Laird, and Rubin (1977) called the procedure the *estimate maximize* algorithm. We recommend Brown's dissertation (1987) for clarity on the subject and Poritz (1988) for a thorough bibliography and historical outline.

[9] For our application, $\mathbf{y}$ is a sequence of observations $(y(1), y(2), \ldots, y(T))$ and $\mathbf{s}$ is a sequence of discrete hidden states $(s(1), \ldots, s(T))$.

[10] For discrete s, this notation means $\langle f \rangle_{s|\mathbf{y},\theta} = \sum_s P_{s|\mathbf{y},\theta}(s|\mathbf{y}) f(s)$.

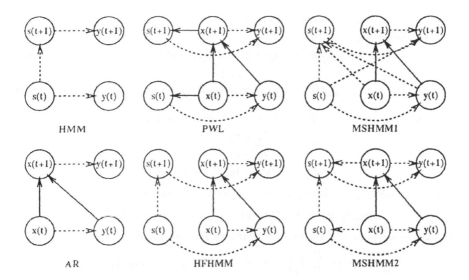

FIGURE 1   Chains of dependency in various model classes. Solid lines indicate deterministic dependence and dotted lines indicate stochastic influence. In each sketch time advances one step with the subsequent conditions appearing above the prior conditions. Given the depicted influences on a node, earlier values of all variables are irrelevant; thus the HMM sketch indicates $P(y(t)|s_1^t, y_1^{t-1}) = P(y(t)|s(t))$ and $P(s(t)|s_1^{t-1}, y_1^{t-1}) = P(s(t)|s(t-1))$. An AR model can be written as $P(y(t)|\mathbf{x}(t)) = (1/\sqrt{2\pi\sigma^2}) \exp -((y(t) - \mathbf{a} \cdot \mathbf{x}(t))^2)/2\sigma^2$ with $\mathbf{x}(t) = F(x(t-1), y(t-1))$ defined by $\mathbf{x}_1(t) = y(t-1)$ and $\mathbf{x}_i(t) = \mathbf{x}_{i-1}(t-1) : 1 < i \leq m$. In the sketch, the solid lines indicate the function F and the dotted lines indicate $P_{y(t)|\mathbf{x}(t)}$. In a piecewise linear (PWL) model, the history vector $\mathbf{x}$ determines the partition element s and thus the linear rule $\hat{y}(t) = \mathbf{a}_s \cdot \mathbf{x}(t)$, then $P_{y(t)|\mathbf{x}(t),s(t)}$ is Gaussian with mean $\hat{y}(t)$ and variance σ_s^2. HFHMMs are discussed in Sections 4 and 6 and type 1 mixed-state hidden Markov models (MSHMM1) are discussed in Section 6.

3. Set $\theta = \hat{\theta}$. If not converged, go to 2.

This procedure will work if,

$$Q(\theta, \hat{\theta}) > Q(\theta, \theta) \Rightarrow P_{\hat{\theta}}(\mathbf{y}) > P_\theta(\mathbf{y}) \tag{6a}$$

and

$$\max_{\hat{\theta}} Q(\theta, \hat{\theta}) = Q(\theta, \theta) \Rightarrow \max_{\hat{\theta}} P_{\hat{\theta}}(\mathbf{y}) = P_\theta(\mathbf{y}). \tag{6b}$$

The truth of the first implication (6a) is shown as follows:

$$P_{\hat{\theta}}(\mathbf{y}) = \frac{P_{\hat{\theta}}(\mathbf{s}, \mathbf{y})}{P_{\hat{\theta}}(\mathbf{s}|\mathbf{y})}$$

$$\log P_{\hat{\theta}}(\mathbf{y}) = \log P_{\hat{\theta}}(\mathbf{s}, \mathbf{y}) - \log P_{\hat{\theta}}(\mathbf{s}|\mathbf{y}).$$

Note $\langle \log P_{\hat{\theta}}(\mathbf{y}) \rangle_{\mathbf{s}|\mathbf{y},\theta} = \log P_{\hat{\theta}}(\mathbf{y})$ because $\mathbf{s}$ does not appear inside the $\langle \rangle_{\mathbf{s}|\mathbf{y},\theta}$. So

$$\log P_{\hat{\theta}}(\mathbf{y}) = \langle \log P_{\hat{\theta}}(\mathbf{s}, \mathbf{y}) \rangle_{\mathbf{s}|\mathbf{y},\theta} - \langle \log P_{\hat{\theta}}(\mathbf{s}|\mathbf{y}) \rangle_{\mathbf{s}|\mathbf{y},\theta}. \tag{7}$$

The Gibbs inequality[11] for two distributions P_{θ_1} and P_{θ_2} says

$$\sum_x P_{\theta_1}(x) \log \frac{P_{\theta_2}(x)}{P_{\theta_1}(x)} \leq 0 \quad \text{or} \quad \langle \log P_{\theta_2}(x) \rangle_{\theta_1} \leq \langle \log P_{\theta_1}(x) \rangle_{\theta_1}.$$

So

$$\langle \log P_{\hat{\theta}}(\mathbf{s}|\mathbf{y}) \rangle_{\mathbf{s}|\mathbf{y},\theta} \leq \langle \log P_\theta(\mathbf{s}|\mathbf{y}) \rangle_{\mathbf{s}|\mathbf{y},\theta}.$$

Now, if $\hat{\theta}$ is chosen so that

$$\langle \log P_{\hat{\theta}}(\mathbf{s}, \mathbf{y}) \rangle_{\mathbf{s}|\mathbf{y},\theta} > \langle \log P_\theta(\mathbf{s}, \mathbf{y}) \rangle_{\mathbf{s}|\mathbf{y},\theta}, \tag{8}$$

Eq. (7) yields the implication (6a)

$$P_{\hat{\theta}}(\mathbf{y}) > P_\theta(\mathbf{y})$$

and the algorithm steps uphill.

The second implication, (6b), does not always hold. Since

$$\left[\frac{\partial}{\partial \hat{\theta}} \langle \log P_{\hat{\theta}}(\mathbf{s}, \mathbf{y}) \rangle_{\mathbf{s}|\mathbf{y},\theta} \right]_{\hat{\theta}=\theta} = \left[\frac{1}{P_{\hat{\theta}}(\mathbf{y})} \frac{\partial}{\partial \hat{\theta}} P_{\hat{\theta}}(\mathbf{y}) \right]_{\hat{\theta}=\theta},$$

and

$$\left[\frac{\partial^2}{\partial \hat{\theta}^2} \langle \log P_{\hat{\theta}}(\mathbf{s}, \mathbf{y}) \rangle_{\mathbf{s}|\mathbf{y},\theta} \right]_{\hat{\theta}=\theta} =$$

$$\left[\frac{1}{P_{\hat{\theta}}(\mathbf{y})} \frac{\partial^2}{\partial \hat{\theta}^2} P_{\hat{\theta}}(\mathbf{y}) - \left\langle \left(\frac{\partial}{\partial \hat{\theta}} \log P_{\hat{\theta}}(\mathbf{s}, \mathbf{y}) \right)^2 \right\rangle_{\mathbf{s}|\mathbf{y},\theta} \right]_{\hat{\theta}=\theta},$$

the critical points of $\langle \log P_\theta(\mathbf{s}, \mathbf{y}) \rangle_{\mathbf{s}|\mathbf{y},\theta}$ and $P_\theta(\mathbf{y})$ are the same, but maxima of the former may be saddle points of the latter. But since their basins of attraction are low-dimensional stable manifolds, it is unlikely that the algorithm will get stuck at a saddle point of $P_\theta(\mathbf{y})$.

[11] While many information theory texts attribute this inequality to Kullback and Leibler (1950), it appeared 50 years earlier in chapter XI Theorem II of Gibbs (1902).

4. HIDDEN FILTER HMMS (HFHMMS)

HFHMMs are a class of time series models to which the EM algorithm can be applied. They are diagrammed in Figure 1 and consist of a hidden first-order Markov process on a set of discrete states $\{s_1, s_2, \ldots, s_{\text{nstates}}\}$ and associated with each discrete state is a linear autoregressive output process. The conditional transition probabilities from any state k are given by $P(s(t{+}1)|s(t){=}k)$, and the output distribution at time t given state $s(t) = k$ and history $\mathbf{x} = (y(t{-}1), y(t-2), \ldots, y(t-m))$ is

$$P_{y(t)|s(t),\mathbf{x}(t)}(y|k,\mathbf{x}) = \frac{1}{\sigma_k \sqrt{2\pi}} \exp -\frac{(y - \bar{y}_k - \mathbf{a}_k \cdot \mathbf{x})^2}{2\sigma_k^2}.$$

A model θ consists of all of the parameters for all of the states; for each state k the parameters are: the transition probabilities $P_{s(t{+}1)|s(t)}(j|k)$, and the output distribution parameters $\bar{y}_k$, $\mathbf{a}_k$, and σ_k. We impose the following constraints:

■ Discrete state transitions are Markov and independent of prior outputs[12]

$$P(s(t)|s_1^{t-1}, y_1^{t-1}) = P(s(t)|s(t{-}1)), \tag{9}$$

and $s(1)$ and $\mathbf{x}(1)$ are independent

$$P(s(1), \mathbf{x}(1)) = P(s(1)) P(\mathbf{x}(1)). \tag{10}$$

Using a notation in which $s_q(t)$ is the tth element in the sequence of states q, i.e., $q \equiv (s_q(1), s_q(2), \ldots, s_q(T))$, we can write

$$P(q) = P(s_q(1)) \prod_{t=2}^{T} P(s_q(t)|s_q(t{-}1)).$$

■ Given $\mathbf{x}(t)$ and $s(t{-}1)$, earlier values of s and y are irrelevant,

$$P(y(t), s(t)|y_1^{t-1}, s_1^{t-1}) = P(y(t), s(t)|s(t{-}1), \mathbf{x}(t)), \tag{11}$$

which with Eq. (9) implies

$$P(y(t), s(t)|s(t{-}1), \mathbf{x}(t)) = P(y(t)|s(t), \mathbf{x}(t))P(s(t)|s(t{-}1)). \tag{12}$$

Given y_1^T, a sequence of observations for training, the EM algorithm adjusts the model parameters θ to maximize the likelihood which can be evaluated as

$$P_\theta\left(y_1^T\right) = \sum_q P_\theta\left(y_1^T, q\right),$$

where the variable q runs over all possible state sequences. Step 2 of the algorithm prescribes selecting new parameters $\hat{\theta}$ to maximize the expected log likelihood $\left\langle \log P_{y_1^T, q, \hat{\theta}}(y_1^T, q) \right\rangle_{q|y_1^T, \theta}$, where the expectation is with respect to the conditional distribution $P_\theta(q|y_1^T)$ based on the old parameters θ. We assume $\mathbf{x}(1)$ is available,[13] and use the assumptions to write:

$$P_{\hat{\theta}}(y_1^T, q) = P_{\hat{\theta}}(y(1), s_q(1)|\mathbf{x}(1)) \prod_{t=2}^{T} P_{\hat{\theta}}(y(t), s_q(t)|s_q(t-1), \mathbf{x}(t-1))$$

$$= P_{\hat{\theta}}(s_q(1)) \prod_{t=2}^{T} P_{\hat{\theta}}(s_q(t)|s_q(t-1)) \prod_{t=1}^{T} P_{\hat{\theta}}(y(t)|s_q(t), \mathbf{x}(t))$$

$$\log P_{\hat{\theta}}(y_1^T, q) = \log P_{\hat{\theta}}(s_q(1)) + \sum_{t=1}^{T-1} \log P_{\hat{\theta}}(s_q(t+1)|s_q(t))$$

$$+ \sum_{t=1}^{T} \log P_{\hat{\theta}}(y(t)|s_q(t), \mathbf{x}(t))$$

$$= \log P_{\hat{\theta}}(s_q(1)) + \sum_{t=1}^{T-1} \log P_{\hat{\theta}}(s_q(t+1)|s_q(t))$$

$$- \sum_{t=1}^{T} \left\{ \log \sigma_{s_q(t)} + \frac{1}{2} \log 2\pi + \frac{\left(y(t) - \bar{y}_{s_q(t)} - \mathbf{a}_{s_q(t)} \cdot \mathbf{x}(t)\right)^2}{2\sigma^2_{s_q(t)}} \right\},$$

To proceed with the optimization formally, we need $P_\theta(q|y_1^T)$. If we use the notation $w(q) \equiv P_\theta(y_1^T, q)$, then $P_\theta(q|y_1^T) = w(q)/W$, where $W \equiv \sum_{q'} w(q')$. The number of terms in $\sum_{q'} w(q')$ depends exponentially on T, precluding a direct evaluation for T's large enough to be interesting, but the sum *can* be evaluated by the *forward-backward algorithm* which is *linear* in T. We will describe the algorithm in Section 5, but first we write out expressions for the required optimization.

$$W \left\langle \log P_{\hat{\theta}}(y_1^T, q) \right\rangle_{q|y_1^T, \theta} = \sum_q w(q) \log P_{s(1), \hat{\theta}}(s_q(1))$$

$$+ \sum_{q, t=1}^{T-1} w(q) \log P_{\hat{\theta}}(s_q(t+1)|s_q(t)) - \frac{W}{2} \log 2\pi$$

$$- \sum_{q, t=1}^{T} w(q) \left\{ \log \sigma_{s_q(t)} + \frac{\left(y(t) - \bar{y}_{s_q(t)} - \mathbf{a}_{s_q(t)} \cdot \mathbf{x}(t)\right)^2}{2\sigma^2_{s_q(t)}} \right\}$$

[12]This may be appropriate if "the noise scale is at least as large as the discrete states," but it is a bad approximation for noise-free deterministic dynamics.

[13]We also drop $P_{\mathbf{x}(1)}(\mathbf{x}(1))$ in all calculations; i.e., set $P_{\mathbf{x}(1)}(\mathbf{x}(1)) = 1$.

and the maximization can be done separately by maximizing the term

$$F(\hat\theta) \equiv \sum_q w(q) \log P_{\hat\theta}(s(1){=}s_q(1)) + \sum_{q,t} w(q) \log P_{\hat\theta}(s_q(t{+}1)|s_q(t)) \qquad (13)$$

and minimizing the term

$$G(\hat\theta) \equiv \sum_{q,t} w(q) \left\{ \frac{\left(y(t) - \bar y_{s_q(t)} - \mathbf{a}_{s_q(t)} \cdot \mathbf{x}(t)\right)^2}{2\sigma^2_{s_q(t)}} + \log \sigma_{s_q(t)} \right\} \qquad (14a)$$

$$\equiv \sum_{q,t} w(q) g(\hat\theta_{s_q(t)}, y(t), \mathbf{x}(t)) \qquad (14b)$$

In Eq. (14) $\hat\theta_s$ refers to the parameters in the model that are associated with the state s, i.e., $\bar y_s$, $\mathbf{a}_s$, and σ_s. We convert the sum over q and t to a sum over s and t:

$$G(\hat\theta) = \sum_{s,t} \left\{ g(\hat\theta_s, y(t), \mathbf{x}(t)) \sum_q w(q)\delta_{s,s_q(t)} \right\} \qquad (15a)$$

$$\equiv \sum_{s,t} g(\hat\theta_s, y(t), \mathbf{x}(t)) w(s,t) \qquad (15b).$$

The function $w(s,t)$ introduced in Eq. (14) is the total probability, considering all possible paths q, that the system is in state s at time t and that the sequence of outputs y_1^T is produced by the model, i.e.,

$$w(s,t) \equiv P_\theta(s(t), y_1^T).$$

In Section 5 we describe how to calculate $w(s,t)$ using the forward-backward algorithm. Given $w(s,t)$, finding new values for $\bar y_s$, $\mathbf{a}_s$, and σ_s is fairly standard linear fitting; solve for $\mathbf{a}_s$ and $\bar y_s$ by using the SVD method[14] to minimize

$$\chi^2 = \sum_t \left\{ y(t)\sqrt{w(s,t)} - (\bar y_s + \mathbf{x}(t) \cdot \mathbf{a}_s)\sqrt{w(s,t)} \right\}^2,$$

and, defining $W(s) = \sum_t w(s,t)$, set

$$\sigma_s = \sqrt{\frac{1}{W(s)} \sum_t w(s,t)\,(y(t) - \hat y(t))^2}.$$

Introducing the notation

$$w(i,j,t) \equiv P_{s(t+1),s(t),y_1^T,\theta}(i,j,y_1^T),$$

[14] See equation 14.3.16 on page 535 of Press et al. (1988).

and denoting the new discrete transition probabilities $f_{ij} \equiv P_{s(t+1)|s(t),\hat{\theta}}(i|j)$, we observe that optimizing Eq. (13) requires maximizing

$$F_{ij} \equiv \sum_{i,t} w(i,j,t) \log(f_{ij}) \text{ subject to}: \sum_i f_{ij} = 1.$$

The Lagrange multiplier method yields

$$f_{ij} \propto \sum_t w(i,j,t).$$

Selecting the new $P_{s(1),\hat{\theta}}(s)$ is a similar problem, and the solution is given in Eq. (16) of the next section.

5. THE FORWARD-BACKWARD ALGORITHM

The forward-backward algorithm is an EM algorithm specifically for time series. The first steps of the algorithm are two passes through the time series: one "forwards" from $t = 1$ to $t = T$ to calculate α's, and the other "backwards" from $t = T$ to $t = 1$ to calculate β's. The factors $w(s,t)$ and $w(i,j,t)$ used in the previous section, can be evaluated in terms of these α's and β's which are defined as follows:

$\alpha(s,t)$. The probability, based on the model, of the observations up to time t and that the system is in state s at time t:

$$\alpha(s,t) \equiv P(y_1^t, s(t))$$

$\beta(s,t)$. The probability, based on the model, of the observations after time t given that the system is in state s at time t and given the previous observations y_1^t:

$$\beta(s,t) \equiv P(y_{(t+1)}^T | s(t), y_1^t)$$

These definitions and Eqs. (11) and (12) yield

$$w(s,t) = \alpha(s,t)\beta(s,t),$$
$$w(i,j,t) = \alpha(j,t)P(y(t+1)|s(t+1)=i, \mathbf{x}(t+1))P(s(t+1)=i|s(t)=j)\beta(i,t+1),$$

and the recursion formulas

$$\alpha(s,t) = \sum_j \alpha(s_j, t-1)P(s(t)=s|s(t-1)=s_j)P(y(t)|s(t)=s, \mathbf{x}(t)),$$
$$\beta(s,t) = \sum_j \beta(s_j, t+1)P(s(t+1)=s_j|s(t)=s)P(y(t+1)|s(t+1)=s_j, \mathbf{x}(t+1)).$$

Note:

1. For the new model, the initial state probabilities are

$$P_{s(1),\hat{\theta}}(s) \propto P_{s(1),y_1^T,\theta}(s, y_1^T) = \alpha(s,1)\beta(s,1) \tag{16}$$

   subject to normalization.

2. The overall likelihood of the observations can be evaluated after a forward pass via:

$$P_\theta(y_1^T) = \sum_s \alpha(s,T).$$

3. The forward recursion is initialized by:

$$\alpha(s,1) = P(s(1){=}s)\ P(y(1)|s(1){=}s, \mathbf{x}(1))$$

4. The backward recursion is initialized by:

$$\beta(s,T) = 1.$$

5.1 PROGRAMMING TRICKS

If

$$\gamma(t) \equiv \sum_s \alpha(s,t) = P(y_1^t),$$

and the process has an entropy rate h, then

$$\gamma(t) \approx e^{-ht},$$

and something must be done to prevent underflow (overflow) for even moderate values of t. The trick is: at each step in the forward recursion record only $a(t)$ and $c(t)$, and at each step in the backward recursion record only $b(t)$, where

$$c(t) \equiv \frac{\gamma(t)}{\gamma(t{-}1)} \text{ or } \gamma(t) = \prod_{\tau=1}^t c(\tau),$$

$$a(j,t) = \frac{\alpha(j,t)}{\gamma(t)} = P(s(t){=}j|y_1^t),$$

and

$$b(s,t) = \frac{\beta(s,t)\gamma(t)}{\gamma(T)}.$$

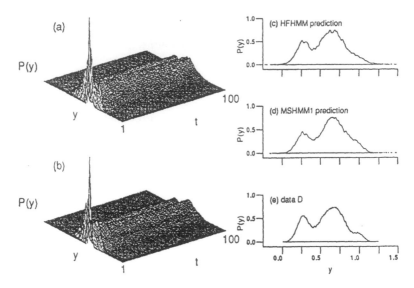

FIGURE 2   Plot (a) is a forecast of Data Set D generated by a HFHMM. The model
has 25 discrete states and the autoregressive filters are eighth order. A MSHMM1
with 20 discrete states and eighth order autoregressive filters generated the forecast
in plot (b). Plots (c) and (d) illustrate the long-term behavior of the models. Each is a
prediction of the distribution of y one hundred steps in the future. The predictions relax
to distributions that are close to the overall distribution of Data Set D, which is plotted
in (e).

Thus for any t

$$a(s,t)b(s,t) = \frac{\alpha(s,t)\beta(s,t)}{\gamma(T)} \propto w(s,t)$$

and

$$\frac{a(j,t)b(i,t{+}1)}{c(t{+}1)} P(y(t{+}1)|s(t{+}1){=}i, \mathbf{x}(t{+}1))P(s(t{+}1){=}i|s(t){=}j) \propto w(i,j,t).$$

In each iteration of the forward-backward algorithm, there is a loop over discrete
states in which the parameters for each state θ_s are reestimated. The new estimates
are in part based on the weights $\{w(s,t) : t = 1,\ldots,T\}$. Because the Gaussians
used in the models have tails that go on forever, $w(s,t) > 0, \forall t$. The times t with
weights below a small threshold $w(s,t) < \epsilon$ have little effect on the new parameters
θ_s, and discarding these times speeds the computations.

6. FORECASTS OF DATA SET D

We have written a family of programs that construct and optimize HFHMMs. To seed the forward-backward algorithm, we construct a HFHMM based on a partition of the space of autoregressive history vectors that we generate by Lloyd iteration.[15] Specifying a quantization vector v_s and metric $d_s()$ for each cell s defines the partition. An autoregressive history vector $\mathbf{x}$ is in the cell s which minimizes $d_s(v_s, \mathbf{x})$. In each cell, we set the metric proportional to the inverse covariance of the data in the cell. To construct the seed HFHMM, we associate a hidden state with each cell of the partition, initialize the parameters of $P_{y(t)|s(t),\mathbf{x}(t)}$ with a linear fit over training data that fall in the cell, and use relative frequencies of transitions between cells to estimate discrete state transition probabilities.

We used this procedure to fit the model that generated Figure 2(a), a forecast of Data Set D. The model is the result of 70 passes of the forward-backward algorithm, each of which required about 45 minutes on a SPARCstation 2. We estimated the probability density that constitutes the forecast using a Monte Carlo method.

As the number of time steps is increased in very long forecasts, probability leaks away to exponentially larger values of y. Although this effect is subtle in Figure 2(a), for simple chaotic systems it is dramatic. Considering the Lyapunov exponents helps explain this defect. A discrete state sequence q, starting at the end of the observed data T and continuing for a forecast of τ steps $q \equiv (s_q(T+1), s_q(T+2), \ldots, s_q(T+\tau))$, specifies a sequence of linear maps, i.e., derivative information. For typical long sequences q, the magnitudes of the eigenvalues of the products of these maps will grow at exponential rates given by the Lyapunov exponents. For chaotic systems, at least one Lyapunov exponent is positive, and the composed maps are linearly, and hence globally, unstable. The problem is that the model is linear in the sense that

$$y_{T,q,\lambda\mathbf{X}(T)}^{T+\tau} = \lambda y_{T,q,\mathbf{X}(T)}^{T+\tau}$$

and

$$P_\theta(\lambda y_{T,q,\lambda\mathbf{X}(T)}^{T+\tau}|q, \lambda\mathbf{x}(T)) = P_\theta(y_{T,q}^{T+\tau}|q, \mathbf{x}(T))$$

where $y_{T,q,\mathbf{X}(T)}^{T+\tau}$ denotes the y sequence that maximizes $P_\theta(y_{T+1}^{T+\tau}|q, \mathbf{x}(T))$. Thus, in a HFHMM there are no nonlinearities to saturate diverging y values. The models described below address this weakness.

---

[15] For details on vector quantization, see Gersho and Gray's recent text (Gersho and Gray, 1992).

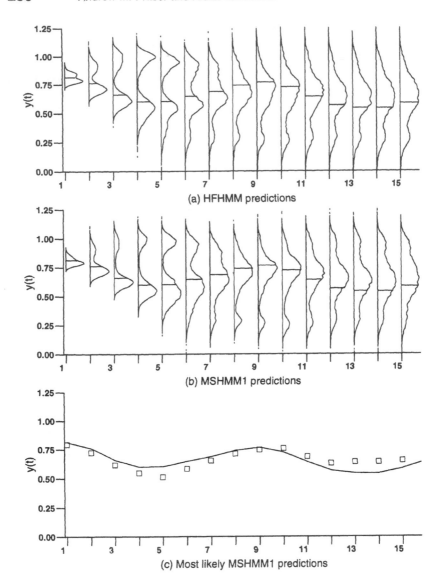

(a) HFHMM predictions

(b) MSHMM1 predictions

(c) Most likely MSHMM1 predictions

FIGURE 3 These plots illustrate the short-range behavior of the models. The first few time steps of the forecasts of Figure 2 appear in plots (a) and (b). For each step, the probability density is sketched and a horizontal bar indicates the *true* continuation data. The squares in (c) denote the peaks of the predictions in (b), and the line connects the points of the true continuation data.

6.1 OUTPUT-DEPENDENT STATE TRANSITIONS

By allowing $\mathbf{x}$ values to influence the transition probabilities between discrete states, one can introduce the nonlinear saturation that HFHMMs miss. In such models, the sequence $(s(t), \mathbf{x}(t))$ still constitutes a Markov process, but the sequence of discrete states $s(t)$ alone does not.

The assumptions we now make are:

a. $P(s(t+1) \mid s_1^t, y_1^t) = P(s(t+1)|s(t), y(t), \mathbf{x}(t))$.

b. $P(y(t) \mid s_1^t, y_1^{t-1}) = P(y(t)|s(t), \mathbf{x}(t))$.

c. $P(y(t), \mathbf{x}(t) \mid s(t), s(t+1))$ has a multivariate normal distribution.

d.   The entire process is stationary.

This type of model is referred to in the figures as MSHMM1. Simpler model types assume that $P(y(t) \mid s(t), \mathbf{x}(t))$ and $P(\mathbf{x}(t+1)|s(t+1), s(t))$ are normal and that $P(s(t+1) \mid s_1^t, y_1^t) = P(s(t+1)|s(t), \mathbf{x}(t+1))$.

Under the above assumptions $\psi(t) = (s(t), \mathbf{x}(t))$ is a Markov process. It is possible to compute all quantities of interest by maintaining, for each transition $s(t) \rightarrow s(t+1)$, the parameters of a multivariate normal distribution $P(\mathbf{x}(t) \mid s(t), s(t+1))$, a normally distributed linear prediction $P(y(t)|s(t), s(t+1), \mathbf{x}(t))$, and the constant $P(s(t+1) \mid s(t))$.

The EM algorithm is applicable to this class of models as well. The transition probabilities in the mixed state space are

$$P(\psi(t+1) \mid \psi(t)) = P(\mathbf{x}(t+1) \mid s(t+1), s(t), \mathbf{x}(t))P(s(t+1) \mid s(t), \mathbf{x}(t))$$

$$= P(y(t) \mid s(t+1), s(t), \mathbf{x}(t))\frac{P(\mathbf{x}(t) \mid s(t), s(t+1))P(s(t+1) \mid s(t))}{P(\mathbf{x}(t) \mid s(t))}$$

The EM algorithm is equivalent to maximizing $\langle \sum_t \log P(\psi(t+1) \mid \psi(t)) \rangle_{q|y_1^T}$. While we do not yet have working code that maximizes this, we have obtained good results using a naive algorithm which maximizes $\langle \sum_t \log P(y(t) \mid s(t+1), s(t), \mathbf{x}(t)) \rangle_{q|y_1^T}$, $\langle \sum_t \log P(\mathbf{x}(t) \mid s(t+1), s(t)) \rangle_{q|y_1^T}$, and $\langle \sum_t \log P(s(t+1) \mid s(t)) \rangle_{q|y_1^T}$, but ignores the denominator term, $- \langle \sum_t \log P(\mathbf{x}(t) \mid s(t)) \rangle_{q|y_1^T}$. Application of this naive algorithm yields essentially monotonic improvement in performance, with each model very close to the true optimal performance for that step in the process.

Predictions made by such a model trained on Data Set D appear in Figures 2(b), 3(b), and 3(c). Even for very long forecasts, there is no leakage of probability to ever larger y's.

ACKNOWLEDGMENTS

This research was supported in part by NSF grant MIP-9113460 and by a contract from Radix Inc. Conversations with many other researchers including Henry Abarbanel, Ronald Hughes, Cory Myers, and Todd Leen have influenced this work.

Eric J. Kostelich† and Daniel P. Lathrop‡

†Department of Mathematics, Arizona State University, Tempe, AZ 85287.
‡Mason Laboratory, Department of Mechanical Engineering, Yale University, New Haven, CT 06520.

Time Series Prediction by Using the Method of Analogues

This paper describes a procedure for making short-term predictions by examining trajectories on a reconstructed attractor that correspond to a dynamical feature of interest, namely, the trajectories near a spiral saddle fixed point in the attractor. Reasonable predictions can be made from short time series records and very good predictions from longer records using local linear approximations of the dynamics, if the dimension of the attractor is not too large. Two methods are described. The first uses nearest-neighbor comparisons similar to E. N. Lorenz' "method of analogues." The second method uses a collection of closely matched trajectories to compute a basis of singular vectors. The observations can be projected onto a subset of the new basis with little loss of information. In the new coordinates, good predictions can be made by finding the slope of a certain least-squares line through the origin.

Times Series Prediction: Forecasting the Future and
Understanding the Past, Eds. A. S. Weigend and N. A. Gershenfeld, SFI Studies
in the Sciences of Complexity, Proc. Vol. XV, Addison-Wesley, 1993   **283**

1. INTRODUCTION

Our strategy in the analysis of chaotic time series data has been to exploit simple ideas and algorithms wherever feasible. In cases where the time series is generated by a low-dimensional dynamical system, a representation of the underlying attractor can be obtained using time delay embedding methods (Sauer et al., 1992; Takens, 1981). The trajectories can be analyzed to find approximate saddle orbits (Lathrop & Kostelich, 1989) that govern the observed behavior, and local linear approximations of the dynamics can be constructed using linear least squares (Eckmann & Ruelle, 1985; Farmer & Sidorowich, 1987; Kostelich & Yorke, 1990). These approximations can be used to predict future values in the time series, at least in the short term. In this paper, we show how to apply these ideas to a laboratory data set (Data Set A of the Santa Fe Institute Time Series Prediction and Analysis Competition), with some modifications to handle the short data record.

Data Set A consists of 1,000 scalar measurements. Figure 1(a) shows the time series. Figure 1(b) shows a projection of an attractor reconstructed using a Takens time delay embedding (Takens, 1981) (the time delay is five time steps). Figure 1(c) is a different time delay embedding, where the coordinates of each point consist of consecutive time series values. (The individual observations are plotted as dots in the latter and are connected by line segments in the former.)

The time delay embeddings suggest that trajectories spiral out from a saddle fixed point. Once they get past a certain distance, they are "reinjected" into a neighborhood of the saddle point. The reinjections occur around $t = 180$ and $t = 600$, and possibly near $t = 500$. The dynamics appear to be relatively low-dimensional chaos, because the reinjections do not occur periodically or with the same amplitude. The time delay plots suggest that the attractor can be embedded in three dimensions; that is, a visual inspection of the data using a three-dimensional graphics workstation suggests that trajectories do not cross themselves and are smooth (differentiable) to observational accuracy. We conjectured that the dynamics are qualitatively similar to those of the Rössler system.[1]

2. ZERO PARAMETER PREDICTION

The prediction problem for Data Set A is to forecast the time series values from $t = 1001$ to $t = 1100$. The main difficulty is that the given time series is so short;

---

[1]The time series comes from a laser experiment, but we did not know this when we entered the contest. The dynamics governing the experiment may be qualitatively similar to those of the Lorenz equations (Hübner et al., this volume; Rössler, 1976).

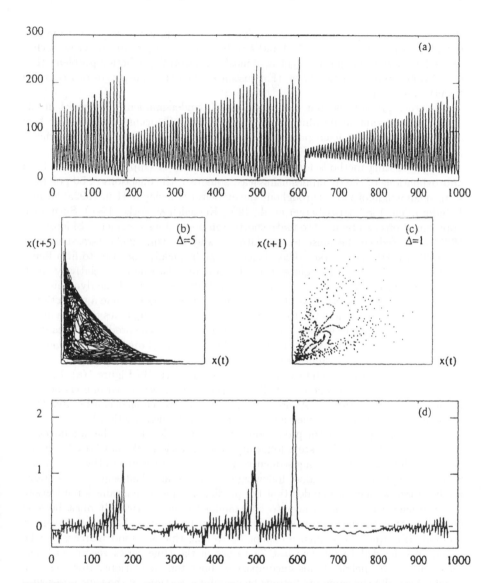

FIGURE 1   (a) Time series plot of Data Set A. (b) Time delay embedding with a delay of five time steps (consecutive points are connected by straight lines). (c) A time delay embedding with a delay of one time step. (d) Local estimates of the expansion rate between nearby initial conditions (in bits per time step). The dotted line indicates the value of the largest Lyapunov exponent. (See Section 4.)

there are only two places where the oscillations have approximately the same amplitude as those between $t = 950$ and $t = 1000$, for example. In our view, such a limited historical record is not typical of most "real world" prediction problems that may involve deterministic chaos. (For instance, even the sunspot record contains twenty or so cycles.)

Another problem is that the time series is undersampled. The peaks in the data are important in deciding when a reinjection will occur (and possibly how close it will be to the fixed point). However, the undersampling makes it difficult to determine the relative maxima accurately.

Most existing methods for analyzing chaotic data attempt to determine local approximations of the dynamics using least squares with collections of 50 points or so in small neighborhoods throughout the attractor (Cawley & Hsu, 1992a; Farmer & Sidorowich, 1987; Grassberger et al., 1991; Kostelich & Yorke, 1990). Such local approximations can be used to make short-term predictions (Farmer & Sidorowich, 1987) or to reduce the noise in the data (Cawley & Hsu, 1992a; Grassberger et al., 1991; Kostelich & Yorke, 1990; Sauer, 1992). Typically one tries to fit a linear model to the data, which requires that the available data lie in a relatively small ball in phase space in order to get an accurate approximation of the dynamics.

Given these limitations, we think that the simplest way to make a prediction is to try to find the section of the time series that most closely matches the data at the end of the record, relative to some norm or other criterion. The prediction consists of the values that follow the most closely matching section. (Lorenz (1963) suggested a similar approach in a paper on weather forecasting.) One can find a good match just by manually overlapping two copies of the time series in Figure 1(a). In some sense, this is a "zero parameter" prediction method because no parameters need to be determined statistically. (A more sophisticated approach that can be used with more observations of the reinjection is described in the next section.)

Our main objective in the prediction contest was to decide whether a reinjection would occur in the 100 time steps following the observations. We have tried to find a good match to a relatively long sequence of the last observations in the time series. Since the sequence of 100 values following each of the matched sequences contains a reinjection, we have concluded that the predicted sequence contains a reinjection.

The length of the sequences to be examined is not critical. It must be long enough so that the last six or so oscillations can be compared with other similar sections. However, the sequence cannot be so long that no good matches can be found. We decided to use the last 75 values of the time series record with the maximum norm metric. (This encompasses about ten oscillations. However, one obtains basically the same predictions by examining sections of the time series with 50 to 85 values. The Euclidean norm can be used instead of the maximum norm, but the maximum norm is more convenient computationally.)

We have written a computer program to compare sections of 75 consecutive time series values and to output the best matches. In other words, the time series from $t = 926$ to $t = 1000$ is treated as a single 75-dimensional vector. We then

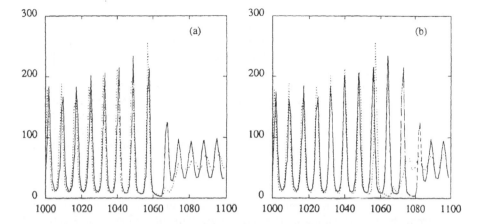

FIGURE 2   Predictions of time series values from $t = 1001$ to $t = 1100$. The solid line is the prediction; the dotted line corresponds to the actual observations. (a) uses the best fit to the last 75 observations; (b) shows the second best fit, which is the same as the solid curve in part (a) but shifted right by 15 time steps.

compute the absolute values of the differences between each component of this vector and every other vector of 75 consecutive values in the data set. The closest match is the vector for which the maximum difference between any two components is smallest. The computation can be done efficiently by arranging the time series values into a binary tree (Bentley & Friedman, 1979) and takes less than a second on a desktop workstation.

There are three vectors of 75 consecutive values that are close to the time series values from $t = 926$ to $t = 1000$. In this paper, distances between vectors will be expressed as a fraction of the time series extent using the maximum norm. (That is, we use the maximum norm and suppose that the time series has been normalized to the interval $[0, 1]$.) The distance between the reference point and its three closest neighbors is 0.13 or less using this convention. Figure 2 shows the predictions corresponding to two of them. The closest match is the time series from $t = 43$ to $t = 117$, with a maximum norm distance of 0.11. The corresponding prediction is the time series from $t = 118$ to $t = 217$, shown as the solid curve in Figure 2(a). The actual values from $t = 1001$ to $t = 1100$ are shown by the dotted line in the figure.

Another close match is the time series from $t = 28$ to $t = 102$, with a maximum norm distance of 0.13. The corresponding prediction is the values from $t = 103$ to $t = 202$, shown in Figure 2(b).

Because the best matching vectors are very close in norm, it was difficult to decide which of them would yield the best prediction. One consideration is that the maximum norm gives only the largest separation between any two components of the vectors, but does not say anything about the size of typical differences. We decided to check the median value of the absolute differences between the components. (This corresponds to the median value of the set $\{|s_{43+j} - s_{926+j}|\}_{j=0}^{74}$ for the closest matching vector.)

The median absolute difference is 0.019 for the closest match (from $t = 43$ to $t = 117$) and 0.016 for the next closest match ($t = 28$ to $t = 102$). Since the latter has the smaller median, we used the 100 values following $t = 102$ as our prediction entry (Figure 2(b)). In retrospect, the other choice is the better prediction, but both of them are reasonably good. This "best match" method determines the reinjection to within 15 time steps, and the amplitude of the oscillations after the reinjection agrees reasonably well with the actual observations, except for some large errors from $t = 1060$ to $t = 1080$.

There are ten vectors altogether whose maximum norm distance is less than 0.20 from the last 75 values of the time series. The 100 values that follow each vector are all plausible predictions. Each predicted sequence contains a reinjection site, but there is some uncertainty where it occurs. (For example, the reinjection occurs after 60 time steps in Figure 2(a) but after 75 time steps in Figure 2(b).)

The different vectors of observations allow one to estimate the pointwise error in the predicted sequence. One way to estimate the error at the jth step is to determine the variance in the jth component over each of the plausible prediction vectors. For example, the pointwise variance in the first 60 predicted values is relatively small, since most of the oscillations coincide, as illustrated by Figure 2. (The variance is larger near the peaks of each oscillation.)

The pointwise variance is largest between the 60th and 70th time steps, corresponding to the uncertainty in the location of the reinjection. Phase differences between the oscillations accumulate, as illustrated in Figure 2, so that the pointwise variance in the 100-step predictions is larger for the last 40 time steps than for the first 60 time steps. The variance among the plausible prediction vectors averages about 0.08% of the time series extent for the first 60 time steps and is about 0.8% for the last 40 time steps.

We emphasize that this variance estimate is computed using only the closest matching prediction vectors relative to the maximum norm. It underestimates the error between the predicted and the actual time series record from $t = 1001$ to $t = 1100$, because the amplitude of the oscillations following the reinjection at $t = 1060$ is not predicted accurately.

3. ONE PARAMETER PREDICTION

A more sophisticated approach to the prediction problem can be taken when more data are available. Suppose one wants to predict the 100 values following the first

1450 observations in the data set. This problem is identical to the original prediction contest problem, except that there are three obvious reinjections instead of two. The time series values leading up to $t = 1450$ are similar to those leading up to $t = 1000$; in fact, the sequence from $t = 924$ to $t = 998$ is the closest match (using the maximum norm) to the 75 values leading up to $t = 1450$. Hence, using the closest match method, the predicted values for $t = 1451$ to $t = 1550$ would be the time series values from $t = 999$ to $t = 1098$. However, the amplitude of the oscillations following the reinjection near $t = 1060$ is smaller than the oscillations than are actually observed near $t = 1525$.

In this example, the closest match method predicts the reinjection time accurately. The amplitude of the oscillations following the reinjection is not predicted with great accuracy, perhaps because it is a function of the amplitude prior to the reinjection. It is not possible to estimate what this functional dependence might be on the basis of the single closest match in the time series.

However, such an estimate can be obtained by examining the three closest matches to the data record from $t = 1376$ to $t = 1450$ and doing a least-squares fit to predict the amplitude of the oscillations after the reinjection. Using a suitable change of coordinates, described below, the problem is equivalent to fitting a straight line through the origin with three points. Thus it is necessary to fit only one parameter (the slope) in order to make the prediction.

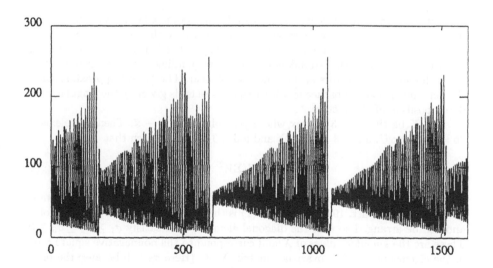

FIGURE 3   The first 1600 values of Data Set A.

The procedure we follow is similar to that outlined in Kostelich (1992) for determining the dynamics near saddle periodic orbits. As in the previous section, we examine the time series to find the best matches to the last 75 values. (Once again, there is nothing special about the use of 75 values; we merely seek comparisons over ten complete oscillations. Somewhat more or fewer values give comparable results.)

This amounts to a 75-dimensional embedding of the attractor, and the time series values from $t = 1376$ to $t = 1450$ are the first through the 75th components, respectively, of a reference vector $\hat{\mathbf{x}}$. The best matches to this sequence correspond to the three nearest neighbors of $\hat{\mathbf{x}}$ in $\mathbf{R}^{75}$ relative to the maximum norm. Since we want to predict the 100 values following $t = 1450$, we also examine the 100 values following each of the best matches. The idea is to compute a linear approximation of the map from each vector of 75 values to the subsequent 100 values. In principle, the map is represented as a 100×75 matrix plus a constant vector, but, of course, it is not possible to estimate 7500 paramaters from these data.

However, there is considerable redundancy in the observations, and the singular value decomposition lets us replace the high-dimensional problem with an approximate one of much lower dimension. In fact, it is possible to project all the observations onto 2 one-dimensional subspaces with little loss of information. In the new coordinates, a least-squares line yields an approximate linear functional relationship between the coordinates of the observations in their respective subspaces. In the new coordinates, predictions of new observations can be made in the usual least-squares sense.

Let the three nearest neighbors of the reference point $\hat{\mathbf{x}}$ be $\mathbf{x}_1$, $\mathbf{x}_2$, and $\mathbf{x}_3$, with mean $\overline{\mathbf{x}}$.[2] The presence of noise virtually assures that the vectors $\mathbf{x}_1, \mathbf{x}_2, \mathbf{x}_3$ are linearly independent, so that the vectors $\mathbf{x}_j - \overline{\mathbf{x}}$, $j = 1, 2, 3$, span a 2-plane through the origin. Because of the redundancy of the corresponding observations in the time series, however, the vectors also are nearly collinear. The "best" approximation of this one-dimensional subspace in a least-squares sense is given by the singular value decomposition of the vectors.

Let X be the 75×3 matrix whose jth column is $\mathbf{x}_j - \overline{\mathbf{x}}$. Then there exist a 75×3 matrix U, a 3×3 matrix Σ, and a 3×3 matrix V such that

$$X = U\Sigma V^T . \tag{1}$$

The columns of U form an orthonormal basis for the columns of X (Dongarra et al., 1979 and Stewart, 1973). Equation (1) is the *singular value decomposition* of X and can be arranged so that the diagonal elements of Σ satisfy $\sigma_1 \geq \sigma_2 \geq \sigma_3 \geq 0$. These are the *singular values* of X and correspond to the nonnegative square roots of the eigenvalues of the covariance matrix $X^T X$. (Here $\sigma_3 = 0$ because the mean has been subtracted from the observations.)

---

[2]The components of the nearest-neighbor vectors are the sequences of 75 values following $t = 49$, $t = 373$, and $t = 924$.

Because there is redundancy in the observations, the second singular value is small relative to σ_1. (Here $\sigma_1/\sigma_2 \approx 10$.) Hence it is possible to project each of the observations $\mathbf{x}_j - \overline{\mathbf{x}}$ onto the first column $\mathbf{u}_1$ of U with little loss of information. (The vector $\mathbf{u}_1$ is the singular vector corresponding to σ_1.) The components of the observations in the remaining orthogonal direction are small, and to a good approximation we can regard them as being composed mostly of the noise. Thus, the projection can be considered as a kind of low-pass filter. In this way, each of the observations $\mathbf{x}_j - \overline{\mathbf{x}}$, which is a sequence of 75 time series values, is replaced by a single scalar $\xi_j = \mathbf{u}_1 \cdot (\mathbf{x}_j - \overline{\mathbf{x}})$.

We apply the same procedure to the vectors $\mathbf{y}_j$ corresponding to the sequences of 100 values following each of the observations $\mathbf{x}_j$ above. (We consider 100-dimensional vectors because we want to predict 100 values at the end of the time series. For instance, the coordinates of the vector $\mathbf{y}_1$ are the time series values from $t = 49 + 75 = 124$ to $t = 223$. The mean of these observations is $\overline{\mathbf{y}}$.) As before, the vectors $\mathbf{y}_j$, $j = 1, 2, 3$, form a linearly independent (yet nearly collinear) set. Let Y be the 100×3 matrix whose jth column is $\mathbf{y}_j - \overline{\mathbf{y}}$, $j = 1, 2, 3$. The singular value decomposition of Y is

$$Y = U' \Sigma' V'^{T},$$

where the columns of U' are an orthonormal basis for the columns of Y. Likewise, the second singular value is small compared to the first because the observations are highly redundant, and the third singular value is zero because the mean has been subtracted from the observations. Thus, with little loss of information, the observations $\mathbf{y}_j$ can be projected onto the singular vector $\mathbf{u}_1'$ corresponding to the largest singular value; that is, $\eta_j = \mathbf{u}_1' \cdot (\mathbf{y}_j - \overline{\mathbf{y}})$, $j = 1, 2, 3$. In this way, the vector pairs $(\mathbf{x}_j, \mathbf{y}_j)$ comprising the relevant sections of the time series can be represented to good accuracy as three ordered pairs (ξ_j, η_j).

A linear approximation of the dynamics along the trajectories can be constructed by fitting a least-squares line through the origin to the pairs (ξ_j, η_j), $j = 1, 2, 3$. That is, we want to find a least-squares estimate of the parameter a in the model

$$\eta_j = a\xi_j, \quad j = 1, 2, 3. \tag{2}$$

(The line goes through the origin, because if one of the observations were the mean $\overline{\mathbf{x}}$, the predicted time series should be the mean $\overline{\mathbf{y}}$ of the subsequent observations.) The reference point $\hat{\mathbf{x}}$ is represented as the scalar $\hat{\xi} = \mathbf{u}_1 \cdot (\hat{\mathbf{x}} - \overline{\mathbf{x}})$ in the new coordinates. We use Eq. (2) to obtain the predicted value $\hat{\eta}$, which in the original coordinates is the vector of 100 values given by

$$\hat{\eta}\mathbf{u}_1' + \overline{\mathbf{y}}.$$

Finally, we remark that an estimate of the error can be obtained by considering the difference between the reference point $\hat{\mathbf{x}}$ and its one-dimensional approximation $\hat{\xi}\mathbf{u}_1 + \overline{\mathbf{x}}$ relative to the singular vector basis. Another way to estimate the error is to repeat these calculations over all collections of nonoverlapping sections of the time series that are reasonably close matches to the reference section, then compare the differences between the resulting predictions. As noted at the end of the previous section, however, the actual error in the prediction may lie outside the bounds given by this method, particularly when the number of observations is limited.

Figure 4 shows the prediction of the time series from $t = 1451$ to $t = 1550$ obtained in this way. Both the location of the reinjection and the amplitude of the oscillations following the reinjection are determined quite accurately.

The singular value decomposition can be computed using standard numerical software (Dongarra et al., 1979 and Stewart, 1973) and requires only a second or so on a desktop workstation. Other sections of the time series can be predicted in a similar manner.

Higher dimensional analogues to Eq. (2) can be tried as more data are used for the prediction. For example, if one has, say, 20 sections that closely match the time series leading up to the section to be predicted, then each observation might be projected onto the first two singular vectors in the corresponding singular value decomposition. Each observation is replaced by a 2-vector in the singular vector coordinates. Multivariate linear regression is used to determine the best 2×2 matrix that minimizes the corresponding sum of squares, so the prediction requires the estimation of four parameters. Such a procedure may yield a better prediction than the one-dimensional version, because it uses more information about each observation.

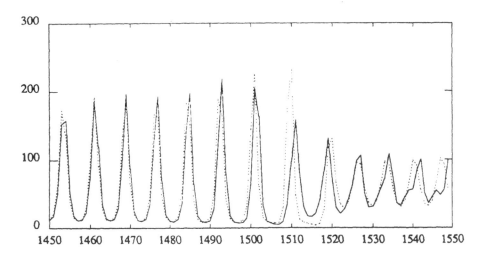

FIGURE 4   Prediction of time series values $t = 1451$ to $t = 1550$. The solid line is the prediction; the dotted line shows the actual observations.

4. VARIABLE PREDICTABILITY

The Lyapunov exponents λ_j, $j = 1, 2, \ldots$, associated with the attractor are an average measure of the rate of growth of distances between nearby initial conditions on the attractor (Eckmann & Ruelle, 1985). By convention, $\lambda_j \geq \lambda_{j+1}$. The distances grow as $2^{\lambda_1 t}$ in the case of one positive Lyapunov exponent, so if $\lambda_1 = 1$ bit/sec, then they double every second. Such an attractor exhibits sensitive dependence on initial conditions, because small uncertainties grow exponentially fast. An attractor is called chaotic if it has at least one positive Lyapunov exponent.[3]

The length of time that one can predict the evolution of a chaotic process is determined by the positive Lyapunov exponents of the underlying attractor. In some average sense, one can predict twice as long into the future when $\lambda_1 = 1/2$ bit/sec as when $\lambda_1 = 1$ bit/sec.

The Lyapunov exponents can be estimated from a time delay reconstruction of the attractor using local linear approximations of the dynamics (Eckmann et al., 1986). More precisely, at each point $\mathbf{x}_i$ on the reconstructed attractor, we approximate the next point $f(\mathbf{x}_i)$ as

$$f(\mathbf{x}_i) = A_i \mathbf{x}_i + \mathbf{b}_i,$$

where A_i is a matrix and $\mathbf{b}_i$ is a vector (Kostelich & Yorke, 1990 and Farmer & Sidorowich, 1987). The matrix A_i is an approximation of the Jacobian matrix of partial derivatives of the flow at $\mathbf{x}_i$. The Lyapunov numbers are the geometric means of the norms of the eigenvalues of the product $A_n A_{n-1} \cdots A_1$. The Lyapunov exponents are the corresponding logarithms (see Eckmann et al., 1986, for details). We estimate $\lambda_1 = 0.087$, $\lambda_2 = -0.033$, and $\lambda_3 = -1.72$ bits/time step from the full 10,000-point time series.[4]

The Lyapunov exponents are averages over the entire attractor. The discussion above therefore suggests that the distance between typical initial conditions on the attractor grows on the average by a factor of $2^{100\lambda_1} = 2^{8.7} \approx 420$ after 100 time steps. Such a large expansion means that it should be effectively impossible to predict 100 time steps into the future, since an uncertainty in the initial data of one laboratory unit (the values are integers between 0 and 255) will expand to the size of the entire attractor within 100 time steps.

Nevertheless, it is usually possible to make good predictions 100 time steps into the future. This consideration means that our estimate for λ_1 probably is too large.

[3] The exponential separation between nearby points does not continue forever, because the attractor is bounded. Eventually the separation process saturates, and trajectories starting from nearby points appear to evolve independently of one another. For convenience, we express the Lyapunov exponents as base 2 logarithms.

[4] The complete data set consists of 10,000 values and is available from the Santa Fe Institute computer server. See the Appendix to this volume. The computation was done using a three-dimensional embedding of the attractor and a least-squares procedure similar to that described in Kostelich & Yorke (1990) and Eckmann et al., (1986).

Regardless of the exact value of λ_1, it is important to note that when the expansion rates are averaged over short sections of each trajectory (a "local" estimate of the Lyapunov exponents), we find considerable variation in their values.

Figure 1(d) shows the "local" expansion rates, defined at the ith time step to be the geometric mean of the norms of the eigenvalues of $A_{i+6}A_{i+5} \cdots A_i$. That is, we take a geometric mean of the expansion rates on the attractor over seven consecutive time steps. (Seven time steps correspond roughly to one complete oscillation.) The geometric means are plotted in Figure 1(d) for the first 1000 values in the time series. The expansion rates are largest where the trajectory is reinjected into a neighborhood of the saddle fixed point; these are regions of large local uncertainty. For example, the distance between nearby points grows by a factor of 20 between $t = 590$ and $t = 600$. In contrast, there is no net expansion between $t = 620$ to $t = 820$.

Thus, the time series exhibits *variable predictability*. Some parts are much easier to predict than others. Our results suggest that predictions of 150 to 200 time steps should be feasible, as long as one avoids intervals where more than one reinjection occurs. The time series from $t = 1001$ to $t = 1100$ contains only one reinjection site and so is relatively predictable. Accurate predictions over certain longer intervals, such as regions similar to that between $t = 700$ and $t = 900$, should be possible. However, it probably is not possible to predict regions like the one between $t = 480$ and $t = 600$, for example, because it encompasses more than one reinjection.

CONCLUSIONS

The data in Data Set A arise from a system whose underlying behavior can be regarded as a low-dimensional chaotic dynamical system. Trajectories spiral outward from a saddle fixed point and eventually are reinjected into a neighborhood of the saddle point.

Predictions of subsequent values can be made by examining sections of trajectories (corresponding to sections of consecutive time series values) in a region of interest on the reconstructed attractor. The simplest procedure locates the section of the time series that most closely matches the last several values in the observed record. The prediction consists of the 100 values following the best fitting section. This method, which is perhaps the most straightforward given the paucity of data, does not require the statistical determination of any parameters.

A more sophisticated procedure can be used if more than two observations are available of the reinjection near the saddle fixed point. The prediction requires the statistical determination of a single parameter in a suitable coordinate system.

The simplicity of our approach should be contrasted to the neural network that Eric Wan used for his prediction (Wan, this volume), which incorporated some

1,105 parameters in a nonlinear system of equations. (Data Set A contains only 1,000 observations; see Gershenfeld and Weigend, this volume.)

Data Set A exhibits variable predictability. Intervals of the time series that correspond to the reinjection of trajectories near the saddle fixed point are associated with high uncertainty (large local expansion rates). Only one reinjection site is in the interval for which the prediction contest was run. This interval is relatively predictable.

The method outlined here should be applicable to almost any low-dimensional chaotic attractor, provided that one can find reasonably close matches to the section of the time series leading up to the interval to be predicted. We have not investigated the data requirements in detail, but the method appears to work well even on short data records like Data Set A. However, the dimension of the underlying attractor cannot be too large (the dimension of attractor for Data Set A is approximately two). This procedure cannot be expected to work on the other time series in this competition, because the attractor dimension is much higher (e.g., about ten for Data Set D). Nevertheless, most standard analysis methods assume that the attractor dimension is tractable (typically four or less for laboratory data sets), and this method should be applicable in such cases.

ACKNOWLEDGMENTS

E. K. is supported by the NSF Applied and Computational Mathematics Program under grant No. DMS-9017174. The authors thank an anonymous referee for several helpful suggestions.

P. A. W. Lewis,[†] **B. K. Ray,**[‡] **and J. G. Stevens**[†]

†Deptartment of Operations Research, Naval Postgraduate School, Monterey, CA 93943.
‡Department of Mathematics, New Jersey Institute of Technology, Newark, NJ 07102

Modeling Time Series by Using Multivariate Adaptive Regression Splines (MARS)

Multivariate Adaptive Regression Splines (MARS) is a new methodology, due to Friedman, for nonlinear regression modeling. MARS can be conceptualized as a generalization of recursive partitioning that uses spline fitting in lieu of other simple fitting functions. Given a set of predictor variables, MARS develops a nonparametric regression model in the form of an expansion in product spline basis functions. MARS can produce continuous nonlinear models for high-dimensional data with multiple partitions and predictor variable interactions. By letting the predictor variables in MARS be lagged values of a univariate time series, one obtains an adaptive spline threshold autoregressive (ASTAR) model, which is a new method for nonlinear modeling of time series. In the semi-multivariate time series extension of ASTAR, which we call SMASTAR, the univariate time series to be modeled has not only its own lagged variables as predictors, but also lagged variables of other related time series. The approach seems well suited for taking into account the complex interactions among multivariate, cross-correlated, lagged predictor variables of a time series system. In addition, one can model the effect on the output series of categorical-predictor time series to obtain a CASTAR model. We illustrate the MARS methodology

Times Series Prediction: Forecasting the Future and
Understanding the Past, Eds. A. S. Weigend and N. A. Gershenfeld, SFI Studies
in the Sciences of Complexity, Proc. Vol. XV, Addison-Wesley, 1993

by applying it to a series of tick-by-tick Swiss franc/U.S. dollar exchange rate returns using lagged returns and lagged times between ticks as predictor variables, as well as series denoting the day of the week and the hour after the start of trading, which we treat as categorical series.

1. INTRODUCTION

Regression modeling is a popular statistical approach that serves as a basis for studying a system of interest. We use regression modeling to develop a mathematical model of the relationships that exist between the dependent (output) variable and the independent (explanatory) variables of the system. Classical methods for developing the functional form of the regression model are based on previous knowledge of the system and on considerations such as smoothness and continuity of the response variable as a function of the predictor variables. Splines provide one functional form that is useful for developing the regression model. Excellent surveys and discussions of splines in statistics are available in the literature (Silverman, 1985; Wegman & Wright, 1983). The objective of our paper is to examine an extension of adaptive regression splines (Friedman, 1991c; Lewis & Stevens, 1991) for developing autoregressive models of univariate and semi-multivariate time series systems, with the additional possibility of including the influence of a categorical time series. We analyze and apply the MARS methodology, when appropriate, to several data sets made available to us through the Santa Fe Institute and discussed at the 1992 NATO Workshop on Comparative Time Series Analysis. One such data set, for example, includes records over time of the heart rate, chest expansion, and blood oxygen level of a sleep study patient, as well as a categorical variable denoting his sleep-wake state at any time point.

To provide a framework for a regression modeling methodology, let y represent a single univariate response variable that depends on a vector of p predictor variables $\mathbf{x}$ where $\mathbf{x} = (x_1, \ldots, x_v, \ldots, x_p)$. Assume we are given N samples of y and $\mathbf{x}$, namely $\{y_i, \mathbf{x}_i\}_{i=1}^N$, and that we can describe y with the regression model,

$$y = f(x_1, \ldots, x_p) + \epsilon \tag{1}$$

over some domain $D \subset \mathbb{R}^p$, which contains the data. The function $f(\mathbf{x})$ reflects the true but unknown relationship between y and $\mathbf{x}$. The random additive error variable ϵ, which is assumed to have mean zero and variance σ_ϵ^2, reflects the dependence of y on quantities other than $\mathbf{x}$. Quoting Friedman (1991b) "$f(\mathbf{x})$ is taken to be that component of y that varies smoothly with changing values of $\mathbf{x}$, whereas the noise is taken to be the leftover part that does not."

The goal in regression modeling is to formulate a function $\hat{f}(\mathbf{x})$ that is a reasonable approximation of $f(\mathbf{x})$ over the domain D. If the correct parametric form of

$f(\mathbf{x})$ is known, then we can use parametric regression modeling to estimate a finite number of unknown coefficients. However, in this paper the approach is nonparametric regression modeling (Friedman, 1991b). We only assume that $f(\mathbf{x})$ belongs to a general collection of functions and rely on the data to determine the final model form and its associated coefficients. Again quoting Friedman (1991b), "The effectiveness of a nonparametric procedure is determined by how well it can gauge the (local) smoothness properties of $f(\mathbf{x})$ and exploit them so as to filter out most of the noise without altering too much of the signal."

In Section 2, we briefly discuss MARS (Friedman, 1991c), a new, computer intensive method of flexible nonparametric regression modeling that appears to be an improvement over existing methodology when using moderate sample sizes N and predictor spaces with dimension $p > 2$. A difficulty with applying existing multivariate nonparametric regression modeling methodologies to problems of dimension greater than two has been called the *curse of dimensionality* (Bellman, 1961). The *curse of dimensionality* describes the need for an exponential increase in sample size N for a linear increase in p, the number of predictor variables, in order to densely populate higher dimensional spaces. MARS attempts to overcome the *curse of dimensionality* by exploiting the localized low-dimensional structure of the data used in constructing $\hat{f}(\mathbf{x})$. Note that the discussion of MARS in this section of the paper is a simple introduction that is only complete enough to motivate the extension to time series modeling with MARS. Further details on MARS can be found in the literature (Friedman, 1991c; Lewis & Stevens, 1991, 1992).

In Section 3, we review briefly the use of MARS for autoregressive modeling and analysis of a univariate time series, x_τ for $\tau = 1, 2, \ldots N$; i.e., the predictor variables in MARS (Lewis & Stevens, 1991) are now the lagged values of the response variable x_τ. The result is an adaptive spline threshold autoregressive (ASTAR) model that is an extension of the threshold autoregressive (TAR) model developed by Tong (1983). One significant feature of both ASTAR and TAR models, when modeling univariate time series data with periodic behavior, is their ability to produce models with underlying sustained oscillations (limit cycles). Time series that seem to possess such behavior include the well-known Wolf Sunspot data (Tong, 1990), the Canadian Lynx data (Tong, 1990), and the Sea Surface Temperature data (Altman, 1987; Lewis & Ray, 1992) to name a few.

Section 4 extends the ASTAR methodology to the semi-multivariate autoregressive modeling of a time series system (Lewis & Stevens, 1992). This builds upon semi-multivariate TAR modeling by Tong, Thanoon, & Gudmundson, 1985. We now seek to use MARS to model a single response variable of a time series system using predictor variables that are the lagged values of both the response and input time series. For example, for $\tau = 1, 2, \ldots, N$, let $\{y_\tau\}$ and $\{z_\tau\}$ be time series that represent system inputs and let $\{x_\tau\}$ be a times series representing the system output response. The set of possible predictor variables for this semi-multivariate time series system are $x_{\tau-1}, \ldots, x_{\tau-d_1}$; $y_\tau, \ldots, y_{\tau-d_2}$ and $z_\tau, \ldots, z_{\tau-d_3}$, where the maximum lags d_1, d_2, and d_3 are not necessarily equivalent. Also, $d_1 + d_2 + d_3 + 2 = p$, the total number of predictor variables. If we apply MARS to this system, the result is a *semi-multivariate* ASTAR model (SMASTAR) that is well suited for modeling the complex interactions that occur between the multivariate, cross-correlated, lagged predictor variables of a time series system.

Section 5 discusses briefly the introduction of categorical data into the time series modeling. This was introduced by Lewis & Ray (1992), who analyzed wind-direction data as a function of present and past wind speed and wind direction. The circular wind-direction data is conveniently modeled as a categorical-valued time series, although there are other ways this could be handled.

Section 6 gives the results of applying the MARS methodology to a series of tick-by-tick quotes of the Swiss franc/U.S. dollar exchange rate. The data is divided into ten sections, each spanning a period of about two weeks. We report the results of modeling one section of the data, using lagged exchange rates, as well as the time between ticks, the day of the week, and the hour of trading as predictor variables in the MARS algorithm. Prediction results are also discussed.

Finally, in Section 7, we discuss the similarities and differences between the MARS-based time series modeling methodology and other methods of time series analysis. Included in our discussion are neural network modeling strategies and methods using, for example, radial basis functions in the estimation of the regression function $f(\mathbf{x})$. Some thought is also given to problems of homogeneity in the part of the observed data used to estimate the model and the time series as a whole, as well as to possible assumptions on the marginal distribution or other structure of the time series which may decrease or increase the utility of the MARS approach. We use the sleep study patient data to illustrate some of these points.

2. MULTIVARIATE ADAPTIVE REGRESSION SPLINES (MARS)

In the regression context, MARS can be conceptualized as a generalization of a recursive partitioning (RP) strategy (Breiman et al., 1984; Morgan & Sonquist, 1963) that uses spline fitting in lieu of other simple fitting functions. The central idea in MARS is to formulate a modified recursive partitioning model as an additive model of functions from overlapping, instead of the usual disjoint, subregions. We initially present recursive partitioning (RP) as a basis for developing the MARS methodology.

2.1 RECURSIVE PARTITIONING (RP)

As in Eq. (1), we have N samples of y and $\mathbf{x} = (x_1, \ldots, x_p)$, namely $\{y_i, \mathbf{x}_i\}_{i=1}^N$. Let $\{R_j\}_{j=1}^S$ be a set of S disjoint subregions of D such that $D = \bigcup_{j=1}^S R_j$. Given the subregions $\{R_j\}_{j=1}^S$, recursive partitioning estimates the unknown function $f(\mathbf{x})$ at $\mathbf{x}$ with

$$\hat{f}(\mathbf{x}) = \sum_{j=1}^S \hat{f}_j(\mathbf{x})B_j(\mathbf{x}), \qquad (2)$$

where

$$B_j(\mathbf{x}) = I\left[\mathbf{x} \in R_j\right],$$

and $I[\cdot]$ is an indicator function with value 1 if its argument is true and 0 otherwise. Each indicator function, in turn, is a product of univariate step functions

$$H[\eta] = \begin{cases} 1 & \text{if } \eta > 0, \\ 0 & \text{otherwise}, \end{cases} \qquad (3)$$

that describe each subregion R_j. Thus $B_j(\mathbf{x})$ is a basis function with value 1 if $\mathbf{x}$ is a member of the R_jth subregion of D. Also, for $x \in R_j$, we have usually $\hat{f}(\mathbf{x}) = c_j$, the sample mean of the y_i's whose $\{\mathbf{x}\}_{i=1}^N \in R_j$.

In general, the recursive partitioning model is the result of a two-step procedure that starts with the single subregion $R_1 = D$. The first, or forward, step uses a combination of straightforward *exhaustive search* and recursive splitting of established subregions, to iteratively produce a large number of disjoint subregions $\{R_j\}_{j=2}^M$, for $M \geq S$, where M is chosen by the user to be substantially larger than the number of basis functions that seems adequate for the final model. The second, or backward, step reverses the first step and trims the excess $(M - S)$ subregions from the model using a criterion that evaluates both the model fit and the number of subregions in the model. The goal of the two-step procedure is to use the data to select a good set of nonoverlapping subregions $\{R_j\}_{j=1}^S$ together with the functions $\hat{f}_j(\mathbf{x})$, so that the estimate $\hat{f}(\mathbf{x})$ is "close" to $f(\mathbf{x})$ over each subregion of the domain.

In recursive partitioning the subregions $\{R_j\}_{j=1}^S$ *are disjoint.* Each data point $\mathbf{x}$ is only a member of one subregion R_j. Therefore, the estimate of $f(\mathbf{x})$ over subregion R_j is restricted to the functional form for $\hat{f}_j(\mathbf{x})$. However, as we will address in Section 2.2, MARS *has overlapping subregions*; the estimate of $f(\mathbf{x})$ over subregion R_j may be obtained as a sum of multiple functional forms.

Recursive partitioning is a very powerful methodology that is rapidly computed. However, in general, several aspects of recursive partitioning present serious difficulties during its application in a high-dimensional setting, i.e., $p > 2$. First, recursive partitioning models have disjoint subregions and are usually discontinuous at subregions boundaries. Second, the iterative division and elimination of parent regions during the forward step algorithm causes recursive partitioning to poorly estimate linear and additive functions. Finally, the form of the recursive partitioning model (2), an additive combination of functions of predictor variables in disjoint regions, makes estimation of the true form of the unknown function $f(\mathbf{x})$ difficult for large p.

2.2 FRIEDMAN'S INNOVATIONS AND MARS

To overcome recursive partitioning's difficulty in estimating linear and additive functions, Friedman (1991c) proposes first that the parent region is not eliminated (as in recursive partitioning) during the creation of its sibling subregions. Thus, in future iterations both the parent and its sibling subregions are eligible for further partitioning. An immediate result of retaining parent regions is that one obtains overlapping subregions of the domain. Also, each parent region may have multiple sets of sibling subregions. With this modification, i.e., with the repetitive partitioning of the initial region R_1 by different predictor variables, the modified recursive partitioning algorithm can produce linear models. Additive models with functions of more than one predictor variable can result from successive partitioning using different predictor variables. This modification also allows for multiple partitions of the same predictor variable from the same parent region.

Maintaining the parent region in a modified recursive partitioning algorithm results in a class of models with greater flexibility than permitted in recursive partitioning. However, the modified approach is still burdened with the discontinuities caused by the step function $H[\eta]$ and the fact that $\hat{f}_j(\mathbf{x})$ has a constant value c_j. To alleviate this difficulty, Friedman proposes to replace the step function $H[\eta]$ with linear (i.e., $q = 1$ order) regression splines in the form of left ($-$) and right ($+$) truncated splines, i.e.,

$$T^-(\mathbf{x}) = [(t - x)_+]^{q=1} = (t - x)_+ \text{ and } T^+(\mathbf{x}) = [(x - t)_+]^{q=1} = (x - t)_+ , \quad (4)$$

where $u_+ = u$ if $u > 0$ and 0 otherwise, the scalar x is an element of $\mathbf{x}$ and t represents a partition point in the range of x. P. L. Smith (1979) uses this notation for adaptive splines in the univariate setting and Friedman (1991c) extends the notation to the multivariate setting to explain MARS. *Note that the truncated spline functions act in only one dimension although their argument is a vector of predictor variables.*

Using linear truncated splines Eq. (4) creates a continuous approximating function $\hat{f}(\mathbf{x})$ with discontinuities in the first partial derivative of $\hat{f}(\mathbf{x})$ at the partition points of each predictor variable in the model. The argument for using *linear* truncated splines (4) is that there is little to be gained in flexibility, and much to lose in computational speed, by imposing continuity in the derivatives of the function $\hat{f}(\mathbf{x})$. Linear truncated splines allow rapid updating of the regression model and its coefficients during each exhaustive search for the next partition of an established subregion. The placement of additional partitions may be used to compensate for the loss of flexibility in using linear truncated splines, as opposed to higher order splines, to estimate $f(\mathbf{x})$ over a subregion of the domain.

Implementation of the modifications proposed above to the recursive partitioning algorithm avoids its identified difficulties and results in the MARS algorithm. The MARS algorithm produces a linear ($q = 1$) truncated spline model with overlapping subregions $\{R_j\}_{j=1}^S$ of the domain D. Each overlapping subregion of a MARS model is defined by the partition points of the predictor variables from an ordered sequence of linear truncated splines that we call a product basis function $K_j(\mathbf{x})$. The MARS estimate of the unknown function $f(\mathbf{x})$ is

$$\hat{f}(\mathbf{x}) = \sum_{j=1}^{S} c_j K_j(\mathbf{x}), \qquad (5)$$

where $\hat{f}(\mathbf{x})$ is an additive function of the product basis functions $\{K_j(\mathbf{x})\}_{j=1}^{S}$ associated with the subregions $\{R_j\}_{j=1}^{S}$. Since, for a *given* set of product basis functions, the values of the partition points, which of course are parameters of the model, are *fixed*, the MARS model (5) is a linear model whose coefficients $\{c_j\}_{j=1}^{S}$ may be determined by straightforward least-squares regression.

As in recursive partitioning, the objective of the *forward step* MARS algorithm is to iteratively adjust the vector of coefficient values using an exhaustive search to identify the subregions $\{R_j\}_{j=1}^{M}$, for $M \geq S$, whose product basis functions "best" approximate $f(\mathbf{x})$ based on data at hand. And again, as in the recursive partitioning procedure, it makes sense to follow the forward step procedure with a backward step trimming procedure to remove the excess $(M - S)$ subregions from the model whose product basis functions no longer sufficiently contribute to the accuracy of the model fit.

MARS uses residual-squared-error, because of its attractive computational properties, in the forward and backward steps of the algorithm to evaluate model fit and compare partition points. The actual backward fit criterion that is used for final model selection is a modified form of the generalized cross-validation criterion (Craven and Wahba, 1979) (GCV). This criterion requires the entire modeling procedure to be reapplied N times, with one of the N observations removed at each application. Friedman (1991c) proposed an approximation to the GCV criterion that requires only one evaluation of the model. The modified generalized cross-validation criterion (GCV^*) used to evaluate a MARS model with subregions $\{R_j\}_{j=1}^{M}$ is

$$GCV^*(M) = \frac{1/N \sum_{i=1}^{N} [y_i - \hat{f}_M(\mathbf{x}_i)]^2}{[1 - C(M)^*/N]^2} = \frac{\hat{\sigma}_\epsilon^2}{[1 - C(M)^*/N]^2}. \qquad (6)$$

The numerator in GCV^* is the average residual-squared-error and the denominator is a penalty term that reflects model complexity. The difference in GCV^* and GCV is in the computation of $C(M)^*$, a model complexity penalty function that is increasing in M. In MARS this modification is necessary to account for the heavy use of the data in determining both the partition points and the coefficients of a final model. The use of other criteria, perhaps more suitable to time series applications, has also been examined (Stevens, 1991). The Schwarz-Rissanen criterion (SC), defined as

$$SC(M) = \ln\left(\hat{\sigma}_\epsilon^2\right) + \frac{\ln(N)C(M)^*}{N}, \qquad (7)$$

is a time series model selection criterion based on stochastic complexity analysis. The extensive simulation results of Stevens (1991) find it to be more appropriate when using MARS in a time series setting. The justification for the SC criterion in linear time series analysis is hard to extend to nonlinear time series anlysis, but, besides the simulation results of Stevens, it has been found in practice to give more parsimonious time series models for real data (Lewis & Stevens, 1992).

3. UNIVARIATE TIME SERIES MODELING USING MARS (ASTAR)

Most research in applications of time series modeling and analysis is focused towards linear models. This is due to the maturity of the theory for linear times series, and the numerous studies and statistical packages that exist to facilitate the use of linear time series models. However, more frequently than not, nonlinear time-dependent systems abound that are not adequately handled by linear models. The use of linear models during the analysis of these nonlinear systems may require invalid assumptions that could lead to misleading conclusions. For these systems we need to consider general classes of nonlinear models that more readily adapt to the precise form of a nonlinear system of interest (Priestly, 1988; Tong, 1990).

By letting the predictor variables for the τth value in a time series $\{x_\tau\}$ be $x_{\tau-1}, x_{\tau-2}, \ldots, x_{\tau-p}$, and combining these predictor variables into a linear additive function, one gets the well-known linear AR(p) time series models. What happens if we use the MARS methodology to model the effect of $x_{\tau-1}, x_{\tau-2}, \ldots, x_{\tau-p}$? The answer is that we still obtain autoregressive models. *However, these models, called Adaptive Spline Threshold Autoregressive (ASTAR) models* (Lewis and Stevens, 1991), *can be nonlinear models in the sense that the lagged predictor variables can have threshold terms, in the form of truncated spline functions (4), and can also interact with the nonlinear terms of other lagged predictor variables.* The form and analysis of univariate ASTAR models are the subject of a recent paper (Lewis & Stevens, 1991).

Threshold models (models with partition points) are a general class of nonlinear models that emerge naturally as a result of changing physical behavior. Within the domain of the predictor variables, different model forms are necessary to capture changing relationships among the predictor and response variables. Tong (1983) provides one particular threshold modeling methodology for this behavior, Threshold Autoregression (TAR), that identifies piecewise linear pieces of nonlinear functions over disjoint subregions of the domain D of the time series $\{x_\tau\}$, i.e., the identification of linear autoregressive models within each disjoint subregion of the domain. TAR modeling methodology has tremendous power and flexibility for modeling of many times series. However, unless Tong's TAR methodology is constrained to be continuous, it creates disjoint subregion autoregressive models that are discontinuous at subregion boundaries. Nor is its implementation systematic.

With MARS, by letting the predictor variables be lagged values of a time series, one overcomes the limitations of Tong's approach and admits a more general class of continuous nonlinear threshold models than permitted in TAR models. In fact, the MARS methodology provides a systematic procedure for deriving nonlinear threshold models that are naturally continuous in the domain of the predictor variables, allows interactions among lagged predictor variables and can have multiple lagged predictor variable thresholds. In contrast, Tong's methodology creates nonlinear threshold models from piecewise linear models whose terms are restricted to the initial sets of candidates of the ASTAR algorithm. Tong's threshold models do not allow interactions among lagged predictor variables and the models are usually limited to a small number of thresholds due to the difficulties associated with the threshold selection process.

Numerous simulation studies have been conducted to evaluate the ability of the ASTAR methodology to identify and evaluate simple linear and nonlinear times series models (Lewis & Stevens, 1991). They also obtained an ASTAR model for the widely studied yearly Wolf sunspot numbers, a univariate nonlinear time series with periodic behavior. The predictions achieved using the ASTAR model were as good as those achieved very recently using a neural network model (Weigend et al., 1990) and are better than those of any other procedure reported in the statistical literature (Tong, 1983).

4. SEMI-MULTIVARIATE TIME SERIES MODELING USING MARS (SMASTAR)

While ASTAR models of univariate time series certainly have widespread applicability, the identification of semi-multivariate threshold autoregressive models of times series systems that account for the dependence of x_τ on $x_{\tau-i}$, $i = 1, \ldots, p$, as well as on other cross-correlated, lagged predictor variables, would have even greater applicability. Note that we are *not concerned* with complete (vector-valued) multivariate models (Box and Tiao, 1977; Tiao & Tsay, 1989), although a recent paper (Lewis & Stevens, 1992) does contain a discussion of complete multivariate modeling using MARS.

There are numerous semi-multivariate times series systems that appear well suited for analysis using the MARS methodology. Sea surface temperatures using lagged temperature, surface winds, and time as predictor variables (Altman, 1987; Lewis & Ray, 1992), or riverflow using lagged river flow, temperature and precipitation as predictor variables (Gudmundsson, 1970; Tong, Thannoon, & Gudmundsson, 1985) are just two such examples. One possible source of nonlinearity in the riverflow system might occur due to the change in temperature above and below freezing. Below freezing, precipitation (snow) does not "run off" as rapidly as precipitation (rain) at higher temperatures. Other applications exist for any multivariate times series system with suspected nonlinear behavior, if the objective is to model a single output stream given multiple system inputs. Thus the heart-rate data from the sleep study patient mentioned in Section 1 and shown in the top panel of Figure 1 could be modeled using lagged heart rates, chest volumes, and blood oxygen levels.

To provide a framework for the semi-multivariate time series model, suppose that for $\tau = 1, 2, \ldots, N$, $\{y_\tau\}$ and $\{z_\tau\}$ denote the input time series and $\{x_\tau\}$ the output time series for a time series system we wish to model. Following Eq. (1), and using the notation (Tong, Thanoon, & Gudmundsson, 1985) $\|$ to separate the predictor variables of each different time series, we can describe x_τ with the semi-multivariate time series regression model

$$x_\tau = f\left(1 \parallel x_{\tau-1}, x_{\tau-2}, \ldots, x_{\tau-d_1} \parallel y_\tau, y_{\tau-1}, \ldots, y_{\tau-d_2} \parallel z_\tau, z_{\tau-1}, \ldots, z_{\tau-d_3}\right) + \epsilon_\tau,$$
(8)

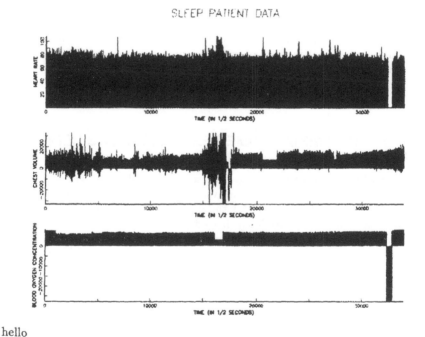

hello

FIGURE 1 Plots of heart rate, chest volume, and blood oxygen concentration for a sleep study patient. Four sleep-wake states were color-coded into the bottom plot but show as different grey.

where 1 denotes a model constant and again the maximum lags d_1, d_2, and d_3 are not necessarily equivalent. Also, y_τ and z_τ, the current values of the predictive time series, may or may not be included in Eq. (8), depending on the time series system.

A similar methodology for semi-multivariate TAR modeling that follows the TAR methodology for a univariate time series, i.e., identification of linear semi-multivariate autoregressive time series models in each disjoint subregion of the predictor variable space, has also been suggested (Tong, Thanoon, & Gudmundsson, 1985; Tsay, 1989). For example, a *simple* two-subregion semi-multivariate TAR model based on a single partition in the space of all the predictor variables at, for example, $z_\tau = 3$ is

$$\widehat{x}_\tau = \begin{cases} f_1 \left(0.5 \parallel 1.1 \parallel -2.7, 1.1 \parallel 4.3, -2.8\right) & \text{if } z_\tau \leq 3, \\ f_2 \left(2.3 \parallel 0.1, -0.2 \parallel 1.7 \parallel -0.1, 2.1\right) & \text{if } z_\tau > 3, \end{cases}$$

which represents the model

$$\widehat{x}_\tau = \begin{cases} 0.5 + 1.1x_{\tau-1} - 2.7\,y_\tau + 1.1\,y_{\tau-1} + 4.3z_\tau - 2.8z_{\tau-1} & \text{if } z_\tau \leq 3, \\ 2.3 + 0.1x_{\tau-1} - 0.2x_{\tau-2} + 1.7y_\tau - 0.1z_\tau + 2.1z_{\tau-1} & \text{if } z_\tau > 3. \end{cases} \quad (9)$$

Other semi-multivariate TAR methodologies (Tong, Thanoon, & Gudmundsson, 1985; Tsay, 1989) focus on univariate and bivariate scatter-plot analysis and on the evaluation of empirical percentiles of preselected threshold variable candidates. *These methods are also permitted with* MARS. However, the predictor variables of a time series system may possess physical behavior not readily apparent when we restrict our modeling methodology to the above approach. The key point is that Tong's and Tsay's methods (Tong, Thanoon, & Gudmundsson, 1985; Tsay, 1989) are time consuming, generally limited to one or two dimensions, and may not be sufficient for identifying changes in the physical behavior of a nonlinear time series system. Thus, a semi-multivariate TAR model is still burdened with the limitations of a univariate TAR model, i.e., a threshold model created with the piecewise linear models from each disjoint subregion of a domain D of the predictor variables. Also the TAR model is usually discontinuous at each subregion boundary (threshold) and is limited to a small number of thresholds, most often using only one variable, due to the difficulties associated with the threshold selection process.

The MARS methodology supplements the approach of Tong, Thanoon, & Gudmundsson, (1985) by admitting a more general class of continuous nonlinear semi-multivariate threshold models than permitted with the semi-multivariate TAR methodology. We call the methodology for developing this class of nonlinear semi-multivariate threshold models SMASTAR (Semi-Multivariate Adaptive Spline Threshold Autoregression). The general formulation of the SMASTAR model is complex (Lewis & Stevens, 1992) and is not given here; instead, using a simple example, we show that the SMASTAR methodology admits a more *general* class of continuous nonlinear threshold models than permitted with semi-multivariate TAR methodology.

Using Eq. (8), let x_τ be a time series we wish to model with the lagged predictor variables $x_{\tau-1}$, $x_{\tau-2}$, $y_{\tau-1}$, $y_{\tau-2}$, $z_{\tau-1}$, and $z_{\tau-2}$. Also, let the notation $(U - t)_+^\pm$ represent $(t - U)_+$ and $(U - t)_+$, where $(u)_+ = u$ if $u > 0$ and 0 otherwise. Each forward step of the MARS algorithm selects *one and only one* set of new terms for the SMASTAR model from the candidates specified by previously selected terms of the model. For our example problem the sets of candidates in the initial forward step of the MARS algorithm are

$$(x_{\tau-1} - t_x^*)_+^\pm \text{ or } (x_{\tau-2} - t_x^*)_+^\pm \text{ or}$$
$$(y_{\tau-1} - t_y^*)_+^\pm \text{ or } (y_{\tau-2} - t_y^*)_+^\pm \text{ or} \tag{10}$$
$$(z_{\tau-1} - t_z^*)_+^\pm \text{ or } (z_{\tau-2} - t_z^*)_+^\pm,$$

where t_x^*, t_y^*, and t_z^* are unknown partition points (thresholds) in the range of their respective lagged predictor variable. For our example problem, assume that the MARS algorithm selects the lagged predictor variable $x_{\tau-2}$ with threshold value $t_x^* = t_1$; i.e., $(x_{\tau-2} - t_1)_+$ and $(t_1 - x_{\tau-2})_+$ are the initial terms (other than the constant) in the SMASTAR model. The sets of candidates in the second forward step of the MARS algorithm includes *all univariate candidates in* Eq. (10) and the new sets of multivariate candidates (interactions):

$$(x_{\tau-1} - t_x^*)_+^\pm(x_{\tau-2} - t_1)_+, \text{ or } (x_{\tau-1} - t_x^*)_+^\pm(t_1 - x_{\tau-2})_+, \text{ or}$$
$$(y_{\tau-1} - t_y^*)_+^\pm(x_{\tau-2} - t_1)_+, \text{ or } (y_{\tau-1} - t_y^*)_+^\pm(t_1 - x_{\tau-2})_+, \text{ or}$$
$$(y_{\tau-2} - t_y^*)_+^\pm(x_{\tau-2} - t_1)_+, \text{ or } (y_{\tau-2} - t_y^*)_+^\pm(t_1 - x_{\tau-2})_+, \text{ or}$$
$$(z_{\tau-1} - t_z^*)_+^\pm(x_{\tau-2} - t_1)_+, \text{ or } (z_{\tau-1} - t_z^*)_+^\pm(t_1 - x_{\tau-2})_+, \text{ or} \tag{11}$$
$$(z_{\tau-2} - t_z^*)_+^\pm(x_{\tau-2} - t_1)_+, \text{ or}$$
$$(z_{\tau-2} - t_z^*)_+^\pm(t_1 - x_{\tau-2})_+,$$

due to the initial selection of $(x_{\tau-2} - t_1)_+$ and $(t_1 - x_{\tau-2})_+$ as terms in the SMASTAR model. It follows that SMASTAR models could have multiple thresholds on one variable, say $x_{\tau-2}$ in our example, by again selecting the set of terms $(x_{\tau-2} - t_x^*)_+^\pm$ in Eq. (10) for some new partition point $t_x^* \neq t_1$. The forward step algorithm continues at each step by selecting the set of univariate or multivariate terms that, for a given threshold t_x^*, t_y^*, or t_z^* discovered using exhaustive search, most contributes to "improving" model fit. The sets of candidates for each subsequent forward step of the SMASTAR algorithm is nondecreasing in size and is based on previously selected terms of the model. As discussed previously the forward step algorithm is followed by a backward-step algorithm that trims excess terms of the model that no longer sufficiently contribute to the model fit. And again, both the forward and backward steps of the algorithm use the generalized cross-validation criterion or the Schwarz-Rissanen criterion to evaluate model fit versus model complexity.

By modeling a time series system using MARS, we overcome some of the limitations of the TAR modeling approach. The MARS methodology provides a systematic procedure for deriving a model that is naturally continuous in the domain of the predictor variables. As recently shown (Lewis and Stevens, 1991) for the yearly Wolf sunspot numbers, ASTAR models of univariate time series can possess multiple thresholds and high-level predictor variable interactions. This has now been extended to the multivariate setting with SMASTAR models, which can also possess multiple thresholds and high-level predictor variable interactions. Only now the threshold values and predictor variable interactions can take place among the cross-correlated, lagged predictor variables of a time series system.

As an illustration of SMASTAR's ability to model an actual semi-multivariate times series system, Lewis and Stevens (1992) examined the riverflow for the Vatnsdalsa river in Iceland. The resulting model is more parsimonious and gives better predictions than the model derived by Tong and his colleagues (Tong, Thanoon, & Gudmundsson, 1985).

5. CATEGORICAL TIME SERIES AS PREDICTOR VARIABLES (CASTAR)

The latest version of the MARS program (MARS 3.50) allows for categorical input. There is a question of what one means by smoothness for variables which do not have an ordering. In MARS this is taken to mean that a smooth set of possible categorical values is one over which the structure of the (continuous, ordered) part of the process generating the data varies very little. This has been shown to work well for categorical time series (Lewis & Ray, 1992). The need for these kinds of additional modeling capabilities for time series is clear. Intervention analysis, in which the input to a system is either 0 or 1, is one example of time series modeling with categorical variables, and the sleep-study patient data mentioned in Section 1 is another. In the heart-rate data, one of the pieces of information being measured is the sleep state of the patient; for example, the patient can be either awake, in a groggy condition between sleeping and waking, asleep, or in REM sleep. The bottom panel of Figure 1 indicates the sleep state of the patient over time through grey scale. One interesting point though is that there is only one period of REM sleep and it occurs in the middle of the record. It shows as the lightest grey in the lower panel and is associated with a large transient in the Chest volume data. This could be an artifact of the data collection, but the initial indication is that that the heart rate and the chest volume of the patient changes when he is in REM sleep.

The MARS model involving categorical variables has the following form:

$$x_\tau = \hat{f}(\mathbf{z}) = a_0 \sum_{m=1}^{M} a_m \prod_{k=1}^{K_{cm}} I(z_{v(k,m)} \in A_{km}) \prod_{k=1}^{K_{om}} (s_{km(z_{v(k,m)}-t_{km})})_+^q, \qquad (12)$$

where K_{om} is the number of factors in the mth ordinal basis function, $s_{km} = \pm 1$ and indicates the (left/right) sense of the truncation, and t_{km} is a knot location on each of the corresponding variables. See Friedman (1991c) for a more complete discussion of the computational and interpretational aspects of the ordinal-categorical model.

Other related methods for modeling categorical time series data are the Markov chain-driven TAR described in Tong (1990, pp.102–103) and methods given by Tyssedal & Tjostheim (1988).

6. APPLICATION TO SWISS FRANC/U.S. DOLLAR DATA

In this section, we apply MARS to a series of tick-by-tick quotes of the Swiss franc/U.S. dollar exchange rate and discuss the resulting models. The exchange rate data is from the Union Bank of Switzerland and was made available to us as part of the Santa Fe Institute's Comparative Time Series Analysis workshop. The data consists of ten sections, each containing 3000 records. Each record contains the exchange rate, an integer denoting the day of the week (e.g., 1 denotes Monday), and the elapsed time since the start of trading that day. Each section contains data for approximately two weeks of trading; however, the ten sections are not contiguous.

There are several features of the data which complicate its analysis. The first is that the quoted exchange rate is recorded on a tick-by-tick basis. We do not model the actual exchange rate, p_t, at tick t, but the return r_t at tick t, which we define (Taylor, 1986) as $r_t = \ln p_t - \ln p_{t-1}$. In examining the returns series for the first section of data, we found that approximately 30% of the returns were 0.00, and the remainder of the data was fairly discretized at several points symmetric about 0.00. This occurs because the change in exchange rate at each tick is always a multiple of 0.0005. An underlying assumption of the MARS algorithm is that the data comes from a continuous distribution, thus the discretization of the returns may cause problems in the application of MARS to this data. Another complication of the tick-by-tick quotes is that the ticks are not equally spaced in time. A variable amount of time may pass before another exchange rate quote is recorded. Rates are usually not quoted overnight; thus, there is a gap of about 13 hours between the last quote for one day and the first quote of the next day. Unequally spaced time intervals are a problem if we want to use the model of the tick-by-tick returns for prediction purposes. We can predict the return at a future tick, *but cannot predict when that tick will occur in real time!*

One method of dealing with this problem is to average the exchange rate quotes occurring in a specified time interval, e.g., five minutes, and model the returns obtained from the averaged exchange rate series. If no quote occurs in a time interval, we assume that the exchange rate is the same in that interval as in the previous interval. Counting the number of quotes in the time interval provides a measure of market activity that may also be useful in modeling the returns. Of course, aggregating the data over successively longer time intervals acts to smooth out some of the structure in the data, unless the data is long-range dependent (Cox, 1984). When we applied MARS to the series of returns from the third section of the Swiss franc/U.S. dollar exchange rate data aggregated over five-minute intervals, the data appeared to have very little dependence structure; i.e., only the first sample correlation value was significantly different from zero at a marginal level. We mention another approach to the problem of unequally spaced time intervals in Subsection 6.3.

A second complication encountered in analyzing the exchange rate data is its apparent nonstationarity. When we applied the MARS algorithm to different sections of the returns data, holding the input parameters to the MARS algorithm fixed, the resulting model was a constant for most sections, indicating that the best model for the exchange rate series was a random walk. In the remaining cases, the resulting ASTAR models varied greatly from one section to another.

6.1 ASTAR MODELS FOR THE SWISS FRANC/U.S. DOLLAR EXCHANGE RATE

We made several attempts at modeling different sections of the data. We present here only the results for modeling the third section of data. Before applying MARS to the returns data, we examined its autocorrelation structure. We found only a very small negative autocorrelation between returns at lag one. We also examined the autocorrelation between the absolute value of the returns, as well as the squared returns. Significant correlation between the absolute returns or the squared returns in the absence of correlation among the actual returns themselves may indicate the presence of nonlinearities or non-Gaussianity in the series (Tong, 1990). The squared returns showed negligible autocorrelation, whereas the absolute returns showed small positive correlation at several lags.

We first applied the MARS algorithm to the tick-by-tick returns series, using 30 lagged returns as predictor variables. The returns were scaled up by a factor of 1,000 to avoid possible underflow problems in the computation of the model. The minimum value of the returns in this section is -128.0, the maximum value is 70.15, with 50% of the values falling between -3.892 and 3.893. We allowed 60 forward steps in the MARS algorithm, with a maximum of three interactions and a minimum span of five. We use the Schwarz-Rissanen criterion to choose the model in each application. The first 2,504 data points were used to estimate the model, with the remaining 465 values saved for validating the model predictions. The prediction problem for the exchange rate data will be discussed in Subsection 6.3. The resulting ASTAR model is

$$\hat{r}_t = \begin{cases} & \textbf{Model 1} \\ 0.096 & -1.739(r_{t-11} - 11.439)_+(r_{t-28} - 3.896)_+ \\ & -0.038(-3.815 - r_{t-1})_+(-11.474 - r_{t-2})_+(11.439 - r_{t-11})_+ \\ & -0.005(r_{t-1} + 3.815)_+(11.439 - r_{t-11})_+(r_{t-25} - 3.111)_+ \\ & +0.369(r_{t-5} - 11.558)_+(11.439 - r_{t-11})_+(-3.890 - r_{t-12})_+ \end{cases}$$
$$(14)$$

The residual sum of squared errors is $\hat{\sigma}_\epsilon = 5.052$ and the value of the SC is 3.311. The model may be interpreted explicitly to give an understanding of how the returns fluctuate over time. For example, the first term of Eq. (14) indicates that when the return 11 ticks ago was greater than 11.439 and the return 28 ticks ago was greater than 3.896, the value of the next return will be pulled down by a factor of -1.739 multiplied by the amount that r_{t-11} is greater than 11.439 and r_{t-28} is greater than 3.896. The fourth term in Eq. (14) acts to bring the value of the next return back up if r_{t-5} was greater than 11.558, r_{t-11} was less than 11.439, and r_{t-12} was less than -3.890.

6.2 SMASTAR AND CASTAR MODELS FOR THE SWISS FRANC/U.S. DOLLAR EXCHANGE RATE

It seems reasonable that the elapsed time between ticks may contain some information useful for modeling the returns. For example, if the market is not active, a longer time may elapse between ticks and the return values may change very little. The estimated cross correlation between the returns series, r_t, and the time between ticks (in seconds), τ_t, indicates a small amount of negative correlation between the time between ticks and the returns one and two ticks later. We use the lagged times between ticks (in seconds) as additional predictor variables in the MARS algorithm to obtain a SMASTAR model for the returns, with the elapsed time between the last quote of the day and the first quote of the next day set to four hours. We also use two categorical series, the first a series consisting of integers from 1 to 5 indicating the day of the week, and the second a series of integers from 0 to 10 indicating the hour after start of trading as additional predictor variables. The use of categorical time series in the MARS algorithm is discussed in Section 5. See Taylor (1986, Ch. 2.5) for a discussion of day-of-week effects in financial time series. The input parameters for the algorithm are the same as in Model 1, with 30 lagged returns, 30 lagged times between ticks, the day of the week, and the hour after start of trading as predictor variables. The resulting SMASTAR model is

$$
\hat{r}_t = \left\{
\begin{array}{l}
\textbf{Model 2} \\
0.588 \quad -0.294(r_{t-5} - 15.516)_+(11.558 - r_{t-11})_+ \\
\quad -1.758(r_{t-11} - 11.558)_+(r_{t-28} - 3.896)_+ \\
\quad -0.034(-3.815 - r_{t-1})_+(-11.474 - r_{t-2})_+(11.558 - r_{t-11})_+ \\
\quad +0.036(r_{t-1} + 3.815)_+(r_{t-5} - 11.662)_+(11.558 - r_{t-11})_+ \\
\quad +0.163(-3.815 - r_{t-1})_+(r_{t-5} - 11.550)_+(11.558 - r_{t-11})_+ \\
\quad -0.095(-3.815 - r_{t-1})_+(r_{t-5} - 7.695)_+(11.558 - r_{t-11})_+ \\
\quad -0.012(-3.815 - r_{t-1})_+(11.558 - r_{t-11})_+(\tau_{t-2} - 545)_+ \\
\quad +0.017(-3.815 - r_{t-1})_+(11.558 - r_{t-11})_+(\tau_{t-2} - 642)_+ \\
\quad -0.000023(r_{t-1} + 3.815)_+(11.558 - r_{t-11})_+(\tau_{t-8} - 650)_+
\end{array}
\right.
$$

$$(15)$$

The residual sum of squared errors is $\hat{\sigma}_\epsilon = 4.960$ and the value of the SC is 3.325. The day of week or hour of day do not appear in the final model, indicating that these variables are not useful in predicting the next return. The time between ticks, τ, does come into the model, indicating different behavior of the returns depending on whether the elapsed time is approximately greater than or less than 11 minutes.

6.3 PREDICTING EXCHANGE RATES USING THE ASTAR MODEL

Model 1 of the previous section may be used to predict the return at future ticks, with the predictions integrated into a trading strategy which uses the predicted change in exchange rate to determine trading actions. Using Model 1 by itself, we cannot predict when the next tick will occur, however. One method of doing this would be to fit a univariate MARS model to the elapsed time between ticks and use the resulting prediction equation to estimate the number of elapsed seconds before the next tick. These predictions could also be used in the prediction equation of the SMASTAR model (Model 2).

We computed the mean-squared k-step-ahead "out-of-sample" prediction errors for the returns series over the last 465 data points (approximately two days of quotes) using Model 1 and compared them to those obtained by using the mean of the series as the predictor at each step. Assuming that the returns series is constant is the same as assuming that the exchange rate series is a random walk (the exchange rate market is efficient). A random walk model was obtained from most other methods used by participants in the conference. At most steps, the MARS model did not perform as well as the random walk model, even though the residual sum of squares, which is the same as the one-step-ahead forecast error over the data used to fit the model, was smaller than that of the sum of squares of the data after subtracting the mean. The residual standard error of the random walk model is $\hat{\sigma}_\epsilon = 7.898$, compared to $\hat{\sigma}_\epsilon = 5.052$ for Model 1. This is another indication that the data may be nonstationary, in that the MARS model estimated using the first 2504 points does not hold for the remainder of the data. The prediction problem may also be caused by the discretization of the data, as discussed at the beginning of this section. Since the random walk model uses the value of the series at time t as the predictor of the series value at time $(t + 1)$, the predictions from this model are also discrete values. This is not true of the predictions generated by the ASTAR model.

7. DISCUSSION AND COMPARISONS TO OTHER METHODS

The following discussion is quite general and probably raises more questions than it even attempts to answer. But if the differences and relative advantages of various of the newer methods of (nonlinear) time series analysis, such as MARS and neural networks, are to be assessed, then some of these questions need to be raised.

7.1 DATA ANALYTIC CONSIDERATIONS

It is tempting to say, as a statistician, that statistical training mandates that one look at the data before leaping into a specific model or methodology, but of course many people do not do this. So assume our job is to analyze the data; why should we look at the data at all? One reason is that, as computers become faster and more accessible, one is likely to be more able to transform problematic data into problematic results. A case in point is the sleep study patient data; it is quite extensive, consisting of 34,000 observations for each of the three series, making even initial analyses about the nature of the data almost overwhelming. However, by simply plotting the data, as we have done in Figure 1, we can get an overall view of its significant characteristics.

Several things stand out that make the analysis difficult. First, there is a large transient near the middle of the data which is reflected in the heart rate, possibly the result of some intervention in the patient's condition. The transient is accompanied by an increase in heart rate and a drop in blood oxygen concentration, which could be an effect of the patient's REM sleep state or could be simply an instrumentation artifact, as is the apparent shut down and resumption of the heart beat at the end of the data set. Clearly this data is structurally inhomogeneous in time, and a model which is based on these assumptions is nonsensical. The data might perhaps be best analyzed in pieces. Figure 1 has the patient's sleep state coded into the bottom panel, making it easier to see what effects the sleep state may have on the heart rate. Consideration of how the model is to be used is also important. Modeling, for example, to predict sleep apnea could be entirely different from modeling to interpret the general oscillations and interactions in the three series.

7.2 ASSUMPTIONS

Closely tied into the admonition to "look at the data" is the issue of the assumptions underlying different methodologies. With regard to MARS, it is clear that the methodology is predicated on having continuous response variables. Does the MARS methodology require normality as well? This issue is important when the time series data is autoregressed on itself; in particular, it is clear that very skewed, positive-valued data is hard to handle with MARS, and data of this type is usually transformed. Thus in modeling the Vatnsdalsa river flow data and the related rainfall data, which are both very skewed, a symmetrizing log transformation is found to be necessary (Lewis & Stevens, 1992) as it is in most regression or autoregression situations. Again for the exchange rate data, the returns are zero about 30% of the time. The discretization of the data suggests that the Mars methodology will not work for that data. Similarly, neural networks which are based on sigmoid functions produce continuous valued output; thus, analogous problems may be encountered with standard neural networks.

The discreteness of the returns data suggests that, as an alternative to MARS and other methodologies, it might be better to model the data using a Markov chain model with, for example, five states: no change in the return from time period $t-1$ to t, a small positive change, a big positive change, a small negative change, and a big negative change. Assuming that the distribution of the return at time t is conditional only on the state of the return in the previous two time periods, one has 25 states with which to determine the best predictor of the state of the return at time t. The model is "autoregressive" but clearly nonlinear. How well would this model work for someone who was trying to make money out of trading in currencies?

Going further, what assumptions are there, if any, concerning the variance homogeneity and independence of the residuals in Eq. (1)? These are standard assumptions in stationary, linear autoregressive modeling, and much work has been expended to accommodate cases in which they are clearly not true. For example, ARCH models (Engle, 1982) have been developed to handle time series which occur in economic contexts and which clearly do not exhibit variance homogeneity. Our experience shows that making a variance-stabilizing transformation before modeling the sea surface temperatures discussed in Section 5 increases the accuracy of the predictions.

7.3 SUBJECTIVITY IN TIME SERIES APPLICATIONS

Another important issue is that of the issue of parameters in the application of time series methods. Neural networks have sometimes been advocated as devices for doing statistics without thinking. We doubt that this is true, especially since simple questions like the number of nodes or the number of hidden layers to have in the network must be addressed by the user.

The same thing is true for the MARS methodology. One must choose the goodness-of-fit criterion, although our research has indicated a strong preference for the Schwarz-Rissanen criterion (Eq. (7)) over generalized cross-validation (Eq. (6)). One must also choose the maximum number of forward steps performed by the algorithm (best chosen to be as large as is feasible), the minimum span (a smoothing parameter), the maximum level of interactions (three seems to be the maximum one can cope with interpretively), the number of auto-predictors (lags) to allow, and the cross-predictors to use. For example, if we were to apply MARS to the sleep-study patient data, we would need to consider whether the heart rate is of interest on its own, or do we need to use a semi-multivariate SMASTAR model and include chest volume and blood oxygen concentration as predictors of heart rate? Choosing the predictors is a problem with any method; with regard to the other parameters, experience has shown that, when using MARS, it is best to allow as extensive a model as possible and let the algorithm do the choosing. This is true partly because it is difficult to project one's expectations of behavior into nonlinear equations, and partly because it is the nature of the MARS algorithm to delineate structure. The caveat here is that MARS is a computer-intensive methodology; thus, allowing more parameters, more interactions, and a finer search can be time consuming.

7.4 RANGE OF APPLICABILITY OF DIFFERENT METHODS

Data Set A consisted of 1,000 values generated as output from a laser. The data shows a high-frequency oscillation whose amplitude builds up rapidly, collapses, and then builds up again. No parameterization of an ASTAR model produced good predictions for this data more than about ten steps beyond the end of the data. In fact, the predicted values then exploded, in complete contrast to the slow buildup in the real data.

What causes this instability in the ASTAR model? It could well be a lack of robustness to extreme values in the MARS methodology which contaminates higher level interactions. The experience with the time series data for the competition seems to confirm this. It would be interesting to know if other methods of time series prediction—for example those based on bilinear models (see Tong (1983), pp. 414-415 and Example 5 on p. 444) or locally linear methods (Farmer & Sidorowich, 1988) show a similar instability, particularly with this data set.

7.5 OTHER BASIS FUNCTIONS

Several other schemes for finding a suitable form for $f(\mathbf{x})$ have been proposed, in particular that of radial basis functions (Casdagli, 1989). Friedman (1991a) has pointed out, however, that radial basis functions give approximations to $f(\mathbf{x})$ which involve the highest possible interaction order. It is therefore possible that they do not work well with functions which are almost additive. They do not have the disadvantage of neural networks of not giving explicit formulae for $f(\mathbf{x})$, but since they do not do input variable subset selection, the resulting equation will probably not be easily interpretable.

7.6 NEURAL NETWORKS

There is, of course, wide interest in using neural networks in a variety of applications, for example, in meteorology (Elsner & Tsonis, 1992). A great disadvantage of neural nets is that the transformation of input to output is an extremely complicated function, making it difficult to elucidate the physical structure of the process generating the input and output. Another point is that there seems to be an implicit assumption of stationarity in the training data and in the response function; it is not clear that neural networks can handle the case in which this assumption does not hold. Using the MARS methodolgy, for example, if there really is an evolutionary trend in the data, e.g., if sea surface temperatures are really continually increasing, then by using a semi-multivariate SMASTAR model in which one of the cross-predictor series is time, one can accommodate this situation. Can this be done using a neural network?

Yet another question arises if the time series under study is generated by a linear, autoregressive process; will a neural network identify this situation, and work well and parsimoniously? We have not seen any simulation results for neural nets like those of Stevens (1991), which validate the ability of the ASTAR procedure to identify and estimate parameters in a linear autoregressive process.

7.7 LOCALLY LINEAR METHODS

We have not compared our empirical results to those obtained using locally linear methods (Farmer & Sidorowich, 1988), though this would seem to be a fruitful question for further research.

8. APPENDIX: PROGRAMS AND TIMINGS

The latest version of the MARS program is 3.5. We have modified Friedman's program in several ways to run more easily with time series and have provided preprocessing and postprocessing programs. The resulting package of programs is available from P. A. W. Lewis at 1526p@NAVPGS.BITNET or from B. K. Ray at borayx@M.NJIT.EDU. The programs are written in FORTRAN-77 and run on microcomputers, workstations, and mainframes with only minor modifications.

With regard to run time, an extreme case is one in which 20 years of daily sea surface temperatures were modelled, allowing 365 predictor variables (365 lagged values of sea surface temperatures) and up to three-way interactions. The required computation time was approximately eight hours on an IBM 3090 mainframe computer. In contrast, complicated models for the Wolf sunspot numbers required only seconds to run on a 25-megaherz 486 microcomputer.

ACKNOWLEDGMENTS

The research of P. A. W. Lewis was supported at the Naval Postgraduate School under an Office of Naval Research Grant. The research of B. K. Ray was carried out while she was a National Research Council postdoctoral fellow at the Naval Postgraduate School. The interest of P. A. W. Lewis in nonlinear threshold models was stimulated by attendance at the Research Workshop on Nonlinear Time Series, organized by H. Tong at Edinburgh, Scotland from 12–15 July 1989. The authors are also grateful to Ed McKenzie for his helpful ideas when applying MARS to the modeling of time series. Finally, we thank Professor Jerome Friedman for supplying us with his MARS program and for help and encouragement.

George G. Lendaris and Andrew M. Fraser
Systems Science Ph.D. Program, Portland State University, Portland, Oregon 97207-0751
email: lendaris@sysc.pdx.edu, andy@ee.pdx.edu

Visual Fitting and Extrapolation

We describe how, matching several segments of the time series plotted on overlaid sheets of vellum tracing paper, we created the forecast of Data Set A that we entered in the Santa Fe Time Series Prediction and Analysis Competition.

INTRODUCTION

We made our forecast of Data Set A by searching the given sequence of 1,000 measurements for segments that nearly match the pattern of the last 21 and then guessing that similar antecedent segments are followed by similar consequent segments. This description fits almost any forecasting procedure, but, while most entries in this competition are interesting because of the computer pattern-matching techniques used, we did our pattern matching by eye—plotting segments of the data on vellum tracing paper, laying the plots on top of each other and shifting them to obtain the best match.

Looking at a plot of the entire data set, such as Figure 1(a), we saw three antecedent segments that roughly match y_{980}^{1000}. We plotted each of these three segments and the corresponding consequent 65 points, namely y_{105}^{190}, y_{430}^{515}, and y_{535}^{620}, on

Times Series Prediction: Forecasting the Future and
Understanding the Past, Eds. A. S. Weigend and N. A. Gershenfeld, SFI Studies
in the Sciences of Complexity, Proc. Vol. XV, Addison-Wesley, 1993   **319**

$17'' \times 22''$ vellum sheets. On each of these plots, we drew a curve through the data points. We made our forecast by plotting y_{980}^{1060} on a fourth vellum sheet. First, we plotted the given points, y_{980}^{1000}, then we laid the forecast plot on top of each of the three roughly matching antecedent segments in turn. We shifted the vellum back and forth to get the best match; in each case $t = 980$ was aligned with a noninteger time (105.5, 429.1, and 532.6). The alignment is depicted in Figure 2. After we had set the alignment with each antecedent segment, we plotted estimates of the forecasts, $\{y(t) : t = 1001 \cdots 1060\}$, by locating the points where the curves connecting the consequent segments drawn on the underlying vellum intersected the vertical grid lines on the upper vellum.

After using each of the matching antecedent segments in this fashion, we had three estimates for each of the values to be forecast. For each time step, $\{t = 1001 \cdots 1060\}$, our contest entry consisted of the mean and sample variance of these three estimates.

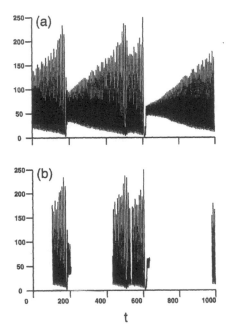

FIGURE 1   The entire contest data set, y_1^{1000}, is plotted in (a). The segment on the right in (b) is y_{980}^{1000}, the antecedent for the forecast, and the three segments on the left were selected because they begin with segments that roughly match the forecast antecedent on the right.

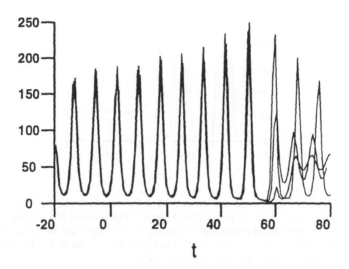

FIGURE 2 The three segments with matching antecedents identified in Figure 1(b), y_{105}^{204}, y_{105}^{204}, and y_{105}^{204}, are superimposed here. We used the vellum overlays to align the t axis of each trace to get the best match with the forecast antecedent, y_{980}^{1000}. On the plot, $t = 0$ corresponds to $y(1000)$, $y(125.5)$, $y(449.1)$, and $y(552.6)$. The divergence of the traces suggests that we cannot forecast more than 60 steps into the future.

Our forecast, along with the true continuation data, appears in Figure 3, and the *score*[1] for each point k

$$\left[\frac{1}{\sqrt{2\pi\,\sigma_k^2}} \int_{d_k-0.5}^{d_k+0.5} e^{-(\xi-y_k)^2/(2\,\sigma_k^2)}\,\mathrm{d}\xi \right]$$

is plotted in Figure 4. The average score for our entry is 8.63. As shown in Figure 4, most of our score is due to errors in the last two cycles we tried to predict. If we had shifted our prediction one cycle, nine sampling periods, to the left, our average score would have improved to 4.84.

[1] In this expression, for time step k, y_k and σ_k^2 are the forecast mean and variance respectively of the entry submitted, and d_k is the true continuation value. Gershenfeld and Weigend justify this choice in the appendix of the first chapter in this volume.

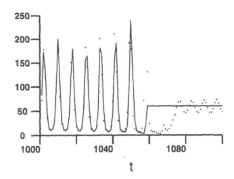

FIGURE 3   The solid line traces our forecast, and the dots indicate the true continuation data. We could not guess the phase or magnitude of the oscillation beyond $t = 1060$, so we forecast a mean that is roughly an average over all three of the consequent segments of a one-cycle moving average, and we selected a variance that is roughly the mean square amplitude of oscillations.

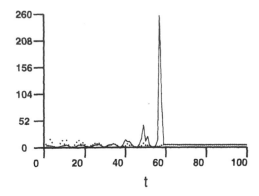

FIGURE 4   The solid trace is the score for our entry, and the dotted trace is the score we would have had if we had shifted our entry one cycle to the left.

Leonard A. Smith
Mathematics Institute, University of Oxford, Oxford, OX1 3LB, United Kingdom;
e-mail: lenny@maths.ox.ac.uk

Does a Meeting in Santa Fe Imply Chaos?

I have no data yet. It is a capital mistake to theorize before one has data.

— Holmes to Watson in *A Scandal in Bohemia*
(see Doyle, 1930)

This chapter compares the success of several nonlinear prediction techniques applied to Data Set A of the Santa Fe Time Series Prediction and Analysis Competition (both A.dat and A.cont. The advantages of a new approach making predictions based on selective use of several different delay reconstructions are illustrated, and a comparison of both local linear and local nonlinear predictions is given. In addition, the phase coherence of the system and the self-consistency of the data is examined using the longer data set A.cont; the latter locates a possible sensor failure in this data set. Limitations due to the amount of data, the sampling rate, and the saturation in the data, in combination with the quality of the predictions achieved with very little information on the value of the initial condition

Times Series Prediction: Forecasting the Future and
Understanding the Past, Eds. A. S. Weigend and N. A. Gershenfeld, SFI Studies
in the Sciences of Complexity, Proc. Vol. XV, Addison-Wesley, 1993

(32 bits or less), suggest that, while the system is clearly nonlinear, evidence from A.dat for sensitive dependence on initial condition, if any, is slight.

---

1. INTRODUCTION

The theme of the workshop was the extraction of information from data. In this chapter, we will approach this task with a variety of nonlinear prediction techniques and an examination of the data itself. For all the prediction results presented, we will construct the predictor from the data in file A.dat and test it on the data in file A.cont which is a continuation of A.dat, thus all error statistics are out of sample. Because the predictors are constructed from A.dat, this data set will be called the learning set.

In addition to global predictors, in which each prediction is influenced by the entire learning set, and local predictors, based on a local (one hopes relevant) subset of points, we shall introduce predictors that take advantage of different reconstructions at different times (i.e., in different regions of phase space). This approach can provide better predictions than any single "optimal" reconstruction.

Many of the results presented at the conference, along with those reported here, illustrate that the system is very predictable. Examination of the data in the continuation set, especially following collapses (defined below), supports this view. We suggest that the dynamics of the underlying system might be modeled simply; while the series is clearly nonlinear, there is little evidence of sensitivity

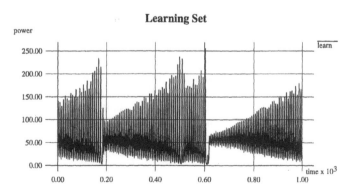

FIGURE 1   The original data set A.dat. These 1,000 points are used as the learning set for the predictors presented below.

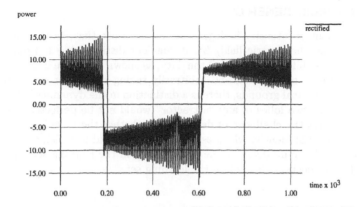

FIGURE 2 An artificially rectified version of A.dat; the regular growth of the oscillations appears more clearly after taking the square root of the data. In the second cycle, the negative square root has been arbitrarily taken to illustrate the possible degeneracy.

to initial condition. Although the physical system may be in fact chaotic, it seems premature to theorize this from the presented analyses based on the data in A.dat alone. Finally, we suggest modifications to the observations that underly this data set which could clarify these issues.

While we shall concentrate on prediction below, related techniques can be employed to detect anomalies in the data stream, which may result from either sensor failure or a major change in the system's dynamics. Such an anomaly is, in fact, detected in A.cont.

2. THE DATA

The initial data set A.dat, shown in Figure 1, consists of 1,000 observations, equally spaced in time and digitized to 8-bit integers in the range 0 to 255 (inclusive). We will call the high-frequency oscillations "cycles," the sudden large decreases in the amplitude of the cycles "collapses," and the packet of growing oscillations between two collapses "an event." The series samples three events, in two of these the collapse is observed; note that if this data set were to arise from motion on a chaotic attractor, it would provide a sparse image of the attractor since only one complete event (collapse through growth to collapse) is available. With an eye on prediction we note that, as only two collapses are observed, the prediction of collapses may be expected to prove difficult, even if the mechanism giving rise to them is straightforward.

2.1 THE PHYSICS: GENERIC?

The measurement of intensity is degenerate in the sense that, as the square of a physical variable (the electric field, E), it does not distinguish $+E$ from $-E$. To illustrate this possibility, we "rectify" the data, as shown in Figure 2, by taking the square root of each observation and arbitrarily taking the center set of cycles as negative. If, on physical grounds, there is a distinction in the dynamics between $+E$ and $-E$, then the two lobes of any "attractor" would not be perfectly symmetric; this would suggest the challenge of determining whether the techniques discussed at this meeting could identify which oscillations correspond to which lobe. If this is indeed the case, it would decrease the data density in the learning set still further and introduce projection effects into the learning set.

2.2 THE STATISTICS: STATIONARY?

We should also consider whether or not the statistics of A.dat are stationary or, more accurately, whether the statistics computed from A.dat have converged to those of an underlying stationary process as, technically, stationarity is defined for a process, not a data set. A necessary condition for such convergence is that the histogram of observed values be representative of that of the system. However, this is not sufficient for the reconstruction methods discussed below. These methods would require that the reconstruction in higher dimensions be "well explored."[1] Given that only two collapses occur in the learning set, this criterion could only be met for a very simple attractor.

The data also appears to be undersampled, judging from the series (the beating of the sampling rate and the cycles is reflected in Figure 4). Thus the times and magnitudes of successive local maxima are poorly approximated; it is suggested below that an excellent model of this system could be constructed from these two parameters. We also note that the signal saturates at 255, this is clearly illustrated in the histogram of observations from the continuation set (not shown, but the effect is seen in Figure 9). This saturation tends to coincide with collapse, making it difficult to judge the true sensitivity to initial conditions.

---

3. THE PREDICTORS

In this section we consider a variety of methods to predict the short-term behavior of the system. The methods considered require that the single observable first

---

[1]This would, of course, depend on the length of the data set, the sampling rate, and the embedding dimension. It is quite reasonable to expect that a given series reconstructed in three dimensions could appear "stationary" while the same data reconstructed in ten dimensions would not. Note, however, the discussion of this issue in Sugihara and May (1990).

be embedded to form a higher dimensional reconstruction. The time ordering of points in this reconstruction is then used to turn the prediction problem into one of interpolation.

3.1 THE METHOD OF DELAYS

The first step in applying these methods of prediction is to *reconstruct* the time series in a geometrical framework. A trajectory, $\mathbf{x}(t)$, is reconstructed in M dimensions from the single observable, $s(t)$, recorded with uniform sampling time, τ_s, by the method of delays to yield a series of vectors

$$\mathbf{x}_i = (s_i, s_{i-j}, \ldots, s_{i-j(M-1)}) \tag{1}$$

where j (or $j\tau_s$) is called the delay time, τ_d. For a deterministic system and a generic observable, this reconstruction preserves many of the characteristics of the original system for sufficiently large M (see, e.g., Casdagli et al., 1991; Packard et al., 1980; Sauer, Yorke, & Casdagli, 1991; and Takens, 1981). In this chapter, we will restrict attention to delay reconstructions of the full series where $M = 4$ with either $\tau_d = 1$ or $\tau_d = 4$. In addition to the prediction time, τ_p, there remains one further time scale to be considered, the width of the "window" used to determine a point in the reconstruction space, τ_w. For the method of delays, $\tau_w = ((M-1)j + 1)\tau_s$. We have observed, however, that multivariate reconstructions of the maximum value and duration of each cycle (i.e., prediction of the value and time of occurrence of the next maxima, given the series of local maxima) provide excellent, if preliminary, results. This approach avoids some of the difficulties discussed below, particularly with respect to collapses. Predictions based on this approach should be significantly improved by additional data with higher resolution (in both time and intensity), so that the maxima are more sharply defined, and may provide the clearest evaluation of sensitivity to initial condition.

3.2 MODES OF PREDICTION

In this section, we clarify several details in the approach to making predictions that must be specified independently of the details of the predictor itself.

3.2.1 DIRECT AND ITERATIVE FORECASTS.
In general, we will employ *direct fore-casts* with a single-step predictor invoked once per prediction. We contrast this single prediction a time τ_p into the future, with *iterative forecast predictions* where the final prediction is based upon a number of predictions made at a smaller time step (e.g., i forecasts, each advancing τ_p/i well into the future).

3.2.2 RUNAWAY AND UPDATE EXTENSION. For the competition, we were asked to take an initial condition and iterate it as far as possible, using the output of the predictor to continue the series. We will denote these as *runaway* extensions in contrast to *update* extensions, where the true value is incorporated after each step of the prediction. While the former are of interest when extending a true series into the future, the latter allow the evaluation of a predictor at a fixed τ_p, which eases parameter determination and interpredictor comparisons. All the results in the paper, with the exception of Figures 6 and 7, are based on update predictors.

3.2.3 ENTRAINMENT. Finally, we note that for quantized data, the system may repeatedly produce the same exact string of observed data values and then diverge. If this string is as long as the window considered by a predictor, τ_w, a deterministic *runaway* predictor cannot reproduce the behavior; it will become entrained within a periodic cycle quite unrelated to the primary dynamics of the system. A related problem distinguishes direct and iterated forecasts, namely that iterated predictors can't get through regions where, for example, the series appears to be constant for a time greater than τ_w; direct predictors can effectively "jump" over such a data segment. Depending upon location in phase space it is possible to combine the advantages of both iterative and direct forecasts by selective usage. This approach, illustrated below, combines two reconstructions with different delay times.

3.3 GLOBAL PREDICTORS

Having produced a four-dimensional reconstruction of the $(n_L = 1000)$ points in A.dat with $\tau_d = 1$ and $\tau_p = 1$, we first consider a global predictor, $F(\mathbf{x}) : R^4 \to R^1$, which estimates s for any $\mathbf{x}$. $F(\mathbf{x})$ is constructed about n_c centers

$$\mathbf{c_j}, \quad j = 1, 2, \ldots, n_c; \quad \mathbf{c_j} \in R^4 \tag{2}$$

chosen from the learning set in such a manner that they are spread out on the reconstruction. We will consider $F(\mathbf{x})$ of the form

$$F(\mathbf{x}) = \sum_{j=1}^{n_c} \lambda_j \phi(||\mathbf{x} - \mathbf{c_j}||) \tag{3}$$

where $\phi(r)$ are radial basis functions (Powell, 1987), in this contribution, either $\phi(r) = r^3$ or $\phi(r) = e^{-r^2/c^2}$ where the constant, c, is based on a multiple of the average distance between data points considered in the fit, d_{nn}. The λ_j are constants determined by a least-squares fit to the observations in the learning set:

$$\mathbf{b} = \mathbf{A}\lambda \tag{4}$$

where λ is a vector of length n_c whose jth component is λ_j and $\mathbf{A}$ and $\mathbf{b}$ are given by

$$A_{ij} = \omega_i \phi(||\mathbf{x_i} - \mathbf{c_j}||) \qquad (5)$$

and

$$b_i = \omega_i s_i \qquad (6)$$

where $i = 1, \ldots, n_L$ and $j = 1, \ldots, n_c$. Traditionally, the weights ω_i associated with each point in the learning set reflects its accuracy; alternatively, the ω_i may be tuned to improve the fit in under-represented regions of the reconstruction (e.g., near the collapses). For global reconstructions, we shall restrict attention to the case where all ω_i are equal (but note the discussion in L. A. Smith, 1992). Details

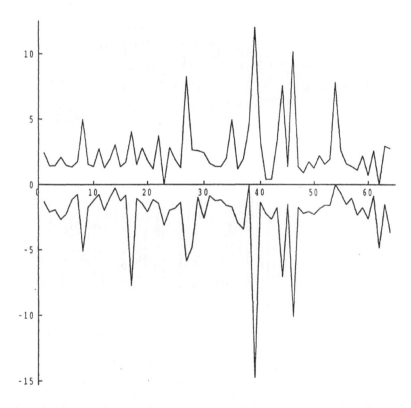

FIGURE 3 Variation of the average error with location in phase space as denoted by the nearest center. Positive and negative errors are averaged separately. Note the great variation at different locations on the reconstruction.

of the construction of this type of predictor may be found in Broomhead and Lowe (1988), Casdagli (1989), and Farmer and Sidorowich (1988). We call function F constructed in this way an RBF predictor.

There are several advantages of global predictors. The reconstruction is smooth and the coefficients are fixed. Thus, since all the computational overhead is done at the outset, large test data sets are easily evaluated. The global reconstruction

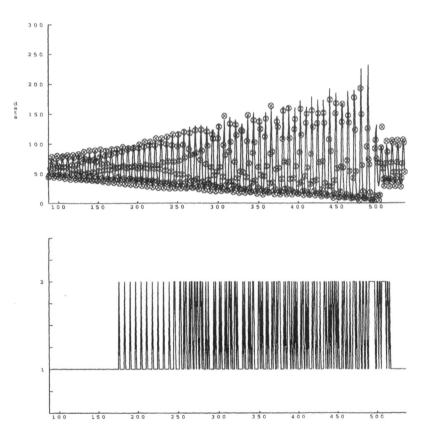

FIGURE 4   Global, multireconstruction predictor results in an update mode. The upper panel showed the observed (solid) and predicted (symbol) values. Note that, as the predictions are fairly accurate, they reveal the beating between the cycles and the sampling rate. The lower panel shows which of the two predictors (denoted by either 1 or 2) was used for the prediction at each given instant.

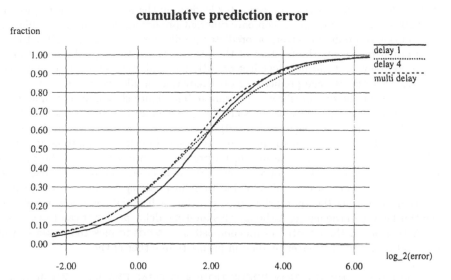

FIGURE 5 The prediction profiles of three global reconstructions with one-step ahead (update) predictions. The curves indicate the fraction of the test set that is predicted to within a given error; thus, for a given value of the error, the highest curve represents the best predictor. The solid and dotted lines reflect the profiles of two different fixed-delay predictors, while the dashed line is that of a (more accurate) third predictor combining the two, as described in the text.

also provides a natural partition of the phase space which can serve many uses. For instance, variation in the quality of predictions with location provides an estimate of the expected error in a given prediction (L. A. Smith, 1992). Figure 3 shows the average positive and negative error in the neighborhood of each center, illustrating the strong variations which occur in practice. In some regions there is a definite bias in the predictions (i.e., they are always too low), this can occur when there is not sufficient data (weight) in these regions and the least-squares fit to the entire data set simply under-fits them; alternatively, one cause of variation in the quality of "mean zero error" regions is the variation of projection effects within a given delay reconstruction. It is this second point we shall discuss here.

Once we have used a particular partition to classify each point in the data series, there is no need for the restriction to a single delay reconstruction for making predictions; given a single, global delay reconstruction to define partitions, we then pick the particular delay reconstruction that works best in each partition and use it.[2] Such a multireconstruction predictor is given in the next section. First, we introduce a graphical method to compare different predictors through the cumulative distribution of prediction error, or predictor profile, for short (L. A. Smith, 1990). This graph represents the fraction of the test set that can be predicted within a given accuracy, and clarifies the effect of outlying "bad" predictions that may bias statistics like the average predictor error. Prediction profiles for two $m = 4, \tau_p = 4$ global RBF predictors are shown in Figure 5. The figure shows, for example, that the predictor corresponding to the solid line predicts 20% of the test set with an error of less than 1 bit ($\log_2(\text{error}) < 0.0$). In one, $\tau_d = 1$, while $\tau_d = 4$ in the other. Note that the $\tau_d = 1$ predictor has more very small errors and more large errors which are large; where it is accurate, its predictions are superior to the $\tau_d = 4$ predictor, yet where it is inaccurate, they are far inferior. The point is that these two predictors work best in different locations in phase space; by recording those locations and using the appropriate predictor for each initial condition, we obtain the third curve in the figure. Although this multiple-delay predictor can't make any prediction better than "both" of its composites, it tends to pick the better of the two and hence gives a better prediction profile than either of them.

3.3.1 MULTIRECONSTRUCTION PREDICTION: A GLOBAL EXAMPLE.
We illustrate what is happening by examining the predictions and which predictor is used where. The upper panel of Figure 4 shows an out-of-sample segment of the observed (solid) and predicted (symbol) time series. The lower panel notes which of the two predictors was used to make each prediction. For the small amplitude oscillations (and near local minima), the $\tau_d = 4$ predictor is preferred; it provides a good estimate of phase and amplitude which is more easily obtained in a longer window. When the oscillations are large and more irregular, the shorter delay contains the more relevant information. This makes sense; it may be a bit naïve to attempt to compute an "optimal delay time" if there is no need to average over the entire reconstruction.

3.4 LOCAL PREDICTORS

An alternative to the single global predictor is to construct a local predictor from the points in the neighborhood of each data point. For local methods other than the nearest neighbor, this involves defining a local neighborhood, usually with a fixed radius or fixed number of k neighbors. Neither is optimal; it would be nice, but computationally expensive, to choose a neighborhood with respect to the local characteristics of the underlying function. Casdagli and Weigend (this volume) investigate the variation of predictions with k using local linear maps in a variety of circumstances.

---

[2]Multiple reconstruction local predictors could also be constructed, utilizing either the error in a withheld subset or the in-sample error to determine the best delay.

3.4.1 NEAREST NEIGHBOR(S). The nearest-neighbor predictor simply chooses the point in the learning set closest to the point to be predicted and uses its image as the prediction. As shown in Figure 6, runaway nearest-neighbor prediction works rather well in **A.dat**. For coarsely quantized data, these predictors are likely to become entrained in the learning set; once a series of predictions produces a point corresponding to a point in the learning set, the predictor will march through the learning set until it reaches the end of the learning set—producing results similar to the striking overlay predictions by Kostelich and Lathrop and others in this volume. When an entrained nearest-neighbor predictor reaches the end of the data

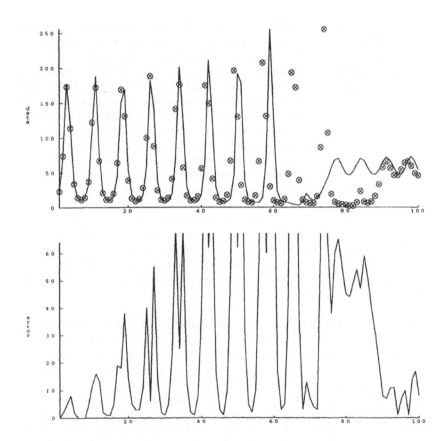

FIGURE 6 The results for runaway (single) nearest-neighbor prediction showing the observed (solid) and predicted (symbol) values. Here the image of the nearest neighbor of the point to be predicted was chosen where $M = 4, \tau_d = 1$. The lower frame shows the error as a function of time.

set, a sudden, rather disorganized-looking behavior may ensue as it suddenly finds a sharp increase in the distance to nearby points; this behavior may continue until the predictor "drifts" near a region where the data density is higher and, perhaps, becomes re-entrained. Other "overtrained" predictors may exhibit similar behavior. In the instance shown here, the nearest-neighbor predictor became entrained in less than ten steps.

Note that this predictor does collapse (two cycles late) and that the phase of the observed and predicted data nearly coincide *after* the collapse. This suggests that phase information is preserved through the collapse, a question investigated below.

3.4.2 LOCAL LINEAR AND QUADRATIC PREDICTION.

Given the coordinates of points within a neighborhood, a local linear predictor finds the best linear combination of these distances for interpolating the future observation. A local quadratic predictor works similarly but includes the quadratic terms. For small data sets, determining the correct size for each neighborhood is crucial; if it is too large, higher order nonlinear effects will be included while, if it is too small, the quadratic and RBF predictors may overfit the data. (Techniques to avoid overfitting with neural nets are discussed in Weigend and Rumelhart (1991).) A major difficulty is that "large" will vary with location over the reconstruction (in terms of either the k nearest neighbors or some fixed distance d in reconstruction space). While considering the average ratio of the observed error to the expected (local, in-sample) error can provide an idea of whether the local predictor is overfitting the data, it also suffers from averaging over the space. A truly local solution based on information theoretic grounds, suggested by A. Mees, currently under investigation in collaboration with K. Judd. In trials based on A.dat, the local quadratic map often yields the better results, although there is enough variation that the best approach might well be to generate an ensemble of predictions (with uncertainty estimates) and form a suitable weighted average. Beven gave an example of this approach in a hydrological context (Beven & Binley, 1992).

3.4.3 LOCAL RADIAL BASIS FUNCTION PREDICTION.

Local RBF predictors are similar to the global RBF predictor above but use only local data. This has the advantage of not assuming that the neighborhood will be small enough to be linear, but care must be taken not to overfit the data in the locally linear case. Runaway predictions using this method are shown in Figure 7; they provide quite reasonable estimates of the behavior until the collapse, but do not predict the collapse. Indeed, it is very difficult to make this predictor collapse; this is due, in part, to the low weight that the two collapses in the learning set have (low because there are only two realizations out of many observed cycles). This can be changed either by taking smaller neighborhoods or, independently, by increasing the weights w_i of the points in the regions of reconstruction space corresponding to a collapse, thereby forcing an improved fit to that region. Alternatively, one could try to detect (by, say, a simple threshold method) when the next collapse should occur and then to extend predictions through the collapse by visual inspection, following Wan's example (Wan, this volume).

We repeat that only Figures 6 and 7 correspond to the requirements of the contest (i.e., runaway predictors starting at the end of A.dat and predicting the first 100 points of A.cont). Both these sets of predictions were based on 32 bits (an initial vector of 4 data points, each of 8 bits). The quality of these predictions and the observation that they, along with the striking *visual pattern matching* predictions presented at the conference, "fail" only at the transition, suggests that a simple model might explain the majority of the dynamics. Such a model would

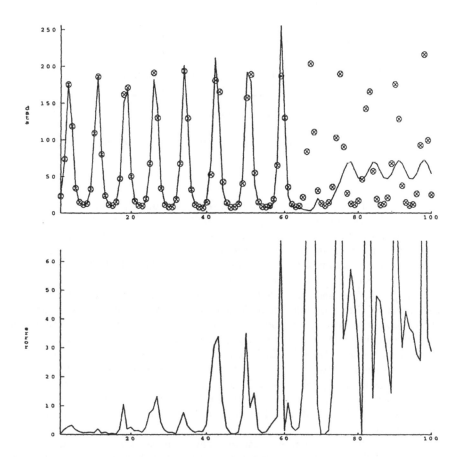

FIGURE 7   Runaway local nonlinear RBF predictions using $\phi = r^3$ with 16 centers distributed between the 32 nearest neighbors of the point to be predicted

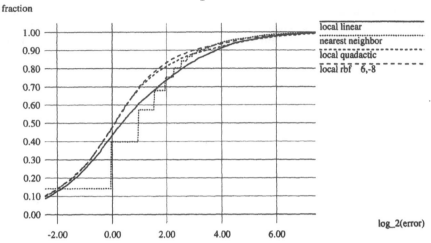

FIGURE 8 Local predictor results for RBF, quadratic, linear, and nearest-neighbor predictors.

be nonlinear, of course (we are measuring $|E^2|$), but quite possibly not chaotic. It would be interesting to determine just how well the amplitude of the cycles after a collapse could be predicted, and also whether those occasions where the amplitude appears to continue to decrease for a few cycles after the collapse could be identified, although the saturation of the signal noted above might make this difficult.

Before pursuing that idea, we compare the local predictors for the case $M = 4$, $\tau_d = 1$, $\tau_p = 4$; the prediction profiles are shown in Figure 8. These graphs reflect the quality of direct update predictions at a fixed step (four samples) into the future. Since the nearest-neighbor predictions are quantized, so are their errors, hence the steplike appearance in the figure. Note that while this predictor gives the fewest accurate predictions (40% of the predictions are accurate to 1 bit or better), it also gives slightly fewer large errors crossing to the left of the other three curves at approximately $\log_2(\text{error}) = 3.5$. The local RBF predictor yields slightly better predictions than the local quadratic, which is significantly better than the local linear. These results are all based on a neighborhood of $k = 32$ nearest neighbors with $n_c = 16$.

We conclude this section by admitting that no motivation for the choice of the particular RBF was given. An advantage of r^3 is that it introduces no additional free parameter. In practice, it is often the case that the exponential, $\alpha \approx 1$ with the constant $c = \alpha d_{nn}$, provides a fit of similar quality and that, by adjusting α and observing the predictor profiles, a better fit can be achieved. (Note that d_{nn}

is a local quantity.) It is also observed that such a fit is often more parsimonious than the original for, while we have introduced an additional parameter, more of the λ_j of the fit are linearly related (due to zeros introduced in a singular value decomposition), thus reducing the total number of degrees of freedom employed in the predictor.

---

4. THE OVERLYING ENGINEERING

4.1 NOISE, SENSOR FAILURE, AND ERROR DETECTION

How similar techniques also can be applied to detect sensor failure and error detection is discussed by L. A. Smith et al. (1991). The basic idea is straightforward and, of course, related to our earlier comments on stationarity. If a deterministic system has explored the phase space of a given reconstruction and a reasonably good predictor has been constructed, then the observation that the expected error is persistently, unexpectedly large would indicate either a change in the systems dynamics or an error in the sensor. When a good predictor cannot be constructed, because the required dimension is too great or the underlying system is stochastic, errors can be suspected when the nearest center distance grows very large. This test is applicable to stochastic systems which, while filling regions in any given reconstruction completely, generally do not fill all regions of the embedding space. In the current application an instance is the detection of the string of zeroes in the continuation set; the series goes to zero for seven time steps near data point 6454.

We also note that, as suggested at the conference (Hubner et al., this volume), if the primary source of noise in this system arises from quantization effects, methods of noise reduction (see Grassberger, Schreiber, & Schaffrath, 1991) should be applied here.

---

5. THE UNDERLYING PHYSICS

The contributions presented have demonstrated that stunningly simple and remarkably complex methods of prediction can be applied to this data set with interesting results. But have we suggested anything about the physics or dynamics of the underlying system? The length of time ahead that we were able to predict based on only 32 bits (4 data points of 8 bits each) makes the system appear very predictable; the difficult predictions appear at the collapses, but even they may not be too complex.

Examining a series of local maxima (not shown), we note that they tend to rise until hitting a threshold near 255 and then suddenly collapse. Figure 9 reveals just how linear this progression is by plotting the local maxima of successive cycles as max_{i+1} against max_i; the bubble at larger values is due, in large part, to poor estimates of the larger maxima resulting from the large sampling time. We have suppressed a scatter of small maxima (as small as four) which usually arise immediately after a collapse, some of which are due to beating between small cycles and the sampling rate. While the majority of values fall above the diagonal (indicating growing amplitude), a significant number fall below, denoting a different dynamical behavior which can be identified in the series. Note that the saturation at 255 is also visible in Figure 9. Plotting the time between maxima as a function of maxima (see Figure 14 in the Appendix) suggests that the frequency of the oscillation is a decreasing function of amplitude; this may be related to the physical processes occurring in the experiment.

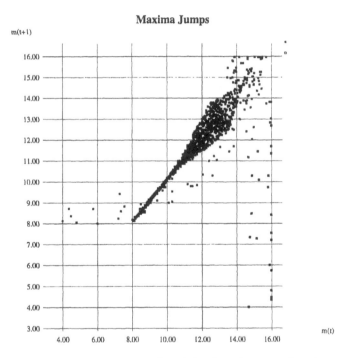

Maxima Jumps

FIGURE 9   A plot of the square root of each local maximum against the one previous showing a clear linear trend. The ballooning at large maxima arises, in part, from variation due to sampling error. The scatter of small local maxima arise just after collapse; however, the interesting sparse band of values just below the diagonal indicates a "slow collapse" ($max_i > max_{i+1}$, are real and can be identified in the next two figures).

Assuming that the growing cycles can be modeled, we consider the collapses where most of the prediction results presented at the workshop (which made it that far) went wrong. Phase coherence and amplitude prediction can be considered separately. To investigate whether or not the phase is randomized during collapse, we compare the evolution of data strings that actually saturate at 255. If we line up 11 consecutive instances where the observed maximum was 255, we see a great organization in both the forward and backward time (see Figures 10, 11, and 12). Figures 10 and 11 reveal that the first maxima after the collapse tend to coincide; further, the relation between larger amplitudes and longer period cycles is revealed in Figure 10 as the larger amplitude cycles are clearly displaced to the right as the time since collapse increases. Figure 12, where the direction of time is reversed, shows an extraordinary coherence in phase 9 τ_s before saturation.

From these observations we conjecture that a predictor based on the values of successive maxima could accurately predict the time and amplitude of the next maximum; initial trails show this breaks down at the collapse, in part, because the saturation at 255 makes these points appear identical.

These observations suggest, as a model, a simple, linearly growing oscillation whose period increases slightly with amplitude. Noting that the signal tends to maintain its phase through the collapse, we have neither determined whether the behavior of the amplitude is random or deterministic, nor explained the interesting segments for which the $\max_i > \max_{i+1}$ just after a collapse. In any event, the

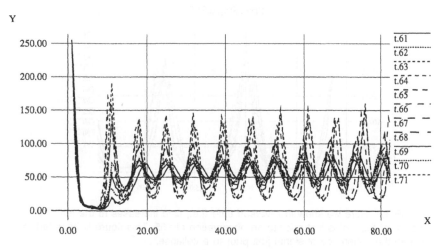

FIGURE 10 Eleven superimposed segments of the time series in A.cont aligned by an observation of value 255. Note the phase coherence after the collapse and that the loss of phase coherence toward the right can be attributed to the larger amplitude oscillations having a longer period (as suggested in the text).

Following a 255

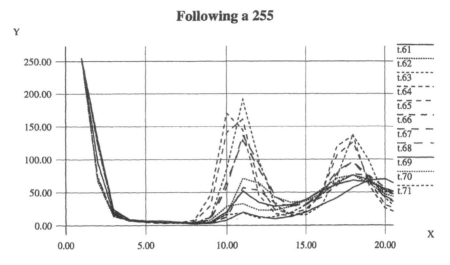

FIGURE 11   An enlargement of the inital section of the previous figure showing the first of two cycles after an observation of 255 in A.cont.

Preceding a 255

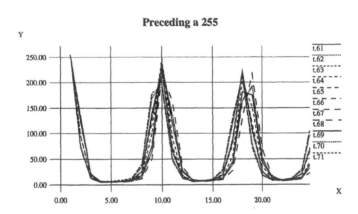

FIGURE 12   Here we show in reversed time (i.e., time increases to the left) the behavior of the series just prior to an observation of 255. This figure is intended to illustrate the coherence of signal just prior to a collapse.

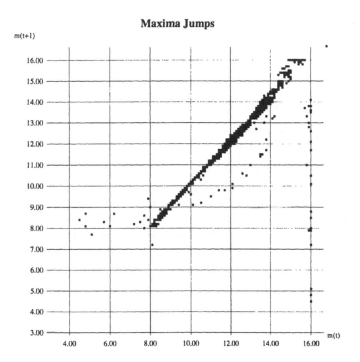

FIGURE 13 A plot of the square root of each local maximum against the previous one for the original (higher resolution) data set. The ballooning at large maxima is decreased. The scatter of small local maxima arise just after collapse; however, the sparse band of values just below the diagonal are real.

phase coherence is interesting and the increase in period may suggest a physical mechanism (such as loss of resonance) that triggers the collapse. It would also be interesting to treat the output of such a simple model (squared and sampled) as surrogate data and to determine whether the actual observations can be distinguished from such series (L. A. Smith, 1992; Theiler et al., 1992a).[3] This approach could be used to discriminate the dynamics of the amplitude after collapses, for instance whether to collapse (retaining phase information) to a random part of the cycle and repeat, or to determine the amplitude after collapse based on the data just before the collapse. While the short-term predictability of this system has been clearly established at this meeting (e.g., the contributions of Sauer (this volume)

[3] Alternatively, one could shuffle the order of segments of the data between collapses; while the majority of statistics computed would remain unchanged (collapses are rare events), long-term determinism would clearly be lost in these surrogates.

and Wan (this volume)), it remains a capital mistake to theorize on the dynamics over the collapse without sufficient data.

APPENDIX: A CLOSER LOOK

After the conference and discussions with James Theiler and Tim Sauer, some doubts were raised about interpreting the balloon of Figure 9 as due to a long sampling time. To address these doubts, the original data (Hübner et al., 1989) were obtained, thus providing a longer data series at twice the sampling rate. Figure 13 corresponds to Figure 9 for the higher resolution data set. Although not resolving the problem of saturation, we see that the balloon is effectively deflated and this suggests that a piecewise linear map could model the series quite well. Then one could then extend such a model to extrapolate true values corresponding to saturated observations. It would be interesting to see if this would provide a method

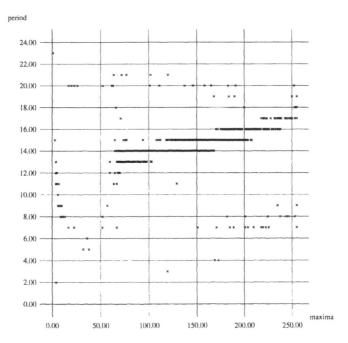

FIGURE 14   A scatter plot of the period between local maximum against the maximum from the full NH_3 data set. The majority of that data clusters near periods 13 to 17, with the greater maxima associated with longer periods.

to determine the height of the first maximum *after* a collapse, in particular, to distinguish the collapses of small, increasing amplitude oscillations from those of intermediate amplitude which decline before growing.

In addition, a reviewer requested an illustration of the claim that the period of a cycle tends to increase with increasing amplitude (see Section 5). This effect, also clearer in the higher resolution series, is reflected in the distributions shown in Figure 14. In this figure, there was no attempt to remove the spurious local maxima due to either beating or noise.

ACKNOWLEDGMENTS

I would like to acknowledge several discussions and ales with M. Muldoon, in particular those regarding stationarity, and an ongoing exchange with J. Theiler vaguely focused on surrogate data. Some of the work reported here was completed during a visit to the UWA during which I enjoyed the hospitality and collaboration of A. Mees and K. Judd. I am grateful for conversations with and data manipulation suggestions by D. Drysdale and D. DeBeer, and many discussions with (as well as the presentations of) other participants at the workshop which contributed to this paper, in particular B. LeBaron's comments on the prediction profiles. Thanks are also due to M. Casdagli, C. Lanone, G. Sugihara, J. Theiler, and A. Weigend for their critical reading of the text. Finally, I owe a special thanks to the organizers, A. Weigend and N. Gershenfeld, for an excellent workshop and for assisting an American in England. This research was supported by a Senior Research Fellowship from Pembroke College, Oxford.

III. Time Series Analysis and Characterization

Martin C. Casdagli† and Andreas S. Weigend‡
†First Boston Corporation, Equity Financial Strategies, 55 E. 52nd St, New York, NY 10055.
‡Xerox PARC, 3333 Coyote Hill Road, Palo Alto, CA 94304.
Address for correspondence: Andreas Weigend, Department of Computer Science and
Institute of Cognitive Science, University of Colorado, Boulder, CO 80309-0430;
e-mail: weigend@cs.colorado.edu.

Exploring the Continuum Between Deterministic and Stochastic Modeling

In order to understand some properties of the system that generated a time
series, we analyze the prediction accuracy as the model class is varied. Our
"Deterministic versus Stochastic Algorithm" can help distinguish between
deterministic and stochastic processes. In this paper, we apply it to Data
Sets A (laser), B (heart), and D (computer-generated data) of the Santa
Fe Time Series Prediction and Analysis Competition.

For each data set, we fit a family of local linear models. After each model is
fitted, its short-term predictive accuracy is assessed on a portion of the data
not used for training. The systematic dependence of these out-of-sample
errors on model parameters—such as the dimension of the reconstructed
state space or the size of the neighborhood for local linear models—can yield
some insight into the dynamics of the underlying system. For example, if
the error is lowest for a highly local model (as for the laser data), we infer
that the time series was generated by a deterministic nonlinear system. If,
on the other hand, the minimum occurs at larger neighborhood sizes (as
for the heart data), we infer that the underlying process was stochastic, or
deterministic of high dimension.

Times Series Prediction: Forecasting the Future and
Understanding the Past, Eds. A. S. Weigend and N. A. Gershenfeld, SFI Studies
in the Sciences of Complexity, Proc. Vol. XV, Addison-Wesley, 1993   **347**

1. INTRODUCTION

In the last decade, powerful computers made it possible to collect time series orders of magnitude longer than any time before in history. Powerful computers also made it possible to simulate nonlinear processes and show that their behavior often violates expectations borne out by the traditional linear paradigm. Putting these two strands together, powerful computers enabled time series prediction and analysis to tackle—often surprisingly successful—long experimental data records by constructing nonlinear models. This has happened in a number of communities, such as in statistics (e.g., Tong, 1990), in artificial intelligence and machine learning (e.g., Weigend, Huberman, & Rumelhart, 1990), in physics (see an earlier volume in this series, edited by Casdagli and Eubank, 1992), and in engineering and control (see, e.g., the volume edited by White and Sofge, 1992).

Nonlinear models with many parameters can be extremely flexible. This gives them the potential to **overfit** the time series: flexible models can extract the features that are genuine to the process but can also fit noise features that are idiosyncrasies of the training data. In statistical jargon, this is the case of "low bias" of the model (since the model can fit almost anything, it is not biased much) and "high variance" of the parameters (since there are so many parameters, they are estimated with large errors). This **bias-variance dilemma** is central to weak modeling, i.e., the case when the modeler does not know first principles that might suggest strong equations. A clear discussion in the context of neural networks is given by Geman, Bienenstock, and Doursat (1992).

The position taken in this dilemma by most statisticians is to severely restrict the number of free parameters in the nonlinear model; this allows them to estimate the parameters quite accurately, i.e., with small errors ("small variance"). However, if the model space is too small (a strong model implies "large bias"), the model might not be flexible enough to emulate the dynamics of the system, resulting in bad out-of-sample predictions due to **underfitting.** For reasons of conservatism, statisticians have tended to favor low-variance/high-bias over low-bias/high-variance. This has resulted in nonlinear models which are either rather small perturbations of linear models, or in models where the nonlinearities are added systematically: statisticians tend to control the *order of the interactions* (e.g., second-order terms, third-order terms etc.).[1] This position (few parameters/limited nonlinearities) works well for time series that actually are generated from simple models where the interesting behavior is mainly due to noise, also referred to as innovations or exogenous shocks, i.e., to outside perturbations of the system.

[1]This is in contrast to neural networks or connectionist modeling where the *number of features* (that are to be extracted) is controlled (for example, by the number of hidden units). Each feature can be a high-order interaction, but the number of features is limited. This is discussed further by Rumelhart et al. (1993) and by Weigend (1993).

A key result of dynamical systems theory is that deterministic and nonlinear systems can also generate complicated or aperiodic behavior, more precisely **deterministic chaos**. Philosophically, this is quite different from assuming that the source of complex behavior is not intrinsic but extrinsic, as described in the previous paragraph, due to **stochastic** influences. In nonlinear deterministic systems the motion in phase space (also called state space) is often restricted to fairly low-dimensional manifolds. Properties of deterministic chaos are discussed in a number of good reviews (Schuster, 1984; Gershenfeld, 1989; Eubank & Farmer, 1990).[2]

For time series prediction and modeling, the existence of low-dimensional chaos suggests considering construction of nonlinear deterministic models for time series. *If* a series is indeed chaotic and low-dimensional, the good news is that it will be possible to obtain precise short-term predictions with such deterministic models, and the bad news is that it will be impossible to obtain good long-term predictions with any model, since the uncertainty increases exponentially in time. This divergence of nearby trajectories is a hallmark of chaotic behavior.

The position recently taken by a number of physicists is to model dynamical systems with fairly general nonlinear models. The problem can be viewed as constructing surfaces in high-dimensional spaces. Functional forms of such models, studied in the past in the field of multivariate function approximation, invariably involve large numbers of parameters.

Often, the explicit functional form for constructing a time series model is not known. For example, our understanding of the underlying dynamics might be too limited and/or we might be able to measure the system adequately. With powerful computers, the range of possibilities for nonlinear models is virtually unlimited. We have developed a tool, the *deterministic versus stochastic algorithm* (DVS) that explores the effect of systematically varying relevant parameters, such as the dimension of the reconstructed state space and the size of the neighborhood for local linear models. We will explain the DVS algorithm in this paper and show how it can lead to some understanding of the underlying dynamics.

Nonlinear models with many parameters are at the heart of connectionist modeling, also known as (artificial) neural network modeling. The Competition showed the power of this recent paradigm. The successful models employed are discussed in the overview article by Gershenfeld and Weigend (this volume).

In this paper, we investigate the middle ground between the extremes of modeling with large numbers of parameters and modeling with a small number of parameters. Similar ideas of varying the model complexity and analyzing the resulting out-of-sample error are presented by Sugihara and May (1990), Weigend and Rumelhart (1991b), Hsieh (1991), and Rubin (1992). We here fit a family of local linear models that range from the deterministic to the stochastic extreme. This is achieved by varying the size of neighborhoods in a reconstructed state space and

---

[2] Although the concepts of chaos and dimension have recently been generalized to stochastic systems (Osborne & Provenzale, 1989; Falconer, 1990; Nychka et al., 1992), we use the term "chaos" to imply *deterministic* chaos.

by computing a linear fit to the dynamics in these neighborhoods. The short-term predictive accuracy of the models is estimated as out-of-sample error on a held-back portion of the time series. The DVS algorithm is defined precisely in Section 2. The results of applying the DVS algorithm to three of the six data sets of the Competition is given in Section 3: we plot the out-of sample errors as a function of the size of the neighborhood in reconstructed state space. (The embedding dimension serves as a parameter in the DVS family.) In each case, these plots allow us to approach two somewhat different objectives:

- First, they allow us to investigate how well a variety of well-known forecasting models would have fared in the Competition, had they been entered. For example, global linear autoregressive models (Yule, 1927) and also simple threshold autoregressive models (Tong, 1990) are located at the stochastic extreme of the DVS family, whereas local linear methods (Farmer & Sidorowich, 1987) and neural networks (Lapedes & Farber, 1987; Weigend, Huberman, & Rumelhart, 1990) are located towards the deterministic extreme.
- Second, qualitative properties of the dynamics of the underlying time series can be inferred by interpreting the DVS plots from the perspective of dynamical systems: If, on the one hand, the underlying dynamics is indeed chaotic and low-dimensional, then models near the deterministic extreme of the DVS family (with small neighborhoods) should give more accurate, short-term forecasts than those at the stochastic (globally linear) extreme. If, on the other hand, the underlying dynamics is indeed stochastic or chaotic of high dimension, then very local models will give less accurate, short-term forecasts since the small neighborhoods give them the flexibility to fit the noise in addition to the signal, thus giving worse out-of-sample forecasts for local models than for global linear models.

The second goal contains more subjective judgment than the first since a number of conflicting alternatives can give rise to similar effects, such as nonstationarity and extreme choices for sampling and prediction times. Some of the complications specific to the DVS method are discussed by Casdagli (1992).

2. THE DVS ALGORITHM

For a scalar time series $\{x_i\} = x_1, x_2, \ldots, x_N$, the DVS algorithm attempts to fit models of the form

$$x_{i+T} \approx f(x_i, x_{i-\tau}, \ldots, x_{i-(m-1)\tau}) \quad .$$

We use a least-squares method to find that (linear) function $f : \mathbb{R}^m \to \mathbb{R}$ that gives the best prediction for x_{i+T} in the sense that that function minimizes the squared error (or quadratic loss) within the model class. The integers T and m define the following quantities:

T : **lead time** or prediction horizon (prediction time into the future),

m : **embedding dimension** or dimension of the reconstructed state space (number of taps of the tapped delay line).

Furthermore, the m past values are combined in the **delay vector** $\mathbf{x}_i$. Here we assume equal spacing of the taps of the delay line, i.e., $\mathbf{x}_i := (x_i, x_{i-\tau}, \ldots, x_{i-(m-1)\tau})$, where τ is the **lag time** or **lag spacing** between each of the taps. Since the series analyzed here have already been sampled fairly coarsely, we choose $\tau = 1$ throughout this paper.

After these definitions, we give the DVS algorithm:

1. Normalize the time series to zero mean and unit variance.
2. Divide the time series into two parts:

    i.   a *training* set or *fitting* set $\{x_1, \ldots, x_{N_f}\}$ used to estimate the coefficients of each model, and

    ii. a *test* set or *out-of-sample* set $\{x_{N_f+1}, \ldots, x_{N_f+N_t}\}$ used to evaluate the model. N_f denotes the number of points in the fitting set, N_t the number of points in the test set.

3. Choose T.
4. Choose m ("outer loop").
5. Choose a test delay vector $\mathbf{x}_i$ for a T-step-ahead forecasting task $(i > N_f)$.
6. Compute the distances d_{ij} of the test vector $\mathbf{x}_i$ from the training vectors $\mathbf{x}_j$ (for all j such that $(m-1)\tau < j < i - T$).[3]

---

[3] For computational efficiency, we use the maximum norm. The algorithm presented here differs from Casdagli's (1992) in that we update the fitting set after each new point is tested rather than holding it fixed. This can be advantageous if the system is nonstationary.

7. Order the distances d_{ij} (using the heapsort algorithm; Press et al., 1992).
8. Find the k nearest neighbors $\mathbf{x}_j^{(1)}$ through $\mathbf{x}_j^{(k)}$ of $\mathbf{x}_i$, and fit an affine model with coefficients $\alpha_0, \ldots, \alpha_m$ of the following form[4]:

$$x_{j+T}^{(l)} \approx \alpha_0 + \sum_{n=1}^{m} \alpha_n x_{j-(n-1)\tau}^{(l)} \quad , \qquad l = 1, \ldots, k.$$

In time series context, this is an autoregressive model of order m fitted to the k nearest neighbors to the test point; i.e., there are k equations. j denotes those times in the training set where the dynamics is similar to the test point. Vary k at several representative values in the range

$$2(m+1) \; < \; k \; < \; N_f - T - (m-1)\tau \quad .$$

9. Use the fitted model from Step 8 to estimate a T-step-ahead forecast $\hat{x}_{i+T}(k)$ starting from the test vector, and compute its robust[5] error

$$e_{i+T}(k) = |\hat{x}_{i+T} - x_{i+T}| \quad .$$

10. Repeat Steps 5 through 9 as $(i+T)$ runs through the test set, and compute the mean absolute forecasting error[6]

$$E_m(k) = \sum_{(i+T)=1}^{N_t} \frac{e_{i+T}(k)}{N_t} \quad .$$

Vary the embedding dimension m (as "outer loop," back to Step 4), and plot the curves $E_m(k)$ as functions of the number of nearest neighbors (k). Such a plot of the family of curves is called a **DVS plot**. These plots will be shown for different prediction times, T (back to Step 3).

Before applying the algorithm to the data sets of the Competition in the next section, a few more general remarks about the DVS algorithm are in order.

[4] Since the coefficients $\alpha_0, \ldots, \alpha_m$ are computed by a least-squares method, they can be interpreted as describing that hyperplane that minimizes the sum of the squared distances to the data points of the k nearest neighbors. We obtain the coefficients by solving the normal equations for this linear system, which can be updated recursively as k is increased and solved by using LU-decomposition for computational efficiency; see Press et al. (1992).

[5] The choice of the error measure is to some extent arbitrary. We use absolute errors in this evaluation step (as opposed to the squared errors used in fitting above) since absolute errors are less sensitive to outliers, i.e., a more robust statistic.

[6] Since we are only interested in the shape of the curve, further normalization (often used for comparison to a simpler forecasting model) is not necessary.

The set of delay vectors $\mathbf{x}_i$, as i varies through the time series, is an example of a *state-space reconstruction* or *embedding* of the time series. If $m = 2$, the embedding can be represented graphically as a *phase portrait* by plotting x_i against $x_{i-l\tau}$.

If the dynamics of the system contains frequencies on several scales, a more parsimonious choice (than keeping all of the taps needed to capture the lowest frequency) might be to space the taps finely in the close past and further apart in the more distant past.[7]

The DVS algorithm as presented here uses a *local linear approximation* in Step 8. The idea of systematically varying embedding dimension and neighborhood size is not restricted to local linear models. More complicated function approximation schemes can be used, such as radial basis functions (Casdagli, 1989; L. A. Smith, 1993), multivariate adaptive regression splines (Friedman, 1991c), and neural networks.

Also, the models can be made considerably more detailed by introducing a variety of noise terms,[8] as is typically done in stochastic modeling (Tong, 1990). We here do not attempt to estimate the probability structure of such noise terms, but focus on properties of the prediction function f. If the noise terms are large, we can only fit functions with few parameters, in order to avoid overfitting. In our setup, these are function that use a large number of neighbors for constructing the fit. Stretching the usual notation, here we refer to such functions with large neighborhoods, resulting in fewer overall parameters as *stochastic models*, even though we are not trying to estimate the noise terms.

Finally, we here always use $N_t = 500$ points for testing, an attempt to balance statistical significance and computational feasibility. For each series, all the remaining points were used as training data; i.e., N_f was initialized to the total length of the series minus N_t.[9] If the data sets are short or nonstationarity, or exhibit confinement (see Section 3.2), a single test set might not be representative enough. The DVS algorithm presented here is easily extended to using several training—test set pairs with subsequent averaging of the prediction errors.

---

[7]Suggestions range from spacing according to Fibonacci numbers to capturing as many non-harmonic frequency components as possible (Tishby, private communication, 1990), to picking the "best" model via subset selection methods for linear regression problems (Tukey, private communication, 1993). Keep in mind, however, that if the system is chaotic, points from too far back in the past (the horizon being determined by the largest Lyapunov coefficient) have lost their predictive information. Takens' theorem assumes the extreme case of a completely deterministic system (i.e., noise-free dynamics). In that (unrealistic) case, only the number but not the location of past values matters.

[8]Here we minimize least squares; i.e., we use a cost function that is quadratic in the errors errors. In a probabilistic (log-likelihood) interpretation, this corresponds to a Gaussian error model. Alternative cost functions are discussed by Weigend, Huberman, and Rumelhart (1992). Furthermore, by taking the sum, we assume the errors to be independent of each other.

[9]The longest computations were required for large values of m for Data Set D. By terminating Step 8 at $k = 1000$ when $m \geq 20$, none of the computations took more than an hour of SPARC2 time.

3. RESULTS

In this section, we report the results of applying the DVS algorithm to three of the
time series of the Competition, Data Set A (laser data), Data Set D (computer
generated data), and Data Set B (medical data). Detailed descriptions of all the
competition data sets are given in the overview article by Gershenfeld and Weigend
(this volume) as well as in the individual papers by the contributors of the data
sets (this volume). We briefly comment on a few salient features and show some
of the phase portraits. Our main focus is on the DVS plots: we interpret them
quantitatively from the point of view of predictive accuracy, and also venture some
qualitative conjectures about the underlying dynamics of the system.

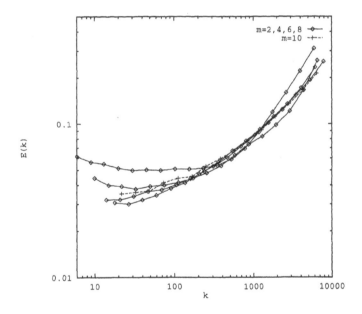

FIGURE 1   DVS plot for the laser data A.con with lead time $T = 1$. The average
out-of-sample prediction error is plotted as a function of the number of neighbors k
(used to construct the local linear models). Models for small values of k (at the left
side of the plot) are local; models for large values of k (at the right) are global. Given
the location of the minimum, the local linear models have lower forecasting errors than
a global linear model. The parameter m denotes the dimension chosen for the state-
space reconstruction.

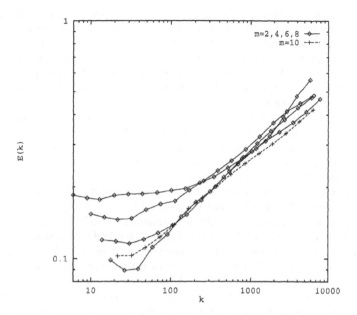

FIGURE 2   DVS plot for the laser data A.con. It differs from Figure 1 that the lead time (prediction time) here is $T = 10$. The the overall error is larger here than for lead time $T = 1$, but the qualitative behavior is the similar to Figure 1.

3.1 LASER ($N = 9,000$)

This univariate time series consists of intensity measurements of a laser believed to be in a chaotic state, with Lorenz-like[10] dynamics (Hübner et al., this volume). We varied the dimension of the reconstructed state space m from 2 to 10, in steps of 2. Figure 1 is the DVS plot for lead time $T = 1$; Figure 2 for lead time $T = 10$. (The lag time is always $\tau = 1$.) The main qualitative result holds for both lead times: models near the local linear extreme (at the left of each plot) yield more accurate forecasts than those near the global linear extreme (at the right of each plot).

By equating the local linear extreme with deterministic dynamics, and the global linear extreme with stochastic dynamics, we infer from the DVS plots that the irregular behavior of this time series is due to an underlying low-dimensional chaotic dynamics rather than stochastic effects. An alternative explanation, also consistent

---

[10] It would be interesting to fit Lorenz-type models to the time series in order to obtain the most accurate predictions and, ultimately, the deepest insights into the underlying dynamics by analyzing the similarities and differences between a fit with such as "strong" model as opposed to a "weak" or "broad" class of models as done in the DVS scheme.

with the DVS plots, could be a deterministic but nonchaotic system. However, such systems usually generate periodic or quasi-periodic series. This turns out to be inconsistent with the frequency spectrum of this data set (not shown).

This interpretation is consistent with the best entries in the Competition where the best contributions for this set were also obtained with deterministic methods, in particular with a neural network (Wan, this volume), and with a local linear method (Sauer, this volume).

For such low-dimensional deterministic chaotic systems, it is of interest to estimate the dimension of the attractor. Our figures here show that embedding dimensions of $m \approx 8$ give the most accurate forecasts. This is an upper limit for the minimal embedding dimension, since there are several confounding effects, as discussed below. Although our optimal m is even low compared to the best entries to the competition (who chose embedding dimensions of 25, 32, 200, and 50), the stereo plot shown by Gershenfeld and Weigend (this volume, Figure 5) reveals that the attractor can be embedded in only three dimensions and that the dimension of its manifold is close to two. A similar conclusion is reached by the simple graphical analysis using Poincaré sections of the phase portrait (not shown).

There are a number of reasons for the fact that more than the theoretical minimal number of lags are required for the best prediction. Examples are (1) the effect of noise (more inputs allow to average out some of the noise), (2) the specific choice for the primitives of the function approximation (we chose simple local linear models, as opposed to more powerful function approximation schemes), and (3) the simple, equidistant lag-space embedding, as opposed to filtered embedding or performing subset selection on the input variables.

3.2 COMPUTER-GENERATED DATA ($N = 50,000$)

This univariate time series consists of observations of a sinusoidally forced particle moving in a four-well potential in four-dimensional space subject to weak damping. (Details are given by Gershenfeld and Weigend, this volume).

Figure 3 shows the phase portrait of D1.dat. Note that the test set of only 500 points (connected by lines) does not cover the phase portrait well. This phenomenon of *confinement*, sometimes also called *barriers to transport*, often occurs in Hamiltonian systems. It follows that the DVS plots shown in Figures 4 and 5 for lead times $T = 2$ and $T = 8$, respectively, will depend on the specific held-back test set. However, it is apparent from both figures that the models somewhat above the deterministic extreme are the most accurate, yielding improvements of between 50% and 100% over linear models. To estimate the dependence on the specific part of the data set, we also analyzed D2.dat (also of length 50,000) and observed similar improvements, although the precise shapes of the DVS plots differed, primarily due to a different test set exploring a different portion of the phase space.

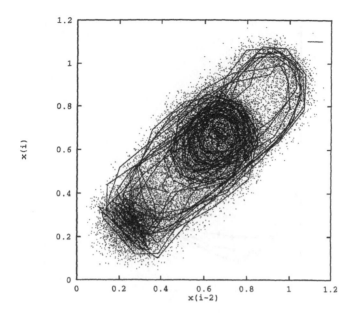

FIGURE 3 Phase portrait of the computer-generated time series D1.dat, plotted with lag time $\tau = 2$. Areas with higher density of points around locations $(0.25, 0.25)$ and $(0.6, 0.6)$ correspond to minima of the potential: the "particle" spends more time at low energies than at high ones. The 500 points of the test set are connected by lines.

Inspecting the DVS plots for Data Set D (Figures 4 and 5), we find that time series D exhibits strong nonlinearity. We also find that the improvement in accuracy over linear models is not strong enough to point to low-dimensional chaos with a few degrees of freedom.[11] Furthermore, the minimum of the family or curves as the dimension of the reconstructed phase space is varied is reached for $m \approx 20$. Here we conclude that the data were generated by a highly nonlinear system of more than a few degrees of freedom.

Before comparing this analysis with the true dynamics, let us discuss two alternative explanations: (1) that there is low-dimensional chaos but the dynamics is not very strongly nonlinear and (2) that the dynamics is essentially linear, but the system is nonstationary. The first hypothesis can be rejected on the grounds that chaotic systems tend to be highly nonlinear, at least for the range of lead times considered here.

[11]Behavior similar to Figures 4 and 5 herein were found for time series of finely sampled pressure fluctuations of speech. The structure there was mainly apparent in the vowels (Casdagli, 1992).

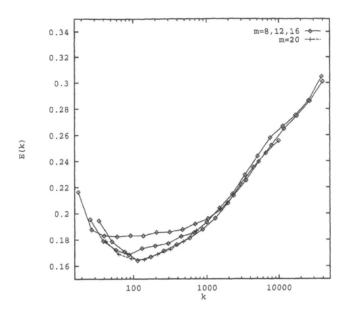

FIGURE 4   DVS plot for the computer-generated data D1.dat with lead time $T = 2$.

The second hypothesis is difficult to address without a precise definition for nonstationarity. In some sense, *all nonlinear systems are nonstationary linear systems* since, in a small enough region of state space, the dynamics is locally linear. However, nonstationarity usually refers to phenomena occurring over much longer time scales than the natural time scale of the dynamical system. With this loose definition of nonstationarity, and a visual inspection of the phase portrait of Figure 3, we also reject the second alternative hypothesis.

We now compare the conclusions reached by interpreting the DVS plots to what is known about the system: it has nine degrees of freedom and is approximately deterministic.[12] Distinguishing nine (deterministic chaos) from infinity (noise) is beyond the reach of a DVS algorithm that employs simple local linear models. At the present stage, it is not clear whether more powerful embedding and/or better suited function approximation techniques could yield much more accurate short-term predictions (in which case evidence for deterministic chaos would have been found), or whether the DVS algorithms predictions are already close to optimal for this system.

[12]There is a small amount of additive Gaussian noise and a small drift in one of the parameters (biased random walk); see Gershenfeld and Weigend (this volume).

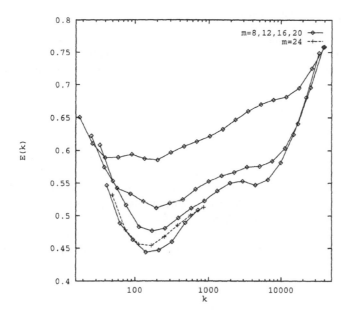

FIGURE 5   DVS plot for the computer-generated data D1.dat with lead time $T = 8$.

In the competition, Zhang and Hutchinson (this volume) used neural networks and predicted the next 10 to 20 data points with moderate success. Unfortunately, they do not give the expected out-of-sample errors as function of the lead time T which would allow us to compare our method to theirs.

An impressive step beyond simple point-predictions for this 100,000-point training data set (Data Set D) was taken by Fraser and Dimitriadis (this volume). They used a hidden Markov model with mixed (i.e., discrete and continuous) states[13] (containing more than 6,000 parameters) and forecast the evolution of the entire probability density.

[13] Fraser and Dimitriadis (this volume), similar to here, used local linear models (of order $m = 8$). However, they allowed the "coefficients" of their models to depend on the hidden states; hidden states can be viewed as a generalization of reconstructed state space. In terms of languages, they moved from regular grammars (as implemented by a simple local linear model or a feedforward neural network) to more powerful context-free grammars. Context-free grammars also can be obtained with recurrent neural networks.

3.3 MEDICAL TIME SERIES ($N = 17,000$)

The last of the three time series analyzed is the medical time series B2.dat of a patient with sleep apnea. We first give the univariate analysis based on the heart rate alone, then turn to a multivariate analysis using also chest volume (breathing) and blood oxygen concentration.

3.3.1 UNIVARIATE ANALYSIS. The first variable of B2.dat consists of the heart rate. Each time step corresponds to 0.5 seconds; see Rigney et al. (this volume). A phase portrait, showing x_i vs. x_{i-4}, is given in Figure 6. As for the computer-generated data (cf. Figure 3), a 500-point test set is not long enough to cover the phase portrait well. Here, however, this is probably due to nonstationarity of the series: the patient was presumably exerting himself when the heart rate increased to over 80 in the fitting set, but no such event occurred during the test period.

The DVS plots are shown in Figures 7 and 8, for lead times $T = 4$ and $T = 16$, respectively. They reveal that nonlinear models near the deterministic extreme yield poor predictions and that the best stochastic nonlinear models yield only about a

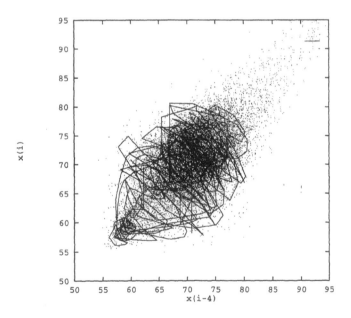

FIGURE 6 Phase portrait of time series B2.dat, for heart rate, with lag time $\tau = 4$, corresponding to 2 seconds. Solid lines connect the points in the test set. (Phase portraits for $\tau = 1$, 10, and 30, corresponding to 0.5, 5, and 15 seconds, are given in the chapter by Glass and Kaplan, Figure 4, this volume.)

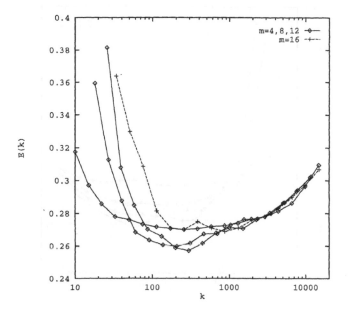

FIGURE 7   DVS plot for the heart data B2.dat with lead time $T = 4$.

15% to 20% improvement in predictive accuracy over linear models.[14] We have not yet carried out a proper statistical test, but we believe this improvement to be statistically significant. Rubin (1992) applies the powerful idea of bootstrapping, essentially substituting assumptions about the distribution by computer time (Efron, 1979) in order to address the question of whether such differences are significant or not.

[14] We obtained somewhat larger improvements in predictive accuracy over a linear model for time series B1.dat where a substantial portion of the testing set was confined to the sparse region where the heart rate was over 80.

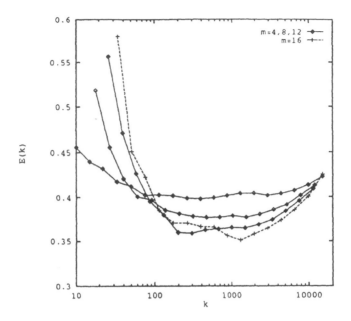

FIGURE 8   DVS plot for the heart data B2.dat with lead time $T = 16$.

3.3.2 MULTIVARIATE ANALYSIS. As a first attempt at a multivariate analysis of
the time series, we considered the bivariate (heart rate, chest volume) time series
(x_i, v_i). These two variables appear to vary over similar time scales (unlike the blood
oxygen concentration which is also available). We analyze here the *first differenced
time series,* defined by

$$(X_i, V_i) := (x_i - x_{i-1} , v_i - v_{i-1}) .$$

An exploratory (although linear!) measure describing how series X and V are coupled is given by the *cross-correlation function* (CCF)

$$\mathrm{CCF}_{XV}(\Delta) = \frac{\mathrm{E}_i \left\{ (X_i - \mu_X)(V_{i+\Delta} - \mu_V) \right\}}{\sigma_X \, \sigma_V} .$$

μ_X is the mean of series X, σ_X the standard deviation, and $\mathrm{E}_i\{\cdot\}$ denotes the
expectation value, i.e., the averaging over the series. $\mathrm{CCF}_{XV}(\Delta)$ thus measures

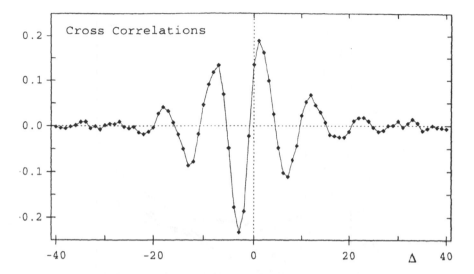

FIGURE 9   Cross-correlation function for the bivariate (heart rate, chest-volume) time series.

how series X, and V delayed by Δ time steps, are linearly correlated. It turns out that the coupling shows up more clearly on the first differenced time series than on the raw time series, possibly due to the first differencing reducing the effects of nonstationarity in the time series. It is plotted in Figure 9, revealing a time-lagged linear coupling between heart rate and chest volume. This effect is known as "sinus arrhythmia" in the medical community.

In the spectral perspective, further analysis includes the cross-spectra (in rectangular in polar coordinates) and also the CCF between one series and the first-differenced of the other (not shown). Here we give the DVS plots for lead times $T = 4$ and $T = 16$ (in Figures 10 and 11) for a bivariate model that predicts the heart rate in the following way:

$$x_{i+T} \approx f(x_i, \ldots, x_{i-(m-1)\tau}, v_i, \ldots, v_{i-(m-1)\tau}) \quad .$$

The normalization (Step 1 of Section 2) is applied to both variables. Figures 10 and 11 should be compared to Figures 7 and 8. As expected from the CCF analysis, the predictive accuracy improves slightly (by about 10%) over univariate models.

Treating the univariate and the multivariate case together, the improvement in accuracy over linear models ranges from about 10% to 25%. This is too small an improvement to conclude that the concept of low-dimensional chaos can explain the irregularities in this data set. There *are* pathological cases that can fool the

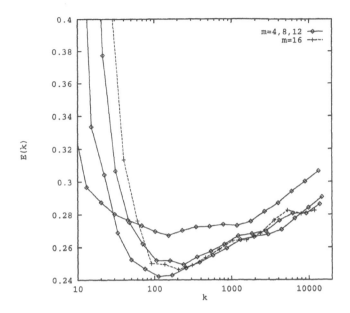

FIGURE 10 DVS plot for the bivariate model for predicting the heart rate from both heart rate and chest volume. Lead time is $T = 4$.

DVS algorithm, such as pseudo-random number generators or, more generally, any deterministic chaotic time series that is sampled much more slowly than its natural time scale. In such cases, low-dimensional time series can indeed reveal DVS plots like those of time series B. However, the sampling time for Data Set B is not slow compared to the intrinsic dynamics. (Actually, is is the most highly sampled of the three we analyzed here; this also influenced our choice of using the longest lag times here.)

Having ruled out low-dimensional chaos, what else could it be? For example, is there evidence for nonlinearity? Let us compare the DVS plots for this series with DVS plots for different problems where the best predictions were also obtained (to within 5% accuracy) at the linear extreme (with large remaining prediction errors). Casdagli (1992) discusses several series that fall into this category: computer-generated data with noise, fully developed turbulence, cellular flames, and EEG data. Also, financial series are notorious for their lack of forecastability; see Briggs (1990) and Hsieh (1991). Compared to these time series, Data Set B seems to reveal some interesting structure that could suggest nonlinear dynamics. However, as pointed out

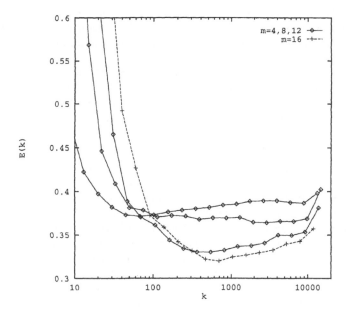

FIGURE 11   DVS plot for the bivariate model for predicting the heart rate from both heart rate and chest volume. Lead time is $T = 16$.

above, the nonstationarity clearly present in this series can also be responsible for this effect. Although the bivariate model is somewhat more forecastable, the main conclusion for Data Set B is the absence of low-dimensional chaos. Incidentally, DVSplots can be used to classify (medical and other) time series data, as an alternative to traditional spectral analysis.

4. CONCLUSIONS

Applying DVS plots to three series of the Competition, we reached the following conclusions:

- A.con: Low-dimensional chaos was clearly identified in the laser time series by obtaining short-term forecasts with deterministic local linear models which were 5 to 10 times more accurate compared to global linear models.

- D2.dat: Nonlinearity, but not low-dimensional chaos, was identified in the computer-generated time series by obtaining short-term forecasts with non-linear stochastic models which were 50% to 100% more accurate compared to global linear models.
- B2.dat: Nonlinearity, but not low-dimensional chaos, was identified in the medical time series by obtaining short-term forecasts with nonlinear stochastic models (univariate and multivariate) which were 10% to 25% more accurate compared to global linear models. The multivariate models revealed that coupling between heart rate and chest volume can be exploited to improve predictions over the univariate models.

The identification of low-dimensional chaos in the laser time series A is not surprising, since this series was collected from a carefully controlled experiment on a physical system with only few degrees of freedom. The results for the computer-generated Data Set D are not surprising either. This time series is generated from an approximately deterministic system with nine degrees of freedom. The DVS algorithm as defined here is powerful enough to detect strong nonlinearity in the time series, but not powerful enough to identify the underlying system as being deterministic. The results for the medical time series B are quite interesting. Although low-dimensional chaos was not identified, some evidence was obtained for the benefits of modeling this system with nonlinear multivariate models. Given the history of the art of dimension estimation, we also would like to point out that the DVS algorithm does not seem to identify low-dimensional chaos when it is not there, unlike some interpretations of dimension calculations (as discussed by Ruelle, 1990).

DVS plots are clearly only one of the many ways that might lead to some insight into the systems that generated an observed time series. They are general enough to be useful as a broad first tack but should be used in addition to other tools, such as testing for (non-)linearity and (non-)stationarity, analyzing recurrence plots, estimating dimensions, etc., since results from one algorithm in isolation are at best incomplete and at worst misleading.

The results of this paper suggest some directions for further study. As mentioned before, the main idea in DVS is the balance between low-variance/high-bias models (few parameters that can be estimated well even for very noisy, i.e., stochastic systems, but that are insufficient for complicated nonlinear mappings), and low-bias/high-variance models (which are flexible with regard to nonlinearities in the model but also with regard to overfitting if there is noise). The deterministic vs. stochastic idea can be extended to embeddings more powerful than delay lines (e.g., to embedding by expectation values; see Gershenfeld and Weigend, this volume), and to function approximation techniques more powerful than local linear approximation (e.g., connectionist networks). In addition to providing more accurate short-term forecasts, such extensions might enable us to identify higher dimensional chaos (e.g., in Data Set D)—another example of understanding the system through prediction.

Fernando J. Pineda and John C. Sommerer
M. S. Eisenhower Research Center and The Johns Hopkins University Applied Physics
Laboratory, Laurel, MD 20723-6099

Estimating Generalized Dimensions and Choosing Time Delays: A Fast Algorithm

We describe a fast, easily coded numerical algorithm for estimating generalized entropies. The box-counting algorithm presented here may be thought of as the natural culmination of a number of incremental improvements to naive box counting suggested over the years. Using this algorithm, the box-counting approach becomes competitive with more conventional correlation and k-nearest-neighbor approaches. We modify the Fraser-Swinney prescription for choosing delay times so that it makes use of the same fast algorithm. We observe that unlike the original Fraser-Swinney approach, the modified algorithm is applicable to deterministic systems.

We illustrate the algorithms by choosing delays and minimal embeddings for selected data sets from the Santa Fe Time Series Prediction and Analysis Competition. For a data set with 10^5 points, we are able to estimate information and correlation dimensions simultaneously in a little over two minutes per embedding on a scientific workstation.

Times Series Prediction: Forecasting the Future and
Understanding the Past, Eds. A. S. Weigend and N. A. Gershenfeld, SFI Studies
in the Sciences of Complexity, Proc. Vol. XV, Addison-Wesley, 1993   **367**

1. INTRODUCTION

When confronted with an experimental time series, it is natural to ask whether it results from a deterministic, low-dimensional dynamical system, or from a very high- or even infinite-dimensional process. To answer this question many investigators have proposed and analyzed techniques for estimating the dimensionality of dynamical systems from experimental time series (Grassberger & Procaccia, 1983a; Takens, 1984; Theiler, 1990b; R. L. Smith, 1992; Gershenfeld, 1992). Much of this work has focused on statistical estimators for precise determination of dimension; however, it is now recognized that for all but systems of very low dimension, the amount of experimental data required for a precise and accurate determination is likely to be unavailable (L. A. Smith, 1988; Ruelle, 1990; R. L. Smith, 1992). In any case, the utility of a precise dimension estimate remains debatable. However, if the dimension question can be answered (even approximately) in favor of the low-dimensional alternative, there are further questions that may reasonably be investigated, even in the case of limited data. These questions focus principally on the degree to which the dynamics can be reconstructed via time-delay or other embeddings.

The Takens embedding theorem (Takens, 1981) states that for a scalar time series $X(t)$ obtained from a d-dimensional deterministic system, the vector with time-delayed coordinates $(X(t), X(t - \tau), \ldots, X(t - E\tau)$, $E \leq 2d + 1$, will trace out a trajectory that is a smooth coordinate transformation of the attractor of the original dynamical system. In particular, if the dynamical system has an attractor of a particular dimension, the embedded trajectory will have the same dimension. In fact, it has been recently shown (Sauer et al., 1992) that, if the attractor underlying the time series is of box-counting dimension D_0, then a time delay embedding into a space whose embedding dimension E is the smallest integer greater than or equal to D_0, will allow accurate estimation of D_0, D_1, and D_2. The dynamical system is said to be *reconstructed*. Reconstruction considerably simplifies the task of prediction and can lead to other more detailed analyses.

The dimensions we calculate are a set of numbers known as generalized dimensions (Renyi, 1970). If the system attractor is covered by E-dimensional boxes of side length ε, then

$$D_q \equiv \frac{1}{q - 1} \limsup_{\varepsilon \to 0} \frac{\log \left(\sum_i p_i^q \right)}{\log(\varepsilon)}, \tag{1}$$

where p_i is the measure of the attractor in box i and the sum is only over occupied boxes. In practice, the limit is not calculable with finite data, and it is usual to assume that for small, finite ε, the sum will scale as ϵ^{-D_q}. For finite data, the measure in the ith box is estimated as $p_i = n_i/n$, where n_i is the number of points in the ith box and n is the total number of points. The generalized dimension D_q is then estimated as the slope of a straight-line region on a log-log plot of the sum in Eq. (1) vs. the box side length, ε. (We ignore the effects of lacunarity in this discussion.)

While Eq. (1) is defined for all real q, the dimension D_q is sensitive to low-probability boxes when $q < 0$. This means that we typically confine our interest to $q = 0$ (capacity dimension), $q = 1$ (information dimension), or $q = 2$ (correlation dimension). We note that the scaling assumptions and limitations due to finite data are not exclusive to the box-counting approach. Similar assumptions are made about the scaling of the directly calculated correlation integral or nearest-neighbor distances used in the usual determination of correlation dimension. In each case, there is an implicit hope that a limit such as that in Eq. (1) will be reached in a "well-behaved" way.

There are three main methods that are used for estimating D_q. These methods can be classified according to whether they use box counting, correlation sums, or k-nearest-neighbor distances. Box-counting techniques use a fixed-size mesh with box size ε and estimate the probability p_i in a box according to $p_i \sim n_i/n$ where n_i is number of points in a box and where n is the total number of points in the attractor. Eq. (1) is used to estimate D_q. Correlation sum techniques count the number of points n_i that fall within a ball of radius ε around m randomly chosen reference points (centers). The point used to define a center is excluded from n_i and p_i is again estimated from n_i/n. For small n it is possible to take $m = n$. For larger n this becomes intractable and one generally uses $m \ll n$ reference points; various proposals have been offered for choice of reference points. Eq. (1) is again used to estimate D_q. Finally, k-nearest-neighbor (or fixed mass) techniques start like correlation sum techniques in that m randomly chosen centers are selected. For the ith center the distances to the remaining $n-1$ points are calculated and sorted in order of increasing magnitude. The distance $r_i(k)$ to the kth nearest neighbor is used as the size δ_i of the ball with measure $p_i = k/n$. The scaling of mean ball size with k is used to estimate D_q (Grassberger, 1985).

Computationally and spatially efficient algorithms for implementing the aforementioned methods have been developed by various investigators. An efficient algorithm for capacity dimension (D_0) by box counting was described by Liebovitch and Toth (1989). In their algorithm they normalize the coordinates of the attractor so that it lies in an E-dimensional cube of size 2^k on a side. Each normalized coordinate is also truncated so it is represented by an integer in the range $(0, 2^k - 1)$. The attractor is then overlaid by a mesh with 2^m boxes on a side. The first $m - k$ bits of each coordinate is masked off and the resulting bit patterns are concatenated together to form a word that is $E \cdot (m - k)$ bits long. This "z" value uniquely labels the E-dimensional box of size 2^{m-k} that contains the given point. All the points with the same z-value are contained in the same box. To efficiently count the number of occupied boxes of size 2^{m-k}, Liebovitch and Toth sort the n z-values and then scan through the sorted list while counting the number of times the z-values change. The mask-concatenate-sort-count steps are repeated for each different box size. They also note that the algorithm can be made more efficient by interleaving the bits of the normalized coordinates and sorting just once. With this modification, the interleave and sort steps need be performed just once while the mask and count steps are combined into a single step that is repeated for each different box size. The memory complexity of this algorithm is $O(n)$. The normalizing, inter-leaving, and counting have computational complexity $O(n)$ while the sorting has computational complexity $O(n \log n)$ if a quicksort or heapsort algorithm is used. Bingham (1992) has discussed the case where box counting is performed with base $b \neq 2$ numerals and has also pointed out that box counting can be performed with $O(n)$ computational complexity. $O(n)$ computational complexity can be obtained through the use of a radix sort algorithm.

For correlation-sum algorithms, Theiler (1987) introduced box-assisted and prism-assisted algorithms. In box-assisted correlation-sum algorithms, a distance cutoff of magnitude r_0 is introduced. A spatial grid with this box size is used to coarsely bin the points. The distance between a pair of points is computed only if the two points are in the same or adjacent boxes. The execution time depends on the choice of r_0; the smaller r_0, the fewer distances that must be calculated. The minimum run time scales like $O(n \log n)$ which is a tremendous improvement over the $O(n^2)$ that is required if all the pairwise distances are calculated. To find neighboring boxes, the boxes are sorted in a linear list and searched. Ultimately the algorithm is limited by the search time. Bingham and Kot (1989) proposed the use of multidimensional trees and range searching as an alternative to linear lists of boxes and prisms in the box-assisted correlation sum algorithms. This latter im-provement results in a state-of-the-art algorithm whose spatial complexity is $O(n)$ and whose computational complexity is $O(n \log n)$. Bingham and Kot implemented their algorithm in two ways: first, recursively with a stack and a linear array to represent the tree data structure and, second, without a stack by using pointers and linked data structures.

For k-nearest-neighbor algorithms, Hunt and Sullivan (1986) published an effi-cient algorithm and data structure. They normalize, interleave, and sort their data to generate an array of box labels. Thus their data structure is the same as that suggested to Liebovitch and Toth by Kaplan. For k-nearest-neighbor algorithms, however, the sorted list of box labels is used to perform efficient range searching as part of a Monte Carlo algorithm. The sort is $O(n \log n)$ or $O(n)$ depending on how it is implemented while the search can be done with $O(n \log n)$ complexity.

The box-counting algorithm for computing D_q to be discussed here is more gen-eral than the box-counting algorithm for computing D_0 described by Liebovitch and Toth. Furthermore, it is simpler than either the k-nearest-neighbor algorithm dis-cussed by Hunt and Sullivan or the box-assisted correlation algorithm discussed by Theiler and extended by Bingham and Kot. It does, however, exploit the same basic data structure used by the modified Leibovitch-Toth and Hunt-Sullivan algorithms.

Before discussing the algorithm in more detail, we should mention that after we developed and implemented this algorithm, we discovered that it was identi-cal to an unpublished algorithm implemented by Városi (1988) at the University of Maryland under the supervision of J. Yorke (private communication). Nevertheless, we feel that it is important to publish this algorithm because few investigators are aware of its significant computational advantages. We have observed that most of the dimension estimation theory centers around the correlation-sum and k-nearest-neighbor techniques. We believe that this is a natural consequence of the mistaken belief that these two techniques are the easiest to implement. As we will show below, dimension estimation by box counting has significant advantages that call for a closer look on the part of the community. First, it is easier and faster to implement than either correlation-sum or k-nearest-neighbor techniques. Second, it makes full use of every data point, in contrast to practical direct correlation dimension estimates (Kostelich & Swinney, 1989). Third, it is just as easy and fast to implement it in higher embedding dimensions as in lower dimensions. Fourth, it is a simple matter to take into account theoretically derived sampling corrections (Grassberger, 1988). Fifth, the incremental cost of computing multiple dimensions simultaneously ($q = 0, 1, 2, 3, \ldots$) is negligible once the data structures have been set up to do the box-counting calculation. Sixth, because the sequence of generalized dimensions D_q is known to be nonincreasing with increasing q (Hentschel & Procaccia, 1983), and because our approach can yield a finite set of generalized dimensions all at once, we obtain an implicit check on the adequacy of the data set; finding $D_q > D_Q$ for $q < Q$ unequivocally indicates an inadequate data set with no additional statistical assumptions. Seventh, the data structures lend themselves to efficient compression and, thus, it is possible to do dimension estimation with very large data sets. Eighth, the box-counting procedure can be used to estimate the scale-dependent mutual information in the time series, provided the dimension analysis warrants proceeding to dynamical reconstruction. This allows a choice of delay to be made for delay-coordinate embedding by following a variant of the prescription put forth by Fraser and Swinney (Fraser & Swinney, 1986; Fraser, 1989b).

Our experience indicates that short codes, written in high-level languages and executed on conventional scientific workstations can estimate, in minutes, dimensions from experimental data sets consisting of hundreds of thousands of data points. Alternatively, we have been able to perform dimension estimation with nearly a half-billion points using somewhat more complex codes and data structures. In this paper we will focus on the basic in-core algorithm and describe its use in reconstructing dynamics from experimentally obtained data sets.

Finally, we wish to editorialize, so that our viewpoint is clear at the outset. We fully realize that approaching dimension estimation via box counting flies in the face of much "standard practice." Further, the statistical properties of box counting are not as well known as those of correlation algorithms. However, we feel that the need for *precise* dimension estimates from experimental data has never really been demonstrated. Very different systems can produce attractors with identical dimensions, so even accurate estimates of dimensions would not distinguish them. The two main practical uses of dimension estimates are (1) to establish whether or not low-dimensional dynamics underlie a time series and (2) to guide phase space reconstruction efforts. In this we agree with Theiler (1990b):

> ...to answer the question 'is it chaos or is it noise?' a robust estimate of dimension is more important than a precise estimate. In these cases the

subtle distinction between information dimension at $q = 1$ and correlation dimension at $q = 2$, say, may not be so important as the more basic issues that arise from experimental noise, finite samples, or even computational efficiency.

We provide the following in the context of a simple and efficient tool for initial investigation of experimental time series, or as a practical tool for detailed analysis of extremely large data sets ($> 10^8$ points) resulting from computer simulations.

We will next review the definitions of generalized dimensions and mutual information; explain the box-counting algorithm and its computational requirements and limitations; and demonstrate its application to several of the competition time series.

2. DEFINITIONS

To use generalized dimension to distinguish between low-dimensional chaos and noise, one essentially plots D_q vs. E for the time series data. Departure of the plot from the $D_q = E$ line, followed by saturation at a value supportable by the number of data, is evidence that the time series results from a relatively low-dimensional system. It is generally believed that the D_q at which saturation occurs provides a good estimate of the value of E required to obtain a good reconstruction of the dynamics. Recent work has provided an alternative and more direct approach (Liebert, Pawelzik, & Schuster, 1991; Kennel, Brown, & Abarbanel, 1992). This latter false nearest-neighbors approach may be less fallible than the former method. In any case, other recent work shows that given enough data, the saturation will occur for $E > D_o$, rather than for the $E \geq 2d + 1$ required for one-to-one reconstruction of d-dimensional dynamics (Sauer et al., 1992).

In theory, for noise-free data, the particular choice of delay time τ plays no role in reconstruction. In practice, the choice of delay can be quite important. Fraser and Swinney (1986) demonstrated that a good choice for the delay corresponds to the first minimum of the mutual information between the time series $X(t)$ and the delayed time series $X(t - \tau)$. At this point it is useful to define more precisely the information theoretic quantities that we intend to estimate.

Assume that the finite time series $X(t)$ is quantized in units of ε. Then, treating X and Y as discrete random variables corresponding to the time series $X(t)$ and $X(t - \tau)$, respectively, we can define discrete probabilities $P_X(x) = \text{Prob}\{X = x\}$ and $P_Y(y) \approx \text{Prob}\{Y = y\}$. Associated with each of these probabilities are scale-dependent information functions $I_X(\varepsilon)$ and $I_Y(\varepsilon)$ given by

$$I_X(\varepsilon) = - \sum_i P_X(x_i) \log_2 P_X(x_i) \qquad (2a)$$

and

$$I_Y(\varepsilon) = - \sum_i P_Y(y_i) \log_2 P_Y(y_i).\tag{2b}$$

In Eqs. (2a) and (2b), the arguments x_i and y_i denote box indices and the sums are over occupied boxes in one-dimensional embeddings. Both because of the information theoretical context and because we will be dealing later with the case $\varepsilon = 2^{-k}$, we naturally use base-2 logarithms. Similarly, we can define joint probabilities in two-dimensional embeddings, i.e., $P_{XY}(x,y) = \text{Prob}\{X = x \text{ and } Y = y\} \simeq n_i/n$ where n_i is the occupancy of the ith two-dimensional box of linear size ε. The corresponding scale-dependent joint information $I_{XY}(\varepsilon)$ between X and Y is then defined as

$$I_{XY}(\varepsilon) = - \sum_i P_{XY}(x_i, y_i) \log_2 P_{XY}(x_i, y_i).\tag{3}$$

The sums, as before, are taken over all the occupied two-dimensional boxes. Finally, we define scale-dependent mutual information $M_{XY}(\varepsilon)$ between the random variables X and Y as

$$M_{XY}(\varepsilon) = I_X(\varepsilon) + I_Y(\varepsilon) - I_{XY}(\varepsilon).\tag{4}$$

In the $\varepsilon \to 0$ limit, the information sums in Eq. (4) become integrals. Fraser (1989b) points out that the information $I_X(\varepsilon)$, $I_Y(\varepsilon)$, and $M_{XY}(\varepsilon)$ diverge in this limit if the system is deterministic and converges if the system is noisy. He also points out that, unlike each separate term in Eq. (4), the mutual information is independent of coordinates. Fraser also stresses the coordinate independence of the mutual information. We believe that for some purposes this may be a useful feature, but for the purpose of finding embedding delays, this feature is irrelevant since a constant added to information sums or integrals does not affect the location of their minima. Fraser concludes that:

> By moving from the entropy of partitions to mutual information integrals we buy the power to describe noisy measurements at the expense of the ability to describe noise-free systems.

It appears to us that the main advantage of using mutual information integrals is that it produces a diagnostic that is independent of box size. The cost is a relatively difficult-to-implement algorithm which makes use of an uneven partition to estimate the densities (small, rectangular boxes where the data are dense, and larger, rectangular boxes where the data are sparse). The smallest size of their boxes is determined by local statistical criteria.

Fraser (1989b) argues that examining the decrease of mutual information with time is equivalent to examining the increase in entropy with time provided that the noise smooths out the distributions inside elements of a generating partition without moving events from one box to another. This may be the case with measurement noise, but it is certainly not the case with dynamical noise. Thus to the extent that dynamical noise is significant, the use of this particular formulation of mutual information may not be entirely justified.

In what follows we advocate the use of generalized dimensions, calculated via simple and fast algorithms, as partition-independent diagnostics for choosing embedding delays. With this approach there is no need to sacrifice the ability to handle deterministic systems. We do, on the other hand, assume that there is scaling behavior that can be used to extract meaningful dimensions.

As noted previously (Fraser & Swinney 1986; Liebert & Schuster 1989), the joint information $I_{XY}(\varepsilon)$ is precisely the quantity used to estimate the information dimension D_1 for a two-dimensional embedding, i.e., the sum in Eq. (1) for $E = 2$ and $q = 1$. Thus, any box-counting algorithm that can calculate D_1 can also, with minor modifications, be used to estimate the scale-dependent mutual information.

Given infinite data and ignoring lacunarity effects, we expect each of the terms on the right-hand side of Eq. (4) to scale linearly with the logarithm of the box size, ε. The proportionality constant can be interpreted as an information dimension. In particular, we expect $I_{XY}(\varepsilon) \sim D_{XY} \log_2 \varepsilon$ and $I_X(\varepsilon) \sim D_X \log_2 \varepsilon$, where D_{XY} is the information dimension (D_1) for $E = 2$ and D_X is the information dimension (D_1) for $E = 1$. We note that D_X is equal to D_Y for a stationary time series; i.e., a simple one-dimensional histogram of the data values in a stationary time series should have no explicit dependence on τ. On the other hand, we expect any calculable estimate of D_{XY} to depend on τ, when the data are finite or noisy. Thus, under the above scaling assumptions, Eq. (4) can thus be rewritten

$$M_{XY}(\tau, \varepsilon) \sim [2D_X - D_{XY}(\tau)] \log_2 \varepsilon. \tag{5}$$

(As a check, we note that for X and Y independent, $D_{XY} = 2D_X$ and the mutual information is identically zero, as expected.) From this expression it is natural to define the mutual information dimension as

$$D_\mu(\tau) = 2D_X - D_{XY}(\tau). \tag{6}$$

To choose an embedding delay, we seek the minimum of $D_\mu(\tau)$ with respect to τ. Some care must be taken in estimating D_X and D_{XY} since, for finite amounts of data, the information sums will saturate for small enough box size ε. The value of ε at which saturation takes place will typically be larger for higher embedding dimensions. Thus we expect the linear scaling in Eq. (5) to break down at different values of ε for D_X and D_{XY}. It is therefore best to estimate D_X from a linear region on a plot of I_X vs. $\log 1/\varepsilon$ (corresponding to $E = 1$) and D_{XY} from a smaller linear region on a plot of I_{XY} vs. $\log 1/\varepsilon$ (corresponding to $E = 2$). If the scaling were perfect for all ε, finding the minimum of D_μ would be equivalent to seeking a minimum in the mutual information itself for any particular value of ε, or indeed in the limit of infinitesimal ε.

This procedure is extremely efficient computationally as we will see later, but can fail if the information sum shows no scaling region in one- and two-dimensional embeddings (for example, if the maximum value of ε where global shape effects ruin scaling is too close to the minimum value of ε where saturation effects ruin scaling). Of course, in this case there is no possibility of estimating generalized dimensions anyway, and we will have explicitly identified the failure before proceeding to that step.

In summary, our variant of Fraser and Swinney's prescription for choosing the delay time consists of calculating the information dimension in one- and two-embeddings for a series of possible time delays. We calculate the corresponding information dimensions and use Eq. (6) to estimate the mutual information dimension. We choose the delay that minimizes the mutual information dimension. In contrast to the Fraser-Swinney approach, we buy the power to describe deterministic systems and noisy measurements at the expense of requiring the observation of scaling regions.

Liebert and Schuster (1989) argue that τ can be chosen on the basis of the correlation sum (i.e., with dimension D_2), rather than the information sum. This is very similar to the procedure we outline above, since if one assumes that D_X is constant over the delays of interest, a minimum of D_μ corresponds to a maximum of $D_{XY} = D_1$ (for $E = 2$). In any case, if one wishes to use this alternative procedure, the arguments in this section still apply, and the box-counting algorithm can be used to efficiently calculate D_2.

Now that we have reviewed the salient concepts of generalized dimension and mutual information, we turn to an explanation of the box-counting algorithm and its computational requirements and limitations; and demonstrate its application to several of the Competition time series. We will not present detailed statistical analyses of the box-counting algorithm, since we do not apply it to precisely estimate dimensions in the time series context. However, we note that in the context of *numerical* experiments to investigate the properties of strange attractors, such analysis is warranted and may be fruitful.

3. THE BOX-COUNTING ALGORITHM

A. BASIC ALGORITHM

In this section we describe the four steps of the basic box-counting algorithm. A pseudocode is given at the end of this section that demonstrates how to implement the algorithm. We note that the bit manipulations required to implement the interleaving in step 2 or the masking in step 4 may be unfamiliar to many scientific programmers. Nevertheless, we know of no standard scientific language that does not support these manipulations as part of the language definition or as part of a standard library.

To be concrete, let us start by assuming that we have a floating point array of E-dimensional vectors. The array is n elements long. The array represents the state of an E-dimensional dynamical system as a function of time or it represents an E-dimensional time delay embedding of a scalar time series.

1. INTEGERIZE. The first step of the basic algorithm is to scale and truncate the floating point values into unsigned r-bit integers. The integer word length, r, determines the finest resolution that can be used to estimate the dimension. The minimum and maximum expected values of each variable are used to scale the floating point numbers. In particular let $x_i(j)$ be the ith component of jth E-dimensional state vector. Then the integerized value is

$$I_i(j) = \frac{(2^r - 1)(x_i(j) - x_i^{\min})}{(x_i^{\max} - x_i^{\min})}.$$

I_i is an r-bit unsigned integer between zero and $2^r - 1$, inclusive.

2. BIT-INTERLEAVE. Next, for each value of j, interleave the r bits of the E r-bit integers to form a single Er-bit-long integer, z. This operation is shown schematically below.

$$\left.\begin{array}{l} I_1(j) = a_1, \ldots, a_r \\ I_2(j) = b_1, \ldots, b_r \\ \vdots \\ \vdots \\ I_E(j) = w_1, \ldots, w_r \end{array}\right\} \longrightarrow z(j) = a_1, b_1, \ldots, w_1, \ldots, a_r, b_r, \ldots, w_r.$$

When the coordinates of the points in the data set are integerized and interleaved, the resulting array of z-values contains all the information necessary to do the box counting or occupation-number counting, if we are content to restrict the value of ε in Eqs. (1)–(4) to be of the form $\varepsilon = 2^{-k}$. The array of Er-bit integers (the z-array) can be scanned to count the number of occupied boxes at each level of resolution. The trick is to examine only the first kE bits of each z-value. At the kth level of resolution, the box count is obtained by scanning through the array and counting the number of unique kE-bit patterns.

3. SORT. The scan through the z-array can be performed efficiently and with very little additional memory by sorting the z-array. It is easy to see that after the sort, all the points that belong in a given box at any level of resolution are adjacent. Thus counting the number of boxes at the kth level of resolution ($\varepsilon = 2^{-k}$) amounts to scanning through the sorted array and counting the number of times that the kE-bit pattern changes.

If the sort is performed with a quicksort algorithm, then it is done in place and requires minimal additional memory. The time spent sorting is $O(n \log n)$. Alternatively, one can use a radix sort which, if properly implemented, requires $O(n)$ time. A radix sort, however, requires more memory than a quicksort.

4. ACCUMULATE PROBABILITIES AND SUMS. The number of points n_i that are contained in a given box at the kth level of resolution can be deduced by counting the number of times that a given kE-bit pattern is repeated. This is easily done in the sorted z-array since the sort places like patterns next to each other. In the simplest case, probabilities p_i are estimated according to $p_i = n_i/n$. One pass through the entire z-array is required to accumulate the statistics for each of the r levels of resolution, accordingly time spent counting is $O(rn)$.

B. PSEUDOCODE

The pseudocode given below illustrates the basic algorithm. As described below, it is designed to calculate the sums $\sum_i p_i^q$ for $q > 1$. For $q = 1$ the code should be modified to accumulate $\sum_i p_i \log(p_i)$ instead of $\sum p_i^q$. The naive estimate $p_i = n_i/n$ for probability is used for clarity despite the fact that this expression is well known to systematically underestimate the true probability. Grassberger (1988) gives corrections to p_i and to the accumulated generalized entropies that are asymptotic in $1/n_i$. These expressions can be used to obtain increased precision at the expense of somewhat greater computational complexity.

The pseudocode is specialized to a two-dimensional embedding ($E = 2$) with 16 levels of resolution ($r = 16$). The z-values are 32-bit unsigned integers. The most significant bit is designated bit 0 while the least significant bit is designated bit 31. The kth element of the array `bitmask` contains the value 2^k; i.e., `bitmask[k]` $= 2^k$. Array element `boxmask[k]` has bits 0 through $2k$ set to 1 and all other bits set to 0. The function that performs bit shifting is designated `shift()`. It takes as its first argument an unsigned 32-bit integer. Its second argument is the number of bit locations to shift. A shift of -1 corresponds to a right shift by one bit (division by 2) while a shift of $+1$ corresponds to a left shift by one bit (multiplication by 2). The function `shift()` returns a shifted 32-bit unsigned integer.

After the procedure is completed, the array `sumpq[k]` contains the sum in Eq. (1) for the chosen value of q, the logarithm of which can be plotted vs. $\log(2^{-k})$ to determine whether or not a scaling region with slope D_q exists. Several practical factors affect the use of this algorithm for finding dimensions and mutual information. These factors are in addition to the actual fact of whether or not the data result from a process with genuine fractal properties. (See Figure 1.)

First, the process of integerizing the coordinates in the first step of the algorithm implicitly chooses boxes of a specific shape for covering the data set. That shape is governed by the overall aspect ratio of the data set and may not be well tuned to the fine structure of the potential attractor. In the limit of infinite data, one expects the shape of the cover elements to be irrelevant, but with finite data and high-aspect ratio data sets, the algorithm may be impaired.

```
double precision float      x, y, sum, sumpq
unsigned 32-bit integer     lx,ly,lz,temp,currentBox,lastBox,n,ni
unsigned 32-bit integer.    array bitmask[16], boxmask[16], z[n]

BEGIN PROCEDURE
  FOR(j=0 to n-1) BEGIN       main loop for digitizing and interleaving
      getNextPoint(x,y)       read (or calculate) the next point in the sequence
      lx ← (2^16-1)*x         digitize x to 16-bit unsigned integer
      ly ← (2^16-1)*y         digitize y to 16-bit unsigned integer
      lz ← 0
      FOR(k=0 to 15) BEGIN   this loop extracts bits and does the interleaving
          lz ← shift(lz,-1) + shift(bitAnd(ly,bitmask[k]),31-k)   a bit from y
          lz ← shift(lz,-1) + shift(bitAnd(lx,bitmask[k]),31-k)   a bit from x
      END FOR
      z[j] ← lz               save result in ith element of z array
  END FOR

  sort(z, n)

  FOR(k=0 to 15) BEGIN
      ni ← 1
      sum ← 0
      lastBox ← bitAnd(boxmask[k],z[0])
      FOR(j= 0 to n-1) BEGIN
          currentBox ← bitAnd(boxmask[k],z[j])
          IF(lastBox EQUALS currentBox) BEGIN
              ni ← ni + 1                    increment points per box
          ELSE
              lastBox ← currentBox;   change box label
              sum ← sum + (ni/n)^q    accumulate ∑ p_i^q (or ∑ p_i log p_i)
              ni ← 1
          END IF
      END FOR
      sum ← sum + (ni/n)^q                  accumulate last term in sum
      sumpq[k] ← sum
  END FOR
END PROCEDURE
```

FIGURE 1 Pseudocode for box-counting algorithm.

Second, fractal sets often exhibit lacunarity that manifests itself as periodic (in log space) oscillations of the sum in Eq. (1) as ε is varied. The supremum operation in Eq. (1) is required to deal with this fact, essentially by considering the scaling of the envelope of the sum in Eq. (1) with ε. Because the algorithm as described above is restricted to $\varepsilon = 2^{-k}$, such lacunarity may not be resolved. If the generalized information sum has a period of oscillation sufficiently incommensurate with $\log 2$, a limited dynamic range of ε could easily show no scaling region for the information sum, or even an apparently scaling region that yields an incorrect generalized dimension.

Third, it is well known that numerical estimates of generalized dimensions for $q < 0$ do not properly account for sparsely populated boxes. Hence, we do not advocate using this approach for estimating generalized dimensions for $q < 0$.

Fourth, in the above example only boxes with sizes $\varepsilon = 2^{-k}$ with $k = 0$ to 15 are counted. In some cases, for example to exhibit lacunarity, it is desirable to sample the scaling curves more finely. This can be accomplished by running the algorithm twice, first with the usual normalization of the points and then again but with an additional factor $1/\sqrt{2}$. The two runs combined produce boxes with sizes $\varepsilon = 2^{-k/2}$ with $k = 0$ to 31.

Finally, finite data sets will usually produce saturation of the generalized information, as ε is made small enough. For a given value of E, at some sufficiently small ε, the boxes will be small enough that there is only one point per occupied box. Making ε even smaller clearly cannot increase the generalized information sum, because no new boxes will be occupied. For example, the value of the information sum must saturate at $\log(n)$, since $p_i = 1/n$.

An important exception to the one-point-per-box saturation level occurs when the data have been digitized at lower resolution than $\log(n)/D_1$; in this case, many of the integerized points will have the same value out to the maximum resolution at which box counting is performed. This will cause the generalized information sums to saturate at a lower value than would be expected on the basis of the size of the data set alone. We also note that this anomalous saturation will be more pronounced in lower embedding dimensions; for example, in order for two 2-d points to be in the same 2-d box, two pairs of coordinates must have the same value and this is less likely than having just two coordinates the same. We further note that if the quantization scale is large compared to the range of the data, the algorithm will "correctly" determine that the embedded attractor is a point set of dimension zero, even with infinite data.

The implications of saturation result from the expectation of some smooth transition for the generalized information sum from the scaling behavior to the saturated behavior. In this transition region, the slope of the $\log(\text{sum})$ vs. $\log(\varepsilon)$ plot is lower than in the scaling region. If saturation takes place for large enough ε, the true scaling region may be suppressed in favor of an apparent scaling region giving an artificially low estimate of generalized dimension.

4. EXAMPLES

To illustrate the use and performance of this algorithm, we apply it to Data Set A, the far-infrared laser, and to Data Set D, the chaotic oscillator. The codes were written in C and written for ease of debugging and portability rather than for speed. In particular, no effort was made to vectorize the code and the standard C library qsort subroutine was used to perform the sorts. Unless otherwise noted, performance figures are for an IBM RS/6000 workstation.

A. FAR-INFRARED LASER

The far-infrared laser data set has 1000 points. To chose a delay τ for dynamical reconstruction, we embed the data in one and two dimensions and minimize the mutual information dimension D_μ with respect to τ. As we have argued in Section 2, this is similar to Fraser and Swinney's approach of finding a minimum in the mutual information. The result is shown in Figure 2 below. To generate the data in the plot required less than one minute on a Macintosh IIfx. Using either the criterion of first minimum in mutual information or correlation dimension, a suitable delay is seen to be $\tau = 2$. The curve labeled "$2 - D_1$" illustrates our comment in Section 3 that minima in mutual information should correspond to maxima in information dimension.

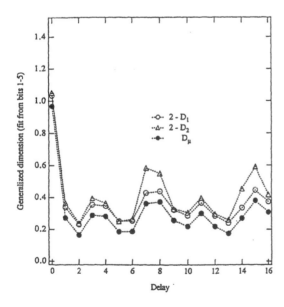

FIGURE 2   Mutual information dimension D_μ as a function of delay for the far infra-red laser. The minimum of D_μ occurs at $\tau = 2$. For comparison we show $2 - D_1$ and $2 - D_2$ for a two-embedding. The fact that the minima and maxima of $2 - D_2$ coincide with those of D_μ indicate that the Liebert and Schuster (1989) prescription of using the maxima of the correlation dimension would result in the same delay as our mutual-information dimension prescription.

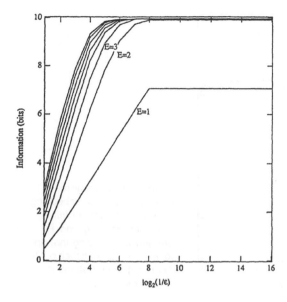

FIGURE 3 Information sums from the far infra-red laser as a function of box size for a delay $\tau = 2$. Each line corresponds to a different embedding. The $E = 1$ information saturates at a lower value due to its enhanced sensitivity to digitization.

To determine a suitable embedding dimension, we calculated the information sums and the correlation sums using the delay $\tau = 2$ in embedding dimensions from 1 through 8 inclusive. The complete calculation over all 8 embedding dimensions and over 16 octaves of resolution required less than 7 CPU seconds. The results for the information sum are shown in Figure 3. The scaling regions occur for $\log_2(1/\varepsilon) < 6$ bits. The $E = 1$ curve saturates at $\log_2(1/\varepsilon) = 8$ bits. This simply reflects the fact that the data were collected with an 8-bit analog to digital converter. Similar curves were obtained for the correlation sums.

In Figure 4 we plot the information dimension and the correlation dimension as a function of embedding dimension. The rapid saturation of these dimensions with the embedding dimension clearly indicates the presence of low-dimensional dynamics in the data set. Because of the limited data, the saturation dimension of about 2.2 for the system's attractor cannot be taken too seriously, but the theoretical requirement that D_1 dominate D_2 is met throughout the saturated region, increasing our confidence that the data are adequate to support the conclusion of low-dimensional dynamics.

Although at this stage of the analysis, we have no right to expect that a 2.2-dimensional set can be embedded in less than 5 dimensions, it is instructive to examine the data in a time delay embedding with $E = 2$, as shown in Figure 5. There are obvious problems with the embedding near the origin; however, the spoke-like structure in the midst of the data is clear. In fact, this structure clarifies the strong maximum in the mutual information dimension around $\tau = 7$ in Figure 2.

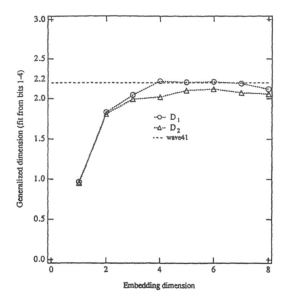

FIGURE 4 Generalized dimension versus embedding dimension for the far infrared laser. Both D_1 and D_2 are shown indicating that up to $E = 8$ the relationship $D_1 \gtrsim D_2$ is satisfied. Saturation occurs for $D_1 \approx D_2 \approx 2.2$.

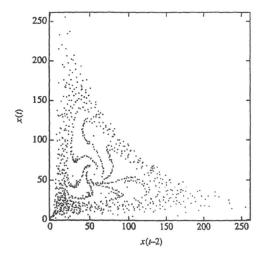

FIGURE 5 The far infra-red laser attractor reconstructed in a two-dimensional embedding with a delay of 2. The seven spokes in the center, corresponding to the peak at delay $\tau = 7$ in Figure 2, are clearly visible.

There are seven spokes evident in the embedded data set. The system trajectory moves out along the spokes gradually, circling the central circular void. Data seven samples apart thus fall very close, leading to high mutual information for that delay.

B. CHAOTIC OSCILLATOR

This data set contained 10^5 points. As in the case of the laser data, we embedded the time series in one and two dimensions and calculated D_1, D_2, and D_μ as functions of delay. The calculation required approximately 1.5 minutes per delay. We found that for $1/\varepsilon > 2^7$, the mutual information was meaningless since its value was merely an artifact of the number of samples and of the limited precision of the data stored in the file (see below). Figure 6 shows the mutual information dimension D_μ as a function of delay. The first minimum occurs at $\tau = 2$. A minor peak occurs at $\tau = 4$. This corresponds to a period of 2 or a frequency of $f = 0.5$. After the time series competition was over, it became known that a drive frequency, $f = 0.6$, was used to generate the data. We hypothesize that this accounts for the peak at $f = 0.5$.

To estimate the dimensionality of the attractor, we used a delay $\tau = 2$ and calculated both D_1 and D_2 for all embeddings from $E = 1$ to $E = 16$. The results are shown in Figure 7. This required approximately 35 CPU minutes or roughly 2.2 minutes per embedding. We found that the $E = 1$ information (and correlation) sums saturated at $I \approx 10$ bits. This is consistent with the fact that only three decimal places were used to save the data in the competition data file. For higher

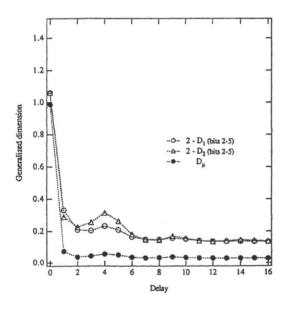

FIGURE 6   Mutual information dimension D_μ as a function of delay for the chaotic oscillator. The minimum of D_μ occurs at $\tau = 2$. For comparison we show $2 - D_1$ and $2 - D_2$. As in the laser data that the minima and maxima of $2 - D_2$ coincide with those of D_μ indicating that the Liebert and Schuster (1989) prescription would result in the same delay as our mutual-information dimension prescription.

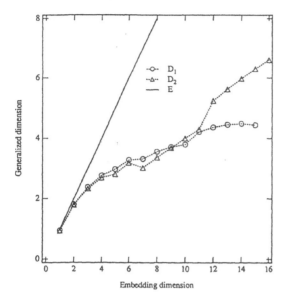

FIGURE 7 Generalized
dimension versus embedding
dimension for the chaotic
oscillator. The large jump in
the D_2 curve is an artifact
of using different numbers
of points to fit straight lines
as the embedding dimension
increases.

embedding dimensions we found that the information (and correlation) sums saturated at $\log_2(100000) = 16.6$ as expected from the number of samples. We estimated D_1 and D_2 by fitting to the scaling regions of the correlation and information sums. For $E \geq 9$ the relationship $D_1 \geq D_2$ is violated yet neither D_1 nor D_2 has saturated. Thus we do not have too much faith in our dimension estimates. It seems reasonable to conclude, however, that the data is consistent with D_1 and D_2 at least 4. Now that the contest is over, it is known that the time series was slightly nonstationary. This may account for the violation of the $D_1 \geq D_2$ relationship.

5. DISCUSSION AND CONCLUSIONS

We have demonstrated that efficiently implemented box-counting algorithms allow convenient initial analysis of time series data. Even large data sets can be quickly assessed as to whether they result from low-dimensional dynamics and, if so, a good delay for time-delay embedding can be determined using the same techniques. The in-core algorithm described in this paper is quite capable of handling around a million points, provided sufficient random access memory is available. Certainly this exceeds the amount of data that is typically available from experiments. For this reason we have chosen not to discuss the extensions[1] to the algorithm that have enabled us to perform calculations with several hundred million points on our workstation.

In Section 2 we confined the discussion to one- and two-dimensional embeddings. Nevertheless, it is clear that the information sums calculated for higher-dimensional embeddings correspond to more general joint-information sums. Conditional-information sums can be estimated by taking the difference between joint-information sums that differ by one embedding dimension. Partition-independent quantities can be extracted by considering the dimensions of the conditional and joint informations. These dimensions measure bits of (joint, conditional, mutual) information per bit of resolution. The same approach can be used to evaluate the marginal and total redundancies discussed by Fraser (1989b).

The fact that we can simultaneously estimate many generalized dimensions at once with little additional computational cost allows us to apply different prescriptions for choosing delays and for finding embedding dimensions. Thus we expect this approach to provide a quite simple and general computation framework for applying dimension estimation to time series prediction.

Finally, lest our computer science colleagues think us remiss, we are compelled to point out that the sorted bit-interleaved array that is the primary data structure necessary to implement this algorithm is quite an old idea whose utility has been discovered and rediscovered over the years. The reader is referred to Samet (1992) for an extensive discussion concerning its pedigree within the computer science community where it is used extensively for representing spatial databases.

ACKNOWLEDGMENTS

We gratefully acknowledge the support of the Department of the Navy, Space and Naval Warfare Systems Command, under contract No. N00039-91-C-0001.

---

[1] The extensions are of two kinds. First, it is necessary to make efficient use of the disk. This is largely an exercise in conventional sort-merge techniques and was discussed by Városi. Second, we note that there is considerable room for improvement in the primary data structure: a sorted bit-interleaved array. In particular, we observe that it is not a particularly compact representation. A considerable amount of memory is wasted storing redundant information. To see that this is the case, it suffices to consider that adjacent entries in the data structure may differ only in their lowest order bits; thus, the higher order bits contain redundant information. We believe that it should not be difficult to perform dimension estimates on data sets with as many as 10^9 points on a conventional scientific workstation with a one-gigabyte disk.

Milan Paluš
Laboratory for Applied Mathematics & Bioengineering, Prague Psychiatric Center, 3rd School of Medicine, Charles University, Ustavni 91, CS-181 03 Prague 8, CSFR.
Present address: Santa Fe Institute, 1660 Old Pecos Trail, Suite A, Santa Fe, NM 87501.

Indentifying and Quantifying Chaos by Using Information-Theoretic Functionals

We present a technique for analyzing experimental time series that provides detection of nonlinearity in data dynamics and identification and quantification of underlying chaotic dynamics. It is based on evaluation of redundancies, information-theoretic functionals which have a special form for linear processes, and on the other hand, when estimated from chaotic data, they have specific properties reflecting positive information production rate. This rate is measured by metric (Kolmogorov-Sinai) entropy that can be estimated directly from the redundancies.

1. INTRODUCTION

The inverse problem for dynamical systems, i.e., determination of the underlying dynamical process based on processing experimental dynamical data (Abraham et al., 1989; Mayer-Kress, 1986), can be considered as a recent alternative to generally used stochastic methods in the area of time series analysis. Algorithms have

Times Series Prediction: Forecasting the Future and
Understanding the Past, Eds. A. S. Weigend and N. A. Gershenfeld, SFI Studies
in the Sciences of Complexity, Proc. Vol. XV, Addison-Wesley, 1993　**387**

been developed which can in principle serve for identification and quantification of underlying chaotic dynamics (Cohen & Procaccia, 1985; Dvořák & Klaschka, 1990; Grassberger & Procaccia, 1983a, 1983c; Wolf et al., 1985). Analyzing experimental and usually short and noisy data, however, ordinary estimators of dimensions or Lyapunov exponents can be fooled, e.g., by autocorrelation of the series under study and can consider as chaotic the process which is in fact linear and stochastic (Osborne & Provenzale, 1989). These complications evoked the necessity of developing methods testing the basic properties of chaotic systems, like nonlinearity, independently of the dimensional or Lyapunov exponents algorithms.

In this paper, we propose an original method suitable for two-step assessment of the character of a time series under study. The first step is testing for nonlinearity in the data dynamics in a general sense, and the second step is identification of chaotic dynamics with the possibility of its subsequent quantification. The whole method is based on evaluation of so-called redundancies, information-theoretic functionals which have a special form for linear processes, and thus a comparison of their linear and general versions can demonstrate the linear or nonlinear nature of the data. On the other hand, when redundancies are estimated from chaotic data, they have specific properties reflecting a positive information production rate. This rate is measured by metric (Kolmogorov-Sinai) entropy that can be estimated directly from the redundancies. In this part of the method, i.e., in estimations of the metric entropy, we follow the original work of Fraser (1989b).

Section 2 introduces the information-theoretic functionals—mutual information and redundancies. The redundancy-based test for nonlinearity is explained and illustrated by numerical examples in Section 3. Its comparison with other methods, testing for nonlinearity or determinism, is given in Section 4. The relation between redundancies computed from time series and metric entropy of underlying dynamical system, giving the possibility of identification and quantification of chaotic dynamics, is explained in Section 5. Section 6 presents the application of the presented methodology on the competition data. The conclusion is given in Section 7. In Appendix 1, remarks concerning the algorithm for estimating redundancies are presented. Details on numerically generated data, used in testing the properties of the presented method, can be found in Appendix 2.

2. MUTUAL INFORMATION AND REDUNDANCIES

In this section we will define basic functionals introduced in information theory. More details can be found in any book on information theory (Billingsley, 1965; Gallager, 1968; Khinchin, 1957; Kullback, 1959; Shannon & Weaver, 1964).

Let x, y be random variables with probability distribution densities $p_x(x)$ and $p_y(y)$. The entropy of the distribution of a single variable, say x, is defined as:

$$H(x) = -\int p_x(x) \log(p_x(x)) dx. \tag{1}$$

For the joint distribution $p_{x,y}(x,y)$ of x and y, the joint entropy is defined as:

$$H(x,y) = -\int\int p_{x,y}(x,y) \log(p_{x,y}(x,y)) dx dy. \tag{2}$$

The conditional entropy $H(x|y)$ of x given y is defined as:

$$H(x|y) = -\int\int p_{x,y}(x,y) \log\left(\frac{p_{x,y}(x,y)}{p_y(y)}\right) dx dy. \tag{3}$$

The entropy of the distribution of the discrete random variable z_i with probability distribution $p_z(z_i)$, $i = 1,\ldots,k$, is defined as:

$$H(z) = -\sum_{i=1}^{k} p_z(z_i) \log(p_z(z_i)). \tag{4}$$

Definitions (2) and (3) for discrete variables can be derived straightforwardly. The average amount of information about the variable y that the variable x contains is quantified by the mutual information $I(x;y)$:

$$I(x;y) = H(x) + H(y) - H(x,y). \tag{5}$$

Clearly, $I(x;y) = 0$ iff $p_{x,y}(x,y) = p_x(x)p_y(y)$, i.e., iff x and y are statistically independent.

Generalization of the definition of the joint entropy for n variables $x_1,\ldots,x_n$ is straightforward: The joint entropy of distribution $p(x_1,\ldots,x_n)$ of n variables $x_1,\ldots,x_n$ is

$$H(x_1,\ldots,x_n) =$$
$$-\int\ldots\int p(x_1,\ldots,x_n) \log(p(x_1,\ldots,x_n)) dx_1 \ldots dx_n. \tag{6}$$

Generalization of the mutual information for n variables can be constructed in two ways:

In analogy with Eq. (5) we define

$$R(x_1;\ldots;x_n) = H(x_1) + \ldots + H(x_n) - H(x_1,\ldots,x_n). \tag{7}$$

This difference between the sum of the individual entropies and the entropy of the n-tuple $x_1, \ldots, x_n$ vanishes iff there is no dependence among these variables. Quantity (7), which quantifies the average amount of common information contained in the variables $x_1, \ldots, x_n$, is called the *redundancy* of $x_1, \ldots, x_n$.

Besides Eq. (7) we define the *marginal redundancy* $\varrho(x_1, \ldots, x_{n-1}; x_n)$ quantifying the average amount of information about the variable x_n contained in the variables $x_1, \ldots, x_{n-1}$:

$$\varrho(x_1, \ldots, x_{n-1}; x_n) = H(x_1, \ldots, x_{n-1}) + H(x_n) - H(x_1, \ldots, x_n). \qquad (8)$$

Clearly, the marginal redundancy $\varrho(x_1, \ldots, x_{n-1}; x_n)$ vanishes iff x_n is independent of all $x_1, \ldots, x_{n-1}$.

The following relations between redundancies and entropies can be obtained by a simple manipulation:

$$\varrho(x_1, \ldots, x_{n-1}; x_n) = R(x_1; \ldots; x_n) - R(x_1; \ldots; x_{n-1}) \qquad (9)$$

and

$$\varrho(x_1, \ldots, x_{n-1}; x_n) = H(x_n) - H(x_n | x_1, \ldots, x_{n-1}). \qquad (10)$$

Now, let $x_1, \ldots, x_n$ be an n-dimensional normally distributed random variable with zero mean and covariance matrix $\mathbf{C}$. In this special case, redundancy (7) can be computed straightforwardly from the definition:

$$R(x_1; \ldots; x_n) = \frac{1}{2} \sum_{i=1}^{n} \log(c_{ii}) - \frac{1}{2} \sum_{i=1}^{n} \log(\sigma_i), \qquad (11)$$

where c_{ii} are diagonal elements (variances) and σ_i are eigenvalues of the $n \times n$ covariance matrix $\mathbf{C}$. (See, e.g., the work of Morgera (1985), two-dimensional case, i.e., mutual information, was derived also by Fraser (1989b).)

Formula (11) obviously may be associated with any positive definite covariance matrix. Thus we use formula (11) to define the *linear redundancy* $L(x_1; \ldots; x_n)$ of an arbitrary n-dimensional random variable $x_1, \ldots, x_n$, whose mutual linear dependencies are described by the corresponding covariance matrix C:

$$L(x_1; \ldots; x_n) = \frac{1}{2} \sum_{i=1}^{n} \log(c_{ii}) - \frac{1}{2} \sum_{i=1}^{n} \log(\sigma_i). \qquad (12)$$

If formula (12) is evaluated using the correlation matrix instead of the covariance matrix, then particularly $c_{ii} = 1$ for every i, and we obtain

$$L(x_1; \ldots; x_n) = -\frac{1}{2} \sum_{i=1}^{n} \log(\sigma_i). \qquad (13)$$

Furthermore, in analogy with Eq. (9), we can define the *marginal linear redundancy* of $x_1, \ldots, x_{n-1}$ and x_n as:

$$\lambda(x_1, \ldots, x_{n-1}; x_n) = L(x_1; \ldots; x_n) - L(x_1; \ldots; x_{n-1}). \qquad (14)$$

3. TESTING FOR NONLINEARITY

Nonlinearity is the necessary condition for deterministic chaos (Schuster, 1988). Thus, in searching for chaotic dynamics in time series, the first step should be assessing its nonlinearity. We can do it using the above-defined redundancies.

In a typical experimental situation, one deals with a time series $Y(t)$. It is usually considered a realization of a stochastic process $\{x_i\}$ which is stationary.

We will study redundancies for variables

$$x_i(t) = Y(t + (i-1)\tau), \; i = 1, \ldots, n, \tag{15}$$

where τ is a time delay and n is the so-called embedding dimension (Takens, 1981). Redundancies of the type

$$R(Y(t); Y(t+\tau); \ldots; Y(t + (n-1)\tau))$$

are, due to stationarity of $Y(t)$, independent of t. We introduce the notation:

$$R^n(\tau) = R(Y(t); Y(t+\tau); \ldots; Y(t + (n-1)\tau)) \tag{16}$$

for the redundancy and

$$L^n(\tau) = L(Y(t); Y(t+\tau); \ldots; Y(t + (n-1)\tau)) \tag{17}$$

for the linear redundancy of the n variables $Y(t)$, $Y(t+\tau)$, $\ldots, Y(t + (n-1)\tau)$; and

$$\varrho^n(\tau) = \varrho(Y(t), Y(t+\tau), \ldots, Y(t + (n-2)\tau); Y(t + (n-1)\tau)) \tag{18}$$

for the marginal redundancy and

$$\lambda^n(\tau) = \lambda(Y(t), Y(t+\tau), \ldots, Y(t + (n-2)\tau); Y(t + (n-1)\tau)) \tag{19}$$

for the marginal linear redundancy of the variables $Y(t), Y(t+\tau), \ldots, Y(t+(n-2)\tau)$ and the variable $Y(t + (n-1)\tau)$.

Relations (9) and (14) can be rewritten as

$$\varrho^n(\tau) = R^n(\tau) - R^{n-1}(\tau) \tag{20}$$

and

$$\lambda^n(\tau) = L^n(\tau) - L^{n-1}(\tau), \tag{21}$$

respectively.

The linear redundancy, according to its definition (13), reflects dependence structures contained in the correlation matrix $\mathbf{C}$ of the variables under study. In the special case considered here, when all the variables are, according to Eq. (15), lagged versions of the series $Y(t)$, each element of $\mathbf{C}$ is given by the value of the autocorrelation function of the series $Y(t)$ for a particular lag. As the correlation is the measure of linear dependence, the linear redundancy characterizes linear structures in the data under study.

We propose to compare the linear redundancy $L^n(\tau)$ with the redundancy $R^n(\tau)$ (or the marginal linear redundancy $\lambda^n(\tau)$ with the marginal redundancy $\varrho^n(\tau)$) considered as the functions of the time lag τ. If their shapes are the same or very similar, a linear description of the process under study should be considered sufficient. Large discrepancies suggest important nonlinearities in links among the variables, or, recalling Eq. (15), among the studied time series and its lagged versions, i.e., in the dynamics of the process under study.

We would like to emphasize that we compare shapes of redundancies as functions of lag τ, not particular values of the redundancies. Estimated values of $R^n(\tau)$ and $\varrho^n(\tau)$ depend on a numerical procedure used ("quantization"—see Appendix 1), while the shapes of their τ-plots are usually consistent for a large extent of numerical parameters used in the redundancy estimations. Therefore each figure, depicting redundancies against time lag τ, is drawn in its individual scale. Redundancies $R^n(\tau)$ and $L^n(\tau)$ are plotted as $R^n(\tau)/(n-1)$ and $L^n(\tau)/(n-1)$. All the redundancies are in bits and time lags in number of samples. In the case of the Rössler and Lorenz systems, time lags are in relevant time units; in the case of EEG, in milliseconds. Different curves in each figure correspond to redundancies of different numbers n of variables (embedding dimension); n is from two usually to five, reading from bottom to top.

Let us recall that the equivalence of redundancy $R^n(\tau)$ and linear redundancy $L^n(\tau)$ can be proved only for a special type of linear processes—the processes with the multivariate Gaussian distribution. In a general case, however, we cannot neglect the possibility that differences between redundancies and linear redundancies are not due nonlinearity, but due a non-Gaussian distribution of the studied data. Nevertheless, after extensive numerical study we can conjecture that the shapes of τ-dependence of redundancy $R^n(\tau)$ (marginal redundancy $\varrho^n(\tau)$) and linear redundancy $L^n(\tau)$ (marginal linear redundancy $\lambda^n(\tau)$) are approximately the same or similar also for different kinds of linear processes. Only for nonlinear processes the differences seem to be of qualitative level. This conclusion is illustrated by the following examples.

We start with a simple torus (two-periodic) time series. (For details about the numerically generated data, see Appendix 2.) Figures 1(a) and (b) illustrate the linear redundancy $L^n(\tau)$ and the redundancy $R^n(\tau)$, respectively, as functions of the time lag τ, computed from the torus series 51,200 samples long containing 50% of uniformly distributed noise. The multivariate distribution of these data is not exactly Gaussian; however the shapes of both the τ-plots of the redundancies are almost the same. All the dependence structures in the data dynamics are detectable on the linear level; i.e., linear description of this data is sufficient.

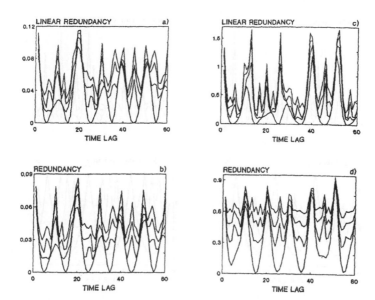

FIGURE 1   (a) $L^n(\tau)$ and (b) $R^n(\tau)$ of noisy torus (two-periodic) time series (50% of noise). (c) $L^n(\tau)$ and (d) $R^n(\tau)$ of the torus series without noise. The embedding dimensions $n = 2$ to 5 (different curves reading from bottom to top).

The same data as above, but without noise, i.e., the "pure" torus, has a multivariate distribution very different from the Gaussian one. This is the cause of differences between $L^n(\tau)$ and $R^n(\tau)$, and Figures 1(c) and (d) are different. Detailed study of redundancies on Figures 1(c) and (d), however, shows that both $L^n(\tau)$ and $R^n(\tau)$ detect the same dependence structures: the maxima and minima of the redundancies are at the same values of the lag τ, and only the relations of their magnitudes are different. These differences we do not consider qualitative and, therefore, in this case, too, we conclude that the linear description of the data should be sufficient.

What we mean by "qualitative differences" of $L^n(\tau)$ and $R^n(\tau)$ is illustrated by the following three data sets, generated by nonlinear dynamical systems.

Figure 2 illustrates redundancies for the series generated by the system with a "strange nonchaotic" attractor, introduced by Grebogi et al. (1984) (51,200 samples of the series length; for details on all numerically generated data, see Appendix 2). Comparing $L^n(\tau)$ and $R^n(\tau)$ on Figures 2(a) and (b), respectively, we can see qualitatively different structures: there are not consistent τ-positions of minima or maxima of the redundancies, there are even different numbers of extrema of $L^n(\tau)$ and $R^n(\tau)$. $R^n(\tau)$ clearly reflects a periodic structure that, unlike that in the above

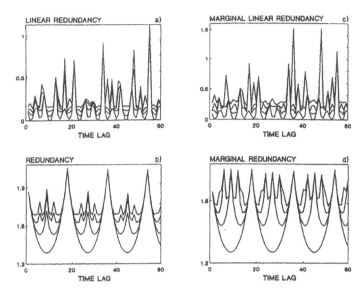

FIGURE 2   (a) $L^n(\tau)$, (b) $R^n(\tau)$, (c) $\lambda^n(\tau)$, and (d) $\varrho^n(\tau)$ of the series generated by the Grebogi strange nonchaotic system.

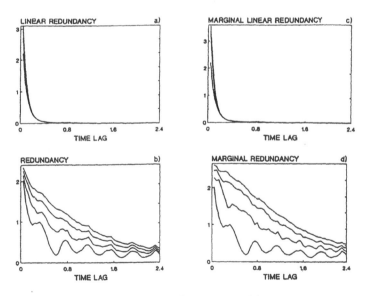

FIGURE 3   (a) $L^n(\tau)$, (b) $R^n(\tau)$, (c) $\lambda^n(\tau)$, and (d) $\varrho^n(\tau)$ of the series generated by the Lorenz system in a chaotic state.

torus data, is not detectable on the linear level. This difference indicates that a linear description of the data is not sufficient and a nonlinear model should be considered.

$L^n(\tau)$ and $R^n(\tau)$ for time series generated by the Lorenz system (Figures 3(a) and (b), respectively) are again different qualitatively: the linear redundancy decreases quickly to values close to zero and detects no dependence for $\tau > 0.4$. On the other hand, the redundancy $R^n(\tau)$ detects nonlinear dependence, the level of which is oscillating with τ and is characterized by the long-term decreasing trend.

In case of time series generated by the Rössler system, both $L^n(\tau)$ (Figure 4(a)) and $R^n(\tau)$ (Figure 4(b)) are of a similar oscillating nature, but the linear redundancies are not able to detect the long-term decreasing trend, clearly reflected in the redundancy $R^n(\tau)$. This difference we again consider as important or qualitative. The importance and nature of this decreasing trend will be discussed in Section 5.

The series length in both the chaotic cases was 1,024,000 samples. These extensive computations were performed in the study of estimation of the metric entropy (Paluš, submitted)—see Section 5—and are not necessary in testing for nonlinearity itself when the nonlinear character of the data can be identified using time series lengths of several thousands or even hundreds of samples.

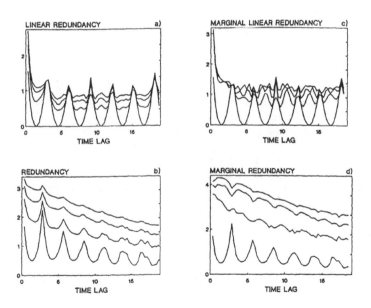

FIGURE 4 (a) $L^n(\tau)$, (b) $R^n(\tau)$, (c) $\lambda^n(\tau)$, and (d) $\varrho^n(\tau)$ of the series generated by the Rössler system in a chaotic state.

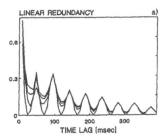 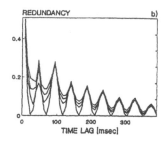

FIGURE 5   (a) $L^n(\tau)$ and (b) $R^n(\tau)$ of normal human EEG, recorded from healthy volunteer in relaxed vigilance with closed eyes, scalp location O_1.

In this section we demonstrated how we can distinguish linear time series from nonlinear ones. At this level we can use either redundancies $L^n(\tau)$ and $R^n(\tau)$ or marginal redundancies $\lambda^n(\tau)$ and $\varrho^n(\tau)$. In Section 5 we will demonstrate the importance of the marginal redundancy for specific detection of chaotic dynamics.

Considering rigorous mathematical results (Eq. 11), we would like to stress that differences between linear redundancy and redundancy cannot be considered as evidence for nonlinearity. We have demonstrated, however, that this difference can be understand as a serious "signature" of nonlinearity. On the other hand, the opposite result, i.e., equality of $L^n(\tau)$ and $R^n(\tau)$ (in the above-considered sense) presents strong support for the hypothesis that the data under study were generated by a linear (stochastic) process. Such a result is valuable—one knows that there is no need to continue analysis of the data by methods of nonlinear dynamics and deterministic chaos; in some cases, it can be also surprising: Figures 5(a) and 5(b) present linear redundancy $L^n(\tau)$ and redundancy $R^n(\tau)$, respectively, computed from a normal human EEG signal (15,360 samples). In spite of many published papers, declaring that "EEG is chaotic" (Başar, 1990), based on results presented in Figure 5, we can conclude that explanation by a linear (stochastic) process is consistent with these data. The same results we obtained from EEG data of ten other subjects recorded in three independent laboratories (Paluš, in press).

4. COMPARISON WITH OTHER TESTS FOR NONLINEARITY AND DETERMINISM

Several authors were led to developing methodology for testing the basic necessary conditions for chaos—i.e., nonlinearity and/or determinism by never-ending discussions on the relevance of "evidences of chaos" in experimental time series by estimating correlation dimension or other dynamical invariants. Our original approach to this problem was introduced in the previous section. Here we present its comparison with two other methods. Both are described by their authors in original papers (Theiler et al., 1992a; Kaplan & Glass, 1992) and also in this volume (Theiler et al., this volume; Kaplan, this volume), so we keep the maximum brevity.

Theiler et al. (1992a). proposed to test for nonlinearity in time series using so called surrogate data.[1] The surrogate data are numerically prepared data with the same statistical properties (mean, variance, spectrum) as the studied experimental data, but the surrogate data are usually created as a linear stochastic process. Then one estimates dimensions and/or other dynamical invariants from both the original and surrogate data. If there are no significant differences between these estimates, the explanation by a linear stochastic process ("null hypothesis") is consistent with the experimental data. Significant differences enable us to reject the null hypothesis of linear stochastic origin of the data and are proposed as signatures of nonlinearity.

We generated surrogate data for all the studied data sets by computing the forward fast Fourier transform (FFT), randomizing the phases and computing the backward FFT. Thus we obtained linear stochastic time series with the same spectra as the original data and computed redundancies from them. We have found that linear redundancies $L^n(\tau)[s]$ of surrogate data are equal to redundancies $R^n(\tau)[s]$ of surrogate data, and both are equal to linear redundancies $L^n(\tau)[o]$ of original data. (Here we again compare time lag plots of redundancies, not their absolute values. [o] means original, [s] surrogate data.) These results are not surprising: The equality $L^n(\tau)[s] = R^n(\tau)[s]$ follows directly from generating the surrogate data as a linear stochastic process. If two series have the same spectra, then they also have the same autocorrelation functions. Recalling the relation between the autocorrelation function and the linear redundancy, the equality $L^n(\tau)[s] = R^n(\tau)[s] = L^n(\tau)[o]$ also can be understood. Then, providing that both the tests, Theiler's and ours, give consistent results, one need not generate surrogate data; the comparison of $L^n(\tau)$ and $R^n(\tau)$ of the studied data is sufficient. But are the tests consistent?

Let us realize that surrogate data are linear and stochastic. When the null hypothesis is rejected, then negation of *"linear AND stochastic"* is *"nonlinear OR deterministic"*; i.e., the following hypotheses are possible:

1. nonlinear and deterministic,
2. nonlinear and stochastic, or
3. linear and deterministic.

On the other hand, in our redundancy test our null hypothesis is linearity. After the results presented in previous sections, we can specify our understanding of linearity: we classify a time series as linear if all the dependence structures can be detected by (auto)correlations and, consequently, by linear redundancy; and

---

[1]The method of surrogate data was used, independently of Theiler et al., by other authors, e.g., Elgar and Mayer-Kress (1989) and others. For simplicity we use the term "Theiler's test."

application of (general) redundancy $R^n(\tau)$ brings no new information. The equality $L^n(\tau)[o] = R^n(\tau)[o]$ also implies the equality

$$L^n(\tau)[s] = R^n(\tau)[s] = L^n(\tau)[o] = R^n(\tau)[o]$$

which means that our test cannot distinguish deterministic linear oscillations from a linear stochastic process. Rejecting the null hypothesis (linearity), the following hypotheses are available:

1. nonlinear and deterministic, or
2. nonlinear and stochastic.

That means, in general, these two tests are consistent in detection of processes (1) and (2). The example for case (3) is our "pure" torus series which we classified as linear (Section 3). Estimations of correlation dimension from this data saturate on 2, but estimations from its surrogate data do not; i.e., the Theiler's null hypothesis is rejected.

Detecting nonlinearity in a time series, we are interested in intrinsic nonlinearity in its dynamics. Influence of a "static" nonlinearity on the nonlinearity tests should be considered. Let us suppose that the underlying dynamics of the studied system is linear—i.e., there is an original (stochastic) linear process $\{x_i\}$, but we can measure series $\{y_i\}$, $y_i = f(x_i)$, where f is a nonlinear function. Such nonlinearity, which is not intrinsic for the dynamics of the system under study but can be caused e.g., by a measurement apparatus, we call static nonlinearity.

A static nonlinearity can influence the results of Theiler's test using linear stochastic surrogate data. Therefore, Theiler et al. (1992a) proposed a more complicated algorithm for generating the surrogate data tailored to this specific null hypothesis including static nonlinearity.

We studied the influence of several types of static nonlinearities on our redundancy test. Figures 6(a) and (b) present linear redundancy $L^n(\tau)$ and redundancy $R^n(\tau)$, respectively, for the noisy torus series (51,200 samples) passed through the quadratic nonlinearity. We can see that $L^n(\tau)$ and $R^n(\tau)$ are not exactly the same, but there is no significant (qualitative) difference in the sense discussed in Section 3; i.e., they reflect the same dependence structures. Redundancies $L^n(\tau)$ and $R^n(\tau)$ for the noisy torus data passed through function tangent hyperbolicus are presented in Figures 6(c) and (d), respectively. They are practically the same, which is the nice result demonstrating the robustness of our test: The function tanh transforms the data to the interval $[-1, 1]$, and the shape of the distribution of the transformed data is approximately like the character M. Two-dimensional distribution of the series and its lagged twin is like a cube with a conic hole. It is very different from the Gaussian "bell."

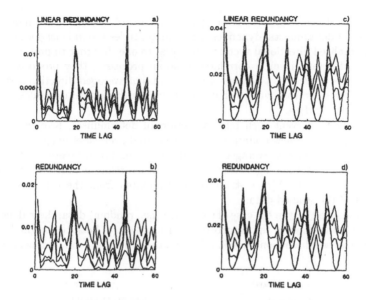

FIGURE 6 (a) $L^n(\tau)$, (b) $R^n(\tau)$ of the noisy torus series passed through the quadratic static nonlinearity. (c) $L^n(\tau)$, and (d) $R^n(\tau)$ of the noisy torus series passed through the function tanh (as another example of the static nonlinearity).

The static nonlinearity effectively means that the distribution of the studied data is not Gaussian. Then the above-demonstrated robustness of our test against the static nonlinearity (i.e., that it indicates only dynamic nonlinearity) is compatible with our original conjecture in Section 3 that the equality of $L^n(\tau)$ and $R^n(\tau)$ can hold not only for Gaussian, but also for different types of (dynamically) linear processes.

We have demonstrated that our redundancy test for nonlinearity has slightly better properties than Theiler's surrogate data test in its basic version. The latter, however, can be adapted for more specific types of surrogate data, depending on the particular problem under study. Such specialized tests can improve one's insight to the origin of the particular data, but lose their general applicability.

Kaplan and Glass (1992) proposed a test for determinism based on computation of average directional vectors[2] in a coarse-grained n-dimensional embedding of a time series. They showed that in the case of random-walk data magnitudes of these vectors decrease with the number N of passes of the trajectory through the box

---

[2]A similar idea was applied in the paper by Cremers and Hübler (1987). We will refer, however, to Kaplan & Glass because we follow their particular implementation.

(i.e., region of coarse-grained state space) as $1/N^{1/2}$. On the other hand, in the ideal case of deterministic dynamics, these magnitudes are independent of N and, using normalized directional vectors, are equal to one. In order to prevent the false detection of determinism in correlated random processes, they proposed the new quantity Λ—the weighted average of magnitudes of averaged directional vectors of both the series decrease with N, but are above the theoretical dependence derived for random-walk data.

In order to prevent such false detection of determinism, Kaplan and Glass (1992) proposed the new quantity Λ—the weighted average of magnitudes of averaged directional vectors, averaged through all the occupied boxes—and compared its dependence on time lag τ with (a) the autocorrelation function of the series under study and (b) $\Lambda(\tau)$ (τ-trace of Λ) computed from the (linear stochastic) surrogate of the studied data.

They showed that in the case of surrogate (i.e., random) data, the τ-dependence of Λ reflects the shape of the autocorrelation function (of both the original and surrogate data—see discussion above). $\Lambda(\tau)$ of a deterministic (and nonlinear, we add)

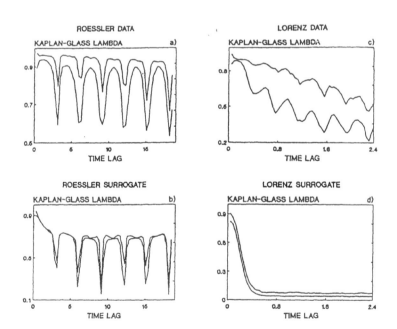

FIGURE 7 The Kaplan-Glass Λ (the weighted average of the averaged directional vectors) as the function of the time lag τ for the Rössler series (a) and its surrogate data (b) and for the Lorenz series (c) and its surrogate data (d). The embedding dimensions are 2 and 3 reading from bottom to top.

time series is different from its autocorrelation function (and $\Lambda(\tau)$ of its surrogate data).

At this step the Kaplan-Glass test becomes similar to our redundancy one and we demonstrate that it also gives very similar results: Figures 7(a) and (b) present the $\Lambda(\tau)$ dependence for the time series generated by the Rössler system and its surrogate, respectively; Figures 7(c) and (d) illustrate the same for the series generated by the Lorenz system and its surrogate. Compare this figures with Figures 3(a) and (b) and 4(a) and (b) illustrating the linear redundancy $L^n(\tau)$ and the redundancy $R^n(\tau)$ for these data. (Remember that $R^n(\tau)$ of the surrogate data is the same as $L^n(\tau)$ of the original data.)

We can conclude that although these tests (that of Kaplan and Glass, and ours) are based on different theoretical approaches, they give practically the same results.

5. DETECTION AND QUANTIFICATION OF CHAOS

In the previous two sections, we discussed the possibility of detecting nonlinearity in a time series. If the series is identified as nonlinear, there are still several possible types of underlying processes: stochastic or deterministic or, more specifically, chaotic processes as a special type of nonlinear deterministic processes. In this section we will demonstrate how the latter can be identified. The tools for identification and also for subsequent quantification of chaotic dynamics are marginal redundancies $\varrho^n(\tau)$, and the basic theoretical concept is the concept of classification of dynamical systems by information rates. The latter was introduced by Kolmogorov (1959) who, inspired by information theory, generalized the notion of the entropy of an information source. In ergodic theory of dynamical systems, it is known as the metric or Kolmogorov-Sinai entropy (Kolmogorov, 1959; Martin & England, 1981; Petersen, 1990; Sinai, 1959, 1976; Paluš, submitted; Walters, 1982).

The rigorous definition of the metric (Kolmogorov-Sinai) entropy requires introduction of basic notions of ergodic theory, which is beyond the content of this chapter. Therefore we refer to any book on ergodic theory (Billingsley, 1965; Cornfeld, Fomin, & Sinai, 1982; Martin & England, 1981; Petersen, 1990; Sinai, 1976, Walters, 1982) as well as to the author's recent paper (Paluš, submitted), and here we will return to an experimental time series considered as a realization of a stationary ergodic stochastic process $\{x_i\}$. We will point out correspondence between a process $\{x_i\}$ and a measure-preserving dynamical system (Cornfeld, Fomin, & Sinai, 1982; Sinai, 1959, 1976; Paluš, submitted; Walters, 1982) for which we can define the metric entropy in terms of entropies introduced in Section 2.

A measure-preserving dynamical system can correspond to a stochastic process $\{x_i\}$ in the sense that an orbit, presenting a single evolution of the system in its state space, is mapped by a measurable map from this space to the set of real numbers. The resulting series of real numbers represents values of physically observable variable measured in successive instants of time—experimental time series.

Conversely, any stationary stochastic process corresponds to a measure-preserving system in a standard way: One can construct a map Φ mapping variables of a stochastic process to a sequence of points $\{z_i\}$ of a measure space Z and define the shift transformation σ on the sequence $\{z_i\}$ as

$$\sigma z_n = z_{n+1}.$$

Due to the stationarity of the original process, such a system is a measure-preserving transformation (dynamical system). (For more details see Petersen's book (1990).)

Considering this correspondence we can express the metric entropy of an underlying dynamical system in terms of entropies of a sequence of random variables x_i. Let us remark first that in practice there is never a continuous probability distribution density, and we consider all the entropies defined using discrete probability distributions on a finite partition of the state space.

Under the above consideration we define the entropy $h(T, \xi)$ of the dynamical system T with respect to the partition ξ as

$$h(T, \xi) = \lim_{n \to \infty} H(x_n | x_1, \dots, x_{n-1}). \tag{22}$$

The metric (Kolmogorov-Sinai) entropy of the dynamical system T is then

$$h(T) = \sup_\xi h(T, \xi), \tag{23}$$

where the supremum is taken over all the finite partitions (Petersen, 1990; Sinai, 1976; Walters, 1982).

Among all the finite partitions, the key role is played by so-called generating partitions. The partition α is generating with respect to the transformation (dynamical system) T if all the measurable sets of the system state space (more precisely σ-algebra) can be generated by countably-fold application of T on α (Petersen, 1990; Sinai, 1976; Walters, 1982). Then the Kolmogorov-Sinai theorem (Petersen, 1990; Sinai, 1976; Walters, 1982) holds:

$$h(T) = h(T, \alpha). \tag{24}$$

The following theorem is important for further consideration here: Let T be a discrete measure-preserving dynamical system. Then

$$h(T^k) = |k| h(T) \tag{25}$$

for every whole number k. For the continuous flow T_t and any real t, the equality

$$h(T_t) = |t| h(T_1) \tag{26}$$

holds (Sinai, 1976).

Now we can proceed to the relation between the metric entropy and the marginal redundancy $\varrho^n(\tau)$: Comparing Eqs. (22) and (10) for $n \to \infty$, we have

$$\varrho^n(\tau) \approx A_\xi - h(T_\tau, \xi),$$

where A_ξ is a parameter independent of n and τ (and, clearly, dependent on ξ) and $h(T_\tau, \xi)$ is the entropy of (continuous) dynamical system T_τ with respect to the partition ξ.

Let ξ be the generating partition with respect to T. Then, considering Eqs. (26) and (24), we have

$$\lim_{n \to \infty} \varrho^n(\tau) = A - |\tau| h(T_1). \tag{27}$$

This assertion was originally conjectured by Fraser (1989c). Application of ideas and methods of information theory in nonlinear dynamics was originally proposed by Shaw (1981).

Let us consider further that the studied time series was generated by an m-dimensional continuous-time dynamical system T_τ fulfilling the conditions of the existence and uniqueness theorem (Arnold, 1973; Kamke, 1959) and the particular trajectory of T_τ is mapped from the state space S to the set of real numbers. There is the unique trajectory passing through each point $s \in S$ so that the evolution on the particular trajectory is fully determined by one m-dimensional point $s \in S$. On the other hand, m-tuples of m successive points $Y(t), \ldots, Y(t + (m-1)\tau)$ can, according to the theorem of Takens (1981), form a mapping of the process $\{Y(t)\}$ to a space Z, so that the sequence $\{z_i\}$ of images of m-tuples $Y(t), \ldots, Y(t+(m-1)\tau)$ is topologically equivalent to the original trajectory $\{s_i\}$ in S. Hence, a particular m-tuple $Y(t), \ldots, Y(t + (m-1)\tau)$ is equivalent to a point from S, and thus it determines the rest of the series $\{Y(t)\}$. It means that only the redundancies $\varrho^n(\tau)$ for $n \leq m$ should be finite and, for $n > m$, redundancies $\varrho^n(\tau)$ should diverge. This is, however, a theoretic consideration providing infinite precision. In experimental and numerical practice, the measurement noise and finite precision emerge, and all the estimated redundancies $\varrho^n(\tau)$ are finite but increasing with n. We can only suppose that the increase of $\varrho^n(\tau)$ for $n > m$ is lower than for $n \leq m$ and is independent of τ.

Intuitively we can explain this supposition by the fact that adding another variable to n variables, $n < m$, the common information measured by $\varrho^n(\tau)$ is increased by specific dynamical information; i.e., the increase $\varrho^{n+1}(\tau) - \varrho^n(\tau)$ depends on n and τ. Addition of another variable when $n > m$ is, considering the increase $\varrho^{n+1}(\tau) - \varrho^n(\tau)$ of the common information, (approximately) adequate to addition of a "noise term" contributing only nonspecific information relevant to noise and finite precision. Therefore, for $n > m$, we expect the curves $\varrho^n(\tau)$ as functions of τ have the same shape, only they are shifted; i.e., $\varrho^{n+1}(\tau) = \varrho^n(\tau) + \text{const}$. Thus the limit behavior of $\varrho^n(\tau)$ for $n \to \infty$, in the case of an m-dimensional dynamical system, is attained for very small n, actually for $n = m + 1, m + 2, \ldots$.

Let us consider that the probability distribution $p(x_1,\ldots,x_n)$ used in the estimation of $\varrho^n(\tau)$ corresponds to the generating partition of the studied m-dimensional dynamical system for a certain extent of τ. Then the limit behavior (27) of the marginal redundancy, i.e., $\varrho^n(\tau) \approx A - \tau h(T_1)$, is attained for $n = m+1, m+2, \ldots$. And this is actually the behavior of $\varrho^n(\tau)$ for low-dimensional dynamical systems. The extent of τ for which marginal redundancies approach the linearly decreasing function is usually bounded by some τ_1 and τ_2, as is discussed in the author's recent paper (Paluš, n.d.).

This phenomenon is illustrated in Figures 3(d) and 4(d) presenting the marginal redundancies $\varrho^n(\tau)$ for the Lorenz and the Rössler system, respectively. $\varrho^n(\tau)$ for $n = 4$ and 5 in both the cases approach linearly decreasing function $A - \tau h$, where the slope h is an estimation of the positive metric entropy of the underlying chaotic dynamical system.

Thus we can consider the linear decrease of the marginal redundancy $\varrho^n(\tau)$ of the studied data as the signature of chaos and the value of its slope—estimation of the metric entropy can serve as the quantitative description of the data under study.

We use the term "signature" and not the popular term "evidence for chaos," because this property, like all so-called evidences, is not a sufficient but necessary condition for chaos (e.g., the implication "chaos $\Rightarrow$ finite dimension" holds, but the inverse does not).

The conditions for successful identification of chaos by $\varrho^n(\tau)$, as we explained above, are n exceeding the dimension of the studied system and a partition fine enough to be generating (redundancy is estimated by box-counting method—see Appendix 1). (Each refinement of a generating partition is generating; i.e., for systems with finite entropy, the boxlike generating partitions exist.) Fine and high-dimensional partitions require a large amount of data (otherwise, estimations of redundancies are heavily biased by the effect of "overquantization"—see Appendix 1); therefore, these conditions cannot always be fulfilled and chaos in time series identified. This is, however, a general property: Using an insufficient amount of data, any method can give biased results.

6. ANALYSIS OF THE COMPETITION DATA

We applied the above-described redundancy method on the Data Sets A, D, and E of the competition data.

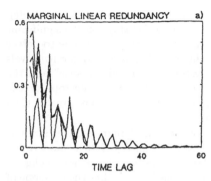
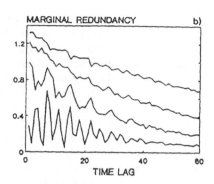

FIGURE 8   (a) $\lambda^n(\tau)$ and (b) $\varrho^n(\tau)$ for the Data Set A—chaotic dynamics of the FIR laser.

DATA SET A

The results, linear marginal redundancy $\lambda^n(\tau)$ and marginal redundancy $\varrho^n(\tau)$, are presented in Figures 8(a) and 8(b), respectively. We can see that $\varrho^n(\tau)$ for $n = 4$ and 5 approach more or less linearly decreasing functions. This is

1. a qualitative difference from the shape of $\lambda^n(\tau)$, i.e., the signature of nonlinearity, and
2. also the signature of chaos—it detects a positive value of the metric entropy of the system underlying this series. The marginal redundancy $\varrho^n(\tau)$ for $n = 5$ is, however, distorted due to "overquantization"; i.e., there is not enough data to obtain "better" results—consistent slopes for $n = 5$ or higher (see Appendix 1).

The result correctly identified the chaotic laser dynamics, which can be described by the system of three ordinary nonlinear differential equations (Hübner et al., this volume), having an attractor of correlation dimension about 2.05 and positive metric entropy. The amount of the data available (10,000 samples) is unfortunately not sufficient to obtain reliable estimation of the value of the metric entropy (Paluš, submitted).

The dynamical system describing this laser dynamics is equivalent to that of the Lorenz system (Hübner et al., this volume). The attractors of these systems have very similar topological properties, including dimension. Considering our results (Figure 8), we can state that the information production or "the levels of chaoticity" of these systems are different. In order to explain this point, let us remember Figures 3(c) and 4(c), depicting $\lambda^n(\tau)$ of the Lorenz and the Rössler systems, respectively. In the case of the Lorenz system, the (marginal) linear redundancy $\lambda^n(\tau)$ for $\tau > 0$ decreases immediately to a close-to-zero level, classifying the series as the sequence of independent random variables—white noise (i.e., $\lambda^n(\tau)$ is not able to detect any dependence at all due to the high rate of information production destroying correlations), while in the case of the Rössler system, $\lambda^n(\tau)$ stays on the significant nonzero level as in the case of regular oscillations (i.e., $\lambda^n(\tau)$ is not able to detect the production of information here). We explain this difference by different "levels of chaoticity" of these systems—the Lorenz system is "strongly chaotic" and the Rössler one is "weakly chaotic." To be more specific, the metric entropy or the positive Lyapunov exponent of the Lorenz system is about ten times greater than that of the Rössler system (Wolf et al., 1985).

According to Figure 8(a) the "chaoticity" of the laser dynamics under study is between those of the Lorenz and the Rössler systems. Examples of such "moderately chaotic systems" in which the linear redundancy is nonzero for a non-negligible extent of lag τ, usually oscillating but decreasing in such a way that the envelope of oscillations of $\lambda^n(\tau)$ reflects the exponential or power-law decrease, are also known (Kot, Sayler, & Schultz, 1992). And this is the case of our chaotic laser (Figure 8(a)).

DATA SET E

Figures 9(a) and (b) present the linear redundancy $L^n(\tau)$ and the redundancy $R^n(\tau)$ computed from the 13th continuous segment (2,568 samples) of Data Set E—the light curve of the variable white dwarf star (Clemens, this volume). We can see there is no significant difference between $L^n(\tau)$ and $R^n(\tau)$, and we conclude that explanation by a linear (stochastic) process is consistent with the data. Analysis of the other two longest segments gave the same results.

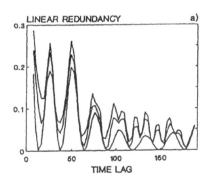 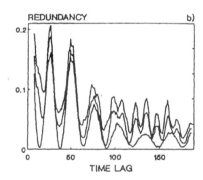

FIGURE 9   (a) $L^n(\tau)$ and (b) $R^n(\tau)$ for the Data Set E (13th continuous segment)— light curve of the white dwarf star.

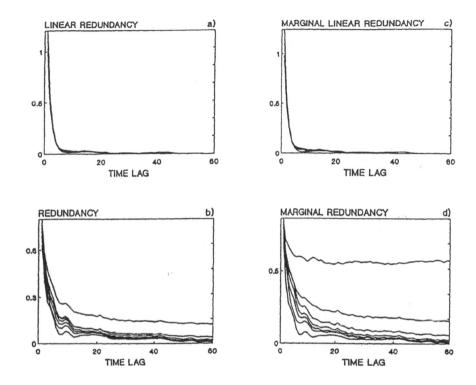

FIGURE 10   (a) $L^n(\tau)$, (b) $R^n(\tau)$, (c) $\lambda^n(\tau)$, and (d) $\varrho^n(\tau)$ for the Data Set D. The embedding dimensions are $n = 2$ to 8 reading from bottom to top. From $n = 6$ the overquantization effect emerges (d).

DATA SET D

We can see all the redundancies in Figure 10. (Redundancies $R^n(\tau)$ and $\varrho^n(\tau)$ for $n = 6, 7, 8$ are distorted due to the overquantization effect—see Appendix 1.) Linear redundancies decrease quickly to near-zero values indicating noiselike dependence level for lags $\tau > 5$ samples, while redundancies $R^n(\tau)$ (or $\varrho^n(\tau)$) are clearly above zero up to lags of approximately 50 samples, and reflect an oscillating phenomenon. This difference we consider as the signature for nonlinearity. There is, however, no specific "chaotic" behavior of marginal redundancies $\varrho^n(\tau)$: They decrease, but a tendency to the linear decrease is not apparent. This long-term trend is closer to an exponential or power-law decrease. Therefore we should assume the following possibilities for the series dynamics:

1. nonlinear stochastic,
2. nonlinear deterministic, but not chaotic, and

3. chaotic, but for identification of its chaoticity, finer partition is necessary and/or its dimensionality is higher than used in analysis. To test these possibilities, longer time series are necessary than the given 100,000 samples.

Above (Figures 1 and 2) we have found that redundancies of deterministic nonchaotic series have no (i.e., zero) long-term trend. A decreasing trend of redundancies of the Data Set D, then, can be explained by the above hypotheses (1) and (3). Can we reject hypothesis (2)?

In the theory of Sections 2, 3, and 5, we required stationarity of the time series under study. Being mathematically strict we should reject application of our redundancy test to nonstationary data (Gershenfeld & Weigend, this volume). But, let us make the exception for a while and study the effect of nonstationarity on time-lag dependence of the redundancies.

We added the Gaussian drift (generated as integration of a Gaussian random variable) to (a) the amplitude of our torus time series, and (b) parameter ω (see Appendix 2) of the system of Grebogi et al. (1984), analyzed above. The parameter was drifted in each iteration step.

The results are presented in Figure 11. Nonstationarity induces the long-term decreasing trend of all the redundancies. (Remember that these data in their original, stationary forms, did not exhibit any long-term decrease of redundancies— Figures 1 and 2.) It is also interesting that the character of these data, i.e., either linear or nonlinear, is not changed (cf. $L^n(\tau)$ with $R^n(\tau)$ in Figures 11(a)–(b) and 11(c)–(d)). We must remark, however, that the roots of this result lie not only in the different nature of the data, but also in different levels of the dynamics influenced by the drift. When we simply added the Gaussian drift to the amplitude of the series generated previously by the system of Grebogi et al. (1984), we found that while the small amplitude drifts do not affect the redundancies, the drifts with larger amplitude "Gaussianize" the series: redundancies $R^n(\tau)$ become similar to the linear redundancies $L^n(\tau)$; i.e., the nonlinear character of the series is hidden by the Gaussian noise added.

Our conclusion based on this experimentation is that hypothesis (2) cannot be rejected. The Data Set D could be generated by a system regularly oscillating in normal conditions, but disturbed by some drift or other nonstationarity in its structural parameters, and in fact this is the case. Using post-competition information, we know that the Data Set D reflects the dynamics of a driven particle in a four-well potential (Gershenfeld & Weigend, this volume). Normally the evolution after short transient period ends in regular nonlinear oscillations in one of the wells: Figures 12(a) and 12(b) present $L^n(\tau)$ and $R^n(\tau)$, respectively, for this case, i.e., for the series obtained by integrating the system without the drift in potential. There is no long-term decreasing trend in these redundancies, and looking closely at them (Figure 12(c) illustrates $L^n(\tau)$ and $R^n(\tau)$ for $n = 2$) we can see that there are several peaks in redundancy $R^n(\tau)$ missing in linear redundancy $L^n(\tau)$, i.e., specific nonlinear dependence structures which are not detectable on the linear

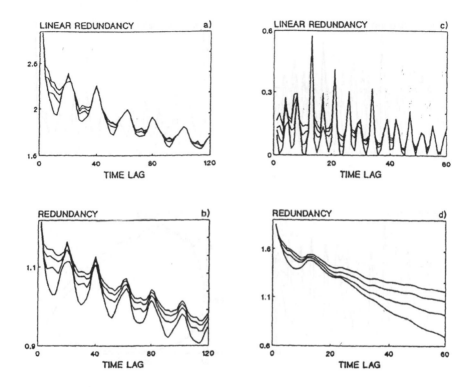

FIGURE 11 (a) $L^n(\tau)$, (b) $R^n(\tau)$ for the torus series with the Gaussian drift in the amplitude, (c) $L^n(\tau)$, (d) $R^n(\tau)$ for the Grebogi system series iterated with the small Gaussian drift in the system parameter ω.

level. Considering the phase portrait of these oscillations, we can see that it is not a simple cycle (Figure 12(d)).

The specific behavior of the Data Set D was induced by a small Gaussian drift in one of the potential parameters. This caused prolonged transient dynamics, consisting of nonlinear oscillations within the wells and random switching among them. Thus the system exhibits a "dynamical loss of memory" (i.e., a long-term decrease of redundancies), but it is of a different origin than this phenomenon in the case of chaotic dynamical systems.

Without information about the origin of this data set, however, we are able to detect only its nonlinearity and none of the above nonlinear hypotheses (1)–(3) can be rejected.

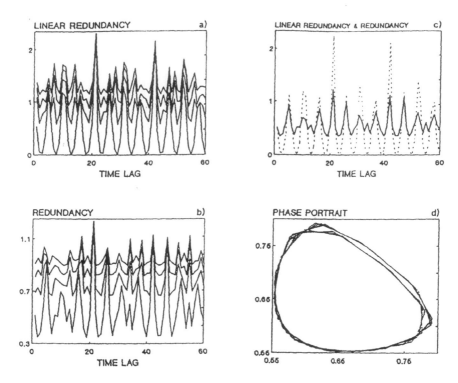

FIGURE 12 Stationary periodic dynamics of the time series obtained by integrating the system used for generation of the data set D, but without the drift in the system parameter: (a) $L^n(\tau)$, (b) $R^n(\tau)$, (c) $L^2(\tau)$ (dashed line) and $R^2(\tau)$ (full line) and, (d) the phase portrait.

7. CONCLUSION

We presented the method for analysis of experimental time series suitable for assessing the nonlinearity of a series and for identification and quantification of the underlying chaotic dynamics. It is based on examination of time-lag dependence of redundancies, the information-theoretic functionals computed from the studied time series and its lagged versions. It was demonstrated in the extensive numerical study that this method is able to discern linear from nonlinear processes and specifically detect low-dimensional chaotic dynamics. It is valuable also for quantitative characterization of chaotic systems by measuring their information production rates in terms of metric entropy.

ACKNOWLEDGMENT

The author would like to thank V. Albrecht, I. Dvořák, A. M. Fraser, A. Longtin, and J. Theiler for stimulating discussions and D. Bensen for careful reading of the manuscript. Special thanks are due to A. S. Weigend and N. A. Gershenfeld.

This study was performed as a part of the projects "Information and entropy properties of the spontaneous brain electrical activity," supported within the framework of the "IBM Academic Initiative in the CSFR," and "Application of EEG complexity measures for quantification of changes of the brain state," supported by the Ministry of Health of the Czech Republic.

APPENDIX 1—REDUNDANCY ALGORITHM

Linear redundancies were computed according to Eq. (13). Eigenvalues of each correlation matrix were obtained by its decomposition using SVDCMP routine (Press et al., 1988).

Practical computations of mutual information and redundancies of continuous variables are always connected with the problem of quantization. By quantization we understand the definition of the finite-size boxes covering the state space. The probability distribution is then estimated as relative frequencies of occurrence of data samples in particular boxes. The naive approach to estimate redundancies of continuous variables should be the use of the finest possible quantization given, e.g., by a computer memory or measurement precision. We must remember, however, that we usually have a finite number N of data samples. Hence, using too fine quantization can cause the estimation of entropies and redundancies to be heavily biased: Estimating joint entropy of n variables using Q marginal quantization levels, one obtains Q^n boxes covering the state space. If Q^n approaches the number N of data samples, or even $Q > N$, the estimate of $H(x_1, \ldots, x_n)$ can be equal to $\log(N)$ or, in any case, it can be determined mainly by the number of the data samples and/or by the number of the distinct data values and not by the data structure, i.e., by the properties of the system under study. We say, in such a case, that the data are *overquantized*. (Even the "natural" quantization of experimental data given by an A/D converter is usually too fine for reliable estimation of the redundancies.)

The emergence of overquantization is given by the number of boxes covering the state space; i.e., the higher the space dimension (the number of variables), the lower the number of marginal quantization levels that can cause overquantization. Recalling definition (7) of the redundancy of n variables, one can see that while the estimate of the joint entropy can be overquantized, i.e., saturated on some value given by the number of the data samples and/or by the number of the distinct data values, the estimates of the individual entropies are not and they increase with finer quantization. Thus the overquantization causes overestimation of redundancy $R^n(\tau)$ and smearing of its dependence on τ.

Recalling $\varrho^n(\tau) = R^{n+1}(\tau) - R^n(\tau)$, one can see that overquantization causes overestimation of marginal redundancy and, moreover, attenuation of its decrease with increasing τ. Further overquantization can lead to a paradoxical, unreal result of $\varrho^n(\tau)$ increasing with τ which formally implies negative metric entropy (Paluš, 1993a).

Therefore, one must be very careful in defining the quantization. Fraser and Swinney (1986) have proposed an algorithm for constructing the locally data-dependent quantization. We have found, however, that one need not develop such a complicated algorithm; a simple box-counting method is sufficient. The only "special prescriptions," based on our extensive numerical experience, concern the method of data quantization:

a. The type of quantization: We propose to use the marginal equiquantization method; i.e., the boxes for box counting are defined not equidistantly, but so that there is approximately the same number of samples in each marginal box.

b. Number of quantization levels (marginal boxes): We have found that the requirement for the effective[3] series length N using Q quantization levels in computation of n-dimensional redundancy is

$$N \geq Q^{n+1};$$

otherwise, the results can be heavily biased due to overquantization as stated above.

APPENDIX 2

Two-periodic noisy data were generated according to the following formula:

$$Y(t) = (R_1 + R_2 \sin(\omega_2 t + \phi)) \sin(\omega_1 t) + \xi,$$

where $R_1 : R_2 = 5 : 4$, $\omega_1 : \omega_2 = 10 : 9$, $\phi = 1.3\pi$, and ξ are random numbers uniformly distributed between $-\Xi$ and Ξ. The term "50% of noise" means that $R_1 : R_2 : \Xi = 5 : 4 : 9$. For the torus series without noise, the same formula with $\xi = 0$ holds.

[3] The effective series length N is $N = N_0 - (n-1)\tau$, where N_0 is the total series length, n is the embedding dimension, and τ is the time delay used in the reconstruction of the n-dimensional embedding.

Chaotic data were generated by numerical integration, based on the Bulirsch-Stoer method (Press et al., 1986), of the Rössler system (Rössler, 1976)

$$(dx/dt, dy/dt, dz/dt) = (-z - y, x + 0.15y, 0.2 + z(x - 10)),$$

with initial values (11.120979, 17.496796, 51.023544), integration step 0.314, and accuracy 0.0001; and the Lorenz system (Lorenz, 1963)

$$(dx/dt, dy/dt, dz/dt) = (10(y - x), 28x - y - xz, xy - 8z/3),$$

with initial values (15.34, 13.68, 37.91), integration step 0.04, and accuracy 0.0001. Component x was used in both the cases.

The time series from the "strange nonchaotic attractor" (Grebogi et al., 1984) was obtained by iterating the system:

$$\Theta_{n+1} = (\Theta_n + 2\pi\omega) mod(2\pi)$$
$$u_{n+1} = \Lambda(u_n \cos(\Theta) + v_n \sin(\Theta))$$
$$v_{n+1} = -0.5\Lambda(u_n \cos(\Theta) - v_n \cos(\Theta))$$

where $\omega = (5^{1/2} - 1)/2$ and $\Lambda = 2/(1 + u_n^2 + v_n^2)$. Component Θ was recorded.

The Gaussian drift was generated by the same algorithm as in the program used for the generation of Data Set D (Gershenfeld & Weigend, this volume).

Daniel T. Kaplan
Department of Physiology, McGill University, Montreal, Quebec, H3G 1Y6 Canada

A Geometrical Statistic for Detecting Deterministic Dynamics

This paper addresses the problem of how to decide whether a time series comes from a deterministic dynamical system. At its root, this problem is ill-posed because any time series of finite length—even one that was generated by a deterministic process—is a possible realization of some random process. In practice, though, one usually has in mind some class of random processes such as linear dynamical systems with Gaussian white noise inputs, and it is worthwhile to pose the problem as deciding whether the time series is a plausible output of this class of random processes. A reasonable approach to this problem is to compute some statistic (or set of statistics) on the time series, and assess the likelihood that the computed values would come from the class of random systems under consideration.

A variety of such statistics have been motivated by developments in the last decade in nonlinear dynamics. Probably the most heavily used has been the correlation dimension (Grassberger & Procaccia, 1983a, 1983c; Theiler, 1990). A widespread belief is that the correlation dimension of a time series from a random process is equal to the dimension of the space in which the time series is embedded. This is not generally the case for time series of finite length, and has been shown to be incorrect even for time series of infinite duration which have power spectra of the form $P(\omega) = \omega^{-\alpha}$ for $\alpha > 1$ (Osborne & Provenzale, 1989; Theiler, 1991).

Times Series Prediction: Forecasting the Future and
Understanding the Past, Eds. A. S. Weigend and N. A. Gershenfeld, SFI Studies
in the Sciences of Complexity, Proc. Vol. XV, Addison-Wesley, 1993   **415**

For these reasons and others (Grassberger et al., 1991; Theiler et al., 1992b), it has been found necessary to apply bootstrapping tests when using the correlation dimension. Such tests involve the construction of "surrogate data" which is generated from the class of random processes under consideration (Theiler et al., 1992b). This bootstrapping approach[1] can be applied to any statistic calculated from a time series, such as the Lyapunov exponent (Bryant, Brown, & Abarbanel, 1990; Theiler et al., 1992b; Wolf et al., 1985). However, although use of the Lyapunov exponent as a statistic is motivated by dynamical considerations, it is unclear that a time series that passes a bootstrapping test does so because its dynamics are deterministic; it may be that the time series simply fails to fall into the particular class of random processes used to construct the bootstrap for reasons that are dynamically irrelevant such as the shape of the time series' histogram.

Another set of statistics that has been proposed to test determinism involves constructing a model of the time series that can be used to make predictions (Casdagli, 1989, 1991; Weigend, Huberman, & Rumelhart, 1990; Sugihara & May, 1990). Such models can be constructed on an ad hoc basis, using, for example, locally linear transformations, radial basis functions, neural networks, etc. Evidence for determinism comes from the ability to make successful predictions; evidence for *nonlinear* determinism comes from the ability to make better predictions with nonlinear models than with linear ones. This is discussed by Casdagi and Weigend (this volume).

The statistic I will describe below directly tests one of the central concepts behind the notion of determinism—that the change of state of a system is a single-valued function of the state. The method, developed with Leon Glass (Kaplan & Glass, 1992; 1993), has certain advantages over the statistics mentioned above: it does not require that a predictive model be constructed, and it does not appear to be sensitive to the details of the specific class of random processes used in bootstrapping. In particular, there is evidence that the statistic is not sensitive to static nonlinear transformations of the time series (Kaplan & Glass, 1993).

To illustrate the use of the statistic, I will apply it to Data Sets A, B, C, and D of the Santa Fe Time Series Prediction and Analysis Competition.

---

[1]An alternative to bootstrapping involves transforming the time series so that the appropriate random process is a simple one, e.g., white noise. This can often be done approximately simply by taking first differences in the time series (Sugihara & May, 1990). Theiler and Eubank (1991), show that "whitening" a time series can make it more difficult to detect determinism.

COARSE-GRAINED FLOW AVERAGES

In a deterministic system, the change of state of the system is a function of the state of the system. In the case of a continuous-time system, one generally has in mind a differential equation

$$\frac{d\underline{x}}{dt} = \underline{f}(\underline{x})$$

where $\underline{x}$ is the "state" of the system, and $\underline{f}(\underline{x})$ takes on a unique value at each $\underline{x}$.

In order to test whether a time series is consistent with such a deterministic mechanism, one needs to construct a state from the time series. This can be done by "embedding" the time series, $z(t)$, i.e., constructing an m-dimensional vector

$$\underline{x}(t) = (z(t), z(t - \tau), \ldots, z(t - (m - 1)\tau))$$

that represents the trajectory of the state of the system (Casdagli et al., 1991; Sauer et al., 1991).

Ideally, one might like to ask whether one can find any pairs of times t_1 and t_2, where $\underline{x}(t_1) = \underline{x}(t_2)$ but $d\underline{x}/dt(t_1) \neq d\underline{x}/dt(t_2)$. If there are no such pairs t_1, t_2, one would like to conclude that $d\underline{x}/dt$ is a single-valued function of $\underline{x}$. At this point, though, one needs to confront the situation encountered in actual practice:

- Exact recurrences of the trajectory do not occur very often. For a chaotic or aperiodic system, the trajectory will never return to exactly the same point in the embedding space (if m is large enough to adequately represent the dynamics). Even for a random walk, it is unlikely (for $m \geq 3$) that the trajectory will ever cross itself, so it is unlikely that one will find a case where $\underline{x}(t_1) = \underline{x}(t_2)$.
- Typically, one deals with time series that are contaminated with noise. One source of noise that is almost universally present in practice is the round-off imposed by digitization. This round-off means that states that were different before digitization may be mapped to the same discrete state by digitization, so that one has no confidence that when the measured states are identical, the actual (noiseless) states are also identical.
  In addition, the noise means that the measured values of $d\underline{x}/dt$ are themselves noisy.

An effective way to deal with these real-world problems is to consider the times $t_1, t_2, t_3, \ldots, t_n$ when the trajectory passes through a small neighborhood of diameter ϵ in the state space. If the dynamical function $\underline{f}(\cdot)$ is approximately constant over this length scale, then for a deterministic system one expects that the vectors $d\underline{x}/dt(t_i), i = 1, \ldots, n$, will point in similar directions. In contrast, for a random walk, one expects the vectors to point randomly.

A simple way to implement and quantify these ideas is to divide the embedding space into boxes, i.e., small hypercubes of edge length ϵ. For a finite-length time series, a finite number of boxes is needed to cover the trajectory. Label each box with an integer, and consider boxes one at a time. Assume that the trajectory enters box j at times $t_1, t_2, \ldots, t_{n_j}$ and exits the box at times $t_1 + b_1, t_2 + b_2, \ldots, t_{n_j} + b_{n_j}$; there are altogether n_j passes through box j. Consider the vectors

$$\underline{v}_{i,j} = \underline{x}(t_i) - \underline{x}(t_i + b_i)$$

and construct the average of the unit length vectors

$$\underline{V}_j = \frac{1}{n_j} \sum_{i=1}^{n_j} \frac{\underline{v}_{i,j}}{|\underline{v}_{i,j}|}.$$

Taking the finite difference $\underline{x}(t_i) - \underline{x}(t_i + b_i)$ rather than the derivative $d\underline{x}/dt$ helps to reduce the influence of high-frequency noise (the finite difference is equivalent to box-car filtering the derivative). It also provides a way to reduce the influence of curvature of the trajectory within a box.

The statistics of $|\underline{V}_j|$ for deterministic systems, random walks, and Gaussian random processes are discussed by Kaplan and Glass (1991; 1993). To summarize:

- The idealized limiting case of perfect determinism will produce

$$\lim_{\epsilon \to 0} |\underline{V}_j| = 1.$$

- If the trajectory is an isotropic random walk,

$$\langle |\underline{V}_j| \rangle_{j \text{ s.t. } n_j = n} = \frac{c_m}{\sqrt{n_j}}$$

where $\langle \cdot \rangle_j$ denotes an expectation value with respect to j and c_m is a constant determined by the embedding dimension, $c_2 = 0.886$, $c_3 = 0.921$, $c_4 = 0.940$, $c_5 = 0.952$, ..., $c_\infty = 1$. The value of $\langle |\underline{V}_j| \rangle_{j \text{ s.t. } n_j = n}$ for an isotropic random walk will be termed $\overline{R}^m_{n_j}$.

- For a deterministic system with finite box size ϵ or with small amounts of noise, $|\underline{V}_j| < 1$, although it is often bigger than $\overline{R}^m_{n_j}$. As a rule of thumb, setting ϵ to be 3–15% of the amplitude range of the time series often gives good results.

- For a Gaussian random process (i.e., Gaussian white noise passed though a linear filter), the value of $\langle |\underline{V}_j| \rangle$ depends on the position of box j in the embedding space and on the autocorrelation function

$$\Psi(\tau) = \langle x(t)x(t - \tau) \rangle_t$$

of the process. By selecting an embedding lag τ appropriately (Kaplan & Glass, 1993), it is possible to bring $\langle |\underline{V}_j| \rangle \to \overline{R}^m_{N-j}$, even though a deterministic chaotic system with the same $\Psi(\tau)$ may have $|\underline{V}_j| > \overline{R}^m_{N-j}$.

■ For discretely sampled data where points that are nearby in time are not necessarily nearby in the embedding space (e.g., the iterates of the logistic map), one often finds that the trajectory is not an isotropic random walk (i.e., $|\underline{V}_j| > \overline{R}_n^m$) even when the data are generated randomly. The reason for this, in a nutshell, is that points near the edge of the embedding cloud are likely to be followed by points nearer the center, even for random data. This creates a directional bias. Thus, the method as described here should not be applied to time series with substantial energy in frequency components near the sampling frequency. A modification of the present method suitable for application to such high-frequency signals is presented in Kaplan (in press).

It is useful to consider two summarizing statistics of $|\underline{V}_j|$. The first constructs a set of numbers $\overline{L}_n^m$ which is the average of $|\underline{V}_j|$ over all the boxes where $n_j = n$. In the limiting case of perfect determinism, $\overline{L}_n^m \to 1$, while in the case of an isotropic random walk, $\overline{L}_n^m = \overline{R}_n^m = c_m/\sqrt{n}$. (Confidence intervals on $\overline{L}_n^m$ are taken as the standard error of the mean of $\langle |\underline{V}_j| \rangle_j$.)

The second summarizing statistic combines the various $\overline{L}_n^m$ with the formula

$$\overline{\Lambda} = \langle \frac{(\overline{L}_n^m)^2 - (\overline{R}_n^m)^2}{1 - (\overline{R}_n^m)^2} \rangle_n$$

where the average is weighted by the number of boxes with n passes. For perfect determinism $\overline{\Lambda} \to 1$, while for an isotropic random walk $\overline{\Lambda} = 0$. (Confidence intervals are constructed by propagating the confidence intervals on $\overline{L}_n^m$ (Kaplan & Glass, 1993).

ANALYSIS OF DATA SETS FROM THE COMPETITION

The time series used in the competition provide examples of a variety of situations: low-dimensional determinism, moderate-dimensional determinism with non-stationarity of the system parameters, and cases where a deterministic mechanism is unlikely.

DATA SET A: LOW-DIMENSIONAL DETERMINISM

Data Set A provides an example of dynamics that appear highly regular during the periods of oscillatory growth, and have irregular-looking resettings from large amplitude to small amplitude. Part of the time series is shown in Figure 1 along with part of a surrogate time series with the same power spectrum. In this case, it is clear to the eye that Data Set A has completely different time domain properties than the surrogate data. For instance, the surrogate data looks qualitatively similar under time reversal, while Data Set A has a marked time asymmetry. The frequency aliasing that occurred because Data Set A was sampled at a rate above the Nyquist frequency is manifest as high-frequency irregularity in the surrogate data.

The visual impression of the difference between Data Set A and the corresponding surrogate data is confirmed in a five-dimensional embedding space by the $\overline{L}_n^5$ statistic, as shown in Figure 2. For an embedding delay of $\tau = 75$, for Data Set A $\overline{L}_n^5$ is approximately 0.9 independent of n, while for the surrogate data $\overline{L}_n^5$ falls off as $c_m/\sqrt{n}$ as expected for a random walk.

By examining the $\bar{\Lambda}$ statistic as a function of the embedding lag τ, one can investigate the time duration over which dynamical correlations persist in the time series. (See Figure 3.) The overall impression is that the $\bar{\Lambda}$ statistic decays from a value near 1 as τ increases. This is consistent with the idea of a positive Lyapunov exponent eliminating dynamical correlations between points in the time series separated by time τ. However, Data Set A is not typical of chaotic dynamical systems I have studied, because the $\bar{\Lambda}$ statistic does not decrease steadily, but shows a second maximum (near $\tau = 400$). In contrast, the Lorenz system—to which the dynamics of Data Set A have been compared (Hübner et al, this volume)—shows no such second maximum. Instead, $\bar{\Lambda}$ for the Lorenz system decays steadily towards 0 as the embedding lag τ is increased (Kaplan & Glass, 1992). The fact that the typical time between amplitude resettings in Data Set A is approximately 400 time units means that an embedding with $\tau \approx 400$ brings together points in the time series that are in the oscillatory growth phase of the time series, although separated by one amplitude resetting. I speculate that the secondary peak in $\bar{\Lambda}$ arises from a lack of divergent

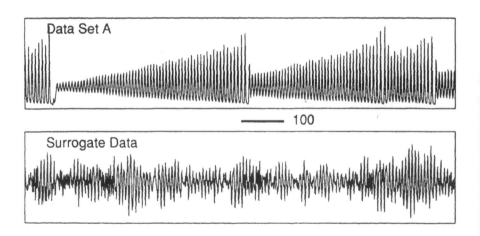

FIGURE 1   A portion of Data Set A and randomly generated surrogate data with the same power spectrum. The bar indicates the time scale in samples.

dynamics during much of the oscillatory growth phase, and that the chaos in the system arises primarily from the irregular dynamics of phase resetting. The Lorenz system shows no secondary maximum in $\bar{\Lambda}$ because there are divergent dynamics even during the oscillatory growth phase.

DATA SET D: MODERATE-DIMENSIONAL DYNAMICS

This time series is measured from the numerical integration of the trajectory of a particle in a four-dimensional potential with four local minima (see Gershenfeld & Weigend, this volume). The particle was forced sinusoidally and damped proportionally to its velocity. The phase space has dimension nine, corresponding to the four position variables, four velocity variables, and the phase of the sinusoidal forcing. The depth of the potential wells was varied as a biased random walk. The measured quantity was the distance of the particle from a fixed point in the phase space, with some additive white noise of small magnitude.

A short segment of the time series is shown in Figure 4, along with a random, surrogate data set with the same power spectrum as the short segment. The

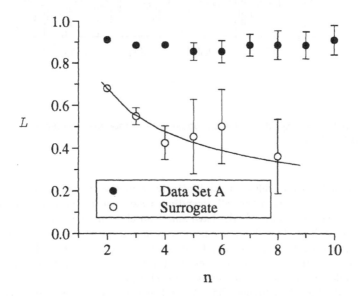

FIGURE 2 $\bar{L}_n^5$ for Data Set A and the surrogate data. (Embedding parameters $\tau = 75$, $m = 5$, box resolution $1/32$.) The theoretical values for a random walk, $c_5/\sqrt{n}$ is shown by the solid curve.

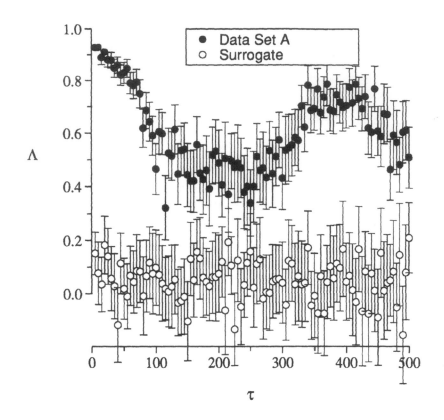

FIGURE 3 $\bar{\Lambda}$ for Data Set A and the surrogate data. (Embedding parameters $m = 5$, box resolution $1/32$.)

surrogate data is visibly different qualitatively from the segment from Data Set D—note how in Data Set D the tall ascending peaks tend to terminiate at similar values, while the descending peaks have small notches in them. This is not true of the surrogate data. This difference is more pronounced when the whole of Data Set D is used to construct the surrogate data; the short-term mean of Data Set D jumps from one value to another (corresponding to jumps between local potential minima), while the surrogate data shows no such jumps.

$\bar{\Lambda}$ for Data Set D is higher than $\bar{\Lambda}$ for the surrogate data. (See Figure 5.) Note that it is easier to distinguish between Data Set D and the surrogate data at $\tau = 5, 10$ than at $\tau = 2$. The dynamics of Data Set D appear to be only slightly better represented with $m = 8$ than with $m = 3$. One reason why $\bar{\Lambda} < 1$ at the

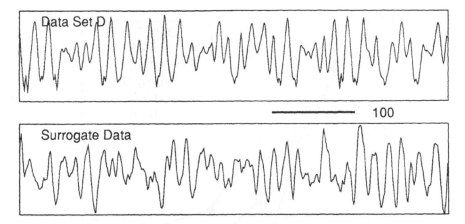

FIGURE 4   A short segment of Data Set D and randomly generated surrogate data with the same power spectrum as the short segment. The bar indicates the time scale in samples.

An important question concerns the amount of data that is required in order to detect the difference between Data Set D and the surrogate data. This question can be addressed by examining short segments of Data Set D and the surrogate data. By examining many such segments, it is possible to calculate the population standard deviation in the estimates of $\bar{\Lambda}$, and compare these to the confidence intervals that are generated internally within each segment. Figure 6 shows $\bar{\Lambda}$ vs. segment length, along with both the population standard deviations and the internal confidence intervals. Three features of this graph are worth emphasizing:

- The population mean of $\bar{\Lambda}$ is independent of time series length, while the confidence intervals tend to decrease with longer time series length.
- For the surrogate data, the internal confidence interval is close to the population standard deviation. For Data Set D, the internal confidence interval underestimates the population standard deviation by a factor of two to five. The reasons for this are the subject of ongoing research. For the purpose of distinguishing Data Set D from the surrogate data, the confidence interval on the surrogate data is the relevant one, since it describes how likely the value of $\bar{\Lambda}$ for Data Set D is if it were actually surrogate data.
- A segment of length 5000 points provides sufficient statistical power reliably to distinguish Data Set D from the surrogate data, if only a single segment were available.

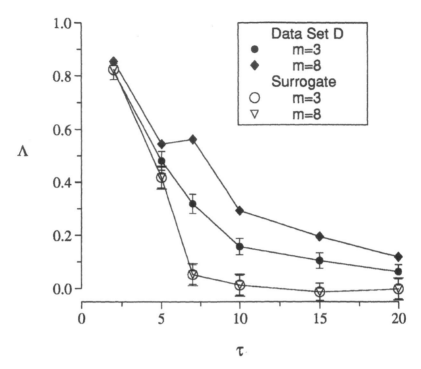

FIGURE 5 $\bar{\Lambda}$ vs. embedding lag τ for Data Set D and surrogate data. (Embedding parameters $m = 3, 8$, box resolution $1/10$.)

smallest embedding lags is that the data set is not perfectly deterministic; one of the parameters in the system is drifting randomly, and there is a small amount of additive noise in the time series.

The autocorrelation function $\Psi(\tau)$ (shown in this volume by Zhang and Hutchinson, Figure 2) suggests that linear correlations in the data are lost over time lags of $\tau \approx 10$. The $\bar{\Lambda}$ statistic suggests that nonlinear correlations are retained over longer times, $\tau \approx 20$.

The important point here is that the internally generated confidence intervals from a single segment can be used to indicate the significance of a detected difference between surrogate and real data, and to suggest whether having more data might help in detecting a difference.

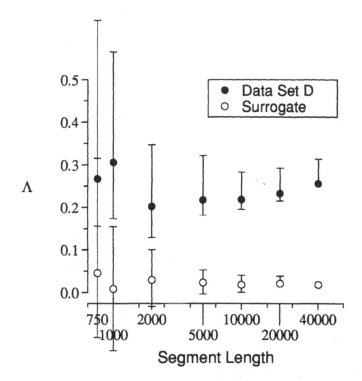

FIGURE 6 $\bar{\Lambda}$ vs. segment length for Data Set D and surrogate data. The upward pointing error bars indicate the standard deviation of $\bar{\Lambda}$ for an ensemble of non-overlapping segments, while the downward-pointing error bars are the internal $1 - \sigma$ confidence interval calculated within each segment (averaged over all segments). (Embedding parameters $m = 6$, $\tau = 10$, box resolution $1/10$.)

DATA SET B: DETERMINISM CONSISTENT WITH LINEAR DYNAMICS

One of the variables of Data Set B contains measurements of heart rate on a sleeping patient. Heart rate is generally not constant even when sleeping, and it has been suggested that heart rate may vary in a deterministically chaotic fashion (Goldberger et al., 1990). When embedded in a three-dimensional space, the $\bar{\Lambda}$ statistic is not able to distinguish between the heart rate data and surrogate data (Figure 7). This suggests that one possibility is that heart rate is consistent with the mechanism that generates the surrogate data—Gaussian white noise passed through a linear filter. Alternatively, it may be that the dynamics of heart rate are too high in dimension to be represented in a three-dimensional embedding. The use of higher dimensional embeddings is precluded by the length of the data set.

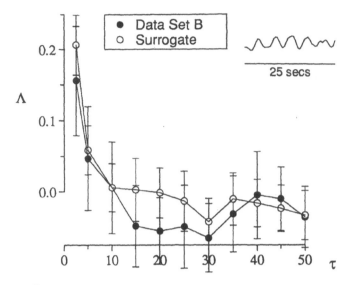

FIGURE 7 $\bar{\Lambda}$ vs. embedding lag τ for the heart rate signal in Data Set B and surrogate data. The inset shows a segment of the heart rate signal. (Embedding parameters $m = 3$, box resolution $1/8$.)

DATA SET C: A RANDOM WALK?

The use of the $\bar{\Lambda}$ statistic in the above examples has emphasized the ability to make a distinction between randomly forced linear systems (as represented by surrogate data) and the data sets themselves. In the case of Data Set C, the exchange rate between the Swiss franc and the U.S. dollar, the first question to answer concerns whether the current exchange rate offers any information about future exchange rates. In terms of an embedded version of the time series, this question can be posed in these terms: "Is the exchange rate a random walk?" If it is, then the current state offers no information about the change in state. If it is not, then one can move on to the more difficult issues of characterizing the dynamics and constructing a predictive model.

The $\bar{\Lambda}$ statistic is well suited to determining whether a given trajectory is consistent with a random walk. In the case of the Swiss franc exchange rate, $\bar{\Lambda} \approx 0$. However, by taking a five-point moving average of the time series, and taking only every fifth point in the filtered time series (to remove the correlations introduced by the filtering), one finds $\bar{\Lambda} = 0.46 \pm 0.03$. Figure 8 shows the box-by-box directions of the flow. The pattern suggests that when the exchange rate is lower at the present

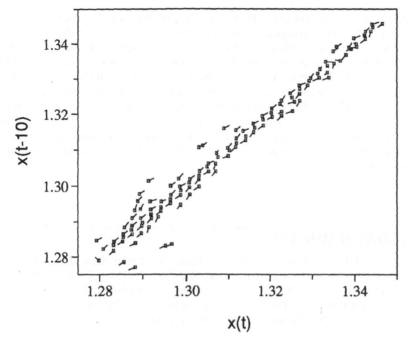

FIGURE 8 Coarse-grained flow averages V_j for the Swiss franc–U.S. dollar exchange rate. The time series was low-pass filtered by averaging over nonoverlapping blocks of five time steps. (Embedding parameters $m = 2$, $\tau = 5$, box resolution $1/32$.)

time than it was five time units ago,[2] the exchange rate will tend to increase, and *vice versa.*

I emphasize that this result does not mean that one can predict future values of the exchange rate with any profit. It simply points out that the low frequency variability (i.e., slower than ≈ 15 minutes) in the exchange rate is not entirely a random walk.

SUMMARY

This paper has described the application of a statistic for detecting determinism in a time series. The statistic indicates whether the trajectory constructed from embedding the time series has a bias towards certain directions as a function of

[2] This time series was sampled once per trade; five time units are roughly 15 minutes.

position in the embedding space. By setting τ appropriately or by using the statistic in conjunction with a comparison to surrogate data generated from a randomly forced linear dynamical system, the statistic provides a test for nonlinear dynamics.

The method may prove useful for screening time series for signs of nonlinear determinism, for evaluating the appropriateness of embedding parameters, and for indicating when useful predictive models can be constructed (as suggested by Eric Kostelich, personal communication). It offers certain technical advantages over other statistics for detecting determinism in time series, particularly in terms of necessary computer resources, the generation of meaningful internal confidence intervals, and its insensitivity to static nonlinear transformations (such as $y(t) = (x(t))^2$).

ACKNOWLEDGMENTS

I thank Leon Glass for helpful discussions, Marc Courtemanche and James Theiler for their comments on the manuscript, and Neil Gershenfeld and Andreas Weigend for organizing the Competition. This work was conducted during the author's tenure as a postdoctoral fellow of the North American Society of Pacing and Electrophysiology, and the Medical Research Council of Canada.

James Theiler,* Paul S. Linsay,† and David M. Rubin‡
*Center for Nonlinear Studies and Theoretical Division, Los Alamos National Laboratory, Los Alamos, NM 87545, and Santa Fe Institute, 1660 Old Pecos Trail, Santa Fe, NM 87501.
†Plasma Fusion Center, Massachusetts Institute of Technology, 175 Albany Street, Cambridge, MA 02139.
‡U. S. Geological Survey, 345 Middlefield Road, Menlo Park, CA 94025

Detecting Nonlinearity in Data with Long Coherence Times

May you have interesting data.

— ancient Chinese curse

We consider the limitations of two techniques for detecting nonlinearity in time series. The first technique compares the original time series to an ensemble of surrogate time series that are constructed to mimic the linear properties of the original. The second technique compares the forecasting error of linear and nonlinear predictors. Both techniques are found to be problematic when the data has a long coherence time; they tend to indicate nonlinearity even for linear time series. We investigate the causes of these difficulties both analytically and with numerical experiments on "real" and computer-generated data.

In particular, although we do see some initial evidence for nonlinear structure in the Data Set E of the Santa Fe Time Series Prediction and Analysis Competition, we are inclined to dismiss this evidence as an artifact of the long coherence time.

Times Series Prediction: Forecasting the Future and
Understanding the Past, Eds. A. S. Weigend and N. A. Gershenfeld, SFI Studies
in the Sciences of Complexity, Proc. Vol. XV, Addison-Wesley, 1993   **429**

1. INTRODUCTION

For time series that arise from chaotic systems, there are certain quantities (e.g., the fractal dimension of the strange attractor, or the spectrum of Lyapunov exponents) that are especially interesting because they characterize intuitively useful concepts (number of active degrees of freedom, or rate of divergence of nearby trajectories) and are invariant to smooth coordinate changes. Algorithms for estimating these quantities are available, but they are notoriously unreliable and often rely heavily on the skill and judgment of their operators. There is an embarrassing lack of consensus, even among the so-called experts, on what constitutes a good estimate of dimension or Lyapunov exponent, or even whether chaos is present in a given time series. To some extent this difficulty may be attributed to inadequate comparison of one algorithm to another (and this conference is aimed at addressing that inadequacy), but to some extent, it is just a hard problem.

The problem is arguably hard enough for long noise-free data sets generated on a computer from low-dimensional maps or differential equations. For "real" data, as the speakers at this conference have repeatedly emphasized, the problem is far more difficult. (And as the organizers have repeatedly reminded us, far more valuable.) Real data is contaminated with noise (which is rarely additive, Gaussian, or white), measured with finite precision, and subject to innumerable external influences in the environment and the measurement apparatus. And, of course, there is never enough of it.

In this article, we will describe (yet) another source of difficulty that arises in the analysis of time series data. The particular problem of detecting nonlinear structure—either by comparison of the data to linear surrogate data, or by comparing linear and nonlinear predictors—is seen to be complicated when the data exhibits long coherence times.

In this section, we define some terms and discuss linear modeling of time series. Section 2 describes the method of surrogate data and compares two approaches to generating surrogate data. We find that both have difficulties trying to mimic data with long coherence time. We illustrate these problems with real and computer-generated time series in Section 3, including the time series E.dat from the the Competition. In the last section, we discuss what it is about the analysis or the data that is problematic.

1.1 TERMINOLOGY

A *time series* is a sequence of measurements $x_1, x_2, \ldots, x_N$ of some physical system taken at regular intervals of time. A time series can be thought of as a particular realization of a stochastic *process*, which we will define as a sequence of random[1] variables $\ldots, X_{-1}, X_0, X_1, X_2, \ldots$. We make this distinction because theorems and formal definitions are available only for *processes*, while the whole purpose of generating this formalism is to assist researchers who are confronted with real experimental *time series*.

We will also distinguish the terms *system* and *model*, by letting *system* refer to the actual underlying physics,[2] and *model* to a hypothetical description of the system. Since the *model* will (for our present purposes) be inferred only from the time series, we cannot expect the model to be expressed in terms of the physics. But, although the model is really nothing more than an operational description of the time series, the hope is that this description—in conjunction with knowledge of the appropriate physics—will actually say something useful about the underlying physical system. When we talk about a *best* or "correct" (always in quotation marks!) model, we will mean the model that—out of a (usually parametric) family of models—has the least root mean squared (rms) error in its one-step-ahead forecast.[3]

Three statistics of particular interest are the *mean* $\mu = \langle X_t \rangle$, the *variance* $\sigma^2 = \langle (X_t - \mu)^2 \rangle$, and the *autocorrelation function* $A(\tau) = \langle (X_t - \mu)(X_{t-\tau} - \mu) \rangle / \sigma^2$. Here $\langle \cdot \rangle$ represents, for the process, an ensemble average. If the process is ergodic, the average could also be over time t, and in that case, good *sample* statistics can be defined from a single *time series*.

When we speak of "coherence time," what we mean is the time beyond which a signal becomes uncorrelated with its past. We can formalize the concept somewhat by defining coherence time as time τ such that the absolute value of the autocorrelation function $|A(T)|$ is smaller than some prespecified value ϵ for all $T > \tau$. This is to be distinguished from the *first* time T that the autocorrelation $A(T)$ drops to a value below ϵ; in other words, we are interested in the "envelope" of the autocorrelation curve. This is not really satisfactory as a formal definition of coherence time—for one thing, it depends on the choice of ϵ—but it is adequate for our current purposes.

In general, however, if the autocorrelation $A(T)$ vanishes exponentially fast as $T \to \infty$, we will say that the coherence time is finite; if it does not vanish at all (if $\limsup_{T \to \infty} |A(T)| > 0$), then we say that the coherence time is infinite.[4]

[1] Note that even a deterministic process is usefully defined as a sequence of random variables. For the logistic process, $X_{t+1} = 4X_t(1 - X_t)$, for instance, each variable X_t has a nontrivial probability distribution $P(X_t)$, but the joint distribution $P(X_{t+1}, X_t)$ reflects the deterministic law.

[2] Mathematically, the *system* is equivalent to the *process*, but the connotation we mean to imply for a *system* is that it is physical.

[3] This is a convenient but basically arbitrary criterion. As Tsay (1992) emphasizes, "it is well-known that the best model with respect to one checking criterion may fare badly with respect to another criterion."

[4] One definition that is consistent with these constraints is $\tau = \sum_{T=1}^{\infty} \sqrt{1 - E^2(T)}$, where $E(T)$ is the rms forecasting error T time steps into the future for the best linear model, normalized

1.2 WOLD DECOMPOSITION

The Wold decomposition is the fundamental theorem of linear time series analysis (e.g., see, Anderson, 1972, Section 7.6.3). This theorem states that any stationary zero-mean process (linear or nonlinear) can be decomposed into the sum of two uncorrelated components: one "deterministic" and one "indeterministic." That is,

$$x_t = z_t + u_t,\qquad(1)$$

where the linearly deterministic z_t can be modeled exactly with a (possibly infinite) linear combination of past values, and where the indeterministic u_t can be modeled by a moving average of uncorrelated innovations.[5]

$$z_t = \sum_{i=1}^{\infty} \alpha_i z_{t-i},\qquad(2)$$

$$u_t = \sum_{i=0}^{\infty} \beta_i e_{t-i}.\qquad(3)$$

It can be shown that the autocorrelation $A_z(T) = \langle z_t z_{t-T}\rangle/\langle z^2\rangle$ of the deterministic component of the time series will be significantly nonzero for arbitrarily large T, whereas the autocorrelation $A_u(T)$ of the indeterministic time series will approach zero as T becomes large. Since u and z are uncorrelated (again, not necessarily independent, but satisfying $\langle u_t z_t\rangle = 0$), it follows that the autocorrelation in the full time series is

$$A_x(T) = \frac{A_u(T)\langle u^2\rangle + A_z(T)\langle z^2\rangle}{\langle u^2\rangle + \langle z^2\rangle}.\qquad(4)$$

From the point of view of the Wold decomposition, then, a process has a finite (resp. infinite) coherence time if and only if its linearly deterministic component is zero (resp. nonzero).

by the standard deviation of the data. We won't actually be using this definition (the informal description in the text will be adequate), but it does seem appropriate to at least write such a definition down.

[5] These innovations are uncorrelated, or "white," but they are not necessarily independent. This means $\langle e_t e_{t'}\rangle = 0$ for $t \neq t'$, but *not* that the joint distribution $P(e_t, e_{t'})$ is equal to the product of the marginal distributions $P(e_t)P(e_{t'})$. The innovations are treated as "noise" in linear analysis, but they may well possess nonlinear deterministic structure. The Wold decomposition is quite general and applies to all stationary processes, including low-dimensional chaos.

1.3 LINEAR MODELING OF TIME SERIES (ARMA)

It follows from the Wold decomposition theorem that any stationary process can be modeled as an autoregressive moving average:

$$x_t = x_o + \sum_{i=1}^{\infty} a_i x_{t-i} + \sum_{i=0}^{\infty} b_i e_{t-i}. \tag{5}$$

For instance, given α_i and β_i from Eqs. (2) and (3), one can take $a_i = \alpha_i$ and $b_i = \beta_i - \sum_{j=1}^{i} \alpha_j \beta_{i-j}$. However, this is not necessarily a unique solution. For indeterministic time series, for instance, it is possible to write the time series as a pure autoregressive (AR)

$$x_t = x_o + \sum_{i=1}^{\infty} a_i x_{t-i} + \sigma e_t, \tag{6}$$

or as a pure moving average (MA)

$$x_t = x_o + \sum_{i=0}^{\infty} b_i e_{t-i}. \tag{7}$$

For time series with infinite coherence time (nonzero linearly deterministic component), however, a full ARMA model is typically required.

In the study of linear Gaussian processes, the innovations are taken to be independent and identically distributed (iid) Gaussian random variables. For indeterministic time series, which can be written as a pure moving average of the Gaussian innovations, this implies that the time series itself will be Gaussian. However, if a deterministic component is present (again, that means an infinite coherence time), then Gaussian innovations do not necessarily imply Gaussian data. For example, a sine wave with added Gaussian white noise can be modeled as a linear process with Gaussian innovations, but the time series is not Gaussian.

Lii and Rosenblatt (1982) have discussed linear (indeterministic) processes with non-Gaussian innovation; they show that these processes are far more complicated than those with Gaussian innovations.

2. SURROGATE DATA

Surrogate data is artificially generated data that is to be used in place of an original data set; the main purpose is to provide a kind of baseline or control against which the original data can be compared. In tests for chaos, for example, one can control against artifacts due to autocorrelation in a time series by generating surrogate data from a random process that mimics the autocorrelation of the original time series. Suppose some algorithm indicates low-dimensional chaos in a time series. If the same algorithm also indicates low-dimensional chaos in the surrogate time series, then one can dismiss the original evidence for chaos as an artifact of the autocorrelation.

More formally, the method provides a mechanism for testing well-formulated null hypotheses. It can be difficult to precisely formulate *interesting* null hypotheses, and often very difficult to prescribe a surrogate data generator that is appropriate for such a null hypothesis. Our work has focused on tests for nonlinearity which take linearly correlated Gaussian noise as the null hypothesis. In this case, one is not looking for chaos *per se*, but for some statistic that is significantly different for the original time series than it is for the linear surrogates. The existence of such a statistic implies that the original time series is inconsistent with the null hypothesis and, therefore, that the original time series is nonlinear.

While the systematic application of this approach to tests of potentially chaotic time series has only recently become fashionable, the basic idea is by no means new. Monte Carlo methods for generating data sets with specified properties are widely used and, in some applications, have reached the status of recipes (Press et al., 1988, Section 14.5). Statisticians have long advocated resampling (so-called "bootstrap") methods, in which new data sets are generated by randomizing the original data set in some prescribed way. We have found the writing of Efron in particular to be enlightening and inspirational (Efron, 1979; Efron & Tsibirani, 1986). The purpose of these methods, however, is usually not to test a hypothesis, but to estimate confidence intervals for some statistic of interest.

The application of these resampling methods to time series is complicated by the temporal dependence of time series data; most of the original bootstrap applications considered individual data points to be independent events. An indirect approach is to remove the linear dependence in the data by considering the innovations (the "residual" time series) of an ARMA model (Efron & Tsibirani, 1986), though the filtering required to produce the residuals can make it harder to "see" the nonlinearity in a chaotic time series (Theiler & Eubank, 1992a). Direct resampling techniques based on temporal "blocks" of data were discussed by Künsch (1989), and an improvement was developed by Leger, Politis, and Romano (1992). While further exploration is certainly called for, it is not clear to us that these methods (at least as they have been applied in Künsch (1989); Politis and Romano (1991) and Leger, Politis, and Romano (1992)) can be used in conjunction with dynamical statistics for the purpose of hypothesis testing.

Parametric bootstraps, instead of resampling the data directly, use the data to set parameter values, and then use these values in a parametric model for generating new data. An incomplete list of authors who have successfully used this approach include: Grassberger (1986), who used a simple linear autoregressive process to generate a time series that mimicked properties of a climate data set originally purported to exhibit low-dimensional chaos; Kurths and Herzel (1987), who compared estimates of dimension and Lyapunov exponent for a time series of solar radio pulsations with those for data from an AR(5) model that fairly accurately matched the spectral properties of the original data; Brock, Lakonishok, and LeBaron (1992), who generated surrogate financial time series to test trading strategies; Ellner (1991), who used this approach to show that a variety of nonchaotic "plausible alternatives" might adequately explain measles and chickenpox data; and TsayTsayR. S. (1992), who provides an excellent overview of the approach with a wide variety of applications.

Kaplan and Cohen (1990) published the first example we are aware of in which the evidence for chaos in a time series was evaluated by comparing against a control data set that was generated by the Fourier transform (FT) method which is described in Section 2.1. Somewhat earlier, Osborne et al. (1986, 1989) inverted $1/f^\alpha$ spectra using an inverse Fourier transform to generate realizations of $1/f^\alpha$ noise, and then showed that dimension estimates of these time series were problematic. (This issue has been further discussed in Provenzale, Osborne, & Soj, 1991; Theiler, 1991.) The use of multiple surrogate data sets for more formal statistical hypothesis testing was suggested in Theiler (1988, 1990) and implemented in Theiler et al. (1992a, 1992b) for a variety of examples. L. A. Smith (1992) has applied the surrogate data methods to fluid dynamical time series, and more recently, to address the issue of inherent periodicities in the climate record.[6] A variant of the surrogate data approach has also been described in Kennel and Isabelle (1992).

The use of formal statistics, in which the null hypothesis is explicitly spelled out and carefully tested against, is only lately gaining popularity in the chaos community. Brock, Dechert, and Scheinkman (1988b) deserve to be singled out for creating perhaps the first statistically rigorous application of the Grassberger-Procaccia (1983c) correlation integral for time series analysis. This work has led to a veritable industry in the economics community involving the application of statistics which incorporate the explicit recognition of chaos (Brock & Dechert, 1989; Brock & Potter, 1992; Brock & Sayers, 1988; Hsieh, 1989, 1991; LeBaron, 1992a; Lee, White, & Granger, in press; Scheinkman & LeBaron, 1989); these complement the more classical approaches taken by the statisticians (Hinich, 1982; Keenan, 1985; McLeod & Li, 1983; Subba Rao & Gabr, 1980; Tsay, 1986, 1991, 1992). Many of these are reviewed in Tong's comprehensive book (1990).

2.1 FT-BASED SURROGATES

To test for nonlinearity, we begin with the presupposition that the time series is linear. A more precise formulation of the null hypothesis is that the data arise from a linear stochastic process with Gaussian innovations.[7]

---

[6]L. A. Smith (personal communication).

[7]An extended null hypothesis which considers that there is an underlying process that is Gaussian, but one is observing a static nonlinear transform of that process, is discussed in Theiler et al. 1992a.

The algorithm we generally use for making linear surrogate data is based on the Fourier transform (FT). Specifically, we compute a discrete Fourier transform of the original data, and replace the phases at each frequency with random numbers in the interval $[0, 2\pi)$ while keeping the magnitude at each frequency (i.e., the power spectrum) intact,[8] and then apply the inverse Fourier transform to produce the surrogate time series.

This is a kind of nonparametric bootstrap which by construction produces surrogates that have the same power spectrum as the original data. In fact, the surrogate time series have exactly the same *sample* power spectrum as the original time series. The Wiener-Khintchine relations assure us that two processes with the same power spectrum will also have the same autocorrelation function, but in comparing the sample statistics, we have to be more careful. Jenkins and Watts (1968) note that there are (at least) three different ways to define a sample autocorrelation (In these definitions, the time series is for convenience assumed to have zero mean):

- Unbiased estimator: $\frac{1}{N-T}\sum_{t=1}^{N-T} x_t x_{t+T}$.
- Biased estimator (lower variance than unbiased estimator): $\frac{1}{N}\sum_{t=1}^{N-T} x_t x_{t+T}$.
- Circular autocorrelation: $\frac{1}{N}\left(\sum_{t=1}^{N-T} x_t x_{t+T} + \sum_{t=N-T+1}^{N} x_t x_{t+T-N}\right)$.

The estimators agree to order $O(T/N)$, and for $T \ll N$ and $N \to \infty$, all three approach the actual autocorrelation of the process. But for finite N they are only approximately equal. And of the three, it is the circular autocorrelation that is exactly preserved in going from the original to the surrogate data sets.

For a Gaussian linear *process*, all of its properties are encoded in the mean, variance, and autocorrelation. But when we say that the "linear" properties of the *time series* are preserved in the surrogate time series, what that means exactly is that the sample mean, sample variance, and circular autocorrelation are preserved.

2.2 ARMA MODEL-BASED SURROGATES

Instead of attempting to exactly preserve some preselected set of sample statistics, an alternative approach for generating surrogate data is to directly fit the data to a constructive parametric linear model, such as a finite-order ARMA(p, q):

$$x_t = x_o + \sum_{i=1}^{p} a_i x_{t-i} + \sum_{i=0}^{q} b_i e_{t-i}. \tag{8}$$

Constructing a parametric model from a finite set of data involves choosing the "correct" values for q and p, and this is an issue of some subtlety; one wants

[8] It is important that the phases be symmetrized in such a way that the inverse Fourier transform is real and the power at each frequency is unaffected; we remark that the recipe for doing this in Theiler et al., 1992b, Section A.1,#4) is incorrect. We are indebted to W. Schaffer for pointing out this error.

enough terms to capture the correct correlations in the data but not so many terms that the data is over-fit. Akaike (1974) and Schwarz (1978) have suggested fairly general criteria; a more recent discussion specific to the ARMA model can be found in Pukkila, Koreisha, and Kallinen (1990). For fixed values of q and p, the optimal parameters (a_i, b_i) depend in principle only on the autocorrelation of the stochastic process.

If there is no deterministic component ($z_t = 0$ in Eq. (1)), then an ARMA(p, q) process can be modeled by a pure autoregressive AR model or a pure moving-average MA model, of appropriately large order.[9] We note that in practice it is much easier to fit coefficients to a pure AR model than to an MA or ARMA model. In that case, assuming a zero-mean process for convenience, the formula is given by Anderson (1971, p. 187):

$$
\begin{bmatrix}
1 & A(1) & \cdots & A(m-1) \\
A(1) & 1 & \cdots & A(m-2) \\
\vdots & \vdots & \ddots & \vdots \\
A(m-1) & A(m-2) & \cdots & 1
\end{bmatrix}
\begin{bmatrix}
a_1 \\
a_2 \\
\vdots \\
a_m
\end{bmatrix}
=
\begin{bmatrix}
A(1) \\
A(2) \\
\vdots \\
A(m)
\end{bmatrix}.
\tag{9}
$$

These are sometimes called the Yule-Walker equations.

Having determined the appropriate ARMA model, one can generate surrogate data by inserting Gaussian iid random numbers into the e_t terms, and then iterating Eq. (8). One is assured that in the long run, the autocorrelation in the surrogate data will approach the autocorrelations used in Eq. (9), but note that this is different from the exact match of sample statistics that is seen in the FT surrogates.

A common alternative practice is to bootstrap the residuals themselves. Having fit the model to the data, one derives a time series of residuals e_t which are then scrambled and reinserted into Eq. (8). This avoids the assumption of Gaussian innovations and, therefore, leads to a broader class of time series, and presumably tests against a looser null hypothesis; however, linear processes with non-Gaussian innovations do not always behave in "linear" ways—for instance, see Tong (1990, pp. 13–14), Lii and Rosenblatt (1982), or Kanter (1979) for examples of some of the pathologies.

It is also worth noting that this AR model is also the optimal linear predictor; that is, the average squared errors,

$$
\langle E_t^2 \rangle = \left\langle \left(X_t - \left[x_o + \sum_{i=1}^{m} a_i X_{t-i} \right] \right)^2 \right\rangle,
\tag{10}
$$

are minimized when the coefficients a_i are chosen according to Eq. (9).

---

[9]But, in general, a pure AR or pure MA will be less parsimonious than the best ARMA(p, q) model; that is, the AR or MA models will usually require more than $p + q$ parameters. Having said this, we should further note that the ARMA formalism doesn't necessarily generate the most parsimonious description of linear Gaussian processes either.

2.3 COMPARISON OF FT AND ARMA SURROGATES

The superficial equivalence of FT and ARMA modeling rests with the notion that both the Fourier spectrum and the AR coefficients depend only on the autocorrelation function of the original time series, which is (at least approximately) mimicked by the surrogates in both cases.

The difference between FT and ARMA surrogates is basically the difference between "fitting the data" versus "fitting the model." FT surrogates exactly match certain sample statistics (mean, variance, and circular autocorrelation) of the original data. ARMA surrogates are generated from a model that is *fit* to the original time series. These surrogates exhibit sample statistics that are usually but not necessarily in approximate agreement with those of the original time series.

Another difference is the way the data sets are generated. The FT method makes a whole new time series all at once, and it necessarily has the same length as the original time series. The ARMA method generates new points iteratively, one at a time, and can generate arbitrarily long or short data sets. This is not necessarily an advantage, though. Generating points sequentially, one is vulnerable to instabilities that may amplify small errors into large effects in the long term. This is a general difficulty with model-based surrogate data methods; two models that are approximately equal (say, have nearly equal ARMA coefficients) can give rise to time series that are markedly different.[10] A second difficulty that arises when modeling processes with long coherence times is that the qualitative long-term behavior, even the overall amplitude of the process, can depend not only on the model parameters but on initial conditions as well.

While the FT method is nonparametric, in the sense that one does not directly fit a model to the data, one can think of it rather as having a very large number of parameters, $N/2$, corresponding to the amplitude of the power spectrum at $N/2$ frequencies. As a model, then, the FT provides an extreme overfit to the data. By contrast, ARMA models are parsimonious, in that the modeler is (usually) careful to choose the minimum number of parameters needed to fit the data.[11]

---

[10] For nonlinear modeling, this can be extremely problematic. A parametric model that exhibits chaotic behavior, for instance, can, with an arbitrarily small change in parameter, give rise to stable periodic behavior. This is sometimes referred to as the genotype/phenotype conundrum. One associates genotype with equations of motion, and phenotype with the long-term behavior of those equations. Small perturbations in the genotype can give rise to huge differences in phenotype. And inferring the genotype from the phenotype is much more difficult than the other way around. We should remark that for linear modeling, the difficulty is not this extreme. In this case, if the roots of the characteristic polynomial of the AR part of the model are well within the unit circle, then a small perturbation of parameters will not grossly affect the overall behavior. (However, we might also remark that for high-order polynomials, small changes in the coefficients can lead to large changes in the roots.)

[11] The problem of parsimony and "effective number of parameters" is much more subtle in the case of nonlinear modeling; see Weigend, Huberman, and Rumelhart (1990), Weigend and Rumelhart (1991b); and Moody (1992) for interesting discussions of this issue.

Which approach is preferable depends on the application. Our view is that FT surrogates are better for testing hypotheses, while ARMA surrogates may be better for estimating confidence intervals. Certainly the FT surrogates will be useless for estimating confidence intervals for estimates of mean, variance, and autocorrelation.

3. APPLICATION TO TIME SERIES DATA

In this section, we will investigate four different time series. The first is a real time series that was part of the Competition. Though we seem to see evidence for nonlinearity in this time series, we give reasons to suspect the results. The second data set is an artificially generated sine wave plus noise. This data is meant to be a caricature of the real data, but a caricature whose underlying process is known. With this second data set, we are able to see the same effects that we observed with the "real" data set and, thereby, confirm our suspicions that the effects we saw were artifacts of the long coherence time. The third data set is also, strictly speaking, a sine wave plus noise, but it is a particularly simple example that permits some analytical discussion. For the third data set, we compare the theoretical efficiency of linear versus nonlinear predictors. Finally, the last data set is a sum of two commensurate sine waves with some added noise; in this case, we see numerically what we described in theory for the third data set: namely, that the prediction error of a nonlinear model fit to the data is smaller than the error of a linear model fit to the data.

3.1 THE INVESTIGATION OF E.DAT

We apply tests for nonlinearity based on the method of surrogate data to Data Set E.dat. These data are observations of the light curve of a variable white dwarf star, and are sampled every ten seconds. We concentrate on a single series #14, chosen more or less arbitrarily.[12] Figure 1(b) shows the first $N = 2048$ points of this time series. The most noticeable feature is the coming and going of an oscillation with a period of 50 time units (500 seconds, or about 8.3 minutes). We computed discrete Fourier transforms on all 17 data sets, using data segments of varying length and location in the time series, and both with and without a Hanning window. (See Figure 2(a) for a particular case.) We see considerable variation, and would not be confident in attempting our own detailed interpretation of the

---

[12]We were partly motivated to use this series because we knew that M. Paluš (this volume) had looked at the same series.

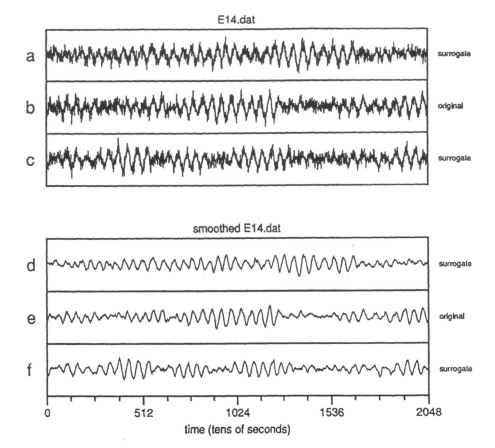

FIGURE 1   (a,b,c) The top three time series are (b) Data Set E.dat #14, and (a,c) two
surrogate data sets. (d,e,f) The bottom three are (e) set E.dat #14 smoothed with a
moving average window of size ten time steps, and (d,f) two of its surrogates. Figures
(b,e) are the first $N = 2048$ points of an approximately 2600-point data set.

power spectrum.[13] However, we do consistently see two peaks in the vicinity of the
dominant frequency (0.002 Hz), suggesting that the signal is quasi-periodic and that
the "coming and going" may be a beating phenomenon. The autocorrelation curve
in Figure 2(b) supports this interpretation and also indicates that the coherence
time is at least on the order of a thousand time steps, and possibly much longer.

[13] We have not attempted to use the information (which was provided) that gave the absolute
starting times for each of the seventeen time series. Combining the data into one long time series
with appropriate gaps should permit much more precise spectral estimation.

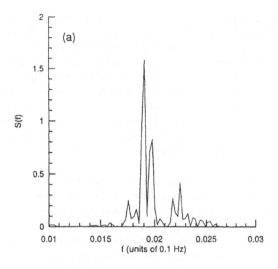

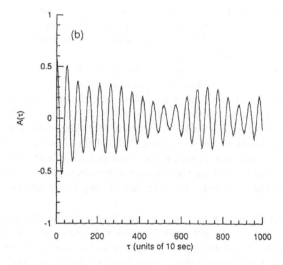

FIGURE 2 (a) Power spectrum, computed by a discrete Fourier transform from all $N = 2602$ points in the time series E14.dat, using a Hanning window. (b) Autocorrelation of the Data Set E14.dat computed with the biased estimator $A(T) = \frac{1}{N} \sum_{t=1}^{N-T} x_t' x_{t+T}'$, where $x_t' = (x_t - \mu)/\sigma$ is the normalized time series value.

In searching for nonlinear structure, any nonlinear statistic in principle is adequate. We used an estimator of fractal dimension, obtained from the slope of a correlation integral at a point r equal to half of the rms amplitude of the time series. While this is not our best shot at what the actual dimension is (in fact, for this data, we do not really even see a hint of low dimensionality), it does provide a nonlinear statistic against which we can compare real data to surrogate. What we see in Figure 3(a) is that—for this statistic—the real and surrogate data are indistinguishable. We quantify significance by counting the number of "sigmas" between the original and surrogate values for the discriminating statistic, where a "sigma" is the standard deviation of all the values of the statistic computed for the surrogate data sets.

Because the data set has a lot of what appears to be high-frequency noise, we also considered a crude low-pass linear filter of the data, based on a moving average (equal coefficients) of ten sample points. That is, $x_t' = (x_t + x_{t-1} + \cdots + x_{t-9})/10$. Figure 1(e) shows how smoothing affects this data set, and in Figure 3(b), we again compare real data to surrogates. At about the four-sigma level, the difference between the real data and surrogates is statistically significant. Inspection of the actual values, however, reveals that the difference is never more than 8%; we are inclined to remark that the difference is "significant," but not very "substantial." When we used nonlinear forecasting error instead of estimated dimension as our discriminating statistic, we did not see any significant evidence for nonlinearity for either the smoothed or the raw data set.

Now, if the surrogate data really is mimicking all the linear properties of the original time series, then any linear statistic computed from both surrogate and original data should give the same value. We plot one such statistic, the in-sample fit error of the best linear model, in Figure 4(a),(b). For both E14.dat and the smoothed E14.dat, there is a small but statistically significant discrepancy. So the surrogate data evidently is *not* mimicking "all" of the linear properties. The technical explanation is that the in-sample fit error is a sample statistic that does not depend precisely and entirely on the circular autocorrelation. That the discrepancy should be systematic, however, is an artifact of long coherence times, as we show in Section 3.2.

In general, as Figure 5 shows, generating surrogate data by the FT algorithm leads to surrogates that do not have the coherent structure of the original sine wave. It *is* possible to generate good surrogates by fortuitous choice of data length. For periodic data, this is only a slight inconvenience (requiring the use of a general discrete Fourier transform (DFT) instead of the fast Fourier transform (FFT) which requires data length to be a power of two); for quasi-periodic data, this is trickier, because one must choose the length of the time series to be (at least approximately) commensurate with *both* periods.

(a)

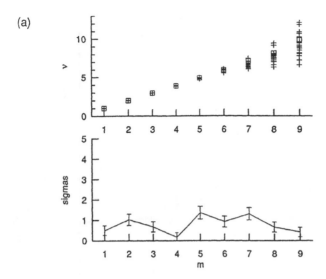

(b)

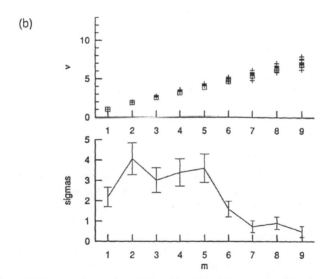

FIGURE 3 Significance of evidence for nonlinearity based on an estimate of the correlation dimension: (a) for Data Set E14.dat and (b) for the smoothed data set. The top panels show estimated dimension for real (□) and surrogate (+) data, as a function of embedding dimension. The bottom panels show "number of sigmas" as an indication of the statistical significance with which one can reject the null hypothesis that the experimental data is linear.

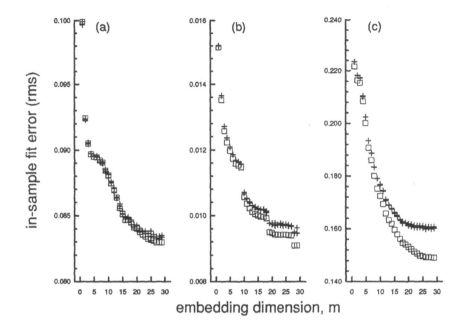

FIGURE 4   A linear statistic, the in-sample rms fitting error, is computed for linear models with embedding dimension m. Here (a) is for E14.dat, (b) is for the smoothed data, and (c) is an artificially generated sine wave with measurement noise. It is apparent, particularly for cases (b) and (c), that the surrogate data is not as linearly predictable as the original data.

3.2 SINE WAVE PLUS NOISE

To investigate the effects of long coherence time in a situation where we know the underlying process, we generated artificial data with infinite coherence time by adding measurement noise to an underlying sine wave. We chose the period and noise level to (very crudely) approximate that of the smoothed E.dat.

There are two effects going on here. The first involves choosing the length so that the periodic continuation is at least continuous (doesn't have a jump). If this is not done, one introduces spurious high frequencies into the data. This effect can be alleviated to some extent by windowing the data, e.g., with a Hanning window (see Theiler et al., 1992b, Section 2.4.2). The second effect involves choosing the length of the time series so that *all* the relevant periods are commensurate with this length. If this is not done, then the DFT takes the power from a single frequency and distributes it to adjacent frequency bins; upon inverting the DFT after randomizing the phases, one sees a beating between the adjacent frequencies instead of the pure frequency in the original time series. In this second case, windowing the data does not help.

Using this sine wave plus noise, we see in Figure 6 that a dimension-based test for nonlinearity is able to distinguish the real and surrogate data with high statistical confidence. Again, the difference is extremely "significant" but not especially

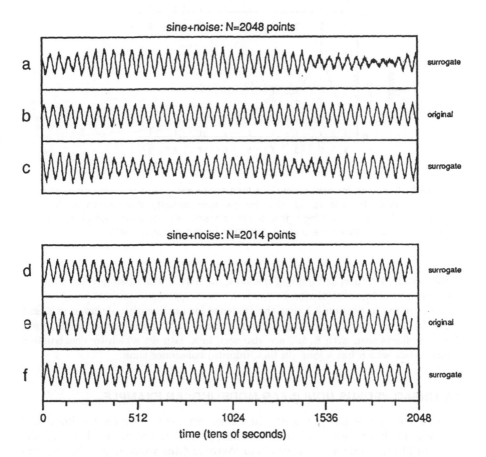

FIGURE 5 Sine signal with white measurement noise, and surrogate data sets generated by the FT algorithm without windowing. In (a,b,c), the length of the time series is a convenient (for FFT purposes) power of two, $N = 2048$; the original data set is in (b), while (a) and (c) are the surrogate data sets. In (d,e,f), we use the same time series, slightly truncated to $N = 2014$ points, so that there is a near-integral number of oscillations in the time series. Again the middle data set (e) is the original, while (d) and (f) are two surrogates.

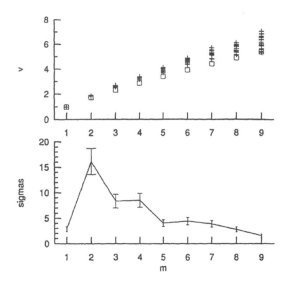

FIGURE 6   Evidence for nonlinearity in a time series composed of a sine wave plus white noise. While there is no indication for low dimensionality (there is too much noise to see that the underlying signal is one-dimensional), the estimated dimension is significantly different for the original data than for the surrogate data.

"substantial." Further, as Figure 4(c) shows, the data is also distinguished from the surrogates by a linear statistic.

We argued in Section 3.1 that the evidence for nonlinearity observed in E.dat might be an artifact of the long coherence time. In this section, we have shown that the effects seen with E.dat are also seen in a data set which by construction is linear, but which has a long (in fact, infinite) coherence time.

3.3 LINEAR VERSUS NONLINEAR MODELING: AN EXAMPLE

Another way to test for nonlinearity in a time series is to compare the linear and nonlinear models to see which more accurately predicts the future. For example, Casdagli (1991, 1992) and Casdagli and Weigend (this volume) describe an "exploratory" approach in which the data is fit with local linear models using k nearest neighbors. The parameter k is swept from $m + 1$, the minimum value required to make a local linear fit in m dimensions, up to the size N of data set itself. For $k < N$, the model is nonlinear, but for $k = N$ it is equivalent to a globally linear model. If the error decreases monotonically with k, then the process is taken to be linear. If the error increases monotonically with k, then the process is taken to be nonlinear and deterministic. If, as most often happens, the error first decreases with increasing k and then increases, the process is taken to be nonlinear and stochastic.

Kanter (1979) has shown the unsurprising result that for an indeterministic linear Gaussian process, the optimum predictor is a linear predictor.[14] We consider a slightly different case than Kanter studied; we look at a *time series* that is generated by a linear *process*, and compare the linear and nonlinear models that are *fit* to the finite length of the time series. What we find for data with long coherence times is that the nonlinear models are often superior.

Consider again a sine wave with additive measurement noise, but to keep the analysis simple, take the sampling rate to be exactly twice the frequency of the signal. The time series is produced by $x_t = s_t + n_t$, where the signal $s_t = (-1)^t S$ is alternating in sign while maintaining a constant amplitude S, and the noise n_t is a white-noise process of amplitude σ. That is,

$$x_t = (-1)^t S + \sigma e_t, \tag{11}$$

where the e_t's are unit variance iid Gaussian random variables. The time series is linear and can be produced by the ARMA model

$$x_t = -x_{t-1} + \sigma e_t + \sigma e_{t-1}, \tag{12}$$

with appropriate initial conditions. In fact, the initial conditions are crucial; notice that the signal amplitude S does not even appear in Eq. (12). This is a general property of processes with infinite coherence times. For a process with a finite coherence time, the amplitude of the signal is determined solely by the coefficients of the ARMA model.[15]

3.3.1 LINEAR MODELING.
Let $M_o(m)$ denote the best order-m autoregressive (AR) linear predictor; here

$$\hat{x}_t = \sum_{k=1}^{m} a_k x_{t-k}, \tag{13}$$

where the coefficients are chosen to minimize the mean squared error of the process:

$$\langle (x_t - \hat{x}_t)^2 \rangle = \langle (x_t - \sum_{k=1}^{m} a_k x_{t-k})^2 \rangle \tag{14}$$

$$= \left[1 - \sum_{k=1}^{m} (-1)^k a_k \right]^2 S^2 + \left[1 + \sum_{k=1}^{m} a_k^2 \right] \sigma^2. \tag{15}$$

---

[14] What *is* surprising in Kanter's paper is that linear non-Gaussian processes *can* be more accurately modeled with a nonlinear predictor. We are grateful to J. Scargle for pointing out this reference to us.

[15] This is another disadvantage of ARMA surrogates compared to FT surrogates; the FT surrogates will by construction possess the same amplitude as the original time series.

With a little algebra, one can show that these coefficients are given by

$$a_k = (-1)^k \frac{S^2}{mS^2 + \sigma^2}, \tag{16}$$

and that the average squared error of this optimal predictor is given by

$$E^2[M_o(m)] = \langle (x_t - \hat{x}_t)^2 \rangle = \sigma^2 + \frac{\sigma^2 S^2}{mS^2 + \sigma^2}. \tag{17}$$

Note that the error decreases monotonically with increasing m, approaching a "floor" of σ^2 as $m \to \infty$. It is basically impossible to beat this error with any model, linear or nonlinear, because it is the noise on the signal which is by definition unpredictable. It is also worth remarking that the convergence of the linear AR model is algebraically slow with embedding dimension m. For chaotic processes (or more generally, for any stochastic processes with a finite coherence time), the convergence is usually faster.

This error assumes that the "correct" model is chosen for a given order m. In practice, one fits a model M_s to a finite sample of N data points. The fit is optimal for the data in the sample set, but in general is not optimal for out-of-sample data. Particularly when m is large, and N is small, the difference between the out-of-sample error for "correct" model and for the *fit* model can be significant.

The effect can be quantified with the aid of the Akaike Information Criterion (AIC) (Akaike, 1974), which provides a measure of the difference between in-sample error E_s and out-of-sample error E_o for the best in-sample model M_s. (See Tong, 1990, Section 5.4) for a modern discussion.) Here,

$$\log(E_o[M_s(m, N)]) = \log(E_s[M_s(m, N)]) + \frac{m}{N}. \tag{18}$$

We have observed numerically that the error of the "correct" model M_o lies roughly half way between the in-sample and out-of-sample error of the in-sample fit model M_s; that is, the difference m/N in Eq. (18) can be split into two roughly equal components[16]:

$$\begin{aligned}
\log(E_o[M_s(m, N)]) - \log(E_s[M_s(m, N)]) = \\
\{\log(E_o[M_s(m, N)]) - \log(E_o[M_o(m)])\} \\
+ \{\log(E_s[M_o(m)]) - \log(E_s[M_s(m, N)])\},
\end{aligned} \tag{19}$$

[16] S. Ellner (personal communication) has provided a heuristic argument for why the terms should be equal. The argument notes that the difference in error between M_o and M_s can be expanded as a Taylor expansion in $\theta_o - \theta_s$ (where θ represents the finite vector of parameters in model M), and that the relevant term is the second derivative of E_o and E_s, respectively, multiplied by $(\theta_o - \theta_s) \cdot (\theta_o - \theta_s)$. Since one expects E_s and E_o to be asymptotically equal (as $N \to \infty$), it follows that their second derivatives should also be asymptotically equal; thus, the expected differences $E_s[M_o] - E_s[M_s]$ and $E_o[M_s] - E_o[M_o]$ should also approach equality for large N.

where

$$\log E_o[M_s(m, N)] - \log E_o[M_o(m)] \approx \frac{m}{2N} \qquad (20)$$

and

$$\log E_o[M_o(m)] - \log E_s[M_s(m, N)] =$$
$$\log E_s[M_o(m)] - \log E_s[M_s(m, N)] \approx \frac{m}{2N} . \qquad (21)$$

For large N and large m (but $m \ll N$), we can combine Eqs. (17) and (21) to write the total squared error as the sum of noise (unavoidable), model inadequacy (m too small), and parameter misspecification (N too small):

$$E^2[M_s(m, N)] = \sigma^2 + \frac{\sigma^2}{m} + \frac{m\sigma^2}{N} . \qquad (22)$$

Thus, for a given finite N, there will be an optimum m for which the total error is minimized. In particular, for large N, the optimum model occurs when $m = \sqrt{N}$, and the total squared error in this case is given by $\sigma^2 + 2\sigma^2/\sqrt{N}$.

3.3.2 NONLINEAR MODELING. By contrast, consider as an example, the following parametric nonlinear model:

$$\hat{x}_t = -S^* \mathrm{sgn}(x_{t-1}) \qquad (23)$$

where sgn is the "signum" or "sign" function; its value is $+1$ or -1, depending on the sign of its argument. Here $m = 1$, and using a learning set of size N, one can estimate the parameter S^* to within an error of $\sigma/\sqrt{N}$. (This assumes $\sigma \ll S$ so that $\mathrm{sgn}(x_t) = \mathrm{sgn}(s_t)$ at almost every time step.) Then, the total squared error is given by

$$\langle (x_t - \hat{x}_t)^2 \rangle = \langle (\sigma e_t + (S - S^*) \mathrm{sgn}(x_{t-1}))^2 \rangle \qquad (24)$$
$$= \sigma^2 + \langle (S - S^*)^2 \rangle \qquad (25)$$
$$= \sigma^2 + \sigma^2/N . \qquad (26)$$

Though both the linear and nonlinear model converge to the same "floor" in the $N \to \infty$ limit, the nonlinear model converges more quickly. For a given N, the nonlinear model (with $m = 1$) beats the best linear AR model (with any m).

One might argue that using this parametric form for the nonlinear model is unfair, since in general one does not know the nature of the model that generated the time series. However, we remark that this model is not far from a local linear approximation that uses $N/2$ nearest neighbors.[17]

---

[17] If σ is small, then the $N/2$ neighbors of a point with $x_t > 0$ will be a cloud of points which all have $x_t > 0$. A linear fit to this data will be of the form $\hat{x}_{t+1} = A + B(x_t - S)$ where $A \approx S$, in particular $A - S \sim \sigma/\sqrt{N}$, and $B \sim 1/\sqrt{N}$. It follows that the reduced squared error $\langle (S - \hat{x}_t)^2 \rangle$ will scale like σ^2/N. By contrast the linear $m = 1$ model achieves $\langle (S - \hat{x}_t)^2 \rangle \sim \sigma^2$. Already, at $m = 1$, the local linear model is better than the best global linear model, which requires an embedding dimension $m = \sqrt{N}$ and achieves a reduced squared error that scales as $\sigma^2/\sqrt{N}$.

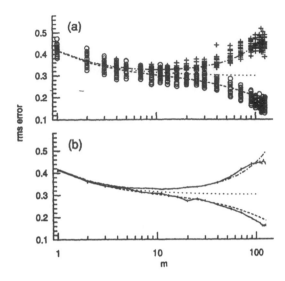

FIGURE 7   In-sample and out-of-sample errors are plotted against embedding dimension, for linear AR models fit to $N = 128$ data points of a signal-plus-noise process defined in Eq. (11), with signal amplitude $S = 1$ and noise amplitude $\sigma = 0.3$. (a) Circles denote in-sample errors and pluses are out-of-sample errors for individual trials. (b) Median (with error bars given as the standard error) of the errors shown in the above panel. In both (a) and (b), we have plotted three theoretical curves. The dotted line corresponds to the expected error $E[M_o(m)]$ of the "correct" order-m model, given in Eq. (17). The dashed line is the theoretical in-sample error of the best-fit model $E_s[M_s(m, N)]$, given in Eq. (21), and the dashed-dotted line is the theoretical out-of-sample error of the best-fit model $E_o[M_s(m, N)]$, given in Eq. (20).

In the example in Figure 7, we consider $S = 1$, $\sigma = 0.3$, and $N = 128$. The ratio of signal-to-noise power is $S^2/\sigma^2 = 11$. Figure 7 shows both the theoretical error and the results of numerical simulations for these parameters.

3.4 NONSINUSOIDAL PERIODIC SIGNALS

The difficulties associated with periodic sine signals seem to be compounded when higher harmonics are added. (Some of these extra difficulties were also addressed in Rubin (1992).) Consider a time series given by $x(t) = \sin(t) + \sin(3t) + \sigma e_t$, where $\sigma = 0.05$ and e_t is uniform noise with unit range. Even for very small σ, a linear model of this data requires four past values to predict the future, because it has to estimate the phase and amplitude of two sine waves—the phase and amplitude are *not* coded into the model itself. One finds that for the same embedding dimension m, nonlinear models fit this data better than linear models.

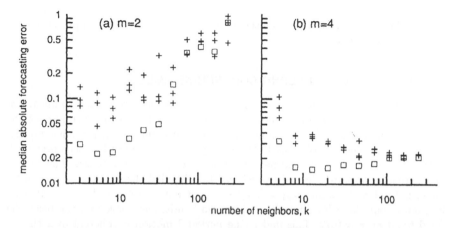

FIGURE 8   DVS plots (Casdagli & Weigend, this volume) for a time series generated by adding two sine waves and a small measure of white noise (□), as well as for some surrogate time series (+). Although this is by construction a linear time series, the plot of forecasting error versus number of neighbors in the local linear predictor indicates that nonlinear models are superior to linear models. Here, $N = 1024$ points are taken from the time series, and the embedding dimension is (a) $m = 2$, and (b) $m = 4$, which is in principle adequate for a linear model of two sine waves. Note that comparison with surrogate data also implies that the time series is nonlinear.

In particular, as seen in Figure 8, DVS plots ("deterministic versus stochastic"; see Casdagli and Weigend, this volume) of forecasting error as a function of number of neighbors in the local linear fit indicate nonlinearity in a time series, even though the system is formally speaking linear. Our intuitive explanation is that the nonlinear models are able to use information that is unavailable to the direct linear model, namely, the amplitudes and relative phase of the two sine waves. So, while the linear model requires four degrees of freedom, the nonlinear model is relatively successful with only one. We should emphasize that the DVS plot was intended as an "exploratory" method of time series analysis, and that it appears very well suited for that purpose. The ambiguity of interpretation that arises when the DVS plot is applied to data with long coherence times is a problem that is not unique to the DVS plot, but is just another artifact of the long coherence time.[18]

And in generating surrogate data sets, one again finds that nonsinusoidal periodicity is even worse than sinusoidal. As well as the usual difficulties, one has the added problem that the FT algorithm does not preserve the phase relation between the harmonics. It is this phase relation that determines the shape of the periodic waveform. ARMA modeling is even worse, because in that case, the model encodes

[18]We are tempted to say that the problem lies not in the analysis but in the data itself!

neither the phase relation between the harmonics *nor* the relative amplitudes of the sinusoidal components.

3.5 ASIDE: CHAOS AND LONG COHERENCE TIMES

Although the situation we have described so far has been restricted to linear systems with noise, we note that fully deterministic chaotic systems can also exhibit long coherence times. While this may seem at first counterintuitive, since positive Lyapunov exponents imply a finite "forgetting" time, the effect has been previously noted (Farmer et al., 1980; Farmer, 1981) in the context of the Rössler flow (1979), and is readily apparent in maps which exhibit "banded chaos." An example of the latter is the logistic map, $x_{t+1} = \lambda x_t (1 - x_t)$, at parameter $\lambda = 3.6$. The attractor is chaotic, but the orbit alternately visits two bands, one above and one below the fixed point at $x = 0.72$. This underlying period 2 motion is coherent over the full length of the trajectory.

---

4. DISCUSSION

We provide three possible interpretations of the basic source of the problems that arise when surrogates of highly coherent time series are generated. The first is technical; the second and third have more of a philosophical, almost existential flavor.

4.1 THE SURROGATE DATA GENERATOR IS FLAWED

One might argue that the inability of the FT algorithm to generate surrogates that mimic the original data indicates a flaw in the algorithm. For coherent signals, the true power spectrum contains instrumentally sharp spikes. However, when estimating the power spectrum from a finite time series, the spike is spread out over several distinct frequency bins with a very specific phase relation between them. When these phases are scrambled, and the FT is inverted, the resulting time series has a shorter coherence time.

One can imagine various *ad hoc* solutions, such as randomizing phases only for frequencies not in the vicinity of the dominant frequency. We have not investigated such modifications, and are hesitant to do so, since they are difficult to automate in a way that would be applicable to all time series. (For example, suppose one has a quasi-periodic time series, and that each of the component frequencies also has higher harmonic frequencies. Distinguishing the peaks from the broadband then becomes a nontrivial task.)

Technical problems arise with ARMA models as well, namely with stability, and the difficulty of choosing the coefficients "just right." For ARMA modeling of the sine wave, even if the correct coefficients are chosen, there is no way to assure that the surrogates will be the same amplitude as the original time series; the amplitude of linear models is coded not in the coefficients but in the initial conditions. One must not only model the coefficients of a linear model; one then must further restrict the initial conditions which are iterated. (Normally, when the coherence time is finite, the amplitude *is* specified by the coefficients, though its dependence becomes increasingly sensitive as the coherence time increases.) Rescaling so that the amplitude of the surrogates matches that of the original time series does not really solve this problem, because if the signal is composed of several sine waves, one must also find a way to maintain all their *relative* amplitudes.

An interesting possibility which we have not pursued is to model the time series not as an ARMA but directly in terms of its deterministic and nondeterministic components, i.e., Eqs. (1)–(3). The surrogates would be generated with new white-noise realizations in Eq. (3) for the indeterministic component, but the deterministic (coherent) component would be kept the same as the original.

4.2 THE TIME SERIES IS NONSTATIONARY

If the fault is not in the algorithm, then perhaps it is in the data. While *stationarity* has a clear-cut meaning for a stochastic *process*, it is a fuzzier concept when applied to a *time series*. The Lorenz flow (1963) is a stationary chaotic process, but if a time series is taken over a short enough segment, it will appear very nonstationary. For a time series, we argue that an useful operational definition of *stationarity* is that the characteristic time scales in the data are much shorter than than the length of the data set itself.

If we think of the coherence time as one of the characteristic time scales, then highly coherent time series are not stationary.

It may seem odd to characterize a sinusoidal signal as nonstationary. But one way to see why this is reasonable is to consider two sine waves whose frequencies are nearly equal. The sum of the two sine waves will exhibit a low-frequency beating as they slowly move in and out of phase with each other. If the length of the time series is shorter than the beating period, then the resulting time series will appear quite nonstationary. So, if we'd like the sum of two stationary time series to itself be a stationary time series, we cannot permit time series with long coherence times to be considered stationary.

4.3 THE TIME SERIES IS NONLINEAR

A third interpretation of the spurious identification of nonlinearity in time series with long coherence times is that the nonlinearity is not spurious at all. Typical linear processes do not produce long coherence times, because their parameters need to be precisely adjusted.[19]

We note that those clean and highly coherent sine waves that come out of signal generators in the laboratory do not arise from RLC circuits, but depend crucially on the nonlinearity of the electronics. In nature, and in the laboratory, nonlinear limit cycles are very common and very robust. Thus, one might argue that a long coherence time is in itself evidence for underlying nonlinear dynamics.

4.4 SUMMARY

Although a time series may be generated by a process that is formally linear, if it has a long coherence time, it can often fool tests for linearity and can be mistaken for nonlinear time series. In particular, it is difficult to generate surrogate data which mimics the linear properties of the process that generated the data.

In testing a time series for nonlinearity, it is a good idea to compare with surrogate data using both nonlinear *and* linear statistics. Good evidence for nonlinearity requires that the nonlinear statistics do distinguish the real and surrogate data, and that the linear statistics do not. Even more important is to plot an autocorrelation curve for the real data and make sure that the autocorrelation $A(T)$ vanishes as T gets large. If there is significant autocorrelation for T on the order of the length of the time series, then one must beware the dangers of long coherence times.

Regarding E.dat, we did see some evidence for nonlinearity, but we note that that evidence is seen only for the smoothed data, and that it is "significant" but not "substantial." Finally, because E.dat has a long coherence time, we are further inclined to discount this evidence for nonlinearity.

ACKNOWLEDGEMENTS

James Theiler acknowledges many useful conversations with Dean Prichard, Steve Ellner, and the LANL/SFI HalfBaked LunchTime NonLinear TimeSeries Working-Group. We are also pleased to express appreciation to Blake LeBaron and many other conference participants whose comments and suggestions have diffused into this manuscript, in one form or another, usually without attribution. Thanks also to Lenny Smith and Andreas Weigend for a critical reading of the manuscript.

[19] For example, to generate a sine wave with additive noise (see Section 3.2) requires an ARMA(2,2) model $x_t = a_1 x_{t-1} + a_2 x_{t-2} + \sigma e_t + b_1 e_{t-1} + b_2 e_{t-2}$, where the roots of $z^2 = a_1 z + a_2$ must lie precisely on the unit circle ($|a_1| < 2$, $a_2 = -1$; or $a_1 = -1$, $a_2 = 0$), and b_1 and b_2 must be precisely equal to $-\sigma a_1$ and $-\sigma a_2$, respectively. If the roots are outside the unit circle, then the time series will diverge to infinity exponentially; and if the roots are precisely on the unit circle but $b_i \neq -\sigma a_i$, the time series will diverge to infinity like a random walk. If the roots are inside the unit circle, then there will be a finite coherence time.

Work by James Theiler was partially supported by National Institute of Mental Health grant 1-R01-MH47184, and performed under the auspices of the Department of Energy. P. S. Linsay thanks the Office of Naval Research and the Department of Energy for support. D. M. Rubin was supported by NASA grant W-17,975.

Finally, we want to thank Neil Gershenfeld and Andreas Weigend for inviting us (James Theiler and Paul S. Linsay) to this unique and fascinating conference, even though we lacked the discipline (or was it the courage?) to actually submit an entry to the contest.

Blake LeBaron
Department of Economics, University of Wisconsin, Madison, WI 53706;
telephone: (608) 263-2516; e-mail: blakel@vms.macc.wisc.edu

Nonlinear Diagnostics and Simple Trading Rules for High-Frequency Foreign Exchange Rates

This paper performs some diagnostic tests on a tick-by-tick foreign exchange series, Data Set C of the Santa Fe Time Series Prediction and Analysis Competition. These tests find some evidence for nonlinearity both in the tick-by-tick data, and on an hourly series. Also, tests are run on the properties of the timing of the quotes in the data set. It is found that the timing and the returns may be connected in interesting ways. Finally, some trading rules are set up. These rules demonstrate the importance of considering both the economic objective of trading profits, and transactions costs in a financial forecasting setting.

1. INTRODUCTION

One of the most challenging applications of modern time series forecasting approaches is the area of financial time series. These series show dependencies which are much weaker and harder to detect than many other time series. The reason for this is obvious. Market prices are not determined independent of trader behavior, and traders will not leave obvious patterns around to be converted into trading profits. Still, the question of whether new techniques may pull hidden structures out of these well-studied series is an interesting one.

Times Series Prediction: Forecasting the Future and
Understanding the Past, Eds. A. S. Weigend and N. A. Gershenfeld, SFI Studies
in the Sciences of Complexity, Proc. Vol. XV, Addison-Wesley, 1993   **457**

Recently, many time series at high frequency have become available for many markets. This paper tests one such high-frequency series from the foreign exchange market. Such series have the potential to reveal the detailed dynamics going on at the actual level of trading, using information which is lost at the daily, weekly, and monthly horizons.[1]

This paper performs several simple diagnostics on these new series. In Section 2 it describes the series used along with some of the problems involved in using them. These are not the ideal series one would want to study. Unfortunately, such series do not exist in the foreign exchange market. Finally, it provides some basic summary statistics and tests for linear dependence on both the price series and the times between recorded bids. Section 3 continues with some nonlinear diagnostic tests. Section 4 concentrates on the relation of the time between quotes to the magnitude of returns. Section 5 deals with the issue of prediction versus actual trading profits. A very simple trading rule is simulated and the results are discussed both in terms of their economic and statistical significance. Some important points are made concerning forecasting and the economic objective of trading profits. Finally, Section 6 concludes and suggests some work for the future.

2. DATA DESCRIPTIONS

The series used in this study all come from Data Set C used in the Competition. This series contains tick-by-tick bid prices for the Swiss franc to U.S. dollar exchange rate along with time stamps for each bid. These prices are offers to buy, submitted by one foreign exchange trading bank.

The foreign exchange market operates in a slightly unusual fashion. Offers to buy and sell, "quotes," are posted electronically and appear on traders screens around the world. However, when a trade is actually made, it is confirmed over the phone, and no electronic record is posted in the quote sequence. This leaves us with only a partial picture of what is happening in the market. Offers can be observed in a detailed way at high frequency, but whether these were prices where actual trades took place is not known.[2] In this situation it is therefore difficult to tell if a quote was "serious," or if traders were just testing a new price range. Also, the

---

[1]There have been many forecasting experiments done at daily and weekly horizons. Several have found little evidence of forecastability (Diebold & Nason, 1990; Meese & Rose, 1990; Mizrach, 1989). A few papers have found some small improvements in out-of-sample forecasts. These include LeBaron (1992b) and Weigend et al. (1992).

[2]This is less of a problem for more centralized markets such as the New York Stock Exchange. For this market some series are available which include information both on quotes and trades made. However, these data sets may still miss some trades which do not get executed on the floor of the exchange.

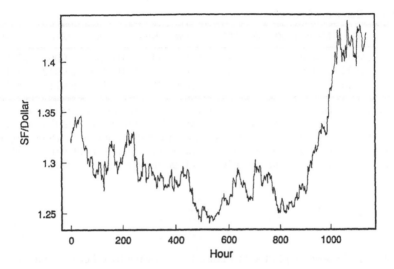

FIGURE 1   Hourly observations of SF/Dollar from August 7, 1990 to April 18, 1991.

fact that Data Set C only gives bid quotes from one bank suggests some caution in interpreting our results. For example, a trading rule tested using this series may not have actually performed as well in the real world since it may not have been able to buy or sell at some of the quotes.

There have been several published studies of high-frequency foreign exchange series. In general, they must deal with the same data limitations faced here.[3] A paper by Bollerslev and Domowitz (1990) gets around the problem of having only quote data in an interesting way. They feed the quotes into an automated trading system which automatically crosses buyers and sellers and generates a pseudo-transaction price series. Obviously, their results depend on the closeness of this algorithm to the trading process, but this is an interesting approach.

The actual time series consists of 30,000 data points containing a time stamp and a bid price. The series covers the period from August 7, 1990 to April 18, 1991. This series will be broken into two equal halves for study. Also, an hourly series is generated using the time stamps as a guide. The hourly series contains large overnight holes when no bids are posted.[4]

[3] Several of these studies are by Baillie and Bollerslev (1991); Engle, Ito, and Lin (1990); Feinstone (1987); and Goodhart and Figliuoli (1991).
[4] The foreign exchange market is one of the few that is actually open around the clock. A continuous series is possible if quotes from different markets (London, New York, Tokyo) are pasted together. See Engle, Ito, and Lin (1990) for details.

TABLE 1 Summary Statistics[1]

Series	N	Mean*1000	Std*10000	Skew	Kurt	Max	Min	% = 0
1	14935	-0.018	4.52	-0.40	11.8	0.0035	-0.0068	27.4
2	14934	0.025	4.45	0.11	8.46	0.0049	-0.0042	28.9
Hour	1140	0.663	27.4	0.700	12.60	0.020	-0.014	8.7

[1] Summary statistics for hourly and tick-by-tick returns constructed from Swiss franc bid prices from August 7, 1990 to April 18, 1991. Series 1 and 2 are the first and second halves of the entire sample of tick-by-tick data with overnight periods removed. Hour is hourly returns including the overnight periods. Skew and Kurt are skewness and kurtosis. % = 0 is the percentage of zeros in each series.

The hourly series is plotted in Figure 1, by pasting together the different days. The picture shows few discernible patterns except for a general upswing at the end of the sample. Some of this movement is probably related to movements in short-term interest rates over the period. The U.S. interest rates dropped from about 8% at the beginning of the sample to close to 6% at the end while the Swiss interest rate stayed at about 8% for the entire sample period. Interest rate movements are clearly connected to exchange rate movements in important ways. Generally, currency with relatively high rates of interest should be depreciating on world markets. The loss due to this depreciation counters the high rate of return from bonds in that currency.[5] This study will concentrate on the foreign exchange series alone since this was all that was available to forecasters in the time series competition. The question of whether additional information can be obtained from interest rate movements awaits future study.

Since the foreign exchange price series are nonstationary, "log"-ged differences will be used. Define P_t as the price at time t and $p_t = \log(P_t)$. Now let

$$r_t = p_t - p_{t-1} = \log\left(\frac{P_t}{P_{t-1}}\right).$$

This will be referred to as the returns series. Also, time differences will be considered. Define s_t as the number of seconds from the beginning of the time series to the time when P_t was recorded. Let the time difference series be

$$d_t = s_t - s_{t-1}.$$

[5] See Baillie and McMahon (1989) for a good survey of the empirical evidence on the connections between interest rates and foreign exchange.

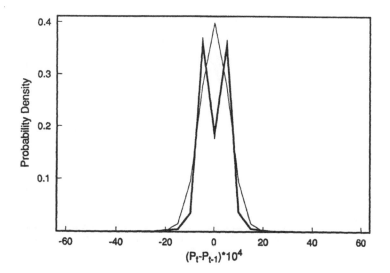

FIGURE 2   Price change empirical density (thick) and normal density (thin).

In other words d_t is the number of seconds that has passed from the previous quote until the quote at time t. The series used in this study for returns and timing have the overnight periods removed. The timing of quotes is also used to build an hourly series. This is constructed by starting at time t and moving through the series until $s_{t+\tau} - s_t > 3600$. After this, a quote is recorded, the timer is set back to zero, and the process repeats. Hourly quotes are not exactly hourly, and they are not even the closest quote to the turn of the hour. For this hourly returns series, the overnight returns remain in the series.

Table 1 presents some summary statistics for the various series. All the series show evidence of leptokurtosis, or fatter tails than a normal distribution.[6] The most important number in Table 1 is in the last column. This number presents the fraction of times the quoted price did not change. For the tick-by-tick data, this fraction is very high, 27% and 29%, for the first and second halves respectively. For the hourly series it is still a sizeable 8.7%. The fact that the returns series contain so many zeros may cause some problems for standard time series analysis for this series.

---

[6] The overnight returns were removed from the tick-by-tick series, but remain part of the hourly series.

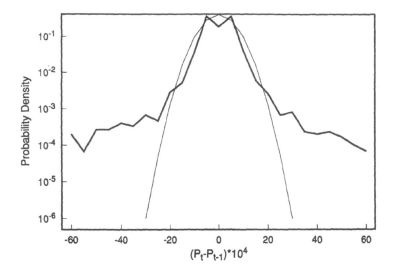

FIGURE 3   Price change empirical density (thick) and normal density (thin), semilog scale.

TABLE 2 Autocorrelations[1]

Series	ρ_1	ρ_2	ρ_3	ρ_4	ρ_5	LBP(5)
1	-0.110*	0.011	-0.008	-0.011	-0.009	186.1*
2	-0.114*	-0.005	-0.025*	-0.013	0.020	212.8*
Hour	0.026	0.022	-0.073*	-0.030	0.002	8.4
d_t (delta time)	0.283*	0.251*	0.224*	0.225*	0.196*	8424*

[1] ρ_j is the autocorrelation at lag j. LBP is the Ljung-Box-Pierce test. Numbers marked with an * are significant at the 99% confidence level using the Bartlett distribution for the autocorrelations and the $\chi^2(5)$ for the LBP test. Series 1 and 2 are the first and second halves of Data Set C (tick-by-tick data) with overnight periods removed. Hour is hourly returns including the overnight periods.

Figures 2 and 3 plot the estimated density of price changes $(P_t - P_{t-1})$ in comparison with an appropriately scaled normal distribution.[7] The price changes

[7] The normal distribution has the same mean and standard deviation as the price change series.

themselves are not continuous random variables since at this frequency the discreteness of actual trading prices becomes an issue. All the price changes are in multiples of 0.0005 SF/dollar. The bins in Figures 2 and 3 are aligned with these discrete changes. The actual distribution of price changes shows relatively fewer price changes at zero than the normal would predict, but this is difficult to tell because of discreteness. The different behavior in the tails of the two distributions is clear in Figure 3 which is plotted on a semilog scale. Here, we see the many large price changes relative to the normal distribution. This a pictorial version of the large kurtosis observed in Table 1.

Table 2 presents linear statistics, the autocorrelations at lags 1 to 5, and the Ljung-Box-Pierce test on the first five autocorrelations.[8] For the tick-by-tick data, there is strong evidence for first-order negative autocorrelation. This was also found in a paper by Goodhart and Figliuoli (1991). The correlation shows up also as a strong rejection of zero autocorrelation using the Ljung-Box-Pierce statistic. The hourly series shows little evidence for any linear dependence in any of the diagnostics.

Tests were also run on the d_t series, which shows strong evidence for linear dependence. All the correlations are large and positive. This indicates that the quote generation process is somewhat clustered. Many quotes come close together followed by periods of relative calm. This could be caused by some simple explanations such as certain periods of the day when activity is lighter (i.e., lunch). Figure 4 presents the estimated density for the d_t series. This density is plotted on semilog paper. It appears that the distribution of d_t is close to exponential. This is an interesting fact and it suggests that the number of bids occurring during a specific time interval may be Poisson.[9]

The diagnostics presented here suggest some form of linear predictability. An autoregressive model for the returns τ of order one,

$$r_t = a + br_{t-1} + \epsilon_t,$$

[8] The Ljung-Box-Pierce test for M lags is defined as

$$Q = n(n+1) \sum_{j=1}^{M} \frac{1}{n-j} \rho_j^2,$$

where n is the length of the series, M is the number of lags in use, and ρ_j is the autocorrelation at lag j. It is a general diagnostic for linear correlations. A good reference is the paper by Box and Jenkins (1976). Under a null hypothesis of white noise, this is distributed as a chi-square with M degrees of freedom.

[9] This connects this to many other physical systems such as equipment failure and radioactive decay.

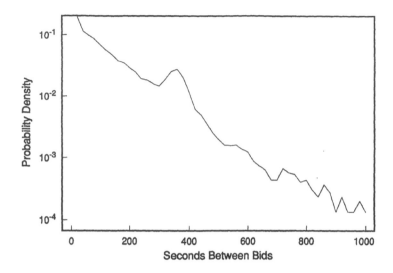

FIGURE 4 Probability density for seconds between bid prices plotted on semilog scale.

was fit to the first half of the series. The point estimates for the parameters using ordinary least squares are $a = -2*10^{-6}$ and $b = -0.110$. The conventional standard errors for a and b are $4*10^{-6}$ and 0.008, respectively. This model was used for out-of-sample forecasting in the second half of the series. The percentage improvement in mean squared error over using the unconditional mean in this section was 1.29%, a very modest improvement, but there is clearly some linear forecastability in high-frequency series.

3. NONLINEAR DIAGNOSTICS

This section examines some nonlinear diagnostics for each series. The first test used is the Brock-Dechert-Scheinkman (BDS) test developed by Brock et al. (1988). This test examines the underlying probability structure of a time series searching for any kind of dependence. It was inspired by the Grassberger-Procacia correlation dimension, but it is a test for any kind of structure in a series, linear stochastic, nonlinear stochastic, or deterministic chaos.

It is set up as follows. Let $\{x_t\}$ be a sequence of scalar observations of length T with cumulative distribution F. Define the embedded subvector as

$$x_t^m = (x_t, x_{t+1}, \ldots, x_{t+m-1}),$$

and the fraction of close pairs,

$$C_{m,T}(\epsilon) = \frac{2}{T_m(T_m - 1)} \sum_{t<s} I_\epsilon(x_t^m, x_s^m),$$

$$T_m = T - m + 1.$$

$I_\epsilon(x_t^m, x_s^m)$ is an indicator function which equals 1 if $\|x_t^m - x_s^m\| < \epsilon$, where $\|\cdot\|$ is the sup norm over the subvector.[10]

Brock, Dechert, & Scheinkman show convergence in distribution for statistics of the form,

$$C_{m,T}(\epsilon) - C_{1,T}(\epsilon)^m \text{ lim dist } N(0, \sigma_m^2(\epsilon)).$$

$$\sigma_m^2(\epsilon) = 4[K^m + 2\sum_{j=1}^{m-1} K^{m-j}C^{2j} + (N-1)^2 C^{2m} - N^2 K C^{2m-2}].$$

C and K can be consistently estimated by

$$C_{1,T}(\epsilon),$$

$$K_T(\epsilon) = \frac{1}{T_m(T_m-1)(T_m-2)} \sum_{t \neq s, t \neq r, r \neq s} I_\epsilon(x_t^m, x_s^m) I_\epsilon(x_s^m, x_r^m).$$

For the tests used here, ϵ will be set to one half the standard deviation of the series. The dimension, m, will be set to 2 and 3. The statistic is divided by the asymptotic standard deviation, so it is distributed asymptotic normal with variance 1 under the null of an independent, identically distributed x_t's. These normalized statistics will be labeled c(2) and c(3) for $m = 2$ and $m = 3$, respectively. Experiments by Brock et al. (1991) suggest that for samples above 500, and $m < 5$, the distribution is close to normality.

The BDS statistic will reject any deviation from independence. In this case, evidence from Table 2 shows that two of the series are not independent. In this situation the statistic can be used as a diagnostic on residuals for a fitted model. An AR(1) is estimated and the statistics are estimated on the fitted residuals for the tick-by-tick series. Rejection of independence here indicates structure beyond the fitted linear model. The hourly series showed little evidence for linear structure so the test is run on the raw series. Table 3 indicates nonlinear dependence for both subsamples of the tick-by-tick series, and the hourly series using the c(2) and c(3) statistics. For example, the hourly series gives a c(2) value of 2.70 which is large compared to a 99% critical value for a standard normal of 2.58. All the series show strong evidence for some kind of dependence beyond linearity using the BDS test.

---

[10] The sup norm over vectors x and y is equal to $\max_j |x_j - y_j|$, or simply the largest absolute difference over the components of the vector.

The second diagnostic run is the Tsay (1986) statistic. This is implemented as follows on a series x_t.

1. Fit an AR(2) to x_t and save the residuals as $\hat{\epsilon}_t$.
2. Regress the vector $(x_{t-1}^2, x_{t-2}x_{t-1}, x_{t-2}^2)$ on $(1, x_{t-1}, x_{t-2})$ and save the residual vectors as $\hat{y}_t$.
3. Regress $\hat{\epsilon}_t$ on $\hat{y}_t$ and run an F-Test on the significance of this regression.

This statistic projects aspects of the time series that could not be forecast using a linear model on nonlinear terms such as $x_{t-2}x_{t-1}$. It can be run using more combinations of cross terms. See Tsay (1986) or Brock et al. (1991) for examples. Tsay shows that the test statistic is asymptotically distributed following an F distribution with 3 and $T-4$ degrees of freedom, $F(3, T-4)$, in the case described above where T is the length of the series and the series follows an AR(2). The critical value at the 99% level for this is 3.78. The Tsay test shows evidence for nonlinearity in the first half of the tick data, but no evidence in the second half, or the hourly data. These differing results suggest possible nonstationarities in the series, and appear consistent with Figure 1 where the series clearly looks different from the first to the second half of the time period.

TABLE 3 Nonlinear Diagnostics[1]

Series	c(2)	c(3)	Tsay	ARCH(1)	ARCH(2)
1	23.90*	26.65*	16.62*	483*	485*
2	20.2*	23.3*	2.20	824*	950*
Hour	2.70*	3.82*	1.39	1.65	3.59

[1] c(2) and c(3) are the BDS statistic at dimensions 2 and 3 using ϵ equal to one half the sample standard deviation. Tsay is the Tsay (1986) test for nonlinearity. ARCH(1) and ARCH(2) are the Engle (1982) test for the presence of ARCH at lags 1 and 2. Numbers marked with an * are significant at the 99% confidence level using the the asymptotic normal for the c tests, the F-test for the Tsay, and χ^2 for the ARCH tests. Series 1 and 2 are the first and second halves of the entire sample of tick-by-tick data with overnight periods removed. Hour is hourly returns including the overnight periods. Series 1 and 2 are AR(1) residuals from the returnss series and Hour is the raw hourly returns series.

The final test run is a test for the presence of autoregressive conditional heteroskedasticity. This term refers to a phenomenon which is common to many financial time series. The series are very hard to forecast, but their absolute magnitudes, $|r_t|$, are predictable. In other words, market activity levels are clustered. This suggests models of the form

$$r_t = \sigma_t \epsilon_t \,,$$

where ϵ_t is an independent, identically distributed disturbance, and s_t follows a stochastic process with some linear persistence, such as

$$\sigma_t = \alpha \sigma_{t-1} + \mu_t \,,$$

where μ_t is also iid. Engle (1982) presents a parametric model, the Autoregressive Conditional Heteroskedastic (ARCH) model, for this type of effect along with a test for its presence. The test is basically a direct test of the predictability of the squares of the series from past squares. Tests of a similar style are given by McLeod and Li (1983) and applied to foreign exchange rates by Hsieh (1988). The Engle test runs the following regression,

$$x_t^2 = a_0 + \sum_{j=1}^{p} a_j x_{t-j}^2 + \epsilon_t \,.$$

The test statistic is TR^2, where T is the sample length, and

$$R^2 = 1 - \frac{\sum \hat{\epsilon}_t^2}{\sigma^2(x_t^2)},$$

where $\hat{\epsilon}_t$ are the estimated residuals for the above regression, and $\sigma^2(x_t^2)$ is the variance of x_t^2. For this test x_t will be AR(1) residuals for the tick-by-tick series and the raw returns for the hourly series. Under the null of no ARCH, TR^2 is distributed $\chi^2(p)$. The two tick-by-tick series show strong evidence rejecting a null hypothesis of no ARCH effects. However, the hourly series is unable to reject.[11]

One possible problem with some of these tests may be the existence of higher order moments. The Tsay and ARCH tests require at least fourth moments to exist, while the BDS test does not make such a requirement. Given the large kurtosis and the high fraction of zeros, some of the series may not have some of these higher moments. This may be part of the cause of the differing results across tests. There are many other types of nonlinear diagnostics that can be used. This paper cannot hope to cover them all. One missing test is the neural-net-based test used by Lee,

---

[11]See Bollerslev et al. (1992) for many examples of this in finance.

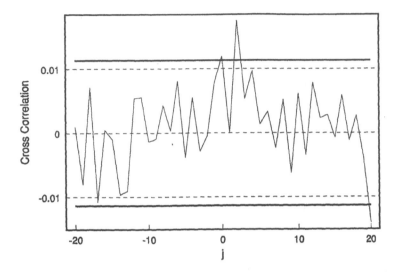

FIGURE 5   Cross correlation between return at t, r_t, and the time between bids at $t + j$, d_{t+j}. Bands are 95% confidence bands using the Bartlett asymptotic standard error.

White, and Granger (1989). This is another available diagnostic which can add to our information from the other tests.

4. DELTA TIME AND PRICE MOVEMENTS

One interesting aspect of these series is that we have information both on the prices and the timing of prices. This timing is not exogenous to the system, and depends on what is going on in the market. This section looks at some simple cross correlations to see if there is any obvious structure in these two series.

Figure 5 presents the cross correlations between returns at time t, r_t, and the time difference over the same period, d_t. This is

$$\left(\frac{1}{T}\right) \sum_{t=1}^{T} \frac{(r_t - \bar{r})(d_{t+j} - \bar{d})}{\sigma_r \sigma_d},$$

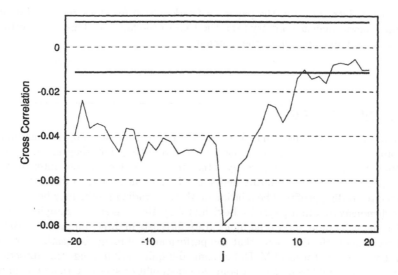

FIGURE 6   Cross-correlation absolute value of return at t, $|r_t|$, and the time between bids at $t + j$, d_{t+j}. Bands are 95% confidence bands using the Bartlett asymptotic standard error.

where $\bar{r}, \bar{d}$ are the means, and σ_r, σ_d are the standard deviations of the two series. The dark lines are the 99% Bartlett confidence bands.[12] The figure shows little linear correlation between the timing of price movements and the price movements themselves.

Figure 6 examines the cross correlations between the absolute value of the return at time t, $|r_t|$, and d_t. This picture shows some interesting connections between the timing of returns and the magnitude of the changes. There is a strong negative contemporaneous correlation ($j = 0$) which spreads over several leads and lags of d_t which is consistent with the fact that d_t itself is correlated. This negative correlation indicates that when the market is moving alot, quotes are coming spaced more closely together. This seems to make sense, but it is not clear exactly what economic theory tells us to expect here. When the speed of information flow into the market increases, it makes sense that quotes will be coming more closely spaced together. However, since the quotes are coming more rapidly, they will have fewer information bits occurring between them. If the quotes came evenly spaced in terms of the number of information events, we would see no connection between the absolute value of returns and d_t. This is related to the idea of economic versus clock

[12]The asymptotic distribution for these estimated correlations is normal with variance $1/T$. See Box and Jenkins (1976).

time which is introduced in the paper by Clark (1973) for stock returns. Finally, the figure shows some asymmetry from the lead to the lag side. This is interesting, and has no easy explanation at the moment.

---

5. TRADING RULES

This paper has, up to this point, dealt with statistical measures of forecastability and nonlinear structure. These measures are interesting, but they are not what an actual trader in a market is really interested in. Traders are not interested in Mean Square Error (MSE) improvements, but in how successfully a forecast can be converted to trading profits. The relation of MSE to trading profits may be tenuous. This is demonstrated in a paper by Levich (1981). He gives some examples of why a trader might be more interested in accurate forecasts of direction rather than the actual movement. He also shows that the performance of many forecasting services was pretty dismal in terms of MSE, but some did quite well in predicting direction.

This section will briefly look at some forecastability issues in terms of a simple dynamic trading strategy to gain some further insights into how certain aspects of forecastability convert into trading profits. This demonstrates how standard measures of predictive accuracy should be modified in an actual trading situation. The trading rules used here are very simple. They assume that the trader is able to switch back and forth between a long (buy) and short (sell) position immediately. A trader starting with a position of $1 would then end up at the end of the period with

$$\prod_{t=1}^{T}(1 + S_t r_t),$$

where S_t is the trading signal, $S_t = 1$ indicates a long position, $S_t = -1$ a short, and $S_t = 0$ would be neutral. For this example, S_t takes only these three values. Also, for a reasonable rule, S_t can only depend on information available up until time $t - 1$. The signal S_t is the final outcome of the processing of forecasts. For this example a very simple rule will be used which tries to make use of the negative correlations in the series (a reversal strategy).[13] If $r_{t-1} > \gamma$ the rule sets $S_t = 1$, a long. If $r_{t-1} < -\gamma$ the rule sets $S_t = -1$, a short. If $|r_{t-1}| \leq \gamma$ then the previous days signal is continued, $S_t = S_{t-1}$.

An important aspect to actually trading is that it is not free. There are transactions costs to be paid every time a trade is executed. For this system transactions

---

[13]Many other rules have been tested in foreign exchange markets. Most of the previous tests have used daily and weekly data. For an early example see Dooley and Schafer (1983), and for some more recent results, see LeBaron (1992b) or Taylor (1992).

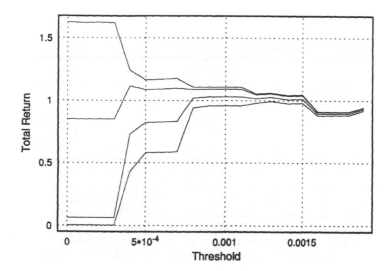

FIGURE 7 Return to reversal strategy for various thresholds and transactions costs. Transactions costs used are 0, 0.01%, 0.05%, and 0.1%. Total return is the return $1 invested at the beginning of the trading period and is decreasing in the level of transactions costs for each threshold value. The trading period covers the first half of the entire sample, a four and a half month period.

costs will be incurred every time s_t changes sign. Many authors have quoted different estimates of transactions costs.[14] General estimates for traders proportionate transactions vary from 0.01% (or 1 basis point) to 0.1%. For this study four costs will be used, 0, 0.01%, 0.05%, and 0.1%.

The trading rules will be estimated to maximize trading profits over an initial training set, the first half of the series. Then they will be applied to the second half of the series. Figure 7 gives the results for different values of the threshold parameter, γ, on the first half of the series. Total return is the total value of $1 at the end of the time period. In a world without transactions costs the return to the simple reversal strategy of over 50% for $\gamma = 0$ seems quite amazing. This number is subject to many of the cautions mentioned in the first section about the source of this series, and whether these prices were tradeable. Also, at this frequency there is

[14] One good example of some estimates is in a paper by Levich (1979). Also, see references in LeBaron's 1991 paper. The cost of trading varies across different foreign exchange series and is usually inversely related to the volume of trade in each currency. For some series, such as the German Mark, transactions costs of 0.01% are reasonable, while for the Swiss franc, 0.1% is probably a better estimate.

TABLE 4 In-Sample and Out-of-Sample Trading Rule Returns[1]

	Zero Cost	0.01%	0.05%
First Half	1.628	1.116	1.029
Trades	6497	1058	145
Second Half	1.681	1.021	0.968
Trades	6804	717	176
Fraction > Original	(0.972)	(0.092)	(0.104)

[1] Numbers are total returns, the amount of money left at
the end if a trader started with $1. The numbers in paren-
theses are the fraction of simulated AR(1)'s with trading
rule profits greater than the original series. The parameter
for the trading rule are optimized over the first half of the
sample. Trades are the number of trades that would have
been executed for a given rule during each half of the sample.

a serious question as to whether a trader could possess the dexterity to flip back and
forth between positions at a minutes notice. On the other hand, these results make
the point that measures of economic forecastability may be very different from the
usual MSE estimates. For a series with very modest improvements in MSE, and
generally low autocorrelation, large returns are observed for the dynamic strategy.

The next part of the figure brings in reality by adding transactions costs. These
two lines show much more modest returns from the dynamic strategy. The maximum
returns are 12% and 3% for the costs of 0.01% and 0.05%, respectively. For the
costs of 0.1%, the returns are not even positive. These more modest returns suggest
caution in the usefulness of this trading rule as an actual real-life strategy. It is
clear that with transactions costs, low values of γ quickly reduce trading profits by
thrashing back and forth between long and short positions.

In-sample and out-of-sample results for these rules are given in Table 4. The
returns are presented for the optimal threshold estimated in the first half of the
series. For the zero-cost case using $\gamma = 0$, the return actually went up to 68%.
However, for the two transaction-cost cases which generated positive returns in the
first half, returns fell dramatically to 2.1% and −3.2% for costs of 0.01% and 0.05%,
respectively.

Table 4 also presents the number of trades generated by each rule. The number
of trades falls as the transaction costs increase. For zero-transaction costs the rule
generates 6,497 trades, which is about 400 trades per week. This might be very

difficult for even the fastest trader to accomplish. Even for the 0.01% transaction-cost case, 1,058 trades are executed which would be about 66 trades per week. For the 0.05% transaction-cost case, a much more reasonable 145 trades yields approximately 9 trades per week. These numbers suggest an important issue to worry about along with the costs of trading. Even if a trader could get zero-transaction costs, it is probably very unlikely that anyone could actual execute the large number of trades required. Any sensible trading rule will have to generate signals which restrict trading in some way.

Trading rules can serve purposes other than as a measure of the profits available to traders. Recently, Brock et al. (1992) and LeBaron (1991) have used trading rules in combination with parametric bootstrapping to test various specifications for stock and foreign exchange price movements. These techniques are similar to the surrogate data approach used by Theiler et al. (this volume), and their origins trace to Efron's (1982) original bootstrap work. In this case an AR(1) is estimated for the exchange rate returns series and then simulated 500 times. Each time the trading rule will be applied to this simulated series. The profits of this strategy are recorded and then compared to the profits from the original series. The results of this experiment are given in the bottom of Table 4. This shows the fraction of simulations giving a value larger than that from the original. For the zero-cost case, we see that this is 97% indicating that the AR(1) generates larger trading profits than the actual series for this level of transaction costs. These fractions change dramatically as we shift to the transaction-cost cases. For example, for the 0.01% transaction-cost case, 9.2% of the simulations generated values as large as those from the actual series. This suggests something unusual in the original series which is not well represented by the AR(1). It is interesting that this result reverses when transactions costs are added.

This section makes two important points to be considered when forecasting financial time series. First, MSE forecast criterion may have little connection with the usefulness of a prediction to traders. Dramatic trading profits may come from rules which only are able to forecast a small fraction of the variability of a financial time series. Second, transactions costs are crucial in converting forecasts into a useful trading strategy. They clearly eliminated almost all of the profits seen here. When financial market forecasts are being converted into trading strategies, they must take these costs into account in their optimization procedures. Finally, trading rules in combination with bootstrap simulations can be used as a specification test. For this case the bootstrap rejects the specification of a simple AR(1) for the foreign exchange series.

6. CONCLUSION

This paper has presented several pieces of evidence on this tick-by-tick series of bid prices on the U.S. dollar/Swiss franc series. Several important facts stand out which may be useful to future forecasters. First, the high-frequency series contains many zero moves. None of the tests run here attempted to deal with this in any clever way. Second, the raw bid series show strong negative serial correlation. This is consistent with previous authors' findings on other series, and an exact explanation for this phenomenon has not yet been found. However, this is one of the most prominent features of this series. Each of the series show some evidence for nonlinearity, but some of the results are mixed across the different frequencies and time periods. The tick-by-tick data show the ARCH features common to many financial time series, but this effect is missing in the hourly series.

The timing of quotes was also studied here and appears to be a very interesting area. First, the arrival process for quotes appears to be close to Poisson because of the exponentially distributed arrival times. Second, there is a inverse relation between the time between quotes and the absolute value of the price change. This relation was not explored beyond simple linear correlation tests. This effect and its relation to the speed of new information arrivals in the market look like interesting areas for future research.

The final results show that trader objectives may yield slightly different results from MSE objectives. Low estimates of MSE improvements may hide the fact that a forecasting rule actually is useful for trading. The connection between the two measures may not be as strong as many forecasters believe. Trading brings in other problems such as transactions costs. A good trading system must take these costs into account if it is going to succeed. It must learn to hold off on trading when its forecasting signals are not strong. Building such costs into some of the forecasting methods used in this volume should be possible. Further enhancements, accounting for the riskiness of such dynamic strategies may be much more difficult.

ACKNOWLEDGMENTS

This research was partially supported by the National Science Foundation (SES-9109671) and the University of Wisconsin Graduate School. The author received helpful suggestions from Andreas Weigend, several anonymous referees, and conference participants at the NATO/ARW conference on Comparative Time Series Analysis held in Santa Fe, New Mexico during May 1992.

Holger Kantz
Physics Department, University of Wuppertal, D-5600 Wuppertal 1, Gauss-Strasse 20,
Germany

Noise Reduction by Local Reconstruction of the Dynamics

We introduce an algorithm for nonlinear noise reduction which is based
on locally linear fits to the nonlinear dynamics. We claim that it is both
robust and flexible enough to apply it to a variety of experimental time
series. In the second part of the paper, we show its performance on Data
Set A. The correlation dimension, Lyapunov exponent, and the Poincaré
map are computed before and after the noise reduction.

1. INTRODUCTION

During the last years, noise reduction methods exploiting nonlinear dynamical properties of a given data set rather than spectral properties have gained large interest.
The reason is that if the unperturbed dynamics of a system is supposed to be
chaotic, statistical methods like filtering techniques fail due to the fact that clean
chaotic signals have broadband spectra and a fast decay of the autocorrelation
function. But exactly this type of data needs noise reduction as a first step of data
analysis, since the concepts used to describe chaotic data are very sensitive to noise.

The basic assumption is that the motion of an unperturbed chaotic system,
which may have a large or even infinite number of degrees of freedom, eventually

Times Series Prediction: Forecasting the Future and
Understanding the Past, Eds. A. S. Weigend and N. A. Gershenfeld, SFI Studies
in the Sciences of Complexity, Proc. Vol. XV, Addison-Wesley, 1993

settles down on an attractor of small dimension. If this is true, not only the dimension itself, but also quanities like Lyapunov exponents (measuring the exponentially fast divergence of nearby trajectories due to chaos) or Kolmogorov-Sinai entropy (which is a measure for gain of information about the state of the system with time, when it is observed) can be computed and characterize the system (Grassberger, Schaffrath, & Schreiber, 1991).

But in many applications the values given for these quantities are quite unsafe, since no clear result of the numerical computation exists. The origin of this dilemma is twofold. On the one hand the assumption of low dimensionality may be wrong or inappropriate due to the existence of different length and time scales. Low in this sense means an attractor dimension of not more than about 5; otherwise, these quantities are hardly computable. But there is a second severe problem: all real data are contaminated with noise on small scales. All the above-mentioned quantities are defined in the limit of infinitely high resolution and generally the scaling behavior by which these quantities are defined sets in only on sufficiently small scales. Therefore, only if the noise level is small enough, the scaling behaviour can be seen. Thus one indispensable step in order to use these tools as a characterization of a system is to estimate the noise level and to do one's best to reduce it. Of course, this is only one motivation for noise reduction. It is obvious that forecasting and model building can be improved by reducing the noise on the input data.

In what follows we first want to review a quite robust scheme for noise reduction and relate it briefly to other methods in the field. In some sense this will be a review of our previous paper (Grassberger et al., 1993), where further details may be found. In the second part we apply it to the Data Set A (laser data) of the Competition. The potentials and limits of the algorithm will be illustrated by this.

2. NOISE REDUCTION FOR CHAOTIC DATA

The problem of noise reduction in some sense is ill-defined. Intuitively it is clear that one wants to subtract the noisy component of a data set, but an a priori definition of this task is lacking.

First of all, in real data it is not at all obvious what the noisy component is. The standard way in nonlinear noise reduction is to assume that the true signal is generated by a deterministic dynamics, whereas the noise is random. For nonchaotic signals, filtering techniques can be very successful, but for chaotic signals these statistical methods mostly fail. Instead, one has to make use of the assumed deterministic dynamics.

Next, what is required for the cleaned signal? Should it be strictly deterministic with respect to the assumed dynamics and close to the noisy signal (as required by Farmer and Sidorowich, 1991), or should one require an even closer new trajectory, allowing a small violation of determinism? This compromise is used by Kostelich and Yorke (1988). One can even ignore the closeness and only use the noisy data as input for a procedure to construct a deterministic orbit (Hammel, 1990). In this paper we define the new trajectory by minimizing the distance towards the original data under the constraint that locally linear fits of the dynamics are fulfilled. We shall show that this prescription has a simple geometrical interpretation.

We shall assume that our data are generated by a low-dimensional deterministic system with chaotic dynamics, superposed by some *additive* noise; i.e., one measures the noisy signal $\vec{y}_t$

$$\vec{y}_t = \vec{x}_t + \vec{\epsilon}_t \qquad \vec{x}_t = \vec{F}(\vec{x}_{t-1}), \tag{1}$$

but there exists a "clean" trajectory $\vec{x}_t$ obeying the unknown deterministic dynamics $\vec{F}$, and some noise $\vec{\epsilon}_t$ with zero average, δ-correlation and some fixed probability distribution. $\vec{F}$ may be a map or the time evolution operator of a first-order ODE.

In many realistic situations this will not be the only perturbation of the data, but, in addition, there may exist *dynamical* noise. This means that the system itself and not only the measurement is disturbed at every moment. Therefore, *a priori* no nearby "clean" trajectory exists, which leads to the shadowing problem (Grebogi et al., 1990). In this case the noise could even be an essential although nondeterministic part of the dynamics, as dynamical noise may change the stability properties of competing attractors. But this problem will not be dealt with in this paper.

In general, measurements are univariate, whereas the underlying system or the supposed chaotic attractor is more dimensional. Therefore, one has to embed the data in the space of delay coordinates of sufficiently high dimension in order to arrive again at a unique deterministic dynamics (Sauer, Yorke, & Casdagli, 1991); i.e., one replaces the state vector $\vec{x}$ by the values of a single observable at adjacent times. In such a space the problem looks

$$y_t = x_t + \epsilon_t, \qquad x_t = f(x_{t-1}, \ldots, x_{t-m}), \tag{2}$$

where the embedding dimension is m. The last equation defines the dynamics in delay coordinates. Rewriting it in an implicit form $\bar{f}(x_t, x_{t-1}, \ldots, x_{t-m}) = 0$ shows that, in an $(m+1)$-dimensional delay coordinate space, the noise-free dynamics is constrained to an m-dimensional hypersurface. For the measured values y_t, this is not true, but the extension of the cloud of data points perpendicular to this hypersurface is of the size of the noise level. Therefore, one can hope to identify this direction and to correct the y_t by simply projecting them onto the subspace spanned by the clean data. Before this, one has to reconstruct this surface from the noisy data. These are the two main ideas underlying different noise-reduction algorithms. They will be exploited in what follows.

We want to concentrate on a class of algorithms where the fit of the local dynamics and the construction of the corrected data is done in one step locally for every point of the time series. In contrast to this in the algorithms by Kostelich and Yorke (1988) and Farmer and Sidorowich (1991), and Davies' (1992b) recent paper, the construction of the corrected data is done in a second step which is different from a single projection, after having determined the underlying dynamics. In a paper by Hammel (1990) a correction scheme is presented which assumes that the underlying dynamics is given and exploits the knowledge of the stable and unstable manifolds of the system. As pointed out by Davies (1992a), the dynamics needs not to be known exactly, but can be estimated from a converging iteration scheme. We do not want to review these algorithms here. Before we finish this paragraph, we should say that all known algorithms that use a dynamics extracted from the noisy data can reduce noise by about one order of magnitude, the differences thus not being dramatic. It seems that the quality of the fit of the dynamics is one limiting factor, and the lack of hyperbolicity another.

But now let us describe in detail one efficient noise-reduction procedure. Schreiber and Grassberger (1991), Cawley and Hsu (1992a 1992b), and Sauer (1992) have designed algorithms which perform the fit of the dynamics and the correction in one step. Here we want to present a formalism which generalizes these works and which in its generality should supply an optimal performance within this class. The algorithms (Schreiber & Grassberger, 1991; Cawley & Hsu, 1992a, 1992b) and the core of Sauer's algorithm (1992) can be deduced from this as special cases. Apart from this, the work of Sauer (this volume) contains a quite interesting framework (prefiltering) to the main step, which may be quite useful for experimental data.

First we want to describe the general formalism, then we shall give a recipe how to implement the algorithm, and finally we shall discuss how the occurring parameters have to be chosen. So, let us go back to Eq. (2) and assume that we use some "overembedding"; i.e., the noise-free dynamics can be represented uniquely already in $m - Q$ dimensions. Then there exists not only one constraint $\tilde{f}(x_t, \ldots, x_{t-m})$, but Q different ones $\tilde{f}^q$. Locally, these $\tilde{f}^q$ can be linearized and then are linearly independent. Let us denote by $\mathcal{U}_n$ the ϵ-neighborhood of the delay vector $\vec{x}_n$ starting at x_n of the univariate time series. For all points contained in $\mathcal{U}_n$, the same set of linear constraints holds, if ϵ is small enough:

$$\vec{a}_q^{(n)} \cdot \vec{x}_k + b_q^{(n)} = 0 \qquad \forall \vec{x}_k \in \mathcal{U}_n, \quad q = 1, \ldots, Q. \tag{3}$$

The Q vectors $\vec{a}_q$ should be linearly independent and properly normalized, which can be achieved by requiring

$$\vec{a}_q^{(n)} \cdot \mathbf{P} \cdot \vec{a}_{q'}^{(n)} = \delta_{qq'} \tag{4}$$

with some diagonal, positive definite matrix $\mathbf{P}$, the role of which will become clear later. For the moment the reader may assume that it is the unit matrix.

Since we do not know the noise-free trajectory, we have to substitute it in Eq. (3) by the noisy one plus the (yet unknown) corrections $\vec{\theta}$:

$$\vec{x}_k = \vec{y}_k + \vec{\theta}_k. \tag{5}$$

The last requirement to arrive at a complete set of equations is that the corrections $\vec{\theta}_k$ are minimal; i.e.,

$$\sigma = \sum_{k:\vec{x}_k \in \mathcal{U}_n} \vec{\theta}_k \mathbf{P}^{-1} \vec{\theta}_k \overset{!}{=} \min . \tag{6}$$

In summary, we have to solve the following minimization problem for each delay vector $\vec{y}_n$ of the noisy time series, where we have to approximate the sums over the clean data by sums over noisy data:

$$L = \sigma + \sum_{k:\vec{y}_k \in \mathcal{U}_n} \sum_{q=1}^{Q} \mu_{kq} (\vec{a}_q^{(n)} \cdot (\vec{y}_k + \vec{\theta}_k) + b_q^{(n)}) + \sum_{q,q'=1}^{Q} \lambda_{qq'} (\vec{a}_q^{(n)} \cdot \mathbf{P} \cdot \vec{a}_{q'}^{(n)} - \delta_{qq'}) . \tag{7}$$

The variation has to be done with respect to the corrections $\vec{\theta}_k$ and the vectors $\vec{a}_q^{(n)}$ and $b_q^{(n)}$, which define the linear constraints. The derivation of the result, which is somewhat tedious, can be found in a paper by Grassberger et al. (1993). Here we want to give only the solution. Since in the end one is only interested in the corrections $\vec{\theta}$, the linear constraints $\vec{a}_q$ and b_q will not be computed explicitly. Therefore, we show only the expression for the corrections $\vec{\theta}_k$.

From the points $\vec{y}_k \in \mathcal{U}_n$, we first construct the mean

$$\eta_i^{(n)} = \frac{1}{|\mathcal{U}_n|} \sum_{k:\vec{y}_k \in \mathcal{U}_n} y_{k+i}, \qquad i = 0, 1, \ldots, m, \tag{8}$$

and the $(m+1) \times (m+1)$ covariance matrix

$$C_{ij}^{(n)} = \frac{1}{|\mathcal{U}_n|} \sum_{k:\vec{y}_k \in \mathcal{U}_n} y_{k+i} y_{k+j} - \eta_i^{(n)} \eta_j^{(n)} . \tag{9}$$

Because of the definition of $\mathbf{P}$, $R_i = (\mathbf{P}^{-1/2})_{ii}$ exists, and we define a transformed version of the covariance matrix

$$\Gamma_{ij}^{(n)} = R_i C_{ij}^{(n)} R_j. \tag{10}$$

The Q orthonormal eigenvectors of the matrix $\mathbf{\Gamma}_{(n)}$ with the smallest eigenvalues are called $\vec{e}_q^{(n)}$, $q = 1, \ldots, Q$. The projector onto the subspace spanned by these vectors is then

$$Q_{ij}^{(n)} = \sum_{q=1}^{Q} e_{q,i}^{(n)} e_{q,j}^{(n)}. \tag{11}$$

Finally the ith component of the correction $\vec{\theta}_n$ is given by

$$\theta_{n,i} = \frac{1}{R_i} \sum_{j=0}^{m} Q_{ij}^{(n)} R_j (\eta_j^{(n)} - y_{n+j}). \tag{12}$$

The result of this minimization is in perfect agreement with the previously introduced intuitive picture: The eigenvectors of a covariance matrix are the semi-axes of the ellipsoid which is the best fit to the cloud of points contributing to the matrix. Therefore, the $m + 1 - Q$ "large" eigenvectors form the subspace on which the noise–free dynamics is supposed to live due to the existence of the constraints Eq. (3). The $\vec{e}_q^{(n)}$ are orthogonal to this hyperplane and are those directions which contribute only due to the presence of noise. Eq. (12) describes the projection onto the hyperplane.

Having computed the corrections $\vec{\theta}_n$ for all delay vectors independently, we construct the corrections ϑ_k of the kth point of the scalar time series by averaging over the corresponding components of the m different $\vec{\theta}_n$,

$$\vartheta_k = \sum_{j=1}^{m} \theta_{k-j+1,j} \, . \tag{13}$$

Because of this, the final correction is not an exact projection onto the hyperplanes but only a compromise between the different possibilities. This is a direct consequence of the fact that we deal with a scalar time series in a time delay embedding. One has to repeat this procedure several times on the whole time series to find convergence; i.e., the average magnitude of the corrections and the violation of the linear constraints decreases.

In principle one can use the correction Eq. (13) with a reduced weight (Sauer, 1992), taking into account that the fit of the dynamics from noisy data is unreliable. This slows down the whole algorithm, since it takes more iterates of the whole procedure to arrive at the final correction. To our experience there is no general rule when this is useful or not; thus, we recommend to take the full weight.

Let us now discuss the use of the matrices $\mathbf{P}$ resp. $\mathbf{R}$. With $\mathbf{P}$ and $\mathbf{R}$ being unit matrices, Eq. (12) describes an orthogonal projection onto the hyperplane. From the naive point of view, this should be the best one can do, as this leads to the smallest possible difference between the original and the corrected time series. With this choice our algorithm is identical to the one by Cawley and Hsu (1992a, 1992b), and the core of Sauer's algorithm (1992). But for chaotic systems this is problematic: infinitesimally close trajectories diverge exponentially fast in time, both in the positive and negative time direction. The rate is given by the positive resp. negative Lyapunov exponents. The corresponding directions in space are transverse to each other (the stable and unstable manifolds). Therefore, in order to correct the location of x_t, one has to use information from both the past and the future. Otherwise, one is left with this exponential uncertainty, at least in one direction. Projecting a delay vector $\vec{y}_n$ orthogonally means to correct in general all components, also those which are at the border of the embedding window, the first and the last component. Because of the above-mentioned effect these corrections are less well defined in a dynamical sense than the central components. To avoid this, one simply should project in a different manner.

TABLE 1 Numerical results for the performance of this algorithm for different choices of the matrix $\mathbf{P}$ and different dynamical systems.[1]

System	T	ϵ	(a)		(b)		(c)	
			r_0	r_{dyn}	r_0	r_{dyn}	r_0	r_{dyn}
Hénon	5000	100%	1.7	3.6	1.7	4.1	1.9	8.3
		10%	2.6	4.5	3.4	7.5	4.2	10.7
		1%	2.9	5.7	3.4	5.3	3.4	6.6
	20000	1%	3.2	6.5	4.1	6.8	4.2	8.4
	100000	0.3%	2.9	4.0	4.7	6.8	5.0	8.3
Ikeda	20000	10%	3.4	—	2.2	—	3.7	—
x-coordinate		1%	1.5	—	1.7	—	2.1	—
Lorenz[a]	20000	10%	3.5	—	2.9	—	3.3	—
x-coordinate	20000	1%	2.3	—	1.8	—	2.3	—
Lorenz[b]	5000	10%	1.8	—	1.5	—	1.8	—

[a] delay $\tau=0.05$
[b] delay $\tau=0.2$

[1] White noise with strength ϵ was added to the numerically generated clean trajectories of length T. Let us denote the clean data by x_i, the noisy data by y_i, and the data after noise reduction by z_i. Then the success of the noise reduction is measured by r_0 and r_{dyn}, which are defined as follows: $r_0 = (\sum(z_i - x_i)^2)/(\sum(y_i - x_i)^2)$ and $r_{dyn} = (\sum(z_i - f(z_{i-1}, z_{i-2}))^2)/(\sum(y_i - f(y_{i-1}, y_{i-2}))^2)$, where f is the known dynamics in delay coordinates (which does not exist for the Ikeda and Lorenz system). Thus r_0 measures the success in reconstructing the unperturbed trajectory, r_{dyn} measures how well the dynamics is recovered. The Hénon map is given by $x_{t+1} = 1 - 1.4x_t^2 + 0.3x_{t-1}$, the Ikeda map reads $(x_{t+1}, y_{t+1}) = (1 + 0.9(x_t \cos\phi - y_t \sin\phi), 0.9(x_t \sin\phi + y_t \cos\phi))$, $\phi = 0.4 - 6.0/(1 + x_t^2 + y_t^2)$, where we have used the x-coordinate as a univariate time series. The Lorenz system is a system of three coupled first-order ODEs (Lorenz, 1963), where again the x-coordinate was taken. The choices of the diagonal matrix $\mathbf{P}$ are: column a, $\mathbf{P} = (m+1) \times (m+1)$ unit matrix; column b, all diagonal elements are 10^{-3} except the center one, which is 1; and column c, all diagonal elements are 1 except the first and the last, which are 10^{-3}. All other parameters where the same in each line of the table, the choices of them were made according to our considerations at the end of Section 2. Detailed information about them is given by Grassberger et al. (1993).

The other extreme, which was used by Schreiber and Grassberger (1991), would be a correction of only the central component of each delay vector. This limit can be obtained, when the matrix $\mathbf{P}$ contains only a single "1" in the middle of the diagonal, the other diagonal elements being very close to zero (say, 10^{-3}).

Now from the dynamical point of view, the correction is optimal, but if the hyperplane happens to be close to tangential to this direction, the correction may be unreasonably large.

Therefore, for chaotic systems an optimal choice of $\mathbf{P}$ consists in small entries at both ends of the diagonals and "1" for the other elements. For chaotic maps this improves the performance of this algorithm tremendously. This is not true for autonomous flows, where the Lyapunov exponent in the direction of the flow is always zero, and one should use a unit matrix. But if one constructs a Poincaré section, one should again use our nontrivial matrix. Numerical results, which support these statements, are presented in Table 1.

We want to finish this section by giving a recipe-like summary and some considerations for the choice of the parameters in this algorithm.

To apply the algorithm, one has to proceed as follows:

- Embed the time series in an $(m + 1)$-dimensional phase space using delay coordinates.
- For each embedding vector $\vec{x}_n$ find a neighborhood containing at least K points. For efficiency it is advisable to use a fast neighbor search algorithm.
- Compute the center of mass $\vec{\eta}_n$, the covariance matrix $\mathbf{C}^{(n)}$, and its transformed form $\mathbf{\Gamma}^{(n)}$.
- Determine all eigenvectors of $\mathbf{\Gamma}^{(n)}$ and compute the correction according to Eq. (12). Note that the ith element x_i of the time series appears as a component of the delay vectors $\vec{x}_{i-m}, \ldots, \vec{x}_i$. Therefore its correction is the (possibly weighted) average of the corresponding components of the corrections $\vec{\theta}_{i-m}, \ldots, \vec{\theta}_i$.
- When all corrections are computed, the time series is replaced by the corrected one and the procedure is repeated.

In the following we present some suggestions for the proper choice of the different parameters of this algorithm.

- For maps, the dimensionality m should not be much smaller than twice the dimension d_{emb}, for which the noise-free dynamics in delay coordinates becomes deterministic (and which is $\leq 2D_f + 1$, when D_f is the fractal dimension of the attractor (Sauer, Yorke, & Casdagli, 1991)). This choice is a compromise. On the one hand, one wants that a constraint of the form $\tilde{f}(x_t, \ldots, x_{t-m}) = 0$ contains all information about the future and the past of at least the central component $x_{t-m/2}$, on which it directly depends due to the forward time evolution $x_t = f(x_{t-1}, \ldots, x_{t-d_{\mathrm{emb}}})$ and the backward time evolution $x_t = f_{\mathrm{back}}(x_{t+1}, \ldots, x_{t+d_{\mathrm{emb}}})$. On the other hand, coordinates which are more than d_{emb} time steps apart are coupled only weakly due to the chaotic dynamics and do not supply useful information to reduce the noise. Not only for numerical convenience, one is interested in having m as small as possible, but the diameter of a neighborhood has to be increased more and more to find a given number of neighboring points in higher dimensions, which deteriorates the linear fit of the dynamics. For flows, especially if the sampling rate is high, m can be much larger because of redundancy of the information, or one should compress the data as suggested by Sauer (1992).

■ The number of constraints can be either kept fixed or increased during the iterations as the separation between signal and noise gets clearer. The remaining subspace has to have a dimension as small as possible but big enough not to destroy the attractor; i.e., it should be about the supposed fractal dimension of the attractor.

■ The choice of the size of the neighborhood has to be a compromise between the need for good statistics (large number of points) and for approximate linearity of the subspace (small radius). It is optimal when these two errors are of equal size. In our 1993 paper, we show that this is the case if the number of neighbors is

$$K \approx [\langle \epsilon \rangle T^{2/D}]^{2D/(4+D)}. \qquad (14)$$

Here T is the length of the time series, D the fractal dimension of the attractor, and $\langle \epsilon \rangle$ the relative noise level.

■ It is obvious that the linear approximation in Eq. (3) induces some error, which causes a deformation of the attractor after several iterations. Therefore, there exists a crossover, when these deformations become of the order of the remaining noise. At latest, at this point one has to stop the iteration. An improvement was suggested by Sauer (1992): Try to get hold of the curvature effects by imposing that the local average of all corrections is zero. But even with this one should shrink the radii of the neighborhoods with increasing number of iterations taking into account the reduction of the noise.

A suboptimal choice of these parameters in not fatal, as our experiences with time series with known clean trajectories show. Generally, the corrections are simply too small. If one is uncertain about the dimension of the unperturbed attractor, it is better to overestimate m. Setting m too small reduces the performance but does not destroy the data, as long as m is larger or equal the minimal embedding dimension derived from Takens' theorem. Nevertheless, for experimental data it is advisable to compare the results with different sets of parameters and to analyse the corrections made. The decay of their autocorrelation function as well as the cross correlation between the corrections and the time series indicate whether one really corrects the noisy component of the data.

3. APPLICATION TO A REAL TIME SERIES: DATA SET A

In this section we want to apply the algorithm presented in Section 2 to a real time series. First, we want to recall the requirements for a time series for the applicability of our noise-reduction scheme: low dimension of the underlying deterministic dynamics, stationarity, additive noise, and a sufficient length of the time series. Furthermore, in order to be superior to a linear filter, the dynamics should be chaotic.

This can be assumed to be valid for Data Set A, the Lorenz-like laser data. The data are taken with a resolution of 8 bits and represented by integers between 0 and 255. The main source of noise on these data seems to be the digitalization, which leads to a noise level of 0.5%. Some of the highest peaks are obviously cut off,

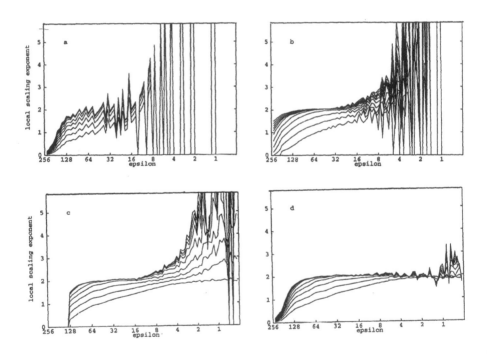

FIGURE 1   The local scaling exponents $d_2(\epsilon)$ computed from the correlation integral $I(\epsilon, d)$ in $d = 2$- to 12-dimensional delay coordinates (curves from below to top). Panel (a) is obtained using the original noisy data and a sup-norm to compute the distance; panel (b): the noisy data, using a Euclidian norm; panel (c): the noisy data plus a half-unit white noise, again sup-norm; panel (d): the data after noise reduction, sup-norm.

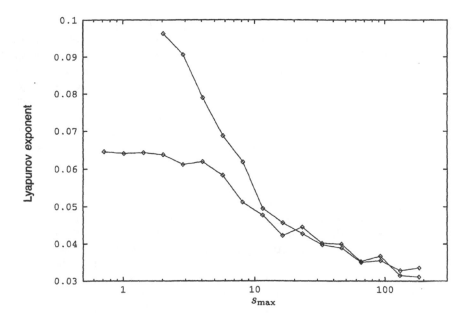

FIGURE 2 The maximal Lyapunov exponent computed with the method of Wolf et al. for the data before (upper curve) and after noise reduction (lower curve) as function of $s_{\max}$.

as the distribution of occurrences of each value decreases to values between 0 and 2 towards 255, whereas in 255 itself there are 12 entries; i.e., all higher values are shifted onto 255. The fractal dimension can be estimated to about 2 (see Figure 2), but the scaling region is rather small, as the small scales are affected by the noise.

If one assumes $D_f \approx 2$, one needs at most a $(d_{\mathrm{emb}} = 5)$-dimensional embedding for a deterministic representation of the dynamics. According to our considerations of the last section, we have chosen $m = 8$. As the Lyapunov exponent is so very small (see below), there is no problem in finding enough neighbors in $m + 1 = 9$ dimensions. The optimal number of neighbors we require is estimated by Eq. (14) to roughly 30, and during the iterations we reduce it to 20. We choose $Q = 7$ constraints such that we project locally onto a two-dimensional subspace. As the data describe a flow, we have chosen $\mathbf{P}$ to be the unit matrix. From the magnitude of the decrease of the average correction, we decided to stop the iteration of the reduction algorithm after the seventh time, using the extended data set of 10,000 points. We want to point out that the choice of these values is not crucial for the results. We obtain nearly the same resulting trajectory, if we choose $m = 10$ or 12 and correspondingly larger Q or iterate the whole scheme two times more.

The following effects of the noise reduction are visible:

■ Dimension estimates: To determine the dimension of the attractor, we use the algorithm by Grassberger and Procaccia (1983c). For various embedding dimensions d, one has to compute the correlation sum $I(\epsilon, d)$, which is the number of all pairs of points in d-dimensional delay coordinates, whose distance is less than ϵ. If d is large enough, for noise-free data one expects to see a scaling range, where $I(\epsilon, d) \propto \epsilon^{D_2}$ independent of d, where D_2 is the correlation dimension. Thus in Figure 1 we show $\log(I(\epsilon, d)) - \log(I(\epsilon', d))/(\log(\epsilon) - \log(\epsilon'))$, where $\log_2(\epsilon) - \log_2(\epsilon') = 1/10$, which is a local scaling exponent $d_2(\epsilon)$. In Figure 1(a) the result for the original time series is shown, where the distance is computed using a maximum norm. Obviously in this plot (as also in Figure 1(b), (c), and (d)), one has to disregard the lowest four curves, for which $d = 2, 3, 4, 5$, as the embedding dimension is too small. But also for the other curves, no scaling region with $d_2(\epsilon) \approx$ const. is visible, but the values of $d_2(\epsilon)$ are more or less independent of d. This indicates that there might be some scaling behavior obscured by the digitalization of the data, which also causes the intervals of $d_2 = 0$ at small scales. Averaging over the local structures of the curves, we find $D_2 = 2.04 \pm .5$, if we restrict the average to $\epsilon \in [90, 16]$, whereas we find $D_2 = 2.9 \pm 7.9$ for $\epsilon \in [16, 1]$ (Note that in Figure 1(a)–(d) the fluctuations around a local mean are anticorrelated). Obviously, for digitalized data the maximum norm is not appropriate to estimate the dimension, as Figure 1(b) shows: with the Euclidian norm the curves smoothen and a nice scaling behavior can at least be seen in the interval $\epsilon \in [64 : 32]$ ($D_2 = 2.02 \pm 0.01$). Whereas for $\epsilon > 64$, one sees the effects of the finite size of the attractor, the steady increase of $d_2(\epsilon)$ and the slight divergence of the different lines for $\epsilon < 32$ indicates the presence of another component of the signal with higher dimensionality, which, as there is no further scaling region, is noise. Instead of using a Euclidian norm, we also have simply added a half-unit white noise to the original data to smear out the gaps in Figure 1(a). The resulting plot is shown in Figure 1(c). Although a scaling region is less clear than in Figure 1(b), for $\epsilon \in [32, 16]$ we find $D_2 = 2.05 \pm 0.01$ ($\epsilon \in [90, 8] : D_2 = 2.04 \pm 0.09$). Summarizing, on a length scale of $\epsilon \approx 32$ the data live on an attractor with dimension close to 2, but they are perturbed on smaller length scales by another process. Finally, in Figure 1(d) we show the local scaling exponents for the data after noise reduction using the maximum norm again. First of all, the curves are very smooth, as the digitalization is removed by this process. But, furthermore and in contrast to Figure 1(c), they are also very flat, giving rise to a much larger scaling region than before. Performing a linear regression for the values of $\log(I(\epsilon, d))$ versus $\log \epsilon$ in the interval $\epsilon \in [64, 2]$ we find $D_2 = 2.02 \pm 0.004$. One might suspect that projecting the noisy data onto a two-dimensional subspace compels the correlation dimension to be two. But this is not true since the hyperplanes are determined individually for each point of the time series and thus generally do not form a global surface. Furthermore, one has to average over the different corrections $\tilde{\theta}$ as described by Eq. (13), such that a resulting dimension 2 surface is a nontrivial effect. The strongest argument is that replacing the data by local averages over their neighborhoods, as it was done by Schreiber (1992), and which corresponds to a projection onto a zero-dimensional subspace, still conserves the main properties of the data and is a zeroth-order noise reduction.

■ The maximal Lyapunov exponent: Since our first goal is not an analysis of the Data Set A but a demonstration of the noise-reduction algorithm, we use here a method by Wolf et al. (1985) to compute Lyapunov exponents which is more sensitive to noise than others (Sano & Sawada, 1985; Eckmann et al., 1986; Holzfuss & Parlitz, 1991). Following again Takens' embedding theorem, we use a five-dimensional embedding space. For the first point of the time series, one looks for a reference point which is a point of the time series very close to the initial point, but at a different time. Then one computes the difference vectors of the sucessive images of these two points. As soon as its length exceeds a bound s_{max}, another point on the trajectory is searched for, which is close to the actual point and whose difference vector points in almost the same direction as the old one. The average increase of these difference vectors is determined by the maximal Lyapunov exponent. The important parameter is the distance s_{max} at which one looks for a new reference point. For $s_{max} > 10$, we find (within some errors) the same exponent for both the data before and after noise reduction (see Figure 2). But this is not the true Lyapunov exponent, as it decreases s_{max}. The values are more and more affected by the finite size of the attractor and thus too small. Only for smaller scales the exponent computed for the cleaned trajectory shows a plateau (below $s_{max} \approx 5$). The value we estimate from this is $\lambda_{max} \approx 0.06$ per time step (base e). This corresponds to about 0.085 bits per time step. In contrast to this the exponents of the noisy trajectory diverge on small scales.

The differences between the two curves can be understood as follows: The parts between subsequent changes of the reference trajectory are pretty long for $s_{max} \approx 10$ (about 20 time steps) such that the effect of noise is averaged out. Only when a new reference trajectory is chosen, there is an error of the order of the noise level for the initial distance between the two trajectories. In contrast to this, for s_{max} approaching the noise level, this error becomes more and more important for the noisy trajectory. Obviously, for s_{max} being of the size of the noise level, the distances fluctuate with the noise. The probability that the reference trajectory is more that s_{max} apart already in the next time step goes to one in this limit and a new reference trajectory is chosen after nearly every time step. In this case we find the maximal value $\lambda_{noise} = 0.097$. For even smaller s_{max} (which is not shown in Figure 2), the result trivially becomes independent of s_{max} and $\lambda(s_{max})$ shows a plateau, since in the numerical computation nothing can change any more. This is a consequence of both the finiteness of the sampling rate of the data and the digitalization. The same saturation effect also exists for the cleaned data (and again is discarded in Figure 2), but the times between choosing a new reference point in the small s_{max} limit are larger than for the noisy data; therefore, it sets in on smaller scales.

Although the noise level is so small for the noisy data a true plateau does not exist. For the cleaned data the scaling range begins below five, where the noisy data are already corrupted by noise. Thus, at least for the Wolf algorithm, noise reduction is an indispensable step to compute Lyapunov exponents.

■ The Poincaré map: To reduce the amount of data and their dimensionality, one can apply the technique of the Poincaré surface of section. One determines all the intersections of the trajectory with a given hyperplane of the phase space, where the trajectory has a given direction of transition (say, from below to above). The mapping from one intersection to the following defines the Poincaré map. A hyperplane is defined by a single constraint. In delay coordinates the simplest and most clear surface in the phase space is the one with velocity zero, i.e., the extrema of the signal. To select a definite direction of the transition through this surface, we choose the minima. They are much better determined than the maxima since the relative sampling rate is higher and there is no cutoff. To determine the points of velocity zero even better, we compute the parameters of a parabula passing through a minimum of the signal and its two neighbors and take its minimum as the point of the Poincaré map. In two-

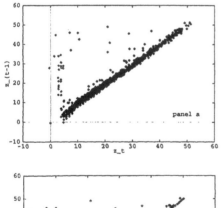

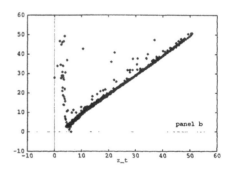

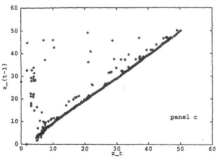

FIGURE 3   The Poincaré map for the surface $\dot{x} = 0$: panel (a) shows the points obtained from the original time series, in panel (b) the points of the corrected time series, and panel (c) shows the result of a noise reduction applied to the points of panel (a). For the latter we used $m = 4$, $Q = 4$ and neighborhoods of at least 28 points. The diagonal elements of **P** were $P_{11} = P_{55} = 0.05$, $P_{22} = P_{33} = P_{44} = 1$.

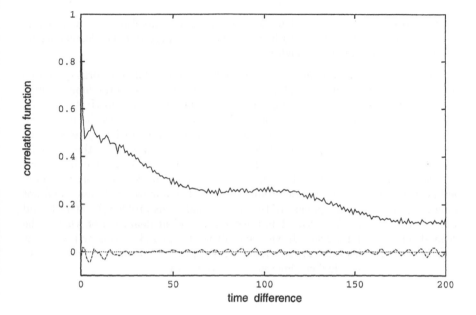

FIGURE 4   Upper curve: part of the autocorrelation function of the total correction,
i.e., of the differences of the points of the time series before and after noise reduction.
Lower curve: The cross correlation between the corrections and the cleaned signal.

dimensional delay coordinates, these points are shown in Figure 3(a) for the
noisy and Figure 3(b) for the data extracted from the cleaned signal. Finally,
in Figure 3(c) we show the effect of noise reduction using the data of Figure
3(a), i.e., cleaning the Poincarè map instead of the flow. In both cases the zig-
zag along the diagonal is reduced a lot. This is remarkable especially for Figure
3(b), since the points there are about seven time steps apart and thus are much
less correlated by the noise reduction than the points in Figure 3(c).

■   Analysis of the noise: We want to finish this section by analyzing the corrections, i.e., the difference between the noisy and the cleaned time series. The average correction is zero, since we have subtracted any trend as it was suggested
by SauerSauerT. (1992). The probability distribution is close to a Gaussian
and has unit variance. Therefore, on the average we correct more than only
the discretization effect. The normalized autocorrelation function consists of a
very narrow peak around zero with quite a high background, which decreases
to 30% of the peak's hight at a time delay of 50 units (Figure 4). Thus what we
subtract is not δ-correlated white noise. When reducing noise on synthetic data
(e.g., the Hénon map), we indeed found a very clear δ-peak and nothing else.
A referee of this article pointed out that noise in a real experiment may well

be colored to some extent. Furthermore, in Figure 4 we also show the properly normalized cross correlations between the corrections and the cleaned signal. They are very small, although being structured.

We think that on this time series "nonlinear" noise reduction works quite well and improves the computation of quantities which are defined in the limit of small scales. As the laser system can be modeled by the Lorenz equations (Hübner et al., this volume), this finding is no surprise, although it demonstrates the high quality of the experimental setup, as the time series is highly stationary, really low-dimensional, and very long. Nevertheless, our results also state that there is more measurement noise than simply the discretization, and the estimated noise level from the corrections is one unit. Nevertheless, since the data stem from a real experiment, it is not possible to determine that one has removed only noise and not a higher-dimensional component of the clean signal. This problem is intrinsic and present in all known noise-reduction schemes, as a clear definition of what is the clean signal cannot be given. In this and all works quoted in this article, a low dimension of the clean dynamics is assumed, such that what one really does is to remove components of higher dimension from the data.

ACKNOWLEDGMENTS

First of all, I want to thank my coauthors (Grassberger et al., 1993) and especially R. Hegger and T. Schreiber, with whom I had many fruitful discussions about the Data Set A. Furthermore, I want to acknowledge the stimulating atmosphere during the SFI conference from which I got many impulses for future work. This work was supported by Deutsche Forschungsgemeinschaft, SFB 237.

IV. Practice and Promise

William H. Press and George B. Rybicki
Harvard-Smithsonian Center for Astrophysics, 60 Garden Street, Cambridge, MA 02138

Large-Scale Linear Methods for Interpolation, Realization, and Reconstruction of Noisy, Irregularly Sampled Data

Various statistical procedures related to linear prediction and optimal filtering are developed for general, irregularly sampled, data sets. The data set may be a function of time, a spatial sample, or an unordered set. In the case of time series, the underlying process may be low-frequency divergent (weakly nonstationary). Explicit formulas are given for (i) maximum likelihood reconstruction (interpolation) with estimation of uncertainties, (ii) reconstruction by unbiased estimators (Gauss-Markov), (iii) unconstrained Monte Carlo realization of the underlying process, (iv) Monte Carlo realizations constrained by measured data, and (v) simultaneous reconstruction and determination of unknown linear parameters. An alternative title for this paper might be "How To Play Connect-the-Dots in a Noisy, Fractal World."

Times Series Prediction: Forecasting the Future and
Understanding the Past, Eds. A. S. Weigend and N. A. Gershenfeld, SFI Studies
in the Sciences of Complexity, Proc. Vol. XV, Addison-Wesley, 1993   **493**

1. INTRODUCTION

This paper, which is closely based on some work published in the astronomical literature (Rybicki & Press, 1992; reprinted here with permission from *The Astrophysical Journal*), is a somewhat "contrarian" contribution to the present volume, since it deals entirely with *linear* methods. Many astronomical data sets, as well as data sets in other observational (as opposed to experimental) sciences, are extremely small, and highly irregular in their sample density. Yet, one wants, on occasion, to draw rather grand conclusions from such data sets!

The issues that arise in doing so are often less concerned with elucidating (or forecasting) the underlying dynamics of a data set, which is the principal focus of the rest of this book, than with simply establishing, in a systematic statistical way, that "apparent" features in the data are in fact there, and, possibly, with the estimation of a small number of parameters from the data set.

In our previous work, for example, Press, Rybicki, and Hewitt (1992a,b; hereafter Papers I and II), we developed a method for determining whether two sets of irregularly sampled data are shifted measurements of the same underlying function and, if they are, for measuring the time lag between them. Papers I and II were focussed on the application to a particular object, gravitational lens 0957+561, and on two particular data sets, the optical data of Vanderriest et al. (1989) and the radio data of Lehár et al. (1992). Those papers thus did not develop the method's more general aspects in any detail.

This paper is more general in scope. We here discuss, in a unified way, a number of related statistical procedures that can be applied to noisy, irregularly sampled data, including data whose underlying physical process is low-frequency divergent (as, e.g., a random walk process). In fact, it is not necessary that the data be a time series or other ordered one-dimensional set. The methods described here apply equally well to spatial data (for example, image reconstruction) or to data measured on an unordered set.

We are interested, primarily, in the problem of estimating the true values of the underlying physical process at points ("times," say) which may or may not be associated with measurements. The methods we discuss all involve solving sets of linear equations over all the data. As such, they are closely related to, or generalizations of, a variety of standard techniques in the literature. Estimation at a measured point is usually called Wiener filtering, or optimal filtering. Estimation at a nonmeasured point is often called linear prediction, or least-squares prediction, or minimum variance estimation. We will see that the issue of statistical bias is an important one; the unbiased case that we discuss is usually called Gauss-Markov estimation (see, e.g., Drygas, 1970; Malley, 1986). We will also be interested in the issue of reconstructing, given a set of measurements, not just the "best" interpolation, but also "typical" realizations, whose statistical properties are as close as possible to the underlying process, which can then be explored with Monte Carlo simulations. This application relates closely to the so-called "missing data" or "data dropout" problem.

To the extent that the methods discussed are standard ones (see, e.g, Rao, 1973; Lewis & Odell, 1971), this paper should be viewed as primarily pedagogical. However, we have found that the existing literature is in practice so fragmented into special cases, and lacking in unified discussion of the above-named techniques, as to be almost irrelevant to the emphasis of this paper. While we do not claim anything in this paper as truly "new," neither are we able, in many cases, to recommend anything in the literature as worth consulting (at least by astrophysicists).

We hope also to be clear about where, along the way, certain statistical assumptions need (or don't need) to be made. Do we assume that a process is stationary? Do we assume that it is Gaussian? At what point is the Bayesian bargain entered into?

In the interests of a practical emphasis, we will illustrate the discussion by application to a particular (artificial) data set. Unfortunately, none of the data sets from the Competition described in this volume are suitable for illustrating

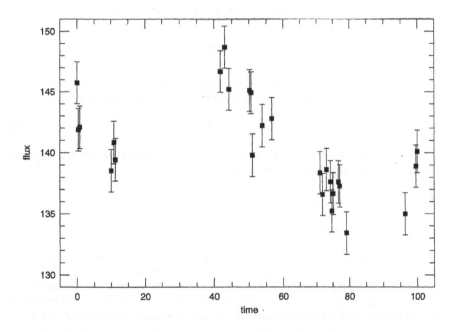

FIGURE 1 Artificial data set used as the example throughout this paper. The data is irregularly sampled, with gaps of various sizes. The process underlying this data is low-frequency divergent, something between $1/f$ noise and random walk.

the specific issues that we want to raise. So, with apologies to the conference organizers, we must introduce a new one. Figure 1 shows a set of 26 irregularly sampled data points, and their error bars. In a nutshell, the question to be answered in this paper is: What is the function that underlies the measurements in Figure 1?

2. WIENER FILTERING OR LINEAR PREDICTION

Let y_i, $i = 1, \ldots, M$, be a set of M measurements, each of which is equal to the sum of an underlying signal s_i and a noise value n_i. It is convenient to represent the quantities as column vectors of length M, so

$$\mathbf{y} = \mathbf{s} + \mathbf{n}. \tag{1}$$

Suppose we want to estimate the value of the signal at some particular point which may or may not be one of the M points already measured. Call the true value there s_* (asterisk can be thought of as taking a value in the range $1, \ldots, M$, or else having a new value $M + 1$). If our estimate is to be linear in the measured y_i's, then we can write

$$s_* = \sum_{i=1}^{M} d_{*i} y_i + x_* \tag{2}$$

where the d_{*i}'s are coefficients that depend on asterisk. The summation is the linear estimate of s_* and x_* is the *discrepancy* of the estimate, or equivalently

$$s_* = \mathbf{d}_*^T \mathbf{y} + x_*. \tag{3}$$

We obtain equations for $\mathbf{d}_*$ by minimizing the discrepancy in the least-squares sense, i.e., minimizing with respect to $\mathbf{d}_*$

$$\begin{aligned}
\langle x_*^2 \rangle &= \left\langle (\mathbf{d}_*^T \mathbf{y} - s_*)^2 \right\rangle \\
&= \mathbf{d}_*^T \left(\langle \mathbf{s}\mathbf{s}^T \rangle + \langle \mathbf{n}\mathbf{n}^T \rangle \right) \mathbf{d}_* - 2 \langle s_* \mathbf{s}^T \rangle \mathbf{d}_* + \langle s_*^2 \rangle.
\end{aligned} \tag{4}$$

Here angle brackets denote statistical ensemble averages, and we have assumed that signal and noise are uncorrelated, $\langle s_i n_j \rangle = 0$. We suppose that we know enough about the underlying process that generates $\mathbf{s}$ and $\mathbf{n}$ so that the nonvanishing averages can be considered known. (We will discuss this more below.) If we define two symmetric, positive definite correlation matrices and a "correlation vector,"

$$\mathbf{S} \equiv \langle \mathbf{s}\mathbf{s}^T \rangle, \quad \mathbf{N} \equiv \langle \mathbf{n}\mathbf{n}^T \rangle, \quad \mathbf{S}_* \equiv \langle (s_*)\mathbf{s} \rangle, \tag{5}$$

then Eq. (4) can be written ("completing the square") as

$$\langle x_*^2 \rangle = (\mathbf{d}_* - \widehat{\mathbf{d}_*})^T [\mathbf{S} + \mathbf{N}](\mathbf{d}_* - \widehat{\mathbf{d}_*})$$
$$- \mathbf{S}_*^T [\mathbf{S} + \mathbf{N}]^{-1} \mathbf{S}_* + \langle s_*^2 \rangle \qquad (6)$$

where

$$\widehat{\mathbf{d}_*} = [\mathbf{S} + \mathbf{N}]^{-1} \mathbf{S}_*. \qquad (7)$$

Since $\mathbf{S} + \mathbf{N}$ is positive definite, the value of $\mathbf{d}_*$ that minimizes Eq. (6) is seen to be $\mathbf{d}_* = \widehat{\mathbf{d}_*}$. The minimum variance estimate $\widehat{s_*}$ for s_* (Eq. (3)) is then

$$\widehat{s_*} = \widehat{\mathbf{d}_*}^T \mathbf{y} = \mathbf{S}_*^T [\mathbf{S} + \mathbf{N}]^{-1} \mathbf{y} \qquad (8)$$

and the mean square residual, i.e., the variance of s_* about $\widehat{s_*}$, is

$$\left\langle (s_* - \widehat{s_*})^2 \right\rangle = \langle x_*^2 \rangle_{\min} = \langle s_*^2 \rangle - \mathbf{S}_*^T [\mathbf{S} + \mathbf{N}]^{-1} \mathbf{S}_*. \qquad (9)$$

Notice that the correlation quantities $\mathbf{S}$, $\mathbf{N}$, and $\mathbf{S}_*$ do not, in principle, depend on the observed data values $\mathbf{y}$, but only on the locations ("times") of the values, and on the underlying process. For example, if the data is an irregularly sampled time series, y_i observed at time t_i, then a typical component $S_{ij} = \langle s_i s_j \rangle = \langle s(t_i) s(t_j) \rangle$ depends only on t_i and t_j, but not on a particular realization of s_i or s_j, since s is the quantity that is ensemble averaged by the angle brackets.

Note also that, in most practical cases, the noise values are uncorrelated, so that the noise correlation matrix $\mathbf{N}$ is diagonal $\mathbf{N} = \mathrm{diag}(\langle n_i^2 \rangle)$. However, our formulation will allow for the more general case of self-correlated noise, with a general correlation matrix. Inclusion of a known (or estimatable) signal-noise correlation can also be done by a simple extension of the present theory, although we shall not give the details here.

One sees in Eq. (8) the connection with Wiener filtering as it is more conventionally presented (see, e.g., Press et al., 1986), generally in the context of a stationary process with a regularly or continuously sampled time series, analyzed in the Fourier domain. In that case, the formula usually given for the optimal estimator is

$$\widehat{s(\omega)} = \frac{\langle |s(\omega)|^2 \rangle}{\langle |s(\omega)|^2 \rangle + \langle |n(\omega)|^2 \rangle} y(\omega). \qquad (10)$$

This is exactly Eq. (8) in the special case that the matrices $\mathbf{S}$ and $\mathbf{N}$ are both diagonal, as is indeed the case for the special assumptions made, since correlation matrices of stationary processes on equally spaced grids are diagonal in a Fourier basis.

In practice, one is often in the position of not having independent statistical information about the process $s(t)$ to estimate $\mathbf{S}$ *a priori*. In that case, one may choose to make the additional assumption of *stationarity*, so that S_{ij} is a function only of the time difference $t_i - t_j$, and not of t_i and t_j separately. Then, every (i, j) pair of data points furnishes a one-point estimate of the correlation function $S(t_i - t_j)$, and one can implement various fitting or smoothing procedures to estimate S and S_*, the latter depending on the time differences $t_* - t_i$ (see Paper I; Edelson & Krolik, 1988; Hjellming & Narayan, 1986; for discussion of the two-dimensional case, see Cressie, 1991). Throughout this paper, the only reason to assume stationarity (in the present sense of time-translation invariance) is if there is no other way to estimate the required correlation quantities.

Indeed, there is an algebraic justification for our somewhat casual attitude about how S is estimated: Suppose that small errors in S lead one to use slightly wrong values $\widehat{d_*}$ in the estimation equation (8). Then, the variance of the estimate for s_* is always larger than Eq. (9). In particular, Eq. (6) can be rewritten as

$$\langle x_*^2 \rangle = \langle x_*^2 \rangle_{\min} + (\widehat{d_*} - d_*)^T [S + N](\widehat{d_*} - d_*). \tag{11}$$

Noting that S and N are both positive definite, one sees that the change in the variance is a pure quadratic form. Thus, first-order errors in N or S, leading to first order errors in $\widehat{d_*}$, lead only to second-order increases in the variance of s_* about $\widehat{s_*}$.

One additional statistical quantity, χ^2, is defined by

$$\chi^2 \equiv y^T [S + N]^{-1} y. \tag{12}$$

Since we have not made any assumption that the underlying processes are Gaussian, the only knowable property of χ^2 is its expectation value, which is the number of data points M:

$$\begin{aligned}
\langle \chi^2 \rangle = \langle \mathrm{tr}(\chi^2) \rangle = \langle \mathrm{tr}([S + N]^{-1} yy^T) \rangle = \mathrm{tr}([S + N]^{-1} \langle yy^T \rangle) \\
= \mathrm{tr}([S + N]^{-1}[S + N]) = M.
\end{aligned} \tag{13}$$

Here we have used the facts that the trace of a scalar is itself, while the trace of a matrix product is invariant under cyclic permutation of its factors. Without further assumptions we cannot say anything about the distribution of χ^2 around its mean.

Equations (8) and (9) are the principal results of this section, giving a prescription for estimating the underlying value s_*, and an uncertainty of the estimate, at any point. Figure 2 shows the application of these formulas to the data of Figure 1. The estimates $\widehat{s_*}$ are shown as the solid curve, while the 1-σ standard deviations (square roots of Eq. (9)) are shown as the grey "snake."

The estimates of N and S that underlly Figure 2 are obtained as follows: N is taken as diagonal, with components equal to the square of the given error bars, n_i^2. For the component S_{ij} of S, we write

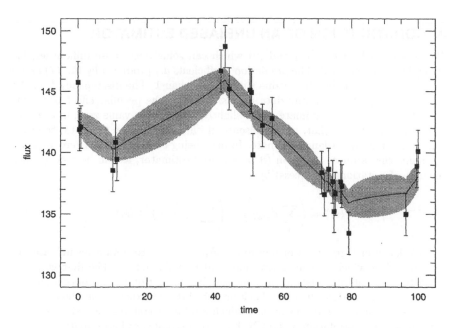

FIGURE 2 Results of applying Eqs. (9) and (10) to the data of Figure 1. The minimum variance prediction for the underlying process is the solid curve. The "snake" indicates 1-σ error bars on the prediction. Notice that the snake narrows where the density of data is highest, and widens (at a rate determined by the data's correlation function) in data gaps.

$$S_{ij} = \langle s_i s_j \rangle = \langle y_i y_j \rangle - n_i^2 \delta_{ij} = \langle y_i^2 \rangle E_i E_j - \frac{1}{2} \langle (y_i - y_j)^2 \rangle - n_i^2 \delta_{ij}. \qquad (14)$$

Here, as a notational convenience, we define a vector $\mathbf{E}$ with all unit components, $E_i \equiv 1$. The trick embodied in Eq. (14) is general, replacing expectation quantities of s (which is unmeasurable) with corresponding quantities on y (which is measurable), and replacing the correlation matrix $\langle y_i y_j \rangle$ by a single population mean square, $\langle y_i^2 \rangle$, and a *structure function*

$$\langle y_i y_j \rangle = \langle y_i^2 \rangle - V_{ij} \qquad (15)$$

$$V_{ij} \equiv \frac{1}{2} \langle (y_i - y_j)^2 \rangle \qquad (16)$$

which is frequently much easier to estimate from the data than is $\langle y_i y_j \rangle$ directly. We then estimate the population mean square by the sample mean square, and the population structure function by a fitting method similar to that described in Paper I.

3. CONSTRUCTION OF AN UNBIASED ESTIMATOR

There is a quirk in Eqs. (8) and (9) which can sometimes cause difficulties, and which is easily remedied. The prediction coefficients d_{*i} produced by Eq. (7) do not in general sum to 1, but to a value slightly less than 1. The discrepancy from 1 is greatest in gaps far from any data points. This has the peculiar effect of making the minimum variance estimate "sag" slightly, towards the value zero, in the gaps. The reason for this is that, in the absence of information from the data, the value zero is a minimum variance estimate. In fact, being constant, it has zero variance! Formally, the estimate equation (8) is a biased estimator, as can be seen (e.g., in the case of a stationary process) by

$$\langle \hat{s}_* \rangle = \left\langle \sum_i d_{*i} y_i \right\rangle = \left(\sum_i d_{*i} \right) \langle s_i \rangle \neq \langle s_i \rangle. \tag{17}$$

We know of three ways of modifying Eq. (8) so as to obtain an unbiased estimator, all of which end up giving identical formulas for $\hat{s}_*$. The first way is to recall the conventional wisdom that one should subtract off the mean of a data set before fitting it (adding the mean back into the fitting predictions at the end). The only question here is *what* mean, i.e., which kind of weighted mean, since we have not only a noise correlation matrix $\mathbf{N}$, but also a signal correlation matrix $\mathbf{S}$ at our disposal.

A cute way to answer the question is to find that value $\bar{y}$ that, when subtracted off, causes χ^2 (Eq. (13)) to be minimized. We seek to minimize with respect to $\bar{y}$

$$\chi^2 = (\mathbf{y} - \mathbf{E}\bar{y})^T [\mathbf{S} + \mathbf{N}]^{-1}(\mathbf{y} - \mathbf{E}\bar{y}). \tag{18}$$

The solution is

$$\bar{y} = \frac{\mathbf{E}^T [\mathbf{S} + \mathbf{N}]^{-1}\mathbf{y}}{\mathbf{E}^T [\mathbf{S} + \mathbf{N}]^{-1}\mathbf{E}}. \tag{19}$$

This is the generalization of the usual "inverse-variance weighted mean" formula. One sees that a data value y_i gets a small weight *either* because its variance is large, *or* because it is highly correlated with other data values (in which case it adds little new information).

In terms of this $\bar{y}$, Eq. (8) is now replaced by

$$\hat{s}_* = \mathbf{S}_*^T [\mathbf{S} + \mathbf{N}]^{-1}(y - \bar{y}\mathbf{E}) + \bar{y}. \tag{20}$$

That is, we subtract $\bar{y}$ from the data, and then add it back to the estimate.

The formula for $\mathbf{d}_*$ implied by Eq. (20) is

$$\widehat{\mathbf{d}_*} = \left(1 - \frac{[\mathbf{S} + \mathbf{N}]^{-1}\mathbf{E}\mathbf{E}^T}{\mathbf{E}^T [\mathbf{S} + \mathbf{N}]^{-1}\mathbf{E}} \right) [\mathbf{S} + \mathbf{N}]^{-1}\mathbf{S}_* + \frac{[\mathbf{S} + \mathbf{N}]^{-1}\mathbf{E}}{\mathbf{E}^T [\mathbf{S} + \mathbf{N}]^{-1}\mathbf{E}}. \tag{21}$$

One can show that $\mathbf{E}^T\widehat{\mathbf{d}_*} = 1$, so the estimator is unbiased. With $\bar{y}$ given by Eq. (19), Eq. (18) can be shown (again completing the square) to be equivalent to

$$\chi^2 = \mathbf{y}^T \left([\mathbf{S}+\mathbf{N}]^{-1} - \frac{[\mathbf{S}+\mathbf{N}]^{-1}\mathbf{E}\mathbf{E}^T[\mathbf{S}+\mathbf{N}]^{-1}}{\mathbf{E}^T[\mathbf{S}+\mathbf{N}]^{-1}\mathbf{E}} \right) \mathbf{y}. \qquad (22)$$

One here sees a projection operator that renders the value of χ^2 independent of *any* constant value added to all the components of the data vector $\mathbf{y}$. It is easy to find the expectation value of χ^2 given by Eq. (22). Defining $\mathbf{Z} = [\mathbf{S}+\mathbf{N}]^{-1}E$ and using Eq. (13), we have

$$\langle\chi^2\rangle = M - \mathrm{tr}\left\langle \frac{\mathbf{y}^T\mathbf{Z}\mathbf{Z}^T\mathbf{y}}{\mathbf{E}^T\mathbf{Z}} \right\rangle = M - \frac{1}{\mathbf{E}^T\mathbf{Z}}\mathrm{tr}\left(\mathbf{Z}\mathbf{Z}\langle\mathbf{y}\mathbf{y}^T\rangle\right)$$

$$= M - \frac{1}{\mathbf{E}^T\mathbf{Z}}\mathrm{tr}\left(\mathbf{Z}\mathbf{E}^T\right) = M - 1. \qquad (23)$$

A second, completely different, way of getting the same result is often computationally more convenient, and also addresses more directly the issue of low-frequency divergent processes. In a low-frequency divergent (sometimes called *weakly nonstationary*) process like a random walk, the *population* mean square $\langle y^2\rangle$ may be infinite or undefined, while the sample mean square is of course finite. In Eq. (15) the first term on the right-hand side will thus not, in general, be estimatable, while the second term (structure function) remains well-behaved. The solution is to substitute Eq. (15) into Eq. (7) analytically, and then take the limit $\langle y^2\rangle \to \infty$ using the Sherman-Morrison formula (see, e.g., Press et al., 1986, sec. 2.10) and the fact that the infinite term is a matrix of rank 1. One finds that Eq. (8) is now transformed exactly to Eq. (20).

Likewise we find that Eq. (22) follows directly from Eq. (12) if the replacement $\mathbf{S} \to \mathbf{S}+\langle y^2\rangle\mathbf{E}\mathbf{E}^T$ is made and $\langle y^2\rangle \to \infty$. (The Appendix discusses a generalization of this result that is used in Section 7.) Armed with this knowledge of equivalence, it is often computationally convenient not to calculate $\bar{y}$ at all, but simply to use the unmodified Eqs. (8) and (9), however substituting for $\langle y_i^2\rangle$ in Eq. (14) a value sufficiently big as to make the d_{*i}'s (as determined by Eq. (7)) sum close enough to 1. In practice, it is adequate to choose $\langle y_i^2\rangle$ to be 10 or 100 times the *sample* variance. This simple trick renders most of the rococo matrix formulas in this section supernumerary, while guaranteeing equivalent results.

Third, finally, as Cressie (1991) notes, one can simply constrain the sum of the d_{*i}'s to 1 by minimizing not Eq. (4) but rather

$$\langle x_*^2\rangle = \left\langle (\mathbf{d}_*^T\mathbf{y} - s_*)^2 \right\rangle + 2\lambda(\mathbf{E}^T\mathbf{d}_* - 1) \qquad (24)$$

where λ is a Lagrange multiplier that enforces the desired constraint. One gets straightforwardly

$$\widehat{\mathbf{d}_*} = [\mathbf{S}+\mathbf{N}]^{-1}[\mathbf{S}_* - \mathbf{E}\lambda] \qquad (25)$$

where

$$\lambda = \frac{\mathbf{E}^T[\mathbf{S} + \mathbf{N}]^{-1}\mathbf{S}_* - 1}{\mathbf{E}^T[\mathbf{S} + \mathbf{N}]^{-1}\mathbf{E}}. \tag{26}$$

Not surprisingly, these equations are algebraically equivalent to Eq. (21). This is the approach usually taken in defining so-called Gauss-Markov estimators.

For the data shown in Figures 1 and 2, removal of the bias, while important in principal, makes a negligible effect in practice. In the remainder of this paper, we will assume (cf. Eq. (14)) that $\mathbf{S}$ always has the form of a rank-one matrix proportional to $\mathbf{E}\mathbf{E}^T$ minus a structure function V_{ij} (Eq. (16)) whose maximum absolute value is very much smaller than the rank-one piece. If $\mathbf{S}$ does not start out having this form, it can be forced into this form by the addition of a constant times $\mathbf{E}\mathbf{E}^T$, as described above. In either case, $\mathbf{S}$ will now have one largest eigenvalue whose eigenvector is close to $\mathbf{E}$ (corresponding to adding a constant to the process) and the estimator $\widehat{\mathbf{d}_*}$ will be close to unbiased.

4. GAUSSIAN PROCESSES

There is something disconcerting about the reconstruction shown in Figure 2: it is too smooth. While that reconstruction is in fact "closest to true" in the minimum variance sense, one has the impression that it is not, itself, a very plausible realization of the process that gave the data points that are shown.

Merely thinking this thought involves, however, some additional assumptions about the process s. Up to now we have assumed nothing about its full probability distribution, but only knowledge of its second moments, in $\mathbf{S}$. To make further statements about "likely" or "unlikely" realizations, we need a full distribution. Absent any additional information, one generally makes the *Gaussian (or Normal) assumption*, that the probability that a vector $\mathbf{s}$ of values is generated is

$$P(\mathbf{s}) \propto \exp\left[-\frac{1}{2}\mathbf{s}^T\mathbf{S}^{-1}\mathbf{s}\right] \tag{27}$$

where the proportionality constant is determined by normalizing the total probability to unity, and similarly for the noise process,

$$P(\mathbf{n}) \propto \exp\left[-\frac{1}{2}\mathbf{n}^T\mathbf{N}^{-1}\mathbf{n}\right]. \tag{28}$$

One must also make a stronger assumption about the uncorrelatedness of $\mathbf{s}$ and $\mathbf{n}$, not just the expectation $\langle \mathbf{s}\mathbf{n}^T \rangle = 0$, but true independence of probabilities,

$$P(\mathbf{s} \text{ and } \mathbf{n}) = P(\mathbf{s})P(\mathbf{n}). \tag{29}$$

One now calculates the probability that the two processes will generate a given set of observations $\mathbf{y} = \mathbf{s} + \mathbf{n}$ as

$$
\begin{aligned}
P(\mathbf{y}) &= \int P(\mathbf{s})P(\mathbf{n})\delta[\mathbf{y} - (\mathbf{s} + \mathbf{n})]\, d^M\mathbf{s}\, d^M\mathbf{n} \\
&= \int P(\mathbf{s})P(\mathbf{y} - \mathbf{s})\, d^M\mathbf{s} \\
&\propto \int \exp\left\{ -\frac{1}{2}[\mathbf{s}^T\mathbf{S}^{-1}\mathbf{s} + (\mathbf{y} - \mathbf{s})^T\mathbf{N}^{-1}(\mathbf{y} - \mathbf{s})] \right\}\, d^M\mathbf{s} \qquad (30) \\
&\propto \exp\left\{ -\frac{1}{2}\mathbf{y}^T[\mathbf{S} + \mathbf{N}]^{-1}\mathbf{y} \right\} \\
&\quad \times \int \exp\left\{ -\frac{1}{2}(\mathbf{s} - \hat{\mathbf{s}})^T[\mathbf{S}^{-1} + \mathbf{N}^{-1}](\mathbf{s} - \hat{\mathbf{s}}) \right\}\, d^M\mathbf{s},
\end{aligned}
$$

where $\hat{\mathbf{s}} = \mathbf{S}[\mathbf{S} + \mathbf{N}]^{-1}\mathbf{y}$. Changing to $\mathbf{s} - \hat{\mathbf{s}}$ as the variable of integration, the integral is seen to be independent of $\mathbf{y}$, so that finally,

$$
P(\mathbf{y}) \propto \exp\left\{ -\frac{1}{2}\mathbf{y}^T[\mathbf{S} + \mathbf{N}]^{-1}\mathbf{y} \right\}. \qquad (31)
$$

Comparing Eq. (31) with Eq. (12), one sees explicitly that, as one might expect, the combined process of signal plus noise has a probability density $\propto \exp[-\chi^2/2]$. That is, it has a classical χ^2-distribution, for which all of the usual probability interpretations apply. As we shall see later, this implies that for Gaussian processes the minimum square discrepancy results of the preceding sections are equivalent to maximum likelihood estimation.

5. UNCONSTRAINED REALIZATIONS OF THE UNDERLYING PROCESS

Having made the Gaussian assumption, we may now generate random realizations of the process $\mathbf{y}$. This is most easily done by first diagonalizing the "covariance" matrix $\mathbf{S} + \mathbf{N}$ appearing in Eq. (31). The resulting "normal modes" are then statistically independent, and the problem is reduced to choosing M independent Gaussian random deviates.

We proceed as follows: First, find the eigenvalues $\lambda_1, \ldots, \lambda_M$ and eigenvectors $\mathbf{v}_1, \ldots, \mathbf{v}_M$ of the positive definite, symmetric matrix $\mathbf{S} + \mathbf{N}$, equivalent to the factorization

$$
[\mathbf{S} + \mathbf{N}] = \mathbf{V}\,\mathrm{diag}\,(\lambda_1, \cdots, \lambda_M)\,\mathbf{V}^T \qquad (32)
$$

where $\mathbf{V}$ is the orthogonal matrix formed out of the eigenvectors by columns,

$$\mathbf{V} = (\mathbf{v}_1 \mathbf{v}_2 \cdots \mathbf{v}_M). \tag{33}$$

Second, identify the large eigenvalue whose eigenvector is close to $\mathbf{E}$ and *set it to zero*. Third, let $\mathbf{r}$ be a vector of M independent Gaussian random deviates of zero mean and unit variance. Then a realization of $\mathbf{y}$ is

$$\mathbf{y} = \mathbf{V}\mathrm{diag}(\lambda_1^{1/2}, \cdots, \lambda_M^{1/2})\mathbf{r} + \overline{y} \tag{34}$$

where $\overline{y}$ is any mean value that you wish to give to the realization.

Alternatively, though less efficiently, we could find the eigenvalues $\xi_1, \ldots, \xi_M$, and eigenvectors $\mathbf{U}$, of $\mathbf{S}$ alone, and find the eigenvalues $\zeta_1, \ldots, \zeta_M$, and eigenvectors $\mathbf{Z}$, of $\mathbf{N}$ alone (these are trivial when $\mathbf{N}$ is diagonal). Setting the large eigenvalue whose eigenvector is close to $\mathbf{E}$ to zero, we could then construct

$$\mathbf{y} = \mathbf{s} + \mathbf{n} = \mathbf{U}\mathrm{diag}(\xi_1^{1/2}, \ldots, \xi_M^{1/2})\mathbf{r} + \mathbf{Z}\mathrm{diag}(\zeta_1^{1/2}, \ldots, \zeta_M^{1/2})\mathbf{r}' \tag{35}$$

where $\mathbf{r}$ and $\mathbf{r}'$ are independent random vectors. The equivalence of procedures (34) and (35) is guaranteed by Eq. (30).

Figure 3 shows two independent realizations of the process underlying Figures 1 and 2, on a set of $M = 250$ equally spaced points and with $\overline{y} = 140$. For clarity, we have set the noise to zero in these realizations. Adding nonzero noise would simply "fuzz" the curves, independently randomly at each point, by a Gaussian of standard deviation equal to the error bars of Figure 1 or 2. We have not here plotted the data from Figure 1, because these realizations are unconditioned by that data. Such unconditioned realizations are frequently useful in Monte Carlo experiments that address questions of how probable are particular features of an observed data set (see, e.g., Papers I and II).

6. REALIZATIONS CONSTRAINED BY THE MEASURED DATA

Section 2 (and Figure 2) constructed minimum variance predictions for values of a process s. If we take the step of making the Gaussian assumption, Eqs. (27) and (28), then the minimum variance prediction is also the maximum likelihood prediction. However, the maximum likelihood process is not itself a very typical realization of the process. Section 4 (and Figure 3) generated realizations that were typical, but were not constrained by the measured data. In this section we show how to combine these techniques and generate an ensemble of realizations, each of which is typical of the underlying process, but also consistent with the measured

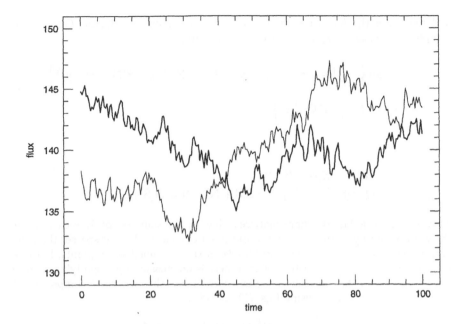

FIGURE 3 Two different unconstrained realizations of the process $s(t)$ underlying the data shown in Figure 1 and 2. Note that these "typical" realizations do not resemble the minimum variance prediction of Figure 2, which is "too smooth."

data. Such an ensemble characterizes ones full knowledge of the actual instance of the process that took place.

Although we could disguise the fact in various ways, our treatment becomes slightly Bayesian at this point (see, e.g., Loredo, 1992): "Bayesian," because we are going to assign probabilities (not likelihoods) to different hypotheses about unmeasured quantities, and "slightly," because, in making the Gaussian assumption, we have already postulated a set of measurements drawn from a stochastic process with a well-defined probability measure.

Bayes' theorem gives the probability of an underlying process **s** *given* a set of measurements **y**,

$$P(\mathbf{s}|\mathbf{y}) = \frac{P(\mathbf{s})P(\mathbf{y}|\mathbf{s})}{P(\mathbf{y})} = \frac{P(\mathbf{s})P(\mathbf{n})}{P(\mathbf{y})}$$
$$\propto \exp\left\{-\frac{1}{2}\left[\mathbf{s}^T\mathbf{S}^{-1}\mathbf{s} + (\mathbf{y}-\mathbf{s})^T\mathbf{N}^{-1}(\mathbf{y}-\mathbf{s})\right]\right\}. \qquad (36)$$

Here the second equality follows from the fact that $\mathbf{y} = \mathbf{s} + \mathbf{n}$, so the probability of **y** conditioned on **s** is just the probability of **n**. The proportionality uses Eqs. (27)

and (28), and the fact that $P(\mathbf{y})$ is merely a normalization factor. Once again completing a square, Eq. (36) can be shown to imply

$$P(\mathbf{s}|\mathbf{y}) \propto \exp\left\{-\frac{1}{2}\left[(\mathbf{s} - \mathbf{S}[\mathbf{N} + \mathbf{S}]^{-1}\mathbf{y})^T[\mathbf{S}^{-1} + \mathbf{N}^{-1}](\mathbf{s} - \mathbf{S}[\mathbf{N} + \mathbf{S}]^{-1}\mathbf{y})\right]\right\}$$
$$\propto \exp\left\{-\frac{1}{2}\left[\mathbf{u}^T\mathbf{Q}^{-1}\mathbf{u}\right]\right\} \tag{37}$$

where

$$\mathbf{u} \equiv \mathbf{s} - \mathbf{S}[\mathbf{N} + \mathbf{S}]^{-1}\mathbf{y}, \tag{38}$$
$$\mathbf{Q} \equiv [\mathbf{S}^{-1} + \mathbf{N}^{-1}]^{-1} = \mathbf{S}[\mathbf{S} + \mathbf{N}]^{-1}\mathbf{N} = \mathbf{N}[\mathbf{S} + \mathbf{N}]^{-1}\mathbf{S}. \tag{39}$$

This derivation assumes square matrices, that is, the same set of M locations for the vectors $\mathbf{s}$ and $\mathbf{y}$. If (in the usual case) the set of measured y_i's is sparser than the desired set of s_j's, then Eq. (37) still holds on the combined set of points, but one must let $N_{jj} \to \infty$ for any value j where there is no measured y_j, signifying infinite uncertainty as to the "measured" value there. Then, the "asterisk" component of Eq. (38) can be rewritten using Eqs. (5) and (8) as

$$s_* = u_* + \mathbf{S}_*^T[\mathbf{N} + \mathbf{S}]^{-1}\mathbf{y} = u_* + \widehat{s_*} \tag{40}$$

where, by Eq. (37), u_* is a Gaussian process with correlation matrix $\mathbf{Q}$ (Eq. (38)).

Equation (40) is a powerful and perhaps surprising result. It says that "typical" realizations, in the correct relative probabilities, are obtained by starting with the minimum variance estimator $\widehat{s_*}$ and adding to it a Gaussian process with zero mean and correlation matrix $\mathbf{Q}$ given by Eq. (39). Computationally, one generates the vector $\mathbf{u}$ by finding the eigenvectors and eigenvalues of $\mathbf{Q}$, and proceeding exactly analogously to Eqs. (32)–(34).

It is useful to note how this works in the limiting case $\mathbf{N} \to \infty$: then $\widehat{s_*} \to 0$ by Eq. (8), $\mathbf{Q} \to \mathbf{S}$, and we obtain an unconstrained realization drawn from the probability distribution of Eq. (27). When (in the typical case) only certain diagonal elements $N_{jj} \to \infty$ (those that have no measured values), then the realization becomes, at these points, constrained only by the information propagated through $\mathbf{S}$. Conversely, if a row and column of $\mathbf{N}$ go to *zero*, then the realization is forced to exactly the measured value y at that point.

Figure 4 shows several independent realizations conditioned on the data of Figure 1, generated by the the procedure just described. If we generate a large number of such realizations, they will, at each abscissa t, have a Gaussian distribution of ordinates, centered on the minimum variance reconstruction (Eq. (8)) and with a standard deviation matching the width of the "snake" in Figure 2 (square root of Eq. (9)). However, there is much more information in such an ensemble than there

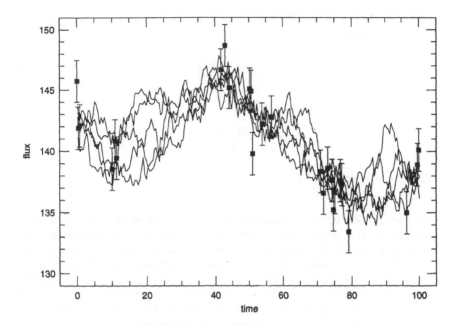

FIGURE 4   Five random realizations conditioned on the measured data points and error bars, generated by Eq. (38). Each realization has the statistics of the true underlying process, is consistent with the measurements, and can be viewed as a plausible reconstruction of the actual process that transpired. The ensemble of such realizations can be used for Monte Carlo exploration of additional statistical questions.

is in Figure 2, since, for each realization, the correlations between different abscissas t are correctly realized. One can thus make use of the ensemble of realizations to answer, via Monte Carlo experiments, many otherwise unaccessible statistical questions. One might ask, for example, "Given the measured data, how often will s be below an upper limit $F = 143$ at time $t = 25$ *and* below an upper limit $F = 145$ at time $t = 30$?" (see Figure 4).

7. SIMULTANEOUS RECONSTRUCTION AND DETERMINATION OF LINEAR FITTING PARAMETERS

In the derivations of Section 2, we relegated χ^2 to a secondary role, since we did not want its Gaussian connotations to be confusing. Now, however, it is safe to point out that Eq. (8) could have been derived simply as a χ^2 minimization, as follows: Let $\tilde{\mathbf{y}}$ be the "augmented" vector

$$\tilde{\mathbf{y}} = \begin{pmatrix} \mathbf{y} \\ s_* \end{pmatrix} \tag{41}$$

whose correlation matrix is

$$\tilde{\mathbf{C}} \equiv \begin{pmatrix} \mathbf{S} + \mathbf{N} & \mathbf{S}_* \\ \mathbf{S}_*^T & \langle s_*^2 \rangle \end{pmatrix} \tag{42}$$

so that the augmented χ^2 is

$$\chi^2 = \tilde{\mathbf{y}}^T \tilde{\mathbf{C}}^{-1} \tilde{\mathbf{y}}. \tag{43}$$

Then, one can readily verify (using the formula for the matrix inverse of a partitioned matrix) that minimizing χ^2 with respect to the value s_* gives exactly Eq. (8). This is much more than simply an interesting alternative derivation, however. Since $\exp(-\frac{1}{2}\chi^2)$ is the probability of a realization of a Gaussian process, we now see that the preceding results using minimum square discrepancy are completely equivalent to *maximum likelihood estimation* for Gaussian processes. Moreover, well-known statistical machinery may be applied to the resulting values of χ^2, leading to confidence limits on the reconstructed signal values, for example.

This notion of generalizing χ^2, and then minimizing it, also yields useful results when applied to the problem of simultaneously reconstructing a process s and fitting for some number N_q of unknown fitting parameters. Suppose that, instead of $\mathbf{y} = \mathbf{s} + \mathbf{n}$, we have

$$\mathbf{y} = \mathbf{s} + \mathbf{L}\mathbf{q} + \mathbf{n} \tag{44}$$

where $\mathbf{q}$ is a vector of unknown parameters (length N_q) and $\mathbf{L}$ is an $M \times N_q$ matrix of known coefficients. From the measured values $\mathbf{y}$, we desire to reconstruct a best estimate of s and, simultaneously, best values for the parameters $\mathbf{q}$.

The generalized χ^2 is

$$\chi^2 = (\mathbf{y} - \mathbf{L}\mathbf{q})^T \mathbf{C}^{-1} (\mathbf{y} - \mathbf{L}\mathbf{q}) \tag{45}$$

where $\mathbf{C} \equiv \mathbf{S} + \mathbf{N}$ is now a convenient abbreviation.

A few examples will clarify Eqs. (44) and (45):

1. If $\mathbf{L}$ is the vector $\mathbf{E}$ (viewed as a $M \times 1$ matrix), and $\mathbf{q}$ is the value $\bar{y}$ (viewed as a 1×1 matrix), then Eq. (45) is exactly Eq. (18), which we used to produce an unbiased estimator for s.

2. If $\mathbf{L}$ and $\mathbf{q}$ are of the form

$$\mathbf{L} = \begin{bmatrix} 1 & 0 \\ \vdots & \vdots \\ 1 & 0 \\ 0 & 1 \\ \vdots & \vdots \\ 0 & 1 \end{bmatrix} \qquad \mathbf{q} = \begin{pmatrix} \overline{y}_1 \\ \overline{y}_2 \end{pmatrix}, \tag{46}$$

then the data consists of two subsets having different (unknown) means. Minimization of Eq. (45) will, in effect, adjust the subsets to a common offset before determining the best reconstruction of the underlying process s. This is precisely what we did in Papers I and II to reduce the two gravitational lens images (which had different, unknown, magnifications) to a common basis.

3. If $\mathbf{L}$ and $\mathbf{q}$ are of the form

$$\mathbf{L} = \begin{bmatrix} 1 & t_1 & t_1^2 \\ 1 & t_2 & t_2^2 \\ \vdots & \vdots & \\ 1 & t_M & t_M^2 \end{bmatrix} \qquad \mathbf{q} = \begin{pmatrix} \alpha_0 \\ \alpha_1 \\ \alpha_2 \end{pmatrix}, \tag{47}$$

then minimization of Eq. (45) has the effect of removing a quadratic trend $\alpha_0 + \alpha_1 t + \alpha_2 t^2$ from the measured data $\mathbf{y}$ before reconstructing a model with correlation matrix $\mathbf{S}$.

Now properly oriented, we can manipulate Eq. (45), once again completing its square, to get

$$\chi^2 = [\mathbf{q} - \widehat{\mathbf{q}}]^T (\mathbf{L}^T \mathbf{C}^{-1} \mathbf{L}) \, [\mathbf{q} - \widehat{\mathbf{q}}] + \mathbf{y}^T [\mathbf{C}^{-1} - \mathbf{C}^{-1} \mathbf{L} \, (\mathbf{L}^T \mathbf{C}^{-1} \mathbf{L})^{-1} \mathbf{L}^T \mathbf{C}^{-1}] \mathbf{y} \tag{48}$$

where

$$\widehat{\mathbf{q}} = (\mathbf{L}^T \mathbf{C}^{-1} \mathbf{L})^{-1} \mathbf{L}^T \mathbf{C}^{-1} \mathbf{y}. \tag{49}$$

From Eqs. (45) and (48), we can now read off the answers to the simultaneous reconstruction of s and $\mathbf{q}$: The parameters $\mathbf{q}$ are estimated by $\widehat{\mathbf{q}}$, which clearly minimizes χ^2 given in Eq. (48). The covariance matrix of this estimate (whose diagonal elements, e.g., are the standard errors for the fitted parameters) is $(\mathbf{L}^T \mathbf{C}^{-1} \mathbf{L})^{-1}$, which is indeed an $N_q \times N_q$ matrix. Once $\widehat{\mathbf{q}}$ is calculated, the vector s is estimated by (cf. Eq. (8))

$$\widehat{\mathbf{s}} = \mathbf{S} \mathbf{C}^{-1} (\mathbf{y} - \mathbf{L}\widehat{\mathbf{q}}), \tag{50}$$

while s_* at any other point is estimated by

$$\widehat{s_*} = \mathbf{S}_*^T \mathbf{C}^{-1} (\mathbf{y} - \mathbf{L}\widehat{\mathbf{q}}). \tag{51}$$

The variance of s_* about this estimate is

$$\left\langle (s_* - \widehat{s}_*)^2 \right\rangle = \left\langle s_*^2 \right\rangle - \mathbf{S}_*^T [\mathbf{C}^{-1} - \mathbf{C}^{-1}\mathbf{L}\,(\mathbf{L}^T\mathbf{C}^{-1}\mathbf{L})^{-1}\mathbf{L}^T\mathbf{C}^{-1}]\mathbf{S}_*. \tag{52}$$

In practice, it is often easier to simply use the formulas for the case of no $\mathbf{q}$ parameter fitting, but add to $\mathbf{S}$ a large scalar multiple of the product $\mathbf{LL}^T$. (The validity of this procedure is proved in the Appendix.) This is a generalization of the analogous technique used in preceding sections for subtracting the mean of the process.

8. DISCUSSION FOR BAYESIANS ONLY

The Bayesian reader may be squirming with displeasure at the procedure just discussed, since the estimated parameters $\widehat{\mathbf{q}}$ are seemingly treated quite differently from the estimated signal $\widehat{\mathbf{s}}$: The former are estimated by maximum likelihood. These maximum likelihood values are then frozen during the estimation of $\widehat{\mathbf{s}}$. From a frequentist viewpoint, this is the only way to proceed; we *have* made the Gaussian assumption that $\mathbf{s}$ has a well-defined probability, but we *have not* made such an assumption about the parameters $\mathbf{q}$. In a spirit of statistical ecumenism, however, we can happily report that, for this particular problem, a straightforward Bayesian calculation, treating $\mathbf{s}$ and $\mathbf{q}$ democratically, gives identical results.

We write the conditional probability of $\mathbf{s}$ and $\mathbf{q}$ given $\mathbf{y}$, by Bayes' theorem, as

$$P(\mathbf{s}, \mathbf{q}|\mathbf{y}) = \frac{P(\mathbf{s}, \mathbf{y}|\mathbf{q})P(\mathbf{q})}{P(\mathbf{y})}, \tag{53}$$

where the conditional probability $P(\mathbf{s}, \mathbf{y}|\mathbf{q})$ is clearly given by

$$P(\mathbf{s}, \mathbf{y}|\mathbf{q}) \propto \exp\left\{ -\frac{1}{2} \left[\mathbf{s}^T\mathbf{S}^{-1}\,\mathbf{s} + (\mathbf{y} - \mathbf{s} - \mathbf{Lq})^T\mathbf{N}^{-1}\,(\mathbf{y} - \mathbf{s} - \mathbf{Lq}) \right] \right\}. \tag{54}$$

To use Eq. (53) to find the most probable estimates of $\mathbf{s}$ and $\mathbf{q}$, we must first consider the factors $P(\mathbf{q})$ and $P(\mathbf{y})$. Actually, $P(\mathbf{y})$ is irrelevant, since it is merely a constant (recall $\mathbf{y}$ *is given*). However, the quantity $P(\mathbf{q})$ poses a more serious problem, since this refers to the probability distribution of the parameters $\mathbf{q}$, which is, in almost all cases, unknown to us. We resolve this Bayesian dilemma, as is usual in such cases, by making the assumption that $P(\mathbf{q})$ is sufficiently broad that it can be considered constant for the purposes of finding the maximum of the probability function $P(\mathbf{s}, \mathbf{q}|\mathbf{y})$. Then, the most probable estimates of $\mathbf{s}$ and $\mathbf{q}$ can be found by minimizing the quadratic expression

$$\mathbf{s}^T\mathbf{S}^{-1}\,\mathbf{s} + (\mathbf{y} - \mathbf{s} - \mathbf{Lq})^T\mathbf{N}^{-1}\,(\mathbf{y} - \mathbf{s} - \mathbf{Lq}), \tag{55}$$

with respect to **s** and **q**. Differentiating with respect to **s** and **q** and setting the results equal to zero yields the simultaneous equations for the minimizing values $\widehat{\mathbf{s}}$ and $\widehat{\mathbf{q}}$,

$$\begin{aligned}(\mathbf{S}^{-1} + \mathbf{N}^{-1})\widehat{\mathbf{s}} + \mathbf{N}^{-1}\mathbf{L}\widehat{\mathbf{q}} &= \mathbf{N}^{-1}\mathbf{y}, \\ \mathbf{L}^T\mathbf{N}^{-1}\widehat{\mathbf{s}} + \mathbf{L}^T\mathbf{N}^{-1}\mathbf{L}\widehat{\mathbf{q}} &= \mathbf{L}^T\mathbf{N}^{-1}\mathbf{y}.\end{aligned} \tag{56}$$

The solution of these equations for $\widehat{\mathbf{s}}$ and $\widehat{\mathbf{q}}$ can be found straightforwardly, yielding precisely the results of Eqs. (49) and (50), showing that this simultaneous procedure is equivalent to our previous separate minimizations.

9. DISCUSSION

Linear estimation is an old subject, and it is perhaps surprising that the principal practical results of this paper (consisting of the progressively more general cases of Eqs. (8)–(9), (19)–(20), (34), (39)–(40), and (49)–(52) are not standard textbook fare. It seems likely that the reason is one of technology, not mathematics: To use the results of this paper you must be able to solve (and possibly diagonalize) linear systems whose size is *the larger* of the size of your data set and the size of the set on which you want to make estimates. For data sets of any interesting size (hundreds or thousands of points), this capability has only recently become readily available in fast desktop workstations. While there may be little in this paper that could not have been written down in the 1940s (if not 40 years earlier!), it is also true that there is little in this paper which could have been *calculated* before the 1980s—and routinely calculated only in the 1990s.

Our principal conclusion is not an equation but an orientation: One now has the capability to solve many significant problems in "classical" data analysis by *global* manipulation of the full data set. Doing so (particularly in conjunction with Monte Carlo methods) can provide unambiguous answers (as in Papers I and II) to otherwise problematic statistical questions.

ACKNOWLEDGMENTS

We thank Jacqueline Hewitt for introducing us to this subject, and for observational insights. This work was supported in part by the U.S. National Science Foundation, grant PHY-91-06678.

APPENDIX

We show here how the minimization of the χ^2 given in Eq. (45) can be expressed as a limit of a certain related χ^2 expression as a parameter approaches infinity. Suppose $\mathbf{C}$, $\mathbf{L}$, $\mathbf{q}$, and $\mathbf{y}$ are defined as in Section 7, and that λ is a real parameter. Our result is then

$$\min_{\mathbf{q}}(\mathbf{y} - \mathbf{Lq})^T \mathbf{C}^{-1}(\mathbf{y} - \mathbf{Lq}) = \lim_{\lambda \to \infty} \mathbf{y}^T \left(\mathbf{C} + \lambda \mathbf{LL}^T\right)^{-1} \mathbf{y}. \qquad (57)$$

To prove this, we first note from Eq. (48) that

$$\min_{\mathbf{q}}(\mathbf{y} - \mathbf{Lq})^T \mathbf{C}^{-1}(\mathbf{y} - \mathbf{Lq}) = \mathbf{y}^T \left[\mathbf{C}^{-1} - \mathbf{C}^{-1}\mathbf{L}(\mathbf{L}^T\mathbf{C}^{-1}\mathbf{L})^{-1}\mathbf{L}^T\mathbf{C}^{-1}\right] \mathbf{y}. \qquad (58)$$

Next, from the well-known Woodbury formula (see, e.g., Press et al., 1986), we have

$$\left(\mathbf{C} + \lambda \mathbf{LL}^T\right)^{-1} = \mathbf{C}^{-1} - \mathbf{C}^{-1}\mathbf{L}(\lambda^{-1} + \mathbf{L}^T\mathbf{C}^{-1}\mathbf{L})^{-1}\mathbf{L}^T\mathbf{C}^{-1}. \qquad (59)$$

Thus

$$\lim_{\lambda \to \infty} \mathbf{y}^T \left(\mathbf{C} + \lambda \mathbf{LL}^T\right)^{-1} \mathbf{y} = \mathbf{y}^T \left[\mathbf{C}^{-1} - \mathbf{C}^{-1}\mathbf{L}(\mathbf{L}^T\mathbf{C}^{-1}\mathbf{L})^{-1}\mathbf{L}^T\mathbf{C}^{-1}\right] \mathbf{y}. \qquad (60)$$

Comparison of Eqs. (58) and (60) proves the result. Note that this result also applies in the case where $\mathbf{q}$ is the scalar $\bar{y}$ and where $\mathbf{L} = \mathbf{E}$ (see Section 3). In that case the parameter λ may be interpreted as the variance $\langle y^2 \rangle$ of the data, as explained in the text.

From the interpretation of Eq. (57) as a fitting of parameters, it seems evident that the result should depend only on the N_q-dimensional subspace spanned by the columns of $\mathbf{L}$ and not on the particular columns themselves. This can be proved by making the replacement $\mathbf{L} \to \mathbf{LR}$, where $\mathbf{R}$ is an arbitrary, nonsingular $N_q \times N_q$ matrix, which changes the columns of $\mathbf{L}$, but maintains the subspace spanned by them. Then

$$\mathbf{C}^{-1}\mathbf{L}(\mathbf{L}^T\mathbf{C}^{-1}\mathbf{L})^{-1}\mathbf{L}^T\mathbf{C}^{-1} \to \mathbf{C}^{-1}\mathbf{LR}(\mathbf{R}^T\mathbf{L}^T\mathbf{C}^{-1}\mathbf{LR})^{-1}\mathbf{R}^T\mathbf{L}^T\mathbf{C}^{-1}$$
$$= \mathbf{C}^{-1}\mathbf{LR}(\mathbf{R})^{-1}(\mathbf{L}^T\mathbf{C}^{-1}\mathbf{L})^{-1}(\mathbf{R}^T)^{-1}\mathbf{R}^T\mathbf{L}^T\mathbf{C}^{-1}$$
$$= \mathbf{C}^{-1}\mathbf{L}(\mathbf{L}^T\mathbf{C}^{-1}\mathbf{L})^{-1}\mathbf{L}^T\mathbf{C}^{-1},$$
$$(61)$$

showing that both results (58) and (60) are invariant under this replacement.

Leon Glass and Daniel Kaplan
Department of Physiology, McGill University, 3655 Drummond Street, Montréal, Québec, Canada H3G 1Y6

Complex Dynamics in Physiology and Medicine

INTRODUCTION

A hallmark of living organisms is that they are not constant in time. Subcellular, cellular, and supercellular processes such as the cycle of cell growth and division, voltage fluctuations in excitable cell membranes, respiration, blood pressure regulation, and the sleep-wake cycle provide spectacular examples of complex rhythms. Interest in these complex fluctuations has been stimulated in recent years by the widespread recognition that deterministic dynamical systems can display *chaotic dynamics*: aperiodic rhythms sensitive to the initial condition (Abraham et al., 1989; Cvitanovic, 1984; Degn et al., 1986; Glass & Mackey, 1988; Krasner, 1990; Mayer-Kress, 1986). Although analyses of theoretical models and experiments in controlled situations have provided good evidence that chaos can sometimes be found in biological systems (Aihara & Matsumoto, 1986; Chialvo et al., 1990; Guevara, et al. 1981, 1990; Hayashi & Ishizuka, 1986), chaos itself serves more often as a motivating idea for research than an unequivocal scientific finding.

In this article we will provide a brief critical summary of time series analysis in biology with emphasis on physiology and medicine. We will first briefly summarize standard methods of time series analysis and indicate how these have been applied.

Times Series Prediction: Forecasting the Future and Understanding the Past, Eds. A. S. Weigend and N. A. Gershenfeld, SFI Studies in the Sciences of Complexity, Proc. Vol. XV, Addison-Wesley, 1993

Then we will indicate some of the ways in which concepts from nonlinear dynamics are being applied. However, time series analysis appears in so many different ways in various fields and subfields that we do not attempt a comprehensive review, but give illustrative examples. We also do not discuss the development of theoretical models for biological systems since other sources provide detailed information about these areas (Glass & Mackey, 1988).

STANDARD METHODS OF TIME SERIES ANALYSIS

The most basic sort of time series analysis is carried out by the human eye, often assisted with calipers to measure distances on a paper chart (corresponding to time intervals in the original data). This is not as primitive as it sounds. The human eye is an excellent pattern recognition device and is capable of carrying out the sophisticated analyses needed to classify time series. For example, in the field of electrocardiography, the interpretation of even exceedingly complex electrocardiograms (ECGs) as carried out, for example, with virtuousity by Pick and Langendorf (1979), requires nothing more than application of several basic concepts in cardiology combined with measurement of timing of the occurrence of beats on comparatively short records. Interpretation of electroencephalograms (EEGs) is carried out in similar fashion by skilled clinicians who have learned how to interpret the frequency, amplitude, and morphology of recordings of electrical activity from different scalp locations (Niedermeyer & Lopes da Silva, 1987). This is sufficient for the identification of a great number of different clinical disorders.

Computer analysis of time series can provide routine diagnosis, such as reading ECGs or EEGs, or can carrry out tasks such as the detection of heart beats (or the lack thereof) in implantable pacemakers or defibrillators. Most medical instrumentation involves signal processing of some sort.

In research, quantitative analysis of physiological time series starts (and often ends) with an analysis of the mean and standard deviation. In some cases, these simple statistics can provide information of physiological importance. For example, the mean heart rate (over tens of seconds) can be used to indicate a level of exertion, and a low standard deviation of heart rate has been shown to be associated with pathology (Kleiger et al., 1987). In respiration, the frequency and the duration of the inspiratory and expiratory phases change with age and are different in different species (Fisher et al., 1982; Mortola, 1991).

The standard deviation by itself often does not provide an adequate characterization of *fluctuations* in physiological systems. For the purpose of characterizing fluctuations, the power spectrum and autocorrelation function, and transfer functions, have successfully been applied. These techniques were introduced to physiology by skilled workers with a background in engineering and have seen significant applications to physiology over the past 30 years. Power spectra and allied techniques have been used in various fields in which the frequencies of oscillations are believed to have functional or clinical significance such as heart rate variability (Akselrod et al., 1981; Kitney & Rompelman, 1981), tremor (Beuter et al., 1989), and electroencephalography (Larsen & Walter, 1970; McEwen & Anderson, 1975). Perhaps because of the systems engineering background of most workers studying spectral analysis, different frequencies are usually associated with different mechanisms that lead to superimposed oscillations.

TIME SERIES ANALYSES USING METHODS INTRODUCED FROM NONLINEAR DYNAMICS

Recognition of the importance of nonlinear phenomena in physiological systems has a long history—its beginnings are perhaps represented by the work of van der Pol and van der Mark (1928) in the early part of the century. In Nobel Prize-winning work, Hodgkin and Huxley (1952) related the dynamics of excitable cell membranes to a system of coupled nonlinear differential equations.

The recent realization that nonlinear dynamical systems can display deterministic chaos has had a strong impact on research in time series analysis in physiology and medicine (Glass & Mackey, 1988; Degn et al., 1986; Goldberger, Rigney, & West, 1990). Whereas a generation ago, researchers were delighted to conjecture that a complex time series from a neuron was well described in terms of random walks (Gerstein & Mandelbrot, 1964), there is now a strong inclination to interpret physiological time series in terms of chaos. Often, this interpretation involves the use of time series analysis techniques motivated by chaotic dynamical systems. In this section we briefly review several of the main concepts in nonlinear dynamics that have been applied to time series analysis.

BIFURCATIONS

One of the most basic concepts in nonlinear dynamics is *bifurcation*. A bifurcation is a change in the qualitative features of the dynamics that arises as some parameter describing the systems changes. Bifurcations may be associated with the onset or the annihilation of oscillations, a sudden change in the period of an oscillation, or the onset or annihilation of chaotic dynamics.

Figure 1 shows an example of bifurcations in a theoretical model of a multi-looped feedback control system represented as a delay differential equation (Glass & Maltag, 1990). Increasing the gain of the control function leads to a cascade of period-doubling bifurcations and, eventually, an aperiodic chaotic rhythm. Examples such as this in model systems abound. It is rarer to find good examples of bifurcations in experimental systems. One experimental system that does show similar phenomena is periodically stimulated chick heart cells. Figure 3 shows a series of traces displaying period-doubling bifurcations and also an aperiodic chaotic rhythm (Guevara et al., 1981).

The occurrence of complex bifurcations is well known in medicine. In cardiology, complex changes of rhythm associated with various arrhythmias are well documented and in some cases may be associated with bifurcations in nonlinear dynamical equations (Glass et al., 1991). For example, the appearance of alternans rhythms in which there is a beat-to-beat alternation of ECG waveforms may in some cases be associated with period-doubling bifurcations (Guevara et al., 1984; Smith et al., 1988).

Time series analysis techniques to detect and characterize bifurcations have not been widely developed. J. M. Smith et al. (1988) have proposed an FFT-based statistic for quantifying alternation and a method for detecting and quantifying

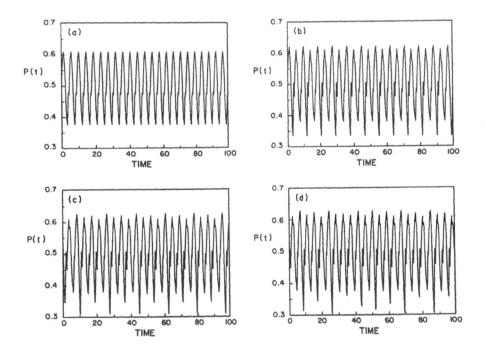

FIGURE 1 Time series generated from equations modeling a multiloop negative feedback system showing bifurcations that occur as the gain of feedback is increased. (a) Periodic orbit, (b) period-2 orbit, (c) period-4 orbit, and (d) chaotic orbit. (Reproduced with permission from Glass and Maltag (1990).)

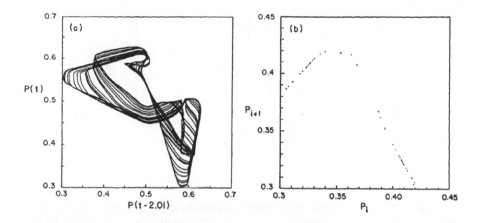

FIGURE 2   Phase plane embedding (a) and Poincaré map (b) for the time series in Figure 1(d). Successive values, designated P_i, of $P(t - 2.01)$ for crossing $P(t) = 0.55$ with $dP/dt > 0$ are determined from the data in (a). (Reproduced with permission from Glass and Maltag (1990).)

alternation that uses an embedding-space formulation is given by Kaplan (1992). In many circumstances, clear examples of bifurcations are difficult to document and a variety of additional measures have been to developed to characterize complex time series.

DYNAMICAL REPRESENTATIONS OF TIME SERIES

Although it is most common to plot time series as a function of time, a variety of other methods suggested by nonlinear dynamics provide powerful insights into dynamics in some circumstances.

One technique, *phase plane embedding* or *phase portrait*, involves plotting $x(t + \tau)$ vs. $x(t)$, giving a trajectory in the phase space. Plots of a variable as a function of its delayed value were first used in theoretical studies of chaos in physiological systems modeled by delay differential equations (Glass & Mackey, 1979), and this technique has since been used widely for systems in which only one variable is easily measured (Sauer et al., 1991). Embeddings of a time series in higher dimensional phase spaces by plotting the current value as a function of several time lagged values can be easily implemented for computational purposes.

An example of a phase plane embedding for the time series in Figure 1(d) is shown in Figure 2(a). In this plot, the trajectory is well confined in a limited region

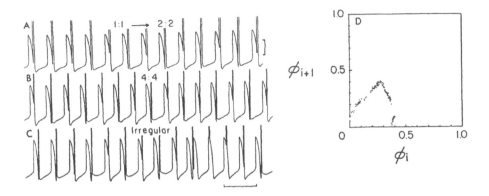

FIGURE 3 Time series showing the effects of periodic stimulation (sharp spikes) on spontaneously beating aggregates of embryonic chick heart cells. (a) Transition from 1:1 phase locking to 2:2 phase locking, (b) 4:4 phase locking, (c) irregular dynamics reflecting deterministic chaos, and (d) return map showing the phase of stimulus $i + 1$ as a function of stimulus i. (Reproduced with permission from Guevara, Glass, and Shrier (1981).)

of phase space. Further insight into the dynamics in this example can be obtained by examining the flow on a cross section to the trajectory of the flow. Successive returns to a cross section to the flow are plotted in Figure (2b). The points fall approximately on a one-dimensional single-humped curve known to give rise to chaotic dynamics. The map, which gives successive returns to the cross section is usually called the *return map* or *Poincaré map*. A recent example in which an unstable cardiac preparation is perturbed and stabilized by feedback calculated from a Poincaré map is described by Garfinkel et al. (1992).

In cases where analysis of the time series suggests a one-dimensional map, it may be possible to derive a form for the map based on a theoretical analysis of the underlying mathematical problem. An example is provided by the periodically stimulated heart cell aggregates (Figure 3). Theoretical analysis (Guevara et al., 1981, 1990) of the effects of periodic stimulation of a limit cycle oscillation, show that (i) if there is rapid relaxation to the limit cycle and (ii) if the stimulus does not alter the properties of the oscillator, then a plot of successive phases of stimulus in the limit cycle follows a one-dimensional map. A map derived from the time series for Figure 3(c) once again falls on a one-dimensional single-humped curve.

In Figures 2 and 3, the time embedding techniques help to identify the underlying deterministic dynamics governing the time evolution. Although such techniques can be readily tested on data sets, there is no guarantee that simple one- (or higher) dimensional maps will be identified. To give an idea of the difficulties that arise in

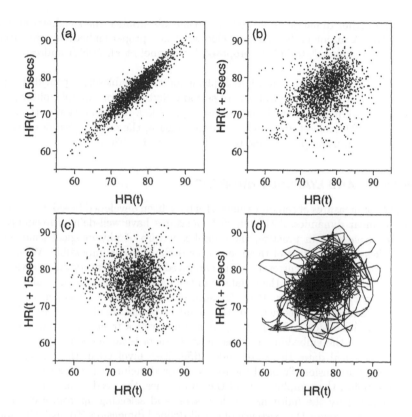

FIGURE 4   Embedding plots of the heart rate time series from Data Set B, with different embedding lags. (a) 0.5 secs, (b) 4 secs, (c) 15 secs, and (d) 5 secs (as in (b)), but now consecutive points have been connected. This shows that the filament structures near the periphery in (b) are the result of single passes of the trajectory through sparsely populated areas.

practical situations, consider Data Set B. Figure 4 shows several two-dimensional embeddings in which heart rate is displayed. The data were sampled at 0.5 sec intervals. The ellipsoidal structure in 4(a)–(b) simply reflects the autocorrelation of the heart rate for small time delays—even random numbers that have been low-pass filtered to produce such correlations will show this structure. Depending on the embedding delay τ, other forms of structure can appear in the embedding. Often, this structure is consistent with passing random white noise through a linear coloring filter. None of these plots give an indication of structure present in the data that would indicate a deterministic origin. Of course, one cannot exclude the possibility that some higher dimensional embedding would disentangle the spaghetti mess in Figure 4(d). A major problem is to decide what is a proper embedding dimension. The "false neighbors" technique proposed by Kennel et al. (1992) addresses this issue.

Two-dimensional representations of physiological variables have provided a useful tool for plotting physiological data of periodic rhythms. For example, phase plane plots of the volume and flow during a breath provide a representation of the respiratory cycle (Mortola et al., 1984). In motor control, there have been descriptions of limb position using phase plane plots (Beuter et al., 1986).

DIMENSION AND LYAPUNOV NUMBERS

A number of measures of complex time series have been developed based on concepts from nonlinear dynamics. Although such measures have well-defined meanings in idealized situations, in practice, the complex nature of physiological time series often makes interpretation of these measures difficult if not impossible. We briefly review several methods currently being employed. An excellent, detailed review of these techniques along with the pitfalls is in the paper by Grassberger et al. (1991).

The *dimension* (Farmer, 1982; Grassberger & Procaccia 1983c) gives a statistical measure of the geometry of the cloud of points. In deterministic chaotic systems the dimension is frequently (but not always!) a fractional number and is independent of the embedding dimension m when m is large enough. In practice, with physiological data, there are many difficulties involved in the analysis of a system using dimension. Some of the issues involved include the stationarity of the dynamics, noise, the sampling rate of the time series, the need to use finite length scales imposed by the finite size of data sets, and selecting appropriate convergence criteria to assert the existence of a well-defined dimension. To deal with some of these problems, computation of the local or "pointwise" dimension has been suggested (Gallez & Babloyantz, preprint; Havstad & Ehlers, 1989; Mayer-Kress, 1988). For example, Skinner et al. (1991) have proposed a technique in which a separate estimate of the correlation dimension is calculated for each point in the time series, based on the several nearest neighbors in the embedding space (Kroll, 1991; Skinner, 1991).

Another set of statistics in wide use are *Lyapunov exponents*. These measure the average local rate of divergence of neighboring trajectories in phase-space embeddings (Wolff et al., 1985). If a system is known to be deterministic, a positive Lyapunov number can be taken as a definition of a chaotic system. There are now a variety of algorithms for estimating Lyapunov exponents from time series. An early algorithm in wide use developed by Wolf (1985) and colleagues estimates the divergence of pairs of neighboring trajectories. Unfortunately, the Wolf algorithm has some serious limitations in the analysis of biological data, but this is still not always recognized. One problem arises when there are large derivatives of approximately

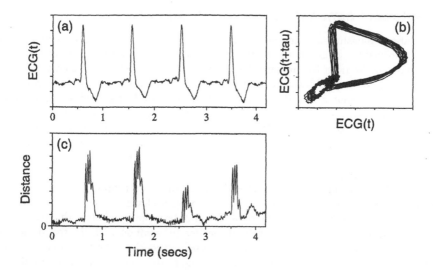

FIGURE 5   (a) A surface electrocardiogram (ECG). The tall, upward spikes (the "QRS complexes") correspond to ventricular activation, the more rounded downward excursions correspond to ventricular relaxation (the "T-wave"). (b) A embedding of the ECG signal. $m = 2$, $\tau = 50$ msecs. (c) The distance between two segments in the embedded ECG trajectory (with $m = 3$, $\tau = 50$ msecs). Initially the segments are quite close, but are rapidly separated during the QRS complexes. Lyapunov exponent calculations that assume that trajectories will separate in an approximately exponential fashion may be mislead by data like this.

periodic functions. An example of an ECG and its embedding in two dimensions is shown in Figure 5. Figure 5(c) is a plot of the distance between two nearby points (in a three-dimensional embedding). The Wolf algorithm would give positive contributions to the Lyapunov number associated with the large increases in the distance in Figure 5(c), but this does not reflect an exponential divergence of the trajectories.

Other algorithms for estimating Lyapunov exponents fit various forms of functions to the embedded data and calculate the local divergence based on local linearization of these functions (Bryant et al., 1990; Nychka et al., 1992; Sano & Sawada, 1985; Zeng et al., 1991). Critical issues in the reliability of the estimation of Lyapunov exponents are the influence of noise, the choice of fitted functions, the size of the neighborhood over which the fitting is done, and introduction of spurious exponents from use of a too high embedding dimension. There has not been a convincing demonstration that current techniques are useful for looking at systems whose attractors are high-dimensional, i.e., $m > 3$.

SURROGATE DATA AND TESTS FOR DETERMINISM

The above measures are frequently difficult to interpret since stochastic systems may also yield a value for the dimension and Lyapunov number. Recent work has emphasized the need to compare the values of these statistics as generated from a time series, to the values generated by suitably constructed stochastic processes. The structure that a statistic is detecting in the cloud of embedded points may possibly be the result of a nonchaotic dynamical mechanism—for example, stochastically forced linear dynamics. We have found, for example, that heart rate signals cannot systematically be distinguished from stochastically forced linear dynamical systems (Kaplan & Glass, 1992). The same is true for the EEGs we have studied (Glass et al., 1992).

Theiler et al. (1991) have proposed a form of bootstrapping that enables any statistic—even those not typically associated with chaos—to be used potentially as a probe of nonlinearity in the data. The bootstrapping involves the generation of "surrogate data" that has certain properties in common with the test data. For example, random surrogate data can be synthesized that has the same autocorrelation function as the test data. The surrogate data can be used informally to evaluate whether a finding of low-dimensional dynamics indicates chaos (see, for example, a study of ventricular fibrillation using a dimension statistic (Kaplan & Cohen, 1990)).

The difficulties above indicate the need for having measures that address whether a given time series is generated by a deterministic process. Our work (Kaplan & Glass, 1992a, 1992b; Kaplan, in press) is based on the observation that deterministic systems will have well-defined vector fields. By embedding a time series, using standard methods and examining the flow in coarse-grained hyperboxes, we can study the extent to which flows through local neighborhoods are locally parallel as would be found for deterministic systems. This work (see, also, the article by Kaplan in this volume) has shown that several short time series showing heart rate variability or electroencephalograms did not show evidence for determinism, but were indistinguishable from a stochastically forced linear system with the same power spectrum (Kaplan & Glass, 1992b).

Another approach to the analysis of determinism is to distinguish between deterministic chaotic systems and stochastic systems based on the extent to which future values of the system can be forecast from past values, i.e., to *test the predictability*.

There are two complementary approaches to using predictability. The fall-off of predictability with increasingly future forecasts can sometimes be used to distinguish different types of dynamics. Although it is sometimes possible to make short-term forecasts for chaotic systems, the predictability is expected to fall off on a time scale governed by the positive Lyapunov exponents. For example, Sugihara and May examined month-by-month records of the numbers of measles and chicken-pox patients, and showed that the measles record was more predictable than white noise, and that the predictability falls off in a manner consistent with chaos (Sugihara & May, 1990). The interpretation of results such as these is somewhat difficult, since stochastically forced linear systems can show similar fall-offs.

A second approach, due originally to Casdagli (1991) and applied to Data Set D by Casdagli and Weigend (this volume), examines whether locally fit linear models perform better at forecasting than globally fit models. If the locally fit models are better, this provides evidence for nonlinear structure in the dynamics—even nonlinear stochastic structure can be detected in this manner. This approach has been adopted, for example, by Longtin (1993) in the analysis of spike trains from sensory neurons.

DIFFICULTIES IN ANALYZING PHYSIOLOGICAL DATA

In this section we consider in more detail some of the properties of physiological data that impose difficulties in the use of many nonlinear dynamics statistics.

NONSTATIONARITY

If statistical characterizations of a time series are not constant in time, the time series is *nonstationary*. One cause of nonstationarity in time series is the constantly changing environment. An obvious but important example is found in the study of 24-hour heart rate variability—in a natural setting, subjects are constantly changing their posture or level of activity. Even when sleeping, changes in sleep stage have a demonstrable effect on heart rate variability.

Other forms of nonstationarity may arise from transients. One role of the various physiological control systems is to respond to changes in environment or other "perturbations." Usually, a physiological control system is not at steady state.

In some cases, the interaction between control systems may lead to very long transients—for example, the time scales over which short-term cardiovascular control systems act ranges from seconds (the parasympathetic nervous system) to hours (renal fluid volume) (Guyton, 1991). Altogether, variability in physiological parameters often has a $1/f$ spectral form, suggesting strong nonstationarity of even long-term physiological recordings (Kaplan & Talajic, 1991; Kobayashi & Musha, 1982).

Several approaches have been used to deal with nonstationarity in physiological time series. The simplest is to use short time series where, it is presumed, nonstationarity is not a severe impediment (e.g., Kaplan & Cohen, 1990). Another technique is to attempt to correct for slow drifts, either by the subtraction of trends or taking the first difference of the time series (e.g., Sugihara & May, 1990). The first step in dealing with nonstationarity is obviously to detect the existence of nonstationarity—visual clues to nonstationarity can be provided by recurrence plots (Eckmann et al. 1987), where the time at which the embedded trajectory returns closest to itself are indicated for each point in the trajectory.

NOISE

Physiological systems often display complex fluctuations that are frequently identified with stochastic "noise." Most physiological processes ultimately are affected by the opening and closing of subcellular ion channels, and most workers believe that the kinetics of the channels is best decribed by stochastic processes. Consequently, stochastic processes or noise are ubiquitous in living systems. Yet conventional wisdom is that the averaging that occurs in going from one hierarchical level to the next, can lead to deterministic models (e.g., the Hodgkin-Huxley equations), being appropriate for cellular and supercellular processes (see Figure 3). There is not now a good mathematical understanding of the properties of nonlinear dynamical systems in the presence of noise, or how the various statistical measures discussed in the previous section are affected by noise.

In addition to this problem, many physiological processes display non-Gaussian noise, and the statistics can be dominated by "outliers." One example, from heart rate variability, concerns the existence of premature ventricular beats. The premature beats introduce very short interbeat intervals into the heart rate record which may have a relatively fixed relationship with the timing of the preceeding and following beats.

An anecdote may illustrate the potential role of such artefacts in nonlinear analysis of physiological time series: in a study of changes in heart rate variability with aging, a comparison was done between the correlation dimension of heart rate and the dimension of a random surrogate time series with the same power spectrum. The purpose was to see if the correlation dimension found evidence for nonlinear dynamics in the heart rate time series. The results were extremely strong: in young people (mean age 28 years), there was no difference between the heart rate time series and the surrogate data. In old people (mean age 75), there was a very distinct and systematic difference: the dimension of the heart rate time series was much less than the dimension of the surrogate data. A careful investigation of the reasons for this revealed that premature beats (which have a different physiological mechanism than normal beats) were producing spikes in the heart rate record, and the randomization technique used in constructing the surrogate data was transforming these highly ordered spikes into random white noise. Although the correlation dimension calculation was not sensitive to the occasional spikes in the heart rate record, it was very sensitive to the white noise in the surrogate data. Since premature beats occur more frequently in old people (something that is clinically well known), the correlation dimension was able to distinguish between heart rate and the surrogate data only in the old people. When the premature beats were removed from the analysis, the difference between heart rate and the surrogate data disappeared.

HIGH DIMENSIONS

Physiological systems are typically high-dimensional and, as such, are difficult to analyze. For example, Data Set B in the Santa Fe Time Series Prediction and Analysis Competition consists of simultaneous measurements of heart rate, respiration force, and blood oxygen concentration. These were only a part of the original data set, which included systolic and diastolic blood pressure. When one realizes that each of these measurements is itself the end result of dynamics coupled through nonlinear delayed feedbacks (Glass & Mackey, 1979; Glass et al., 1991), the possibility emerges of high-dimensional dynamics. Perhaps the intrinsic difficulty of these problems is reflected in the reluctance of most participants in the Competition to analyze Data Set B, in comparison to some of the other sets.

Although it is theoretically possible, under certain conditions in deterministic systems, for a single measured variable to be able to represent the entire system's dynamics, it is not well understood when this is a practical approach—the relationship between, say, respiration force and heart rate is sufficiently complex that it may be necessary to include both measured signals (as well as other coupled signals such as blood pressure) in a dynamical analysis of cardiovascular control. The use of multiple signals can lead to very high embedding dimensions (each signal may need to have several lags used in order to get a meaningful representation). The use of most nonlinear dynamics techniques in high-dimensional embeddings has not been well studied, and there is little knowledge about the best ways of representing multiple signals or of identifying interactions or coupling among signals.

As another example, consider the spread of excitation in heart muscle. This is a problem which is intrinsically infinite-dimensional. Common measurement methods involving surface ECG recordings reflect a projection of this problem to low dimensions. However, the ECG from a single lead is at best a crude indicator of the three-dimensional spread of excitation, particularly for rhythms with complex spatial organization such as ventricular fibrillation and ventricular tachycardia. Current mapping cardiac electrical activity are being made with 512 simultaneous electrodes (Johnson et al., 1992) and optical techniques provide resolutions of upwards of 10,000 pixels (Davidenko et al., 1992). How best to reduce such measurements to a low-dimensional representation is an unsolved problem.

Related issues involve the EEG as a measure of brain activity. Surface EEG recordings reflect averages of electrical activity over millions of cells. The functional significance of the EEG waves are not well understood. Claims that this data reflects low-dimensional dynamics have been numerous (see references in Babloyantz, 1989), but our own preliminary analysis of this problem does not show low-dimensional dynamics or evidence for determinism in data sets of normal EEG activity (Glass et al., 1992).

APPLICATIONS OF TIME SERIES ANALYSIS—CHAOS OR CHARACTERIZATION?

One of the principal goals, to date, of nonlinear dynamical time series analysis in physiology and medicine has been to establish whether time series arise from chaotic dynamical systems. There has been an unfortunate tendency to assume that if a computer program prints out a finite correlation dimension or one that has a non-integer value, or the program gives an estimate of a maximum Lyapunov exponent that is positive, then the time series is chaotic.

The search for deterministic chaos in complex physiological systems has focussed discussion away from other important issues. One possibility is that nonlinear statistics such as the correlation dimension can be effective ways of describing time series from physiological systems *even though these systems may not be chaotic.* Particularly in the EEG literature, there has been an attempt to use the correlation dimension to distinguish between different mental or physiological states (Albano et al., 1986; Babloyantz, 1989; Pritchard & Duke, in press). There have also been attempts to characterize the "complexity" of heart rate and blood pressure variability using statistics motivated by dimension and entropy (Kaplan et al., 1991; Pincus et al., 1993).

Insofar as it is desired to use statistics such as the dimension or entropy as characterizations of the (perhaps stochastic) dynamics of the time series, there are several issues of importance. It needs to be established whether the statistic indicates a physiological quantity (for example, sleep stage or the occurrence of an epileptic seizure or susceptibility to ventricular fibrillation). This can only be demonstrated by showing that the value of the statistic changes in a consistent manner as some physiological condition changes, or by showing differences between populations in physiologically distinct states. Perhaps the most remarkable claims for the application of dimension analysis have been made by Skinner et al. (1993) (see, also, Kroll & Fulton, 1991), who claim that prior to the onset of ventricular fibrillation the pointwise correlation dimension of the heart rate variability falls to a value near 1.

If a nonlinear statistic, such as the dimension, distinguishes time series from different physiological conditions, it is important to know if the statistic reflects information apparent visually in the time series or that can be found from more conventional measures such as the autocorrelation function. For example, EEGs are conventionally classified by the energy in various bands of the power spectrum. Are changes in dimensionality of the EEG correlated with power spectral changes? Questions such as these can perhaps be addressed by systematic use of surrogate data. ("Theoretical" arguments that a given nonlinear statistic is orthogonal to a conventional statistic such as the power spectrum, need to be examined with care. For example, the use of the zero crossing of the autocorrelation function to set the embedding lag τ in correlation dimension calculations may introduce a link between the power spectrum and the calculated dimension.)

In many cases the justification for using a given statistic (such as the correlation dimension) is founded on assumptions that may not be appropriate in physiology, such as that the system is deterministic or that all transients have died out, or that the level of noise is small. In these cases the hope is that the statistic will nonetheless prove to have physiological meaning even when the assumptions do not hold. A better approach might be to use statistics that do not make unwarranted assumptions, for example, statistics that are intended to provide useful dynamical information even for stochastic systems. For example, Pincus (1992) has introduced an "approximate entropy" statistic that can be interpreted for stochastic systems in terms of Markov chains. Nychka et al. (1992) have described an algorithm for calculating local divergence (i.e., Lyapunov exponents) which is based on nonparametric regression (with neural networks, see Gershenfeld and Weigend, this volume) and therefore expressly designed to be resistant to small amounts of noise.

CONCLUSION

In this summary review, we have placed emphasis on examples taken from cardiology. Applications of nonlinear dynamics to cardiology are more advanced than in other areas of medicine reflecting several factors including the comparatively simple anatomy of the heart, the fact that cardiologists deal with complex rhythms on a daily basis, and a tradition of theoretical modeling in cardiology. The other branch of medicine in which nonlinear dynamics is being actively pursued is neurology, but the problems in deciphering dynamics in complex neural systems are much more difficult. Claims that nonlinear dynamics and chaos are of vital importance for neural functioning have appeared (Skarda & Freeman, 1987). A more conservative account is by Milton et al. (1989). Another branch of medicine showing complex dynamics is endocrinology, but here we are still at the very beginning of a quantitative understanding of dynamic phenomena.

The analysis of complex time series requires significant skills in mathematics and computer analysis of data. Since very few physiologists or physicians have such skills, progress in the applications of nonlinear dynamics to physiology and medicine requires interdisciplinary groups. If nonlinear dynamic phenomena turn out to be important in medicine, it will be necessary in the future to offer training in nonlinear mathematics to a subset of physicians. Since complex dynamics cut across all branches of medicine, and it is unlikely that more than a handful of physicians will be interested in strong mathematical training, one can foresee a time when a new medical specialty, a *dynamicist*, will take a place amongst cardiologists, neurologists, and the other "ists" of medicine.

ACKNOWLEDGMENTS

This work is supported by a grant from the Natural Sciences and Engineering Research Council of Canada.

Clive W. J. Granger
Economics Department, University of California at San Diego, La Jolla, CA 92093

Forecasting in Economics

INTRODUCTION

Forecasting is a very serious activity in economics, involving a great deal of effort and money to produce them, and some of these forecasts attract a lot of attention with the press. It is probably fair to say that virtually every forecasting method that has ever been proposed has been applied, or at least carefully considered for use with economic data. I believe that this is true just for stock market data alone, for obvious reasons. It is worth remembering that economics is a decision science, with an objective of trying to explain the decisions of the agents and institutions that make up the economy. Forecasts, or the close cousins called "expectations," are important inputs into these decisions. For example, when a company is considering a large investment in a new factory or piece of machinery, its decision is partly based on a sequence of forecasts of the cash flow that will arise from the investment.

If Y_t is a series to be forecast with a horizon h, so that at time t a forecast is required of Y_{t+h}, based on an information set I_t, which consists of $\underline{X}_{t-j}, j \geq 0$ where $\underline{X}'_t = (Y_t, \underline{W}'_t)$, $\underline{W}_t$ being some set of other explanatory variables, then the random variable Y_{t+h} is characterized by the conditional distribution $g_h(y|I_t)$ where $g_h(y|I_t) = \text{Prob}(Y_{t+h} \leq y|I_t)$. The whole distribution may be a function of

I_t and its quantities may be forecast individually. However, current practice is to concentrate on just two moments, the conditional mean

$$f_{t,h} = E[Y_{t+h}|I_t]$$

and the conditional variance

$$V_{t,h} = E[(Y_{t+h} - f_{t,h})^2|I_t].$$

These are usually viewed as "unconditional forecasts" in that a neutral position is taken towards other future events. A great deal of attention is also paid to "conditional forecasts" in which a scenario is considered corresponding to a particular future event. For example, if the government is thinking of increasing some tax rate next year, what will happen to the sales of my company if this increase does occur? The government may make conditional forecasts about its revenue for the new tax, repeat this for various tax rates, and then decide on the tax by selecting the one that maximizes the forecast revenue. This survey will concentrate on unconditional forecasts, thus, will not be concerned with questions of policy or control.

2. MAIN PROPERTIES OF ECONOMIC VARIABLES AND DATA

The success, or otherwise, of forecasting methods is partly determined by the basic properties of the series to which they are applied. For economics, these properties include the following.

SHORT-LENGTH DATA SERIES

Major macroeconomic series often consist of only 200 to 500 terms as there are about 40 years of monthly data available, at most, since the end of World War II and its immediate recovery period. Many series are much shorter. Spain has only 20 usable annual figures for example—and currently the newly unified Germany has much less data. As a nonexperimental science one cannot just generate extra data, observing the series more frequently is helpful, but very costly, and extending the series back into much earlier periods may be adding irrelevant data because the economy has changed its structure and composition. Some financial data are very long and are available for very short time intervals. An advantage is that complicated models can be attempted and then evaluated quite soon on a "clean," post-sample data set. A problem is that because of the highly speculative nature of most financial markets, they are inclined to be generated in ways that are not found as often elsewhere in economics, including simple nonstationarities, outliers, and dramatic changes in variance. A discussion of the use of alternative forecasting methods with stock market prices is given by Granger (1992).

HIGH LEVELS OF MEASUREMENT ERROR

The noise-to-signal (variance) ratio may well be 1:3 or so for important macroseries. A large, developed economy is very complicated, and the mean variables are estimated by stratified sampling techniques rather than measured directly, which inevitably leads to measurement error. Many of the variables are difficult to define with any precision—exactly what is Gross National Product, what does "unemployed" really mean—and there is always an important hidden economy in which people work, earn, and trade without reporting it to official authorities, to avoid taxes or various constraints, and people often do not know exactly how much they earned or spent over some period. There is, thus, both knowing and unknowing misreporting which produce errors. These errors become embedded into the economic variables and are not simply added to the signal. If inflation is estimated with error and announced one month, this estimate will enter agents decisions about investment and employment next month, say. We have no reason to believe that the noise is Gaussian independent and identically distributed (iid).

NONSTATIONARITY

An assumption of stationarity is often quite unreasonable, there is frequently a clear trend in mean (sometimes with breaks in slope corresponding to structural breaks) and in variance, and there is a seasonality in mean (often not perfectly deterministic, having a stochastic and evolving shape) and possibly in variance. Further, we may expect parameters in any model to be time varying, both slowly changing due to changes in policy, tastes, and technology, and rapidly changing because of structural changes, such as produced by the doubling of oil prices in 1974 or by the political convolutions in Eastern Europe in recent years. Parameters may also change seasonally.

Even if trends in mean are not present or are estimated and removed, the series are often "long memory in mean," defined by

$$f_{t,h} \not\to \text{const} \qquad \text{as } h \text{ becomes large}$$

so that however far ahead you forecast, the best (least-squares) forecast remains a function of the information set. This is not a property of a stationary series, where $f_{t,h}$ tends to the unconditional mean of the series as h becomes large. Many stochastic processes have the long memory in mean property, including random walks and autoregressive unit root processes

$$A(B)Y_t = \varepsilon_t$$

where ε_t is iid, zero mean, and $A(B)$ is a polynomial in the lag operator B with $A(1) = 0$. (There could be several unit roots but this possibility is not considered here.) Many macro series have been found to be apparently long memory in mean processes, and will be denoted $I(1)$ (indicating a single-unit root). It is also possible to define long memory in variance and in distribution, corresponding to nonmixing processes.

NONLINEAR, STOCHASTIC RELATIONSHIPS

My personal beliefs, which I think are widely shared by other econometricians, are that forecasts derived from relationships between several variables are better than from univariate models, that these relationships should be nonlinear, and that the models should be truly stochastic rather than deterministic. This last belief arises because of the observation that the economy is continually being shocked by small innovations and unforecastable news items that affect decisions, as well as the occasional large unexpected shock. If raw macroeconomic data is used in a "dimensional estimator" of the kind occurring in chaos theory, the resulting estimate is often apparently of low dimension, because $\text{corr}(Y_t, Y_{t_h}) \simeq 1$ for these series, but when applied to the differenced series, the dimensions estimated are usually not low.

---

3. DECISIONS TO MAKE WHEN FORECASTING

DATA SET TO USE

One can just use the publicly available data, which is now both plentiful and becoming easily accessible on disks, but some data that one would like is not available, examples being plentiful regional or state data. Occasionally panel data is available, but is expensive to gather and much more is needed. It is possible to gather extra data, such as by running surveys of consumers' confidence, companies investment intentions or the opinions of various forecasters. Some of these prove to be helpful in improving macro forecasts, others are less clearly helpful.

FORECAST HORIZON

Whether to forecast long run (h large) or short run (h small) depends on the purpose for which the forecast is being made. Although most forecasts are fairly short run, there is a great deal of interest in the long run, which perhaps corresponds to an equilibrium much considered by economic theory.

WHAT MODELING TECHNIQUE TO USE

Although most forecasts are made using simple linear models and small information sets, some come from quite complicated nonlinear models. More attention is being given to nonlinear models, including neural networks, flexible Fourier forms, regime switching models, nonparametric techniques such as projection pursuit, and so forth. Tests of linearity are generally used first and, if a null of linearity is rejected, a method of identification of one or a few nonlinear techniques may be attempted, and then the appropriate model(s) estimated and fitted to an "in sample" training data set. The model is usually evaluated by comparing the post-sample forecasting ability of the nonlinear models with alternatives including a linear one. A problem with some of the nonlinear techniques is that, if there is little or no actual nonlinearity in the series, they are inclined to "data mine" and thus overfit in sample but then to perform poorly in the post-sample.

SIZE OF THE SYSTEM MODELED

The series modeled and forecast is generally a vector with m components, but there is considerable difference in strategies about the size of m used, from univariate ($m = 1$) to mid-size vector autoregressive (linear models) ($m = 10$, say) up to large structural econometric models ($m = 100$ to 400, say) usually with simplistic nonlinearities and dynamics. There does seem to be advantages in moving from $m = 1$ to $m = 10$, but going up to $m = 200$ is much more controversial.

HOW TO EVALUATE FORECASTS

A variety of techniques for evaluating models and their forecasts have been evolved, including tests of significance of whether one set of forecasts are better than another, whether there is an advantage in combining the forecasts, either linearly or nonlinearly, and whether a model A "encompasses" model B—which means that A forecasts all of the outputs of B just using the inputs of A. The least-squares cost function is usually used as an evaluation criterion, largely for mathematical and computational convenience, but actual cost functions are probably nonsymmetric. Surprisingly, little work has been done considering actual cost functions in practice. Discussions of the evaluation question can be found in Granger and Newbold (1986) and Clements and Hendry (1993).

4. THE ECONOMY AS A MECHANISM

It should be emphasized that the economy is not a physical or biological mechanism. The U.S. economy, for example, consists of over 80 million families, each of which is making individual decisions designed to optimize their own well-being or "utility," each having different family characteristics and histories and with decisions based largely on different information. The macro economy is then the aggregate of the series produced by the individual families, an example being consumption. Whether or not a nonlinear relationship at the microrelationship produces nonlinearity at the macro level depends on various conditions on the type of nonlinearity and the common features on the various micro information sets.

As an example of a family decision making which is not a rule-driven, autonomous, or deterministic mechanism, consider the use of electricity by a family over some period, such as a month. The family starts the month with a stock of appliances which use electricity, such as an air conditioner, and in most cases this stock changes little over the month. For a longer period, appliances will change and probably become more efficient. The family cannot use electricity without owning these appliances. Some forecasting models of electricity demand, called end-use models, are based just on appliance stocks and forecasts of these stocks. However, throughout the month, the family makes a sequence of decisions about whether or not to use an appliance on any given occasion. These decisions will depend on family characteristics, such as size and age composition, house size and level of insulation, and its economic variables such as the family income and the prices of electricity and other potential purchases. The economists will emphasize income and prices, even though many families do not actually know these variables with any precision. The decisions are made frequently and not independently and differ across families. One aspect of the economy not being a mechanism is that individuals dislike being thought of as being forecastable. If I sit down to dinner in a restaurant and forecast what everyone I am with will order, this will certainly change their decisions and will probably ensure that my forecasts are incorrect.

As has been emphasized, the values taken by many economic variables are strongly influenced by decisions by agents, and it is often true that these decisions are related to forecasts or expectations. A great deal of the modeling effort of econometricians is concerned with these expectations. A simple example is that a farmer decides what crops to plant in his fields according to the expected price he will receive for the various alternative crops. The eventual price he will eventually get depends on what crops were chosen—and how successful they were.

5. AREAS OF CURRENT INTEREST

As with all areas of research there are fluctuations in what is topical and of interest. I would like to emphasize three areas which are currently active in research and have good potential for forecasting.

COINTEGRATION

If X_t, Y_t are both $I(1)$ (long memory in mean) series, but there is a constant A such that the series Z_t given by

$$Z_t = X_t - AY_t$$

is *stationary*, then X_t, Y_t are said to be cointegrated. This is a special property that does not hold for most pairs of $I(1)$ series. It can be shown that it only occurs if there is a decomposition

$$X_t = AW_t + \tilde{X}_t,$$

$$Y_t = W_t + \tilde{Y}_t,$$

where W_t is $I(1)$ and $\tilde{X}_t, \tilde{Y}_t$ are both stationary. Thus, the reason why we use X_t, Y_t, and $I(1)$ is because of the common factor W_t. If X_t, Y_t are cointegrated, it follows that they must appear to have been generated by the "error correction" model:

$$\Delta X_t = \gamma_1 Z_{t-1} + \text{lags}\Delta\gamma_t, \Delta Y_t + e_{xt},$$

$$\Delta Y_t = \gamma_2 Z_{t-1} + \text{lags}\Delta\gamma_t, \Delta Y_t + e_{yt},$$

where e_{xt}, e_{yt} are a joint white-noise vector, and necessarily at least one of γ_1, γ_2 is nonzero. Thus, by testing for the property of cointegration—and several adequate tests now exist—if it is found, this implies that the error-correction model should be fitted, where the number of lags involved are chosen by some model selection criterion, such as BIC as defined in Granger and Newbold (1986). If cointegration is not found, but X_t, Y_t are $I(1)$, the appropriate (linear) model is a vector autoregression in the changes. The error-correction model will provide superior short- and long-run forecasts for X_t and Y_t, and these forecasts for X_t and Y_t, and these forecasts "hang together" in that

$$f_{t,h}^{(x)} = A f_{t,h}^{(y)} \quad \text{for } h \text{ large}.$$

The concept can be generalized to a vector of $I(1)$ series and nonlinearity can be introduced in various ways, with $\gamma_j z_{t-1}$ replaced by $\gamma_j(z_{t-1})$ for $j = 1, 2$ for some functions with $\gamma_j(0) = 0$, in the error-correction model, or with the linear cointegration relationship replaced by

$$z_t = g_1(x_1) - g_2(y_2)$$

with some unbounded functions $g_j(\ ), j = 1, 2$. Improvements in forecasts from such generalizations have yet to be proved in practice.

REGIME-SWITCHING MODELS

A class of model which is now receiving an increasing amount of attention takes the form

$$Y_t = \underline{\beta}_2' \underline{X}_t + \phi(\underline{\gamma}' \underline{X}_t)[\underline{\beta}_2' \underline{X}_t] + e_{yt}$$

where $\underline{X}_t$ is the vector of explanatory variables, possibly including $Y_{t-j}, j > 0$ and $\phi(\ )$ is usually a sigmoid function such as the logistic function

$$\phi(w) = [1 + \exp(-w)]^{-1}.$$

For w large, Y_t obeys the linear model

$$Y_t = \underline{\beta}_1' \underline{X}_t + e_{yt}$$

and for w small, it obeys

$$Y_t = (\underline{\beta}_1' + \underline{\beta}_2') \underline{X}_t + e_{yt}$$

so that

$$W_t = \underline{\gamma}' \underline{X}_t$$

is the switching variable that smoothly moves the generating mechanism from one regime to another. As given, these regimes are linear but could be replaced by specific parametric nonlinear models. The models are fairly easy to specify and to estimate and often have an interpretation that is attractive to an econometrician. For example, W_t may measure the state of the business cycle, as one would expect a different relationship between wages and employment, say, at the peak of the cycle than at a trough.

TIME-VARYING PARAMETERS

For reasons given above, the economy is clearly evolving and so parameters are likely to alter values. For a linear model, these changes can be captured using a Kalman filter, state-space formulation. The resulting model is often difficult to differentiate from some nonlinear models. This approach is not new and there is an accumulation of evidence that superior forecasts are often obtained compared to linear, time constant parameter models. However, the way in which the parameters change do not always have sensible interpretations and may just be proxies for a complicated nonlinearity.

6. LEVELS OF FORECAST UNCERTAINTY

To make a decision in an uncertain environment, which is typically the case in economics, it is necessary to not only have a forecast of the conditional mean but also to have an estimate of the uncertainty level of this forecast. A clear difficulty with economic forecasts is that this uncertainty is high relative to the object being forecast. To illustrate this problem, the following table shows average root mean squared errors (ARMSE) for forecasts for several important macroeconomic variables, using quarterly data for the period 1980:2 to 1985:1 and horizons $h = 1, 4$ and 8 (one quarter, one and two years). The averages are over seven forecasting models (five models for $h = 8$)—mostly large-scale structural models or vector autoregressions. I think that most of these forecasting models can be characterized as being

TABLE 1 Forecast Models

	Average Value Over Period (annual rate)	ARMSE h=1	ARMSE h=4	ARMSE h=8
GNP Price Deflator (Inflation Rate Measure)	~ 6.0	1.67	1.83	2.74
GNP (Nominal) Growth Rate	~ 8.0	4.73	4.09	3.76
GNP (Real) Growth Rate	~ 2.6	3.93	2.76	1.96
Real Nonresidential Fixed Investment	~ 7.4[1]	11.7	9.9	7.5
Civilian Unemployment	~ 7.0	0.3	1.21	1.96

[1] % points per year.

generally considered of good quality and are highly regarded. Their forecasts are issued early in the quarter over which they are forecasting, for $h = 1$. The original data are taken from McNeese (1986) who lists the names of the forecast models.

Let us illustrate some of these numbers.

Interval forecasts for real GNP growth rate three months ahead are

$$2.6 \pm 4 \quad \text{for} \quad 1 \quad 2 \text{ s.d i.e.} \quad -1.4 \quad \text{to} \quad 6.6 \,,$$
$$2.6 \pm 8 \quad \text{for} \quad 2 \quad 2 \text{ s.d i.e.} \quad -5.4 \quad \text{to} \quad 10.6 \,,$$

(s.d. $\equiv$ estimated standard deviation, here taken to be the ARMSE).

For civilian unemployment three months ahead, the bands are

$$\pm 1 \text{ s.d.} \quad 6.7 \text{ to } 7.3$$
$$\pm 2 \text{ s.d.} \quad 6.4 \text{ to } 7.6$$

which are reasonably narrow bands, but, for two years ahead the bands are

$$\pm 1 \text{ s.d.} \quad 5 \text{ to } 9$$
$$\pm 2 \text{ s.d.} \quad 3 \text{ to } 11$$

which are so broad as to be of little value. The worst case is obviously for real fixed investment; for three months ahead the bands are

$$\pm 1 \text{ s.d.} \quad -5.7 \text{ to } 19.1$$
$$\pm 2 \text{ s.d.} \quad -16.0 \text{ to } 30.8$$

which are embarrassingly wide. All of the forecasting models produce large RMSE values, ranging between 10.3 to 15.3 to get an average of 11.7.

In the above tables, the ARMSE values occasionally decrease as h increases, contrary to theory. This can be explained by the short sample sizes and the fact that the $h = 1$ and $h = 8$ are based on slightly different periods, as almost two years of data is lost when forecasting two years ahead rather than just one quarter ahead.

The broad forecast intervals either suggest that the economy quickly becomes difficult to forecast, or that there remains a great deal that can be done to produce superior forecasts.

7. CONCLUSION

The quality of economic forecasts continues to improve steadily as we learn how to use old methods better and which of the new methods are most promising. I believe that nonlinearity will produce a further improvement, but that it is not just a matter of taking any nonlinear technique off the shelf and using it successfully, progress will occur from careful modeling, evaluation, and learning.

V. S. Afraimovich,†‡ M. I. Rabinovich,† and A. L. Zheleznyak†

†Department for Nonlinear Science, Institute of Applied Physics, Russian Academy of Sciences , 46 Uljanov St., 603600 Nizhny Novgorod, Russia.
‡Center for Dynamical Systems and Nonlinear Studies, Georgia Institute of Technology, Atlanta, Georgia, 30332-0190.

Finite-Dimensional Spatial Disorder: Description and Analysis

We discuss a new approach to the analysis of spatial disorder (including the snapshots of the fields evolving in time), taking, as a basis, translational dynamical systems with d times. Space series is used to determine the quantitative characteristics of disorder, such as spatial correlation and pointwise dimensions and others.

1. INTRODUCTION

Spatial disorder, i.e., irregular distribution in space of elements, structures, or fields, is a traditional object for investigations in different branches of science. There are plenty of examples of spatial disorder, such as random distribution of gas molecules or of microcracks in metal before breaking, a picture of rough oceanic surface, distribution of substance in space, and so on.

Spatial disorder is usually analyzed employing methods of statistical physics (see, e.g., Ziman, 1979). Knowledge of standard characteristics such as spatial Fourier spectrum, correlation scale, and others enable us, in particular, to distinguish between short- and long-range order and to determine the size of domains

Times Series Prediction: Forecasting the Future and
Understanding the Past, Eds. A. S. Weigend and N. A. Gershenfeld, SFI Studies
in the Sciences of Complexity, Proc. Vol. XV, Addison-Wesley, 1993   **539**

(a)   (b)

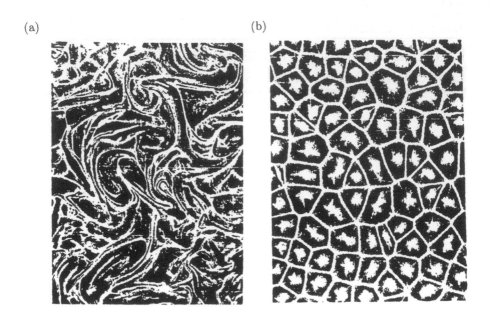

FIGURE 1   (a) Mixed initial state of liquid in Marangoni convection; (b) "half-ordered" lattice observed on the background convective crystal structure.

and orientation of the spins. But traditional approaches do not allow us to answer a natural question: Does pattern formation follow simple laws (rules) or is it random spatial distribution that can be analyzed only statistically? In other words, is spatial disorder of a deterministic origin and, if yes, then what are the properties of the dynamical system generating this disorder? For example, it appears reasonable to suppose that an absolutely random initial (handmade) structure of liquid surface at Bénard-Marangoni convection(Figure 1(a)) evolves into a distinctive cellular structure (Figure 1(b)) according to the laws which have a dynamical (not a statistical) origin.

This issue is highly important. The fact that the disorder is generated by a dynamical system with finite-dimensional phase space already gives us information about its origin, and the changing in time of the system's properties helps us understand evolutionary peculiarities of spatial disorder.

2. THE EXAMPLE OF FINITE-DIMENSIONAL SPATIAL DISORDER

There are many examples illustrating that irregular spatial field distributions generated by finite-dimensional dynamical systems must exist in Nature, and appear to be no less typical than finite-dimensional temporal chaos.

A universal model that is widely used in theory of nonequilibrium media is a one-dimensional generalized Swift-Hohenberg equation

$$\frac{\partial u}{\partial t} = -\varepsilon u + \beta u^2 - u^3 - \left(k_0^2 + \frac{\partial^2}{\partial x^2} \right)^2 u \tag{1}$$

that can be derived, for example, from the Boussinesq equation near the threshold of linear instability in the problem of Rayleigh-Bénard convection (Normand et al., 1977; Swift & Hohenberg, 1977; see also Rabinovich & Sushchik, 1990). It is a gradient equation with a free energy functional (also referred to as the Lyapunov functional):

$$F = \int_{\Omega} \left[\frac{\varepsilon}{2} u^2 - \frac{\beta}{3} u^3 + \frac{1}{4} u^4 + \frac{1}{2} \left(\left(k_0^2 + \frac{\partial^2}{\partial x^2} \right) u \right)^2 \right] dx \,. \tag{2}$$

This equation may be represented in a gradient form

$$\frac{\partial u}{\partial t} = -\frac{\delta F}{\delta u} \,. \tag{3}$$

The functional F can not increase along any trajectory, because:

$$\frac{dF}{dt} = -\int_{\Omega} \left(\frac{\partial u}{\partial t} \right)^2 dx \leq 0 \,, \tag{4}$$

and system (1) can demonstrate only two types of solutions as $t \to \infty$: propagating fronts (if F has no minima) or static distributions along x corresponding to local minima of F. Such distributions $u(x)$ satisfy the ordinary differential equation

$$u_{xxxx} + 2k_0^2 u_{xx} + (k_0^4 + \varepsilon)u - \beta u^2 + u^3 = 0 \tag{5}$$

and can be treated as space series or snapshots, i.e., as functions of the point on the trajectories of the finite-dimensional dynamical system (5) which we refer to as a translational dynamical system.

The dynamics of system(s) may be very complicated. Computer experiments show that a homoclinical structure exists in phase space of the system and contains not only regular (periodic and quasi-periodic) trajectories but also chaotic trajectories (see Figure 2).

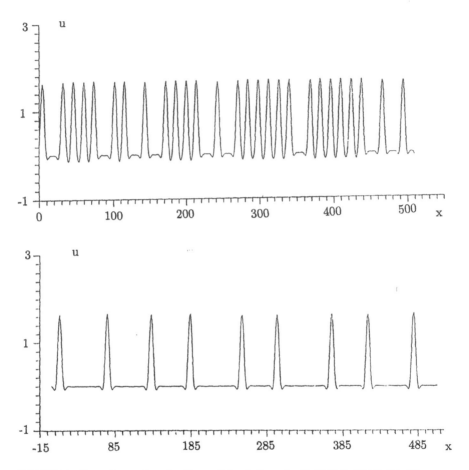

FIGURE 2   Snapshots—the chaotic spatial solutions for Swift-Hohenberg Eq. (1).

3. SPATIO-TEMPORAL ANALOGY AND ANALYSIS OF ONE-DIMENSIONAL SPATIAL DISORDER

Direct spatio-temporal analogy is a handy tool in the investigation of irregular distributions along a single spatial coordinate x, that allows for the description of the properties of spatial disorder in terms of nonlinear dynamics. Thus, advanced mathematical and algorithmic apparatus using the approach proposed by Packard et. al (1980) and Takens (1981) proved to be very fruitful.

Following this approach (see, also, Normand, Pomeau, & Velarde, 1977), the observable $\underline{u}$ that is a sequence of values of the state vector component of the system at discrete moments of time: $\underline{u} = \{u(j), j \in (-\infty, \infty)\}$, where $u(j) = u(t_j)$ can be considered as an element of space B of all infinite sequences with a bounded norm $\|\underline{u}\|_B < \infty$. The shift operator S on B can be defined as a shift of the sequence:

$$S: \{u(j)\} \rightarrow \{(u(j+1))\}.$$

The pair $(B, \{S^t\}_{t \in \mathbf{Z}})$ specifies a universal dynamical system while $\{S^t \underline{u}\}_{t \in \mathbf{Z}}$ is its trajectory that goes through the point $\underline{u}$.

According to the Takens algorithm, m-dimensional clusters

$$U_j^{(m)} = (u(j), u(j+1), \ldots, u(j+m-1))$$

are constructed from points of the observable $\underline{u}$ and the behavior

$$\ldots \xrightarrow{S} U_0^{(m)} \xrightarrow{S} U_1^{(m)} \xrightarrow{S} \ldots \xrightarrow{S} U_j^{(m)} \xrightarrow{S} \ldots$$

of these clusters is analyzed. Formally, the vectors $U_j^{(m)}$ are the projections of the observable $\underline{u}$ onto the m-dimensional subspace, M_m, of space B (we define the so-called natural projection Π_m that retains only m components for each infinite-dimensional sequence). Apparently, if Π_m is invertible on the considered trajectory, then $(M_m, \{\tilde{S}^t\}_{t \in \mathbf{Z}})$, where $\tilde{S}^t = \Pi_m \circ S^t \circ \Pi_m^{-1}$, is a finite-dimensional dynamical system.

If the fractal dimension of the invariant set A_u of a universal dynamical system is finite $(D(A_u) < \infty)$ and the integer $m \geq 2 \bullet D(A_u)$, then, according to the "embedding" theorem (for last results in this direction, see Sauer et al., 1992; on occasion of a usual reference see Mãne, 1980; related to the embedding theorem, see Sauer et al., 1992, remark 4.8), one-to-one bicontinuous projections, $\Pi_m: B \rightarrow M_m$, restricted to A_u are typical among all projections $B \rightarrow M_m$. To put it in other words, in a general case the topological properties of the infinite-dimensional universal dynamical system $(B, \{S^t\}_{t \in \mathbf{Z}})$ and of finite-dimensional system $(M_m, \{\tilde{S}^t\}_{t \in \mathbf{Z}})$, such as the number of fixed and periodic points, the existence of homoclinic trajectories, etc., are the same. If, in addition, Π_m on B and Π_m^{-1} on $\Pi_m(B)$ are Lipschitz-continuous, then the dimensions of invariant sets coincide. The corresponding $\underline{u}$ is called a *finitely generated observable*.

Employing this procedure for the reconstruction of a universal dynamical system by a one-dimensional observable, which is in our case a space series or a snapshot, $\underline{u}(x)$, we can investigate the properties of this system by means of qualitative

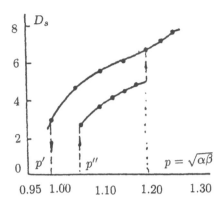

FIGURE 3 "Chaos-chaos" phase transition and hysteresis on the (p, D_s) plane in model (6): $s = \sqrt{\beta/\alpha} = 1.15$ and $R = 10^4$.

characteristics such as dimensions, entropies, or spectrum of Lyapunov exponents of the reconstructed invariant set (see, e.g., Mayer-Kress, 1986).

This approach is extremely important even in a one-dimensional case. Thus, it was used (Bazhenov et al., 1992) to study dimensional characteristics of field distribution within a one-dimensional complex Ginzburg-Landau equation.

$$\frac{\partial A}{\partial t} = A - A|A|^2 + \frac{\partial^2 A}{\partial x^2} + i\alpha\frac{\partial^2 A}{\partial x^2} + i\beta A|A|^2 . \tag{6}$$

The spatial dimension, D_s, versus the parameter $p = \sqrt{\alpha\beta}$ is shown in Figure 3. The plot has two remarkable features: stepwise increase of D_s by almost 1.5 times near the point p'' and hysteresis. The jump in the spatial dimension has the following explanation. The field described by Eq. (6) may have two qualitatively different irregular states. One of them is "phase turbulence" when the phase changes chaotically and the amplitude weakly pulsates near its average value (see Figure 4(a)). The other state is "strong" turbulence when both amplitude and phase are strongly irregular (Figure 4(b)). In the course of formation of spatial disorder, new spatial perturbations (amplitude pulsations) emerge in the transition across the point p'' and this transition can be considered, in a sense, to be a critical phenomenon.

The onset of one or another chaotic regime in the region of the parameters p' and p'' depends on initial conditions (in this case two strange attractors co-exist in the phase space of the dynamical system (6)). This explains the hysteresis phenomenon: in the transition through the critical point p'' from right to left, we remain in the attraction domain of a multidimensional attractor. While starting

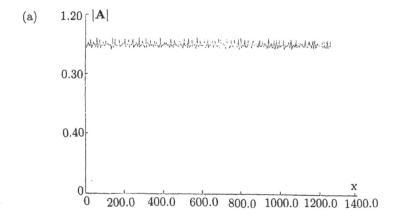

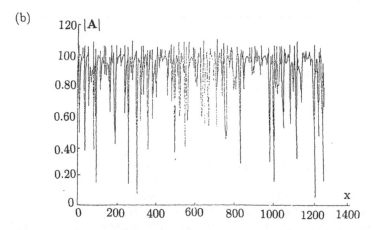

FIGURE 4 Amplitude distribution along x at (a) "phase turbulence" (solution of Eq. (6) for the parameters $\alpha = 1.0$, $\beta = 1.3$, and $R = 10^4$) and (b) strong "amplitude turbulence" ($\alpha = 1.1$, $\beta = 1.44$, and $R = 10^4$).

from regular initial conditions on the interval $[p', p'']$, we enter the regime of low-dimensional chaos.

Thus, the changes in the properties of spatial disorder with the variation of the parameters of system (6) are manifested in the change of spatial dimension D_s. Note that the dimension of time series, D_t, also changes in a jump at the same value of the parameter p''. Perhaps, in this case there is a relation between D_s and D_t and numerical results show that D_s and D_t are related as $D_s \leq D_t$.

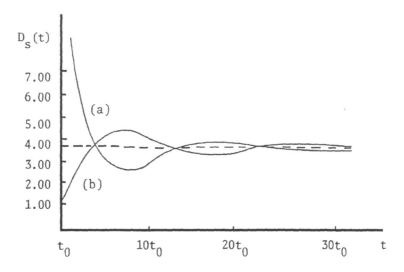

FIGURE 5   Time dependence of the correlation dimension, D_s, of the space series for CGLE under random (a) and generated by Hénon map (b) initial conditions.

Another interesting result that is connected with quantitative analysis of disordered spatial distributions was obtained in computer experiment with CGLE in a very long one-dimensional system (Beilin et al., n.d.). Spatial dimension was continuously evolving in time to one constant value (see Figure 5), irrespective of whether initial spatial distribution had been generated by a infinite-dimensional random process or by a Hénon map with the dimension $D_s = 1.20$. The regime of spatio-temporal chaos whose snapshots are finite-dimensional disorder with a definite value of dimension is established for quite different initial conditions.

Thus, we have confirmed that the dynamical approach is very fruitful for the description of spatial disorder. However, when we have more than one spatial coordinate, we cannot draw direct spatio-temporal analogy.

4. TRANSLATIONAL DYNAMICAL SYSTEMS: FORMAL MATHEMATICAL DESCRIPTION

Consider first the case of two spatial variables, x_1 and x_2. The observable, $\underline{u}$, i.e., the snapshot of the field u, is a function of two variables, $u(x_1, x_2)$. Clearly, the shift operator T on the plane must be determined by the two parameters, α_1 and α_2, corresponding to the shift of the snapshot along each spatial coordinate. Designating $\alpha = (\alpha_1, \alpha_2)$, it is natural to take $T^{\alpha}u(x) = u(x + \alpha)$, where $x, \alpha \in \mathbf{R}^2$. In real experiments the snapshot can often be specified in a discrete form by its values, $u(j)$, at the nodes, $j = (j_1, j_2)$, of a two-dimensional lattice. Then $T^{\alpha}u(j) = u(j + \alpha), j, \alpha \in \mathbf{Z}^2$.

The dynamical systems generated by such shift operators that depend on two parameters or on two "times" (the so-called actions of $\mathbf{R}^2$ or $\mathbf{Z}^2$ groups) are generalization of classical dynamical systems with one "time" (see, e.g., Katok et al., 1985).

We now propose an approach to the investigation of irregular d-dimensional spatial distributions that employs description of the properties of spatial disorder in terms of qualitative characteristics (dimensions, entropies, Lyapunov exponents) of the invariant sets of dynamical systems with "d-times."

Let us introduce formal mathematical definitions. We take interest in isotropic unbounded media whose properties do not depend on the chosen coordinates. The dynamical systems generating such translational invariant distributions are referred to as translational dynamical systems. These systems are reversible (their properties do not change when x is replace by $-x$); consequently, translational dynamical systems are similar to Hamiltonian ones.

Consider a set of continuous (vector) functions $u(x)$, $x \in \mathbf{R}^d (u \in \mathbf{R}^p)$, employing conventional procedures of addition and scalar multiplication. Introducing onto this set a distance, we obtain a metric space B that we refer to as a phase space of the system. To each d-dimensional vector, $\alpha = (\alpha_1, \ldots, \alpha_d) \in \mathbf{R}^d$, we associate a translational map $T^{\alpha}: B \rightarrow B$ that is specified by the expression $T^{\alpha}u(x) = u(x+\alpha)$. In this fashion we determine the translational dynamical system, i.e., the action of the group $\mathbf{R}^d$ on B, or, in other words, we are concerned with a dynamical system with d times.

If the process under study is such that knowing the initial state (initial field distribution) one can unambiguously determine the subsequent states at any moment of time, then a subgroup of evolution operators $\{S^t\}_{t \geq 0}$ also acts on B; i.e., an evolutional dynamical system is determined as well. The behavior of the trajectories of translational and evolutional dynamical systems in the common phase space B gives a full mathematical description of the spatio-temporal properties of the nonequilibrium medium of interest.

Having defined formally the translational dynamical system, we can introduce quantitative characteristics of the snapshot $\underline{u}$ as the characteristics of its invariant set $A_{u(x)} \equiv \{T^{\alpha}u(x)\}_{\alpha \in \mathbf{R}^d}$.

We can rigorously define measure-independent characteristics such as fractal dimension or topological entropy and, if the invariant (with respect to T^{α}) measure μ is introduced on A_u (i.e., $\mu(T^{\alpha}A_u) = \mu(A_u)$), then we can determine measure-dependent characteristics, for example, pointwise and correlation dimensions (see Afraimovich et al., 1992), Kolmogorov-Sinai entropy and, in principle, Lyapunov exponents. All these characteristics will be referred to as the characteristics of the snapshot.

Note that if, for example, a two-dimensional snapshot is periodic with respect to x_1 and x_2, the set A_u is merely a two-dimensional torus and the dimension of the snapshot is $D(A_u) = 2$; if the snapshot has a quasi-periodically repeating structure (like a quasi-crystal), then A_u is also a torus but now of a higher dimension and the dimension of the snapshot is an integer, too; while, for the patterns that are chaotically distributed over the plane, A_u will be a fractal set having a noninteger dimension in a general case. The time evolution of the snapshots corresponds to the motion of the set A_u in the space $B: A_u \to^{S^t} A_{S^t u}$.

5. GENERALIZING TAKENS' APPROACH FOR d-DIMENSIONAL SPACE SERIES

Consider now the generalization of Takens' approach to systems with d-dimensional time. For the sake of simplicity, we will take the space and time to be discrete by analogy with ordinary dynamical systems; i.e., we will take $\mathbf{Z}^d$ instead of $\mathbf{R}^d$ and $\mathbf{Z}_+$ instead of $\mathbf{R}_+$.

Following Takens (1981), we will call the snapshot

$$\underline{u} = \{u(j), j = (j_1, \ldots, j_d) \in \mathbf{Z}^d, u \in \mathbf{R}^p\}$$

to be a finitely generated one if there exist: (1) a dynamical system with d-dimensional time and a finite phase space M_m and (2) a Lipschitz continuous one-to-one conjugate map $h: A_u \to M_m$ such that the inverse map h^{-1} is also Lipschitz-continuous.

Let us introduce the space of observables, B, as a set of all infinite d-dimensional sequences of $\underline{u}$ such that the norm

$$\|\underline{u}\|_B = \sum_j \frac{\|u(j)\|\mathbf{R}^p}{2^{|j|}} < \infty,$$

where $|j| = |j_1| + \ldots + |j_d|$ and $\|\underline{u}\|\mathbf{R}^p$ is a usual norm in $\mathbf{R}^p$. When conventional operations of addition and scalar multiplication are introduced in a standard fashion, B is Banach's space.

Having defined on B the operator of translation, T^α, as

$$T^\alpha: \{u(j)\} \to \{(u(j + \alpha))\},$$

we obtain a universal dynamical system $(B, \{T^\alpha\}_{\alpha \in \mathbf{Z}^d})$ with a trajectory $\{T^\alpha \underline{u}\}_{\alpha \in \mathbf{Z}^d}$ that passes through $\underline{u}$.

Now, like in the case of one spatial coordinate, we will form from neighboring points of the snapshot m^d-dimensional clusters:

$$U_i^{(m)} = \{u(i + j), i \in \mathbf{Z}^d, j \in K_m^d\},$$

where $K_m^d = \{j = (j_1, \ldots, j_d) \in \mathbf{Z}^d, 0 \le j_l < m - 1\}$ is an integral d-dimensional cube having the side m. The set $M_{m^d} = \{U_i^{(m)}\}$ is m^d-dimensional subspace of space B and $\Pi_{m^d} : B \to M_{m^d}$ is natural projection.

Let for a fixed snapshot $\underline{u}$ the dimension of the invariant set be $D(A_u) < \infty$ and m be such that $m^d \ge 2D(A_u)$. Then according to the embedding theorem one-to-one bicontinuous projections are typical $B \to M_{m^d}$ projections restricted to the set A_u, i.e., the topological properties of invariant sets, are, in a general case, the smae for both the dynamical systems, $(B, \{T^\alpha\})$ and $(M_{m^d}, \{\tilde{T}^\alpha\})$, and $\underline{u}$ can be a *finitely generated snapshot*.

6. NUMERICAL ALGORITHMS FOR TWO-DIMENSIONAL SPACE SERIES

The lemmas proved by Afraimovich et al. (1992) allow us to propose the algorithms for the calculation of correlation and pointwise dimensions generalizing the algorithms presented by Grassberger and Procaccia (1983c) and Termonia (1983). Let us take a two-dimensional ($d = 2$) snapshot in the form of a two-dimensional array $\underline{u} = \{u(i,j), i, j \in \mathbf{Z}_+\}$. Actually, the array is limited in size: $i \le N_1, j \le N_2$. For each integer $m \ge 1$, we can construct matrices on the order of $(m \times m)$:

$$A_{K,L}^{(m)} = \{u(k,l), k = K, \ldots, K + m - 1, l = L, \ldots, L + m - 1\}.$$

Let us define the correlation integral in the form

$$C^{(m)}(\varepsilon) = \frac{R^{(m)}(\varepsilon)}{[(N_1 - m)(N_2 - m)^2]},$$

$$R^{(m)}(\varepsilon) = \#\{((K,L),(K',L')) : \text{dist}(A_{K,L}^{(m)}, A_{K',L'}^{(m)}) \le \varepsilon\}; \tag{7}$$

where $\# \{E\}$ is the number of elements on set E. Then, for sufficiently small ε, the ratio $\log C^{(m)}(\varepsilon) / \log \varepsilon$ can be approximately equal to the correlation dimension D_C of the two-dimensional snapshot in the m^2-dimensional embedding space. If D_C becomes constant when $m > m^*$, we can say that the spatial distribution is finitely generated.

Following Ruelle (1990) we can estimate roughly the minimal size of the array $(u(i,j))_{N_1 \times N_2}$ that is needed for correct evaluation of the dimension within the $[\varepsilon', \varepsilon'']$ interval. By virtue of

$$D_C \cong \frac{\log_2 C^{(m)}(\varepsilon'') - \log_2 C^{(m)}(\varepsilon')}{\log_2 \varepsilon'' - \log_2 \varepsilon'}, C^{(m)}(\varepsilon') \ge \frac{1}{N_1^2 \cdot N_2^2}, \text{ and } C^{(m)}(\varepsilon'') \le 1,$$

assuming $\varepsilon'' = 2^\delta \varepsilon'$, we have the following estimate:

$$D_C \leq \frac{2}{\delta} \log_2(N_1 \cdot N_2). \tag{8}$$

Note that for a d-dimensional snapshot, such an estimate has the form

$$D_C \leq \frac{2}{\delta} \sum_{l=1}^{d} \log_2 N_l. \tag{9}$$

Thus, when determining the correlation dimension of a multidimensional snapshot, one must bear in mind that the number of discretization points along each time coordinate may be much smaller than in the case of one-dimensional time.

Construction of the correlation integrals needs a great number of calculations of the distances between the matrices. For this reason we prefer to compute the distance in the form:

$$\text{dist}\left(A_{K,L}^{(m)}, A_{K',L'}^{(m)}\right) = \max_{\substack{k=1,\ldots,m \\ l=1,\ldots,m}} \left(|u_{K+k-1,L+l-1} - u_{K'+k-1,L'+l-1}|\right). \tag{10}$$

The number of calculations reduces significantly if all the matrices $A_{K,L}^{(m)}$ are compared to the set of reference matrices

$$\left\{A_{K_i,L_j}, i \leq i_{\text{ref}}, j \leq j_{\text{ref}}\right\}.$$

In this case the accuracy of calculation of the correlation dimension of a two-dimensional snapshot is determined by the inequality:

$$D_C \leq \frac{1}{\delta} \log_2(N_1 \cdot N_2 \cdot i_{\text{ref}} \cdot j_{\text{ref}}). \tag{11}$$

The testing showed that the behavior of the correlation integrals does not, actually, depend on the turn of the snapshot by an arbitrary angle, which indicates that the algorithm is robust. Figure 6 shows the plots of $\log_2 R^{(m)}$ versus $\log_2 r$ ($r = 2^{10} \varepsilon/\varepsilon_{\text{max}}$) for the two-dimensional snapshot $U(x,y) = \sin x \sin(\sqrt{3}/2y)$ (Figure 7) that is a two-dimensional torus in some phase space. The correlation dimension was calculated to be $D_C \in [1.96; 2.03]$ for $N_1, N_2 = 256, i_{\text{ref}}, j_{\text{ref}} = 4$.

Thus we arrive at the following definition: *Finitely generated disorder* is the disorder that can be considered as a trajectory of a certain dynamical system. In a one-dimensional case, it is an ordinary dynamical system, for instance, an ODE system. In a two-dimensional geometry, we have a dynamical system with two "times."

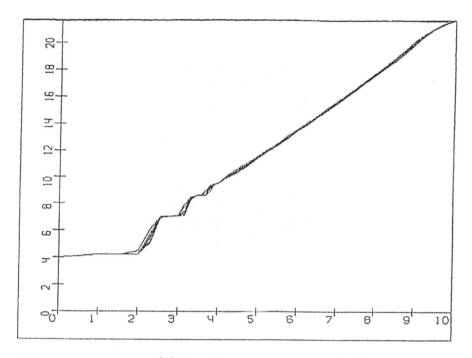

FIGURE 6 The plot $\log_2 R^{(m)}(r)$ versus $\log_2 r$ for two-dimensional field $U(x,y) = \sin x \sin(\sqrt{3}/2y)$ for different values of the dimension of embedding space, m^2.

Such dynamical systems may be specified in different ways. For example, a dynamical system generating a simple field

$$u(x,y) = A\cos(\omega_1 x + \varphi_1)\cos(\omega_2 y + \varphi_2)$$

may be written as a PDE system[1]

$$\frac{\partial u}{\partial x} = u_1 , \frac{\partial u_1}{\partial x} = -\omega_1^2 u , \frac{\partial u_2}{\partial x} = \frac{u_1 u_2}{u} ,$$
$$\frac{\partial u}{\partial y} = u_2 , \frac{\partial u_1}{\partial y} = \frac{u_1 u_2}{u} , \frac{\partial u_2}{\partial y} = -\omega_2^2 u . \tag{12}$$

[1] This example was proposed by L. Glebsky.

FIGURE 7   The picture of field $U(x, y) = \sin x \sin(\sqrt{3}/2 y)$

This system may be considered as a combination of two differential operators, $\mathcal{D}_x$ and $\mathcal{D}_y$, that determine the shift of the field u.

Apparently, when determining the operator of translation, T^α, with two times through the superposition of two shift operators, $T_x^{\alpha_1}$ and $T_y^{\alpha_2}$, a commutation condition along x and y, respectively:

$$T_x^{\alpha_1} \circ T_y^{\alpha_2} = T_y^{\alpha_2} \circ T_x^{\alpha_1} \tag{13}$$

must be imposed.

It can be readily verified that the commutation condition of the operators $\mathcal{D}_x$ and $\mathcal{D}_y$ is fulfilled for system (12); i.e., we have Lee brackets equal to zero:

$$[\mathcal{D}_x, \mathcal{D}_y] \equiv \left[u_1 \frac{\partial}{\partial u} - \omega_1^2 u \frac{\partial}{\partial u_1} + \frac{u_1 u_2}{u} \frac{\partial}{\partial u^2}, u_2 \frac{\partial}{\partial u} + \frac{u_1 u_2}{u} \frac{\partial}{\partial u_1} - \omega_1^2 u \frac{\partial}{\partial u_2} \right] = 0.$$

One can see that even a simple example of translational dynamical system presented above is far from being trivial. Therefore construction of simple model systems with several times and a complex structure of phase space which, at the same time, would be amenable to detailed mathematical analysis is one of paramount conditions for theoretical investigation of spatial disorder.

7. CORRELATION DIMENSION OF CAPILLARY RIPPLES

The procedure for the calculation of the correlation dimension presented above was employed for the description of the spatio-temporal chaos of parametrically excited capillary ripples.

Experiments on the dynamics of capillary waves were performed on a fluid placed in a flat couvette vibrating in the vertical direction. It is known (Faraday, 1831) that, if the amplitude V of vibration is higher than the critical value V_c,

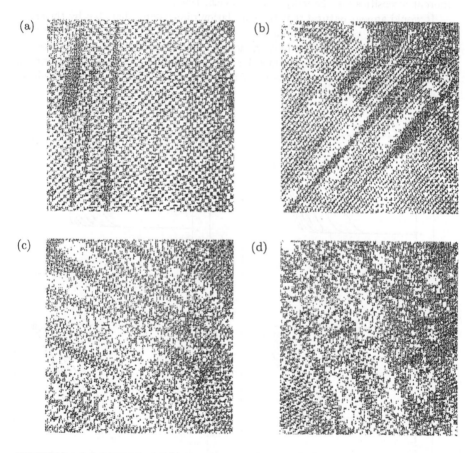

FIGURE 8 Snapshots of capillary ripples for supercriticalities: (a) $s = 0.15$, (b) $s = 0.81$, (c) $s = 1.1$, and (d) $s = 1.85$.

a system of capillary waves is generated on the surface of the fluid. If the size of the couvette is much larger than the wavelength of capillary ripples, then the parametrically excited ripples are a superposition of two mutually perpendicular pairs of counterpropagating waves, irrespective of the shape of the couvette which may be round or square. As the supercriticality $s = V/V_c - 1$ increases, there appear the envelope waves propagating perpendicularly to the originally excited pairs. When the amplitude of vibrations is sufficiently high, the system of envelope waves becomes chaotic—turbulence is formed. This is a quite general scenario of the transition to the wave turbulence although its details may differ in fluids of different viscosities (see Ezersky & Rabinovich, 1990).

It would appear natural that the correlation dimension calculated by the snapshots of capillary ripples would increase with the growth of the supercriticality. To prove these intuitive ideas, we have made quantitative measurements and processing employing the correlation dimension of snapshots. The correlation dimension D_C has been calculated on patterns of capillary ripples for the supercriticalities $s \in [0.15, 1.85]$ (see Figure 8).

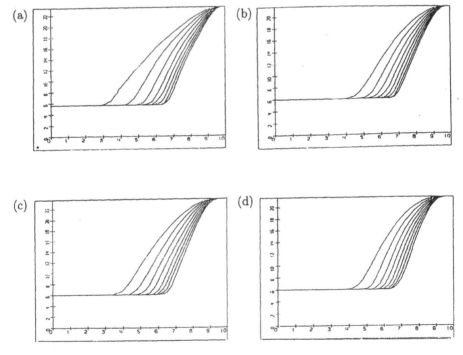

FIGURE 9 Correlation integrals corresponding to Figure 7: (a) $D_C = 6.3$, (b) $D_C = 7.0$, (c) $D_C = 7.4$, and (d) $D_C = 8.0$.

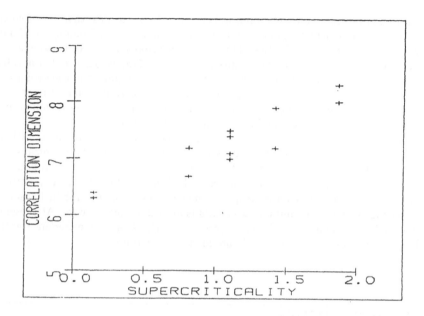

FIGURE 10   Correlation dimension of snapshots of capillary ripples as a function of the supercriticality.

The correlation integrals corresponding to the snapshots are presented in Figure 9. Our facilities (we perform calculations on EC-1037 which is comparable with IBM 4341) allow for the processing of 512×512 matrices that are constructed by the images of capillary ripples, with the number of reference points being i_{ref}, j_{ref}.

Results of calculations show that the snapshots of capillary ripples are finitely generated. As the supercriticality s grows from 0.15 up to 1.85, the correlation dimension of the snapshots increases; for equal supercriticalities the correlation dimensions are calculated to be approximately equal (see Figure 10).

8. CONCLUDING REMARKS

We have considered space series—a new issue in the investigations in the theory of dynamical systems. Space series may be statical (time invariant) spatial pictures of temperature distribution that is established in large-box convective cells, density distributions of population or of economical activity over large areas, instantaneous states of fields rapidly pulsating in time (snapshots), or similar pictures that emerge on a coordinate-time plane. The approach proposed for the investigation of such space series using calculations of spatial dimension and spatial Kolmogorov-Sinai entropy proved to be very effective. Thus, the dimension D_s enables us to answer the fundamental question whether the given "picture" is finitely generated or not. In other words, we are now able to say whether this picture results from the evolution of dynamical system or it is a consequence of random forces distributed in space. Entropy characterizes the degree of spatial disorder and allows one to follow its evolution or degradation in time (or with the variation of parameter).

However, the characteristics of space series studied above are numbers rather than functions; i.e., they give little information on the properties of dynamical spatial disorder. Therefore characteristics that would be functions, like spatial Fourier spectrum, are also highly desirable. An example of such a function is a spectrum of generalized dimension (see the temporal analog by Grassberger, 1983). Perhaps, the spectrum of spatial Lyapunov exponents is even more informative. At present we have only a general idea of such a processing. However, our experience with fields that depend only on one coordinate looks very optimistic.

ACKNOWLEDGMENTS

We are grateful to H. Abarbanel, M. Bazhenov, A. Ezersky, L. Glebsky, A. Fabrikant, L. Korzinov, M. Shereshevsky, and L. Tsimring for fruitful discussions.

Harry L. Swinney
Department of Physics, The University of Texas at Austin, Austin, TX 78712
e-mail swinney@chaos.utexas.edu

Spatio-Temporal Patterns: Observations and Analysis

THE PROBLEM

Algorithms have been developed for computing a variety of dynamical systems properties, including generalized dimensions D_q, Lyapunov exponents, entropies, the $f(\alpha)$ spectrum of singularities, unstable periodic orbits, and mutual information. A number of such algorithms are described in this volume. However, none of these techniques for analyzing dynamical systems provide information on the *spatial* character of spatially extended systems. Until recently the lack of tools for analyzing spatio-temporal behavior has not been a severe limitation in interpreting laboratory studies because most experiments provided primarily temporal information, such as the velocity or temperature measured at a single point in a fluid flow, or the concentration of a particular species measured at a point in a chemical system. In some cases photographs provided detailed spatial information at an instant of time, but the temporal evolution of the spatial structure was not generally available.

Times Series Prediction: Forecasting the Future and
Understanding the Past, Eds. A. S. Weigend and N. A. Gershenfeld, SFI Studies
in the Sciences of Complexity, Proc. Vol. XV, Addison-Wesley, 1993   **557**

The situation has changed dramatically as inexpensive PC-based digital imaging systems have become widely available. Now enormous amounts of information can be quickly obtained about time-dependent two-dimensional spatial patterns. For example, a standard, digital video system yields images with 512×512 pixels of 8-bit resolution at a rate of 30 frames/s (8 Mbyte/s), and in the not too distant future, systems will provide images with 1024×1024 pixels of 12-bit resolution at a rate of 120 frames/s (2 Gbyte/s). The problem is how to interpret this enormous amount of information.

How can one compare two different complex disordered spatio-temporal patterns? (For example, compare the patterns in Figure 4.) How similar or how different are two patterns? How can the complexity of a pattern be quantified so that it can be determined whether a given system is becoming more or less complex as a control parameter is varied? The *universality* question: what properties of patterns and of the bifurcations between different patterns are common in diverse systems? What role does symmetry play in pattern selection? How does turbulence in Navier-Stokes systems (where the nonlinearity is in the $\mathbf{u} \cdot \nabla \mathbf{u}$ term and there is a wide range of spatial and temporal scales) differ from disordered spatio-temporal patterns in reaction-diffusion systems (where the nonlinearity is in the chemical kinetics and there does not seem to be a wide range of spatial and temporal scales)? The answers to these questions are largely unknown at the present. Some of these issues are discussed in a recent review (Cross & Hohenberg, 1993).

In this paper, we first propose a possible classification scheme for spatio-temporal patterns, and then we present some examples of patterns and stress the need for new techniques for analyzing patterns.

CLASSIFICATION OF SPATIO-TEMPORAL PATTERNS

There is no accepted system for classifying spatio-temporal patterns, but patterns can certainly be divided into two major classes, large and small patterns, where the relative size is characterized by the aspect ratio Γ, which is the ratio of the size of a system to the smallest characteristic length scale. For small patterns, say, with an aspect ratio $\Gamma \ll 100$, it is not possible to distinguish behavior that is quasi-periodic in space from spatial chaos, just as it is not possible to determine whether a short time series is quasi-periodic or chaotic. The dividing line between small and large aspect ratio is of course not sharp; the greater the spatial complexity, the larger Γ must be in order to characterize the spatial character of a pattern. The following scheme classifies patterns roughly in terms of increasing complexity:

SMALL PATTERNS ($\Gamma \ll 100$).

■ temporally ordered (time-independent, periodic, or quasi-periodic in time)
■ temporally chaotic

For small patterns, simple periodic spatial variations can be recognized with only a small number of spatial cycles, but more complex spatial variations can only be characterized as nonperiodic. The term spatio-temporal chaos should be reserved for large patterns where it is possible to distinguish spatial chaos from spatial quasi-periodicity. The attractors for small patterns are generally low-dimensional.

LARGE PATTERNS ($\Gamma \gg 100$).

■ spatio-temporally ordered (crystalline)

  ■ in space: periodic or quasi-periodic (e.g., quasi-crystals)
  ■ in time: time-independent, periodic, or quasi-periodic global oscillations; traveling or standing waves

■ spatially disordered but time-independent

  ■ low-dimensional (with a spatial variation like the time variation of a homogeneous low-dimensional chaotic dynamical system)
  ■ high-dimensional attractors with a narrow range of length scales (e.g., patterns with no long-range correlations, just small domains containing ordered patterns)
  ■ high-dimensional attractors with a wide range of length scales (e.g., patterns in self-organized criticality; diffusion-limited aggregation clusters)

■ spatio-temporally disordered (*spatio-temporal chaos*)

  ■ low-dimensional in both space and time
  ■ high-dimensional with a narrow range of length scales (e.g., some patterns in cellular automata, coupled lattice maps, and reaction-diffusion systems; Ginzburg-Landau phase turbulence)
  ■ turbulent, i.e., very high-dimensional with a wide range of length and time scales (e.g., Navier-Stokes systems at large Reynolds number)
  ■ ergodic (purely random)

We are concerned here with patterns formed in dissipative systems far from equilibrium, but these patterns, whether ordered or disordered, may be indistinguishable from patterns formed by equilibrium systems. For example, some disordered reaction-diffusion patterns appear quite similar to patterns in magnetic garnet thin films.

Dynamical systems methods have proved to be valuable in characterizing low-dimensional attractors of homogeneous systems but have failed to provide significant insight into high-dimensional dynamics. Thus there is doubt that these methods will be useful in studying complex large aspect ratio spatio-temporal patterns, although there is some hope that these methods might ultimately be able to identify large-scale *coherent structures* that would be the "particles" of a low-dimensional dynamical system. At the present this is simply a fond hope.

There is some arbitrariness in the proposed scheme for classifying patterns, and the scheme is clearly not exhaustive. Other categories and subcategories could be added, but the proposed classes include a significant fraction of the patterns that have been observed in spatially extended dynamical systems.

A SMALL ONE-DIMENSIONAL SYSTEM

Our group has examined pattern formation in a particular small one-dimensional reaction-diffusion system in both experiments (Tam & Swinney, 1990) and numerical simulations (Vastano et al., 1990). The behavior of this system was studied as a function of the concentrations imposed at the two ends of the one-dimensional reactor. These experiments were conducted on the Belousov-Zhabotinsky reaction: one end of the reactor was in contact with a continuously refreshed reservoir of the oxidizer (bromate), while the other end of the reactor was in contact with a reservoir of the substrate and catalyst (a glucose-acetone substrate and manganese sulfate catalyst). The acetone-glucose substrate was chosen because this chemistry exhibits simple oscillations but no complex oscillations or chaos in a homogeneous reaction. Thus any more complex behavior, if observed in the spatially extended system, must arise from the spatial coupling.

Simulations (Vastano et al., 1990) were conducted using the Tyson-Fife reduced form of the Oregonator kinetics for the Belousov-Zhabotinsky reaction. This model involves only two species, $u(z,t)$ and $v(z,t)$, corresponding to the concentrations of the oxidizer and catalyst, respectively. The resultant reaction-diffusion equation is:

$$\partial_t u = D\partial_{zz}u + \left(\frac{1}{\varepsilon}\right)\left[u(A-u) + qBv\frac{(pA-u)}{(pA+u)}\right],$$

$$\partial_t v = D\partial_{zz}v + Au - Bu,$$

where ε is a constant, D is the diffusion coefficient, and the parameters $A(z)$, $B(z)$, and $q(z)$ are assumed to be known and to vary linearly from one end of the reactor ($z = 0$) to the other ($z = 1$). No flux boundary conditions are imposed on u and v: $\partial_z u = \partial_z v = 0$ at $z = 0$ and $z = 1$. The experiment was designed so that D, which is the same for both species, could be continuously varied in the range

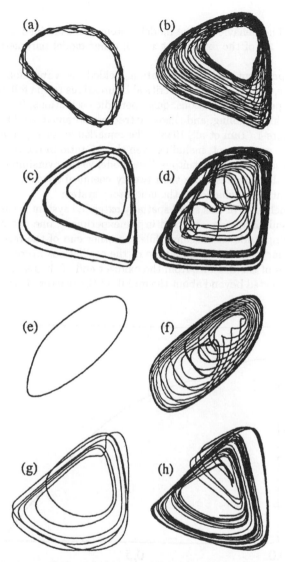

FIGURE 1 Attractors for a one-dimensional reaction-diffusion system obtained in laboratory experiments, (a)–(d), and in numerical simulations of a model, (e)–(h): (a) and (e), limit cycles; (b) and (f), 2-tori, (c) and (g), frequency-locked 7:5 limit cycles; (d) and (h), strange attractors. From Tam et al. (1988).

$0.1 - 10 \ cm^2/s$. The behavior of the model system was examined as a function of the concentration A_o of the oxidizer at $z = 0$; other model parameters were held fixed.

Both the experiment and the simulation yielded, as a function of increasing control parameter, a sequence of supercritical bifurcations to the following regimes: stationary state, periodic oscillations, quasi-periodic oscillations, frequency locking at a ratio 7:5, period doubling, and chaos. Attractors obtained for different regimes are shown in Figure 1 (Tam et al., 1988). The remarkable correspondence between the model and the experiment, including even the 7:5 ratio in the frequency-locked state, must be to some degree fortuitous. The model vastly oversimplifies the chemistry, which actually involves more than twenty chemical species, but it certainly captures the qualitative behavior of the laboratory system.

The spatial behavior of this small spatially extended system is simple but nontrivial. In the oscillatory state observed in the experiment, one end of the reactor exhibited large amplitude oscillations while the other end of the reactor appeared to be in a stationary state: the amplitude of the oscillations decreased monotonically as a function of the distance from the oxidizer end of the reactor and became too small to be detected beyond about the middle of the reactor. These observations

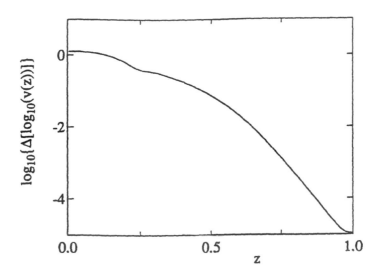

FIGURE 2   The amplitude of the periodic oscillations in $v(z, t)$ versus spatial position z for the model one-dimensional reaction-diffusion systems (with $A_o = 0.022$). Oscillations of amplitude less than 10^{-3} were unobservable in the experiment. From Vastano et al. (1990).

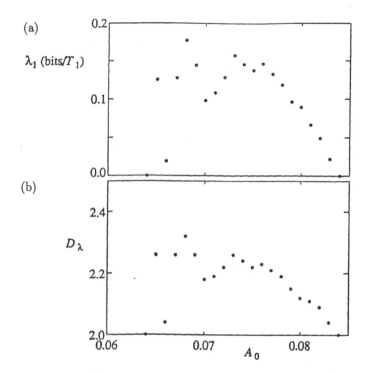

FIGURE 3 (a) The largest Lyapunov exponent λ_1 and (b) the Lyapunov dimension D_λ as a function of the parameter A_o in the reaction-diffusion model in a range where the system is chaotic. (λ_1 is in bits of information per T_1, where T_1 is the mean period of oscillation at one end of the one-dimensional system.) From Vastano et al. (1990).

are supported by results from the simulations, which are shown in Figure 2—the amplitude of oscillations decreases by more than 10^4 in going from the oxidizer end of the reactor to the other end.

The chaos found in this continuum system was found to be low-dimensional: only the largest Lyapunov exponent λ_1 was positive for any value of $A_o(\lambda_2 = 0)$. The third Lyapunov exponent was sufficiently negative so that the dimension of the attractors remained less than three; see Figure 3. (The dimension was determined from the Kaplan-Yorke formula, which in the present case is $D_\lambda = 2 + \lambda_1/|\lambda_3|$.) There is no fundamental reason why the dimension for this spatio-temporal system might not be significantly larger, but extensive explorations of chaos in the model (varying A_o, ε, D, and B_1) generally yielded attractors with dimensions in the range 2.0–2.3.

The behavior of this one-dimensional system is similar to that of a single, periodically forced damped oscillator: increasing the forcing leads from periodic oscillations to quasi-periodicity, frequency locking, and chaos. The results from the model and the experiment indicate that the local kinetics in the one-dimensional reactor are averaged to create, in effect, two oscillators localized near the two ends of the reactor. This can be understood from a consideration of the solutions for the kinetics of a homogeneous (well-stirred) reaction, as discussed elsewhere (Vastano et al., 1990).

This laboratory and numerical study of a reaction-diffusion system provides an example of complex spatio-temporal dynamics in a *small* system where a fairly detailed understanding of the behavior has been achieved. Such understanding is far more difficult to achieve even in moderately large two-dimensional systems, as described in the following section.

EXAMPLES OF COMPLEX TWO-DIMENSIONAL PATTERNS

Six examples of disordered two-dimensional patterns are shown in Figure 4. In each case the pattern was obtained for a system sustained in a state far from equilibrium. Figure 4(a) is a chemical pattern formed in a thin inert gel layer that is in contact with continuously refreshed reservoirs of chemicals on both sides of the thin layer (Ouyang & Swinney, 1991a, 1991b). The reservoirs on the two sides of the thin gel layer contain different reagents that are components of a reacting system that is known to exhibit bistability and oscillations in a homogeneous system. The gel is 2 mm thick and 25 mm in diameter. The wavelength of the pattern is 0.2 mm (only part of the pattern is shown), and the thickness of the pattern in the gel is about 0.2 mm (Ouyang et al., 1992). The pattern is viewed with light transmitted through the gel. The reaction occurs in the gel, which is used to prevent convection; thus any patterns that arise must be a consequence not of convection but of competition between diffusion and the nonlinear chemical kinetics. This particular pattern continuously evolves in time but never develops long-range order. If a regular pattern is imposed by illuminating this photosensitive system with a grid pattern, then, as soon as the light perturbation is removed, the regular pattern relaxes to the type of pattern shown in Figure 4(a) (the relaxation time is about one hour) (Ouyang et al., 1991a).

Figure 4(b) is a pattern formed by surface waves in a 1-cm-deep, 8-cm^2 layer of n-butyl alcohol that is oscillated vertically at 320 Hz (Tufillaro et al., 1989). At small driving amplitudes, patterns form with long-range order, but at large driving amplitudes, as in the figure, the pattern becomes disordered and the correlation range decreases to only a few wavelengths.

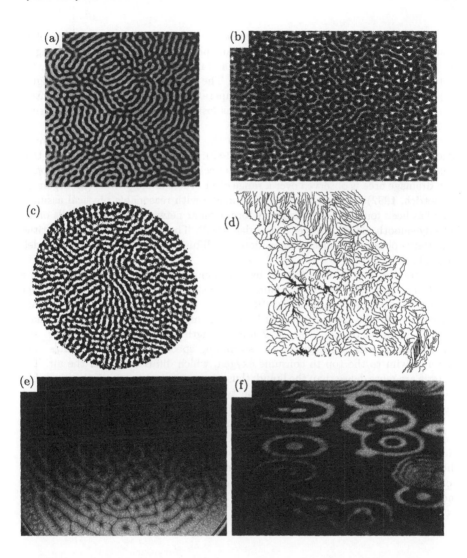

FIGURE 4   Disordered spatio-temporal patterns in (a) a reaction-diffusion system (Ouyang & Swinney, 1991a); (b) capillary wave ripples (Tufillaro et al., 1989); (c) a rotating Rayleigh-Bénard convection system (Y. Hu, R. Ecke, & G. Ahlers, private communication; Zhong et al., 1991); (d) the network of rivers in the state of Missouri (Schroeder, 1982); (e) bioconvection in a suspension of bacteria (J. Kessler, private communication; Pedley & Kessler, 1992); and (f) the oxidation of carbon monoxide on a platinum catalyst (Jakubith et al., 1990).

Figure 4(c) is a pattern formed by convection in carbon dioxide in a Rayleigh-Bénard cell rotating at 25 rad/s (Hu et al., personal communication; Zhong et al., 1991). Near the onset of convection in a stationary (nonrotating) system, the pattern consists of parallel convection rolls. The rotating convection cell in Figure 4(c) is at about 10% above the critical temperature difference at which a convection pattern first forms, and the pattern has lost long-range order. The pattern is continuously evolving in time.

Figure 4(d) is the network of rivers in the state of Missouri (the right-hand edge is the Mississippi River) (Schroeder, 1982). Analyses of landscapes show that there is no characteristic length scale; fractal scaling of the length of a river basin with drainage area is observed over a range of 11 orders of magnitude (Montgomery & Dietrich, 1992)! Recently, a dynamical model with reasonable physical assumptions has been found to spontaneously form similar networks when rain falls on an initially smooth plane (Kramer & Marder, 1992). These patterns have similarities to patterns obtained in electrodeposition and diffusion-limited-aggregation models (Argoul et al., 1988).

Figure 4(e) is a pattern formed by swimming bacteria (*Bacillus subtilis*) in a pan 3 mm deep and 70 mm in diameter (Kessler & Pedley, 1992). The average concentration of bacteria is about $10^9/cm^3$, so there are many bacteria in each convection cell (the concentration of bacteria is high in white regions, low in dark regions). Each bacterium is about 2 μm in length and 0.7 μm in diameter, and is propelled by rotating flagella. The swimming speed is typically 20 μm/s. The bacteria swim to the top to consume oxygen, which diffuses in from the air. The bacteria are more dense than the liquid; thus, the concentration gradient is unstable. The principal variables determining the type of pattern that forms are the cell concentration and the depth of the chamber.

Figure 4(f) is a pattern formed by the oxidation of carbon monoxide on a single-crystal platinum catalyst (Jakubith et al., 1990). Bright regions are covered with carbon monoxide, dark regions with oxygen. The pattern was imaged using photoemission electron spectroscopy; the region shown is about 0.2 by 0.3 mm. Concentric elliptically shaped waves emanate periodically from nucleation centers and propagate with anisotropic velocities. The pattern evolves nonperiodically on a time scale of the order of seconds.

DISCUSSION

How can the patterns in Figure 4 be characterized and compared? If dynamical systems methods are inadequate, what alternative tools are available? Fourier transforms and correlation functions provide insight for systems that are ordered in space and time, but the inadequacy of power spectra even for analyzing time series in spatially homogeneous systems became widely recognized more than a decade ago—it was realized that the same broadband power spectrum could arise from a low-dimensional dynamical system (e.g., the Lorenz model) or from a very high-dimensional system. Experimenters were already conscious of the limitations of Fourier spectra when dynamical systems methods for analyzing attractors constructed from time series became available in the early 1980s, and these techniques were rapidly adopted for characterizing temporal behavior.

Newer techniques including wavelet transforms (Chui, 1992) and proper orthogonal decomposition (Loeve, 1977; Sirovich, 1987) techniques provide some quantitative characterization of patterns but can fail to distinguish different types of dynamics. There is a need for a systematic examination of these and other new techniques. These techniques should first be applied in studies of patterns formed by model systems such as cellular automata (Gutowitz, 1990) and coupled map lattices (Kaneko, 1989), where the system is precisely known and controlled. Ultimately, of course, the goal is to understand physical systems, but laboratory systems are inappropriate for the development of analysis techniques because of long characteristic times (e.g., hours in convection and reaction-diffusion systems) and the difficulty of achieving high accuracy.

Here we have focused on one- and two-dimensional patterns, leaving the more difficult problem of three-dimensional patterns for the future. Three-dimensional data should become widely available as high-speed holography, laser Doppler, acoustic Doppler, and other techniques are refined. Thus experiments will continue to stimulate the development of techniques for analysis of spatio-temporal patterns. It is hoped that in the next decade as much progress will be achieved in the analysis of spatio-temporal patterns as has been achieved in the analysis of time series in the past decade.

ACKNOWLEDGMENTS

The author gratefully acknowledges continual stimulating interactions with the members of the Center for Nonlinear Dynamics at the University of Texas at Austin. This research is supported by the Office of Basic Energy Sciences of the Department of Energy and the Nonlinear Dynamics Program of the Office of Naval Research.

Appendix: Accessing the Server

The data used for the "Santa Fe Time Series Prediction and Analysis Competition," and programs, results of analyses, visualization, sonifications, as well as other data sets, are available from the **Time-Series** archive at **ftp.santafe.edu**. Retrieving and depositing material is straightforward by ftp over the Internet. In the following example session, we show **user input** in bold, **system response** (some parts are deleted here) in teletype font, and *comments* in italics. If you cannot resolve problems with ftp, send e-mail to **ftp@santafe.edu**.

From your computer type:
```
rak_pop% ftp ftp.santafe.edu
Name:       anonymous
Password: your_email_address
ftp>  cd pub/Time-Series
ftp>  dir
-rw-rw-r-- 1 tserver 113 3522 Jun 12 1992 README
```
This file describes the available directories, and it contains the necessary forms to submit data sets and programs to the archive.

```
drwxrwxr-x 2 tserver 113 512 Jun 12 1992 competition
```
This directory has the competition data sets (along with their description) and the original instructions for the competition.

```
drwxrwxr-x 2 tserver 113 512 Jul 28 1750 Bach
```
This directory contains the data of the Bach fugue and its continuations.
See Dirst and Weigend (this volume).

```
drwxrwxr-x 2 tserver 113 512 Jun 12 1992 results
```
This directory is for comparative evaluations of time series techniques and
data sets.

```
drwxrwxr-x 2 tserver 113 512 Jun 12 1992 programs
```
This directory is for programs related to time series analysis.

```
drwxrwxr-x 2 tserver 113 512 Nov 16 1992 data
```
This directory contains data sets that have been added following the close
of the competition, as well as some standard data sets such as the sunspot
series. Please note that some of the data sets are large (e.g., some of the
financial data sets are more than 3MB compressed). Please transfer large
files only at night.

```
drwxrwx-wx 3 tserver 113 512 Apr 13 05:17 incoming
```
New material should be deposited here and timeseries@santafe.edu notified.

```
ftp>  cd competition
ftp>  dir
```

```
-rw-rw-r-- 1 tserver 113     54558 Feb  1 1992 A.cont
-rw-rw-r-- 1 tserver 113      6000 Aug  3 1991 A.dat
-rw-rw-r-- 1 tserver 113    291238 Aug  3 1991 B1.dat
-rw-rw-r-- 1 tserver 113    295321 Aug  3 1991 B2.dat
-rw-rw-r-- 1 tserver 113      1778 Feb  1 1992 C.cont
-rw-rw-r-- 1 tserver 113    285090 Aug  1 1991 C1-5.dat
-rw-rw-r-- 1 tserver 113    285091 Aug  1 1991 C6-10.dat
-rw-rw-r-- 1 tserver 113      3000 Feb  1 1992 D.cont
-rw-rw-r-- 1 tserver 113    300000 Aug  3 1991 D1.dat
-rw-rw-r-- 1 tserver 113    300000 Aug  3 1991 D2.dat
-rw-rw-r-- 1 tserver 113    217946 Aug  3 1991 E.dat
-rw-rw-r-- 1 tserver 113     55687 Dec 29 1991 F.dat
-rw-rw-r-- 1 tserver 113     21196 Feb  1 1992 data_information
-rw-rw-r-- 1 tserver 113     29231 Dec 30 1991 old_instructions
```

```
ftp>  get A.dat
```

```
150 Opening ASCII mode data connect for A.dat (6000 bytes).
226 Transfer complete.
```

```
ftp>  bye
```

Bibliography

This bibliography compiles all references of this book in alphabetical order by first author. (If there is more than one entry for an author, we first list "solo"-works of that author in chronological order, and then works with co-authors in alphabetical order of the first co-author.) Following each reference, we give [in square brackets] the chapter(s) of this book in which the work is cited. (The chapters of this book are indicated by the name of the first author of the corresponding chapter.)

Abarbanel, H. D. I., and J. Kadtke. 1990. "Information Theoretic Methods for Determining the Minimum Embedding Dimension for Strange Attractors." Preprint, May 1990. Unpublished.
[Fraser]

Abboud, F. M., and M. D. Thames. 1983. "Interaction of Cardiovascular Reflexes in Circulatory Control " In *Handbook of Physiology*, edited by J. T. Shephard, F. M. Abboud, and S. R. Geiger, Sect. 2, Vol. 3, Part 2, 675–753. Bethesda, MD: American Physiological Association.
[Rigney]

Abdel-Malek, A., C. H. Markham, P. Z. Marmarelis, and V. Z. Marmarelis. 1988. "Quantifying Deficiencies Associated with Parkinson's Disease by Using Time Series Analysis." *Electroencephal. & Clin. Neurophysiol.* **69**: 24–33.
[Rigney]

Abelson, H. 1990. "The Bifurcation Interpreter: A Step Towards the Automatic Analysis of Dynamical Systems." *Intl. J. Comp. & Math. Appl.* **20**:13.
[Gershenfeld]

Abraham, N. B., A. M. Albano, B. Das, and M. F. H. Tarroja. 1987. In *Fundamentals of Quantum Optics II*, edited by F. Ehlotsky, 32–46. Berlin: Springer-Verlag.
[Glass, Hübner]

Abraham, N. B., A. M. Albano, A. Passamante, and P. E. Rapp, eds. 1989. *Measures of Complexity and Chaos.* New York: Plenum Press.
[Paluš]

Adolph, E. F. 1961. "Early Concepts of Physiological Regulations." *Physiol. Rev.* **41**: 737–770.
[Rigney]

Afraimovich, V. S., A. B. Ezersky, M. I. Rabinovich, A. L. Zheleznyak, and M. A. Sherechevsky. 1992. "Dynamical Description of Spatial Disorder." *Physica D* **58**: 331–338.
[Afraimovich]

Afraimovich, V. S., M. I. Rabinovich, and A. L. Zheleznyak. 1993. "Finite-Dimensional Spatial Disorder: Description and Analysis." This volume.
[Gershenfeld]

Afraimovich, V. S., and A. M. Reiman. 1990. "Dimensions and Entropies in Multidimensional Systems." In *Nonlinear Waves, Dynamics, and Evolution*, edited by A. V. Gaponov-Grekhov, M. I. Rabinovich, and J. Engelbrecht, 2–29. Berlin & Heidelberg: Springer-Verlag.
[Afraimovich]

Ahmed, A. K., S. Y. Fakhouri, J. B. Harness, and A. J. Means. 1986. "Modeling of the Control of Heart Rate by Breathing Using a Kernel Method." *J. Theor. Biol.* **119**: 67–79.
[Rigney]

Aihara, K., and G. Matsumoto. 1986. "Forced Oscillations and Routes to Chaos in the Hodgkin-Huxley Axons and Squid Giant Axon." In *Chaos in Biological Systems*, edited by H. Degn, A. V. Holden, and L. F. Olsen, 1211–132. New York: Plenum Press.
[Glass]

Akaike, H. 1970. "Statistical Predictor Identification." *Ann. Inst. Stat. Math.* **22**: 203–217.
[Gershenfeld]

——. 1974. "A New Look at the Statistical Model Identification." *IEEE Trans. Auto. Control* **19**: 716–723.
[Theiler]

——. 1979. "A Bayesian Extension of the Minimum AIC Procedure of Autoregressive Model Fitting." *Biometrika* 66: 237.
[Rigney]

Akselrod, S., D. Gordon, F. A. Ubel, D. C. Shannon, A. C. Barger, and R. J. Cohen. 1981. "Power Spectrum Analysis of Heart Rate Fluctuations: A Quantitative Probe of Beat-to-Beat Cardiovascular Control." *Science* 213: 220–222.
[Glass, Rigney]

Albano, A. M., N. B. Abraham, G. C. De Guzman, M. F. H. Tarroja, R. S. Gioggia, P. E. Rapp, I. D. Zimmerman, N. N. Greenbaum, and T. R. Bashore. 1986. "Lasers and Brains: Complex Systems with Low-Dimensional Attractors." In *Dimensions and Entropies in Chaotic Systems: Quantification of Complex Behavior*, edited by G. Mayer-Kress, 231–240. Berlin: Springer-Verlag.
[Glass]

Albano, A. M., J. Muench, C. Schwartz, A. I. Mees, and P. E. Rapp. 1988. "Singular-Value Decomposition and the Grassberger-Procaccia Algorithm." *Phys. Rev. A* 38: 3017–3026.
[Hübner]

Alesić, Z. 1991. "Estimating the Embedding Dimension." *Physica D* 52: 362–368.
[Gershenfeld]

Altman, N. S. 1987. "Smoothing Data with Correlated Errors." Technical Report 280, Department of Statistics, Stanford University, Stanford, CA, October.
[Lewis]

Amato, I. 1992. "Chaos Breaks Out at NIH, but Order May Come of It." *Science* 256: 1763–1764.
[Rigney]

Ames, C. 1989. "The Markov Process as Compositional Model: A Survey and Tutorial." *Leonardo* 22: 175–187.
[Dirst]

Anderson, T. W. 1971. *The Statistical Analysis of Time Series*. New York: Wiley.
[Theiler]

Angelone, A., and N. A. Coulter. 1964. "Respiratory Sinus Arrhythmia: A Frequency-Dependent Phenomenon." *J. Appl. Physiol.* 217: 479–482.
[Rigney]

Anrep, G. V., W. Pascual, and R. Rossler. 1936. "Respiratory Variations of the Heart Rate. II. The Central Mechanism of the Respiratory Arrhythmia and the Interrelations Between the Central and Reflex Mechanisms." *Proc. Roy. Soc. Lond. B. Biol. Sci.* 119: 218–232.
[Rigney]

Appel, M. L. 1992. "Closed-Loop Identification of Cardiovascular Regulatory Mechanisms." Ph.D. Thesis, Harvard University, Cambridge, MA.
[Rigney]

Argoul, F., A. Arneodo, G. Grasseau, and H. L. Swinney. 1988. "Self-Similarity of Diffusion-Limited Aggregates and Electrodeposition Clusters." *Phys. Rev. Lett.* 61: 2558–2561.
[Swinney]

Arnold, V. I. 1973. *Ordinary Differential Equations*. Cambridge, MA: MIT Press.
[Paluš]

Babloyantz, A. 1989. "Some Remarks on Nonlinear Data Analysis of Physiological
Time Series." In *Measures of Complexity and Chaos*, edited by N. B. Abraham
et al., 51–62. New York: Plenum Press.
[Glass]

Bachrach, J. 1988. "Learning to Represent State." Master's Thesis, University of
Massachusetts, Amherst, MA.
[Mozer]

Baillie, R. T., and T. Bollerslev. 1991. "Intra-Day and Inter-Day Volatility in Foreign
Exchange Rates." *Rev. Econ. Stud.* **58**: 565–585.
[LeBaron]

Baillie, R. T., and P. McMahon. 1989. *The Foreign Exchange Market: Theory and
Econometric Evidence*. Cambridge: Cambridge University Press.
[LeBaron]

Bainbridge, F. A. 1920. "The Relation Between Respiration and the Pulse Rate."
J. Physiol. Lond. **54**: 192–202.
[Rigney]

Ballard, D. H. 1986. "Cortical Connections and Parallel Processing: Structure and
Function." *Behav. & Brain Sci.* **9**: 67–120.
[Mozer]

Barron, A. R. 1992. "Universal Approximation Bounds for Superpositions of a Sigmoidal Function." *IEEE Trans. Info. Theory* **38**.
[Gershenfeld]

Bartoli, F., G. Baselli, and S. Certutti. 1985. "AR Identification and Spectral Estimate
Applied to the R-R Interval Measurements." *Intl. J. Biomed. Comp.* **16**: 201–215.
[Rigney]

Başar, E., ed. 1990. *Chaos in Brain Function*. Berlin: Springer.
[Paluš]

Baselli, G., D. Bolis, S. Cerutti, and C. Freschi. 1985. "Autoregressive Modeling and
Power Spectral Estimate of R-R Interval Time Series in Arrhythmic Patients."
Comp. Biomed. Res. **18**: 510–530.
[Rigney]

Baum, L. E., T. Petrie, G. Soules, and N. Weiss. 1970. "A Maximization Technique
Occuring in the Statistical Analysis of Probabilistic Functions of Markov Chains."
Ann. Math. Stat. **41(1)**: 164–171.
[Fraser]

Bazhenov, M. V., M. I. Rabinovich, and A. L. Fabrikant. 1992. "The 'Amplitude'-
'Phase' Turbulence Transition in a Ginzburg-Landau Model as a Critical Phe-
nomenon." *Phys. Lett. A* **163**: 87–94.
[Afraimovich]

Beck, C. 1990. "Upper and Lower Bounds on the Renyi Dimensions and the Uniformity
of Multifractals." *Physica D* **41**: 67–78.
[Gershenfeld]

Beilin, K. E., M. I. Rabinovich, and A. L. Zheleznyak. n.d. "Stability of Finite-Dimensional Spatial Disorder in a One-Dimensional CGLE." To be published.
[Afraimovich]

Bellman, R. E. 1961. *Adaptive Control Processes.* Princeton: Princeton University Press.
[Lewis]

Benchetrit, G., P. Baconnier, and J. Demongeot, eds. 1987. *Concepts and Formalizations in the Control of Breathing.* Manchester: University Press.
[Rigney]

Bendixen, H. H., G. M. Smith, and J. Mead. 1964. "Pattern of Ventilation in Young Adults." *J. Appl. Physiol.* **19**: 195–198.
[Rigney]

Bengio, Y., R. De Mori, and R. Cardin. 1990. "Speaker-Independent Speech Recognition with Neural Networks and Speech Knowledge." In *Advances in Neural Network Information Processing Systems II*, edited by D. S. Touretzky, 218–225. San Mateo, CA: Morgan Kaufmann.
[Mozer]

Bengio, Y., P. Frasconi, and P. Simard. 1993. "The Problem of Learning Long-Term Dependencies in Recurrent Networks." *Proc. IEEE Intl. Conf. Neur. Net.*: in press.
[Mozer]

Bentley, J. L., and J. H. Friedman. 1979. "Data Structures for Range Searching." *ACM Comp. Surv.* **11**: 397.
[Kostelich]

Bergel, E. 1985. *Bachs letzte Fuge.* Bonn: Max Brockhaus Musikverlag.
[Dirst]

Berger, R. D., S. Akselrod, D. Gordon, and R. J. Cohen. 1986. "An Efficient Algorithm for Spectral Analysis of Heart Rate Variability." *IEEE Trans. Biomed. Engr.* **BME-33**: 900–904.
[Rigney]

Bernardi, L., F. Keller, M. Sanders, P. S. Reddy, F. Men, and M. R. Pinsky. 1989. "Respiratory Sinus Arrhythmia in the Denervated Human Heart." *J. Appl. Physiol.* **67**: 1447–1455.
[Rigney]

Bernstein, L. 1979. *The Unanswered Question: Six Talks at Harvard.* Cambridge, MA: Harvard University Press.
[Dirst]

Bertholon, J. F., E. Labeyrie, G. Testylier, and A. Teillac. 1987. "In Search of the Attractors in the Control of Ventilation by CO_2." In *Concepts and Formalizations in the Control of Breathing*, edited by G. Benchetrit, P. Baconnier, and J. Demongeot, 9–22. Manchester: University Press.
[Rigney]

Beuter A., H. Flashner, and A. Arabyan. 1986. "Phase Plane Modelling of Leg Motion." *Biol. Cyber.* **53**: 273–284.
[Glass]

Beuter A., D. Larocque, and L. Glass. 1989. "Complex Oscillations in a Human Motor System." *J. Motor Behav.* **21**: 277–289.
[Glass]

Beven, K., and A. Binley. 1992. "The Future of Distributed Models: Model Calibration and Uncertainty Prediction." *Hydrol. Proc.* **6**.
[Smith]

Bharucha, J. J., and P. M. Todd. 1989. "Modeling the Perception of Tonal Structure with Neural Nets." *Comp. Music J.* **13(4)**: 44–53. Reprinted in *Music and Connectionism*, edited by P. M. Todd, and D. G. Loy, 128–137. Cambridge, MA: MIT Press.
[Dirst]

Billingsley, P. 1965. *Ergodic Theory and Information.* New York: Wiley.
[Paluš]

Bingham, S. 1992. "Box-Counting with Base b^n Numerals." *Phys. Lett. A* **163**: 419–424.
[Pineda]

Bingham, S., and M. Kot. 1989. "Multidimensional Trees, Range Searching, and a Correlation-Dimension Algorithm of Reduced Complexity." *Phys. Lett. A* **140**: 327–330.
[Pineda]

Bodenhausen, U., and A. Waibel. 1991. "The Tempo 2 Algorithm: Adjusting Time Delays by Supervised Learning." In *Advances in Neural Information Processing Systems 3*, edited by R. P. Lippmann, J. Moody, and D. S. Touretzky, 155–161. San Mateo, CA: Morgan Kaufmann.
[Mozer]

Bogert, B. P., M. J. R. Healy, and J. W. Tukey. 1963. "The Quefrency Alanysis of Time Series for Echoes: Cepstrum, Pseudo-Autocovariance, Cross-Cepstrum and Saphe Cracking." In *Proceedings of the Symposium of Time Series Analysis*, edited by M. Rosenblatt, 209–243. New York: John Wiley. Reprinted in *The Collected Works of John Tukey*, Vol. I, p. 455–493, Wadsworth.
[Dirst]

Bohlin, T. 1977. "Analysis of EEG Signals with Changing Spectra Using a Short Word Kalman Estimator." *Math. Biosci.* **35**: 221–259.
[Rigney]

Bollerslev, T., R. Y. Chou, N. Jayaraman, and K. F. Kroner. 1992. "ARCH Modeling in Finance: A Review of the Theory and Empirical Evidence." *J. Econometrics* **52** **(1)**: 5–60.
[LeBaron]

Bollerslev, T., and I. Domowitz. 1990. *Trading Patterns and the Behavior of Prices in the Interbank Deutschemark/Dollar Foreign Exchange Market.* Evanston, IL: Northwestern University Press.
[LeBaron]

Bourgoin, M., K. Sims, S. J. Smith, and H. Voorhees. 1993. "Learning Image Classification with Simple Systems and Large Databases." *IEEE Trans. Pat. Anal. & Mach. Intel.*: submitted.
[Gershenfeld]

Box, G. E. P., and F. M. Jenkins. 1976. *Time Series Analysis: Forecasting and Control,* 2nd ed. Oakland, CA: Holden-Day.
[Gershenfeld, Rigney]

Box, G. E. P., and G. C. Tiao. 1977. "The Canonical Analysis of Multiple Time Series." *Biometrika* **64**: 355–365.
[LeBaron, Lewis]

Brachman, R. J., and H. J. Levesque. 1985. *Readings in Knowledge Representation.* San Mateo, CA: Morgan Kaufmann.
[Zhang]

Bradley, E. 1992. "Taming Chaotic Circuits." Ph.D. Thesis, Massachusetts Institute of Technology, September 1992.
[Gershenfeld]

Bradley, E., and G. Gong. 1983. "A Leisurely Look at the Bootstrap, the Jackknife, and Cross Validation." *Am. Stat.* **37(1)**: 36–48.
[Zhang]

Breiman, L., J. Friedman, R. Olshen, and C. Stone. 1984. *Classification and Regression Trees.* Belmont, CA: Wadsworth.
[Lewis]

Bridle, J. 1990. "Training Stochastic Model Recognition Algorithms as Networks Can Lead to Maximum Mutual Information Estimation of Parameters." In *Advances in Neural Information Processing Systems 2,* edited by D. S. Touretzky, 211–217. San Mateo, CA: Morgan Kaufmann.
[Mozer]

Briggs, K. 1990. "Improved Methods for the Analysis of Chaotic Time Series." *Phys. Lett. A* **151**: 27.
[Casdagli]

Brock, W. A. 1991. *Beyond Belief: Randomness, Prediction, and Explanation in Science,* edited by J. Casti and A. Karlovist, Ch. 10. Boca Raton, FL: CRC Press.
[Zhang]

Brock, W. A., W. D. Dechert, J. A. Scheinkman, and B. LeBaron. 1988. "A Test For Independence Based on the Correlation Dimension." University of Wisconsin Press, Madison, WI.
[LeBaron]

Brock, W. A., and W. D. Dechert. 1989. "Statistical Inference Theory for Measures of Complexity in Chaos Theory and Nonlinear Science." In *Measures of Complexity and Chaos,* edited by N. Abraham, 79–98. New York: Plenum.
[Theiler]

Brock, W. A., D. Hsieh, and B. LeBaron. 1991. *Nonlinear Dynamics, Chaos, and Instability: Statistical Theory and Economic Evidence.* Cambridge, MA: MIT Press.
[LeBaron]

Brock, W. A., J. Lakonishok, and B. LeBaron. 1992. "Simple Technical Trading Rules and the Stochastic Properties of Stock Returns." Technical Report 9022, Social Systems Research Institute, University of Wisconsin, Madison, WI. *J. Finance* **47(5)**: 1731–1764.
[LeBaron]

Brock, W. A., and S. M. Potter. 1992. "Diagnostic Testing for Nonlinearity, Chaos, and General Dependence in Time Series Data." In *Nonlinear Modeling and Forecasting*, edited by M. Casdagli and S. Eubank, 137–162. Santa Fe Institute Studies in the Sciences of Complexity, Proc. Vol. XII. Reading, MA: Addison-Wesley.
[Theiler]

Brock, W. A., and C. L. Sayers. 1988. "Is the Business Cycle Characterized by Deterministic Chaos?" *J. Monet. Econ.* **22**: 71–90.
[Theiler]

Brockwell, P., and R. Davis. 1987. *Time Series: Theory and Methods*. New York: Springer-Verlag.
[Sauer]

Broomhead, D. S., R. Jones, and G. P. King. 1987. "Topological Dimension and Local Coordinates from Time Series Data." *J. Phys. A* **20**: L563–L569.
[Sauer]

Broomhead, D. S., and G. P. King. 1986. "Extracting Qualitative Dynamics from Experimental Data." *Physica D* **20**: 217–236.
[Gershenfeld, Sauer]

Broomhead, D. S., and D. Lowe. 1988. "Multivariable Functional Interpolation and Adaptive Networks." *Complex Systems* **2**: 321–355.
[Gershenfeld, Smith, Zhang]

Brown, A. M. 1980. "Receptors Under Pressure: An Update on Baroreceptors." *Circ. Res.* **46**: 1–10.
[Dirst, Rigney]

Brown, P. F. 1987. "The Acoustic-Modeling Problem in Automatic Speech Recognition." Ph.D. Thesis, Carnegie-Mellon University, Pittsburgh, PA.
[Fraser]

Brown, P. F., S. A. Della Pietra, V. J. Della Pietra, J. C. Lai, and R. L. Mercer. 1992. "An Estimate of an Upper Bound for the Entropy of English." *Comp. Ling.* **18**: 31–40.
[Dirst]

Brown, R., P. Bryant, and H. D. I. Abarbanel. 1991. "Computing the Lyapunov Spectrum of a Dynamical System from an Observed Time Series." *Phys. Rev. A* **43**: 2787–806.
[Gershenfeld]

Brusil, P. J., T. B. Waggener, R. E. Kronauer, and P. Gulesian, Jr. 1980. "Methods for Identifying Respiratory Oscillations Disclose Altitude Effects." *J. Appl. Physiol.* **48**: 545–556.
[Rigney]

Bryant, P., R. Brown, and H. D. I. Abarbanel. 1990. "Lyapunov Exponents from Observed Time Series." *Phys. Rev. Lett.* **65(13)**: 1523–1526.
[Glass, Kaplan, Paluš]

Bryson, A., and Y. Ho. 1975. "Applied Optimal Control." New York: Hemisphere.
[Wan]

Buntine, W. L., and A. S. Weigend. 1991. "Bayesian Backpropagation." *Complex Systems* **5**: 603–643.
[Gershenfeld]

Busoni, F. 1912. *Fantasia Contrappuntistica*, edited by W. Middelschulte. Leipzig: Breitkopf und Härtel.
[Dirst]

Butler, G. 1983. "Ordering Problems in J. S. Bach's *Art of Fugue* Resolved." *Music. Qtr.* **69**: 44–61.
[Dirst]

Calaresu, F. R., J. Ciriello, M. M. Caverson, D. F. Cechetto, and T. L. Krukoff. 1984. "Functional Neuroanatomy of Central Pathways Controlling the Circulation." In *Hypertension and the Brain*, edited by G. P. Guthrie and K. A. Kotchen. 3–21. Mount Kisco, NY: Futura.
[Rigney]

Cannon, W. B. 1929. "Organization for Physiological Homeostasis." *Physiol. Rev.* **9**: 399–431.
[Rigney]

———. 1932. *The Wisdom of the Body*. New York: Norton.
[Rigney]

Caprile, B., and F. Girosi. 1990. "A Nondeterministic Minimization Algorithm." Technical Paper, Artificial Intelligence Laboratory, Massachusetts Institute of Technology, Cambridge, MA.
[Zhang]

Caputo, J. G., and P. Atten. 1987. "Metric Entropy: An Experimental Means for Characterizing and Quantifying Chaos." *Phys. Rev. A* **35**: 1311–1316.
[Hübner]

Carley, D. W., E. Onal, R. Aronson, and M. Lopata. 1989. "Breath-by-Breath Interactions Between Inspiratory and Expiratory Duration in Occlusive Sleep Apnea." *J. Appl. Physiol.* **66**: 2312–2319.
[Rigney]

Carley, D. W., and D. C. Shannon. 1988. "A Minimal Mathematical Model of Human Periodic Breathing." *J. Appl. Physiol.* **65**: 1400–1409.
[Rigney]

Carpenter, G. 1981. "Normal and Abnormal Signal Patterns in Nerve Cells." *SIAM-AMS Proc.* **13**: 49–90.
[Rigney]

Casdagli, M. 1989. "Nonlinear Prediction of Chaotic Time Series." *Physica D* **35**: 335–356.
[Casdagli, Gershenfeld, Kaplan, Lewis, Rigney, Sauer, Smith, Zhang]

———. 1991. "Chaos and Deterministic versus Stochastic Nonlinear Modeling." *J. Roy. Stat. Soc. B* **54**: 303–328.
[Gershenfeld, Glass, Kaplan, Smith, Theiler]

———. 1992. "Nonlinear Forecasting, Chaos, and Statistics." In *Modeling Complex Phenomena*, edited by L. Lam and V. Naroditsky, 131–152. New York: Springer-Verlag.
[Casdagli]

Casdagli, M., D. des Jardins, S. Eubank, J. D. Farmer, J. Gibson, J. Theiler, and
N. Hunter. 1992. "Nonlinear Modeling of Chaotic Time Series: Theory and Appli-
cations." In *Applied Chaos*, edited by J. H. Kim and J. Stringer, 335. New York:
Wiley.
[Rigney]

Casdagli, M., S. Eubank, J. D. Farmer, and J. Gibson. 1991. "State Space Reconstruction in the Presence of Noise." *Physica D* **51D**: 52–98.
[Fraser, Gershenfeld, Kaplan, Smith]

Casdagli, M., and S. Eubank, eds. 1992. *Nonlinear Modeling and Forecasting.* Santa
Fe Institute Studies in the Sciences of Complexity, Proc. Vol. XII. Redwood City:
Addison-Wesley.
[Casdagli]

Casdagli, M. C., and A. S. Weigend. 1993. "Exploring the Continuum Between Deterministic and Stochastic Modeling." This volume.
[Dirst, Gershenfeld, Glass, Kaplan, Sauer, Smith, Theiler]

Catlin, D. E. 1989. *Estimation, Control, and the Discrete Kalman Filter.* Applied Mathematical Sciences, Vol. 71. New York: Springer-Verlag, 1989.
[Gershenfeld]

Cawley, R., and G. H. Hsu. 1992a. "A Local Geometric Projection Method for Noise
Reduction in Maps and Flows." *Phys. Rev. A.* **46**: 3057–3062.
[Kantz, Kostelich]

——. 1992b. "SNR Performance of a Noise Reduction Algorithm Applied to Coarsely
Sampled Chaotic Data." *Phys. Lett. A* **166**: 188.
[Kantz]

Chaitin, G. J. 1966. "On the Length of Programs for Computing Finite Binary Sequences." *J. Assoc. Comp. Mach* **13**: 547–569.
[Gershenfeld]

——. 1990. *Information, Randomness & Incompleteness.* Series in Computer Science,
Vol. 8, 2nd ed. Singapore: World-Scientific.
[Gershenfeld]

Chapman, K. R., E. N. Bruce, B. Gothe, and N. S. Cherniack. 1988. "Possible Mechanisms of Periodic Breathing During Sleep." *J. Appl. Physiol.* **64**: 1000–1008.
[Rigney]

Chase, M. H., and E. D. Weitzman, eds. 1983. *Sleep Disorders. Basic and Clinical Research.* New York: S. P. Medical and Scientific Books.
[Rigney]

Chatfield, C. 1988. "What is the Best Nethod in Forecasting?" *J. Appl. Stat.* **15**:
19–38.
[Gershenfeld]

——. 1989. *The Analysis of Time Series*, 4th ed. London: Chapman and Hall, 1989.
[Gershenfeld, Rigney, Zhang]

Chay, T. R., and J. Rinzel. 1985. "Bursting, Beating, and Chaos in an Excitable Membrane Model." *Biophys. J.* **47**: 357–366.
[Rigney]

Chialvo, D., M. C. Michaels, and J. Jalife. 1990. "Supernormal Excitability as a Mechanism of Chaotic Dynamics of Activation in Cardiac Purkinje Fibers." *Circ. Res.* **66**: 525–545
[Glass]

Chui, C. K. 1992. *An Introduction to Wavelets.* San Diego, CA: Academic Press.
[Swinney]

Ciriello, J., C. V. Rohlicek, and C. Polosa. 1983. "Aortic Baroreceptor Reflex Pathway: A Functional Mapping Using Tritiated 2-Deoxyglucose Autoradiography in the Rat." *J. Autonom. Nerv. Sys.* **8**: 111–128.
[Rigney]

——. 1985. "2-Deoxyglucose Uptake in the Central Nervous System During Systemic Hypercapnia in the Peripherally Chemodenervated Rat." *Exp. Neurol.* **88**: 673–687.
[Rigney]

Ciriello, J., C. V. Rohlicek, R. S. Poulsen, and C. Polosa. 1982. "Deoxyglucose Uptake in the Rat Thoracolumbar Cord During Activation of Aortic Baroreceptor Afferent Fibers." *Brain Res.* **231**: 240–245.
[Rigney]

Clark, P. K. 1973. "A Subordinated Stochastic Process Model with Finite Variance for Speculative Prices." *Econometrica* **41**: 135–155.
[LeBaron]

Clemens, C. 1993. "Whole Earth Telescope Observations of the White Dwarf Star (PG1159-035)." This volume.
[Gershenfeld, Paluš]

Clements, M. P., and D. F. Hendry. 1993. "On the Limitations of Comparing Mean Square Forecast Errors." *J. Forecasting*: in press.
[Granger]

Clynes, M. 1960. "Respiratory Sinus Arrhythmia: Laws Derived from Computer Simulation." *J. Appl. Physiol.* **15**: 863–874.
[Rigney]

Cohen, A., and I. Procaccia. 1985. "Computing the Kolmogorov Entropy from Time Signals of Dissipative and Conservative Dynamical Systems." *Phys. Rev. A* **31**: 1872–1882.
[Paluš]

Collet, P., and J.-P. Eckmann. 1980. *Iterated Maps on the Tnterval as Dynamical Systems.* Boston: Birkhäuser.
[Gershenfeld]

Connor, J., L. E. Atlas, and D. R. Martin. 1992. "Recurrent Networks and NARMA Modeling." In *Advances in Neural Information Processing Systems*, edited by J. E. Moody, S. J. Hanson, and R. P. Lippman, Vol. IV, 301–308. San Mateo, CA: Morgan Kaufmann.
[Mozer]

Cope, D. 1991. *Computers and Musical Style.* Madison, WI: A-R Editions.
[Dirst]

Corbalan, R., F. Laguarta, J. Pujol, and R. Vilaseca. 1989. "Lorenz-Like Dynamics in Doppler-Broadened Coherently Pumped Lasers." *Opt. Commun.* **71**: 290–294.
[Hübner]

Cornfeld, I. P., S. V. Fomin, and Ya. G. Sinai. 1982. *Ergodic Theory.* New York: Springer.
[Paluš]

Cover, T. M. 1965. "Geometrical and Statistical Properties of Systems of Linear Inequalities with Applications in Pattern Recognition." *IEEE Trans. Elec. Comp.* 14: 326–334.
[Gershenfeld]

Cover, T. M., and J. A. Thomas. 1991. *Elements of Information Theory* . New York: John Wiley.
[Dirst, Gershenfeld]

Cox, D. R. 1984. "Long-Range Dependence: A Review." In *Statistics: An Appraisal, Proceedings 50th Anniversary Conference*, edited by H. A. David and H. T. David, 55–74. Ames, IA: Iowa State Statistics Laboratory, Iowa State University Press.
[Lewis]

Cox, D. R., and P. A. W. Lewis. 1966. *The Statistical Analysis of Series of Events.* New York: Wiley.
[Rigney]

Cox, J. P. 1980. *Theory of Stellar Pulsation.* Princeton: Princeton University Press.
[Clemens]

Craven, P., and G. Wahba. 1979. "Smoothing Noisy Data with Spline Functions. Estimating the Correct Degree of Smoothing by the Method of Generalized Cross Validation." *Numerische Mathematik* **31**: 317–403.
[Lewis]

Cremers, J., and A. Hübler. 1987. "Construction of Differential Equations from Experimental Data." *Z. Naturforsch.* **42(a)**: 797–802.
[Gershenfeld, Paluš]

Cressie, N. 1991. "Geostatistical Analysis of Spatial Data." In *Spatial Statistics and Digital Image Analysis.* Washington: National Academy Press.
[Press]

Cross, M. C., and P. C. Hohenberg. 1993. "Pattern Formation Outside of Equilibrium." *Rev. Mod. Phys.*: in press.
[Swinney]

Crutchfield, J. P., and B. S. McNamara. 1987. "Equations of Motion from a Data Series." *Complex Systems* **1**: 417–452.
[Gershenfeld, Rigney]

Crutchfield, J. P., and K. Young. 1989. "Inferring Statistical Complexity." *Phys. Rev. Lett.* **63**: 105–108.
[Gershenfeld]

Curtiss, P., J. Kreider, and M. Brandemuehl. 1992. "Adaptive Control of HVAC Processes Using Predictive Neural Networks." Technical Report, Joint Center for Energy Management, University of Colorado, Boulder, CO.
[Mozer]

Cvitanovic P., ed. 1984. *Universality in Chaos*. Bristol: Adam Hilger.
[Glass]

Cybenko, G. 1989. "Approximation by Superpositions of a Sigmoidal Function." *Math. Control, Signals, & Sys.* **2(4)**.
[Gershenfeld, Wan]

Davidenko, J. M., A. V. Pertsov, R. Salomonsz, W. Baxter, and J. Jalife. 1992. "Stationary and Drifting Spiral Waves of Excitation in Isolated Cardiac Muscle." *Nature* **355**: 349–351.
[Glass]

Davies, C. T. M., and J. M. M. Neilson. 1967. "Sinus Arrhythmia in Man at Rest." *J. Appl. Physiol.* **22**: 947–955.
[Rigney]

Davies, M. E. 1992a. "An Iterated Function Approximation in Shadowing Time Series." *Phys. Lett. A* **169**: 251–258.
[Kantz]

———. 1992b. "Noise Reduction by Gradient Descent." Preprint.
[Kantz]

DeBoer, R. W., J. M. Karemaker, and J. Strackee. 1985. "Spectrum of a Series of Point Events, Generated by the Integral Pulse Frequency Modulation Model." *Med. Biol. Eng. Comp.* **23**: 138–142.
[Rigney]

Degn H., A. V. Holden, and J. F. Olsen, eds. 1986. *Chaos in Biological Systems*. New York: Plenum.
[Glass, Rigney]

Dempster, A. P., N. M. Laird, and D. B. Rubin. 1977. "Maximum Likelihood from Incomplete Data via the EM Algorithm." *J. Roy. Stat. Soc. B* **39**: 1–78.
[Fraser]

de Vries, B., and J. C. Principe. 1991. "A Theory for Neural Networks with Time Delays." In *Advances in Neural Information Processing Systems 3*, edited by R. P. Lippmann, J. Moody, and D. S. Touretzky, 162–168. San Mateo, CA: Morgan Kaufmann.
[Mozer]

———. 1992. "The Gamma Model—A New Neural Net Model for Temporal Processing." *Neur. Net.* **5**: 565–576.
[Mozer]

Diebold, F. X., and J. M. Nason. 1990. "Nonparametric Exchange Rate Prediction?" *J. Intl. Econ.* **28**: 315–332.
[Gershenfeld, LeBaron]

DiRienzo, M., G. Mancia, G. Parati, A. Pedotti, and A. Zanchetti, eds. 1992. *Blood Pressure and Heart Rate Variability*. Amsterdam: IOS Press.
[Rigney]

Dirst, M., and A. S. Weigend. 1993. "Baroque Forecasting: On Completing J. S. Bach's Last Fugue." This volume.
[Gershenfeld, Zhang]

Donaldson, G. C. 1992. "The Chaotic Behavior of Resting Human Respiration." *Respir. Physiol.* **88**: 313–321.
[Rigney]

Dongarra, J. J., C. B. Moler, J. R. Bunch, and G. W. Stewart. 1979. *LINPACK User's Guide*. Philadelphia: Society for Industrial and Applied Mathematics. For a description of public domain software for many problems in numerical linear algebra.
[Kostelich]

Dooley, M. P., and J. Shafer. 1983. "Analysis of Short-Run Exchange Rate Behavior: March 1973 to November 1981." In *Exchange Rate and Trade Instability: Causes, Consequences, and Remedies*, edited by D. Bigman and T. Taya, 43–72. Cambridge, MA: Ballinger.
[LeBaron]

Dowell, A. R., C. E. Buckley, R. Cohen, R. E. Whalen, and H. O. Sieker. 1971. "Cheyne-Stokes Respiration: A Review of Clinical Manifestations and Critique of Physiological Mechanisms." *Arch. Intern. Med.* **127**: 712–726.
[Rigney]

Doyle, Sir A. C. 1930 *The Penguin Complete Sherlock Holmes*, 163. London: Penguin.
[Smith]

Drygas, H. 1970. *The Coordinate-Free Approach to Gauss-Markov Estimation.* Lecture Notes in Operations Research and Mathematical Systems, Vol. 40. Berlin: Springer-Verlag.
[Press]

Duda, R. O., and P. E. Hart. 1973 *Pattern Classification and Scene Analysis.* New York: Wiley.
[Gershenfeld]

Dupertuis, M. A., R. E. Salomaa, and M. R. Siegrist. 1987. "Stability of Single-Mode Optically Pumped Lasers." *IEEE J. Quant. Electron.* **23**: 1217–1232.
[Hübner]

Dutta, P., and P. M. Horn. 1981. "Low-Frequency Fluctuations in solids—1/f Noise." *Rev. Mod. Phys.* **53**: 497–516.
[Gershenfeld]

Dvořák, I., and J. Klaschka. 1990. "Modification of the Grassberger-Procaccia Algorithm for Estimating the Correlation Exponent of Chaotic Systems with High-Embedding Dimension." *Phys. Lett. A* **145**: 225–231.
[Paluš]

Ebcioglu, K. 1988. "An Expert System for Harmonizing Four-Part Chorales." *Comp. Music J.* **12(3)**: 43–51.
[Dirst]

Eckberg, D. L. 1983. "Human Sinus Arrhythmia as an Index of Vagal Cardiac Outflow." *J. Appl. Physiol.* **54**: 961–966.
[Rigney]

Eckberg, D. L., Y. T. Kifle, and V. L. Roberts. 1980. "Phase Relationship Between Normal Human Respiration and Baroreflex Responsiveness." *J. Physiol. (Lond.)* **304**: 489–502.
[Rigney]

Eckmann, J.-P., S. O. Kamphorst, D. Ruelle, and S. Ciliberto. 1986. "Lyapunov Exponents from a Time Series." *Phys. Rev. A* **34**: 4971.
[Kantz, Kostelich]

Eckmann, J. P., S. O. Kamphorst, and D. Ruelle. 1987. "Recurrence Plots of Dynamical Systems." *Europhys. Lett.* **4**: 973.
[Glass]

Eckmann, J.-P., and D. Ruelle. 1985. "Ergodic Theory of Chaos and Strange Attractors." *Rev. Mod. Phys.* **57**: 617–656.
[Kostelich, Paluš, Sauer]

Edelman, G. M. 1987. *Neural Darwinism: The Theory of Neuronal Group Selection.* New York: Basic Books.
[Rigney]

Edelman, N. H., and T. V. Santiago. 1986. *Breathing Disorders of Sleep.* New York: Churchill Livingstone.
[Rigney]

Edelson, R. A., and J. H. Krolik. 1988. "The Discrete Correlation Function: A New Method for Analyzing Unevenly Sampled Variability Data." *Astrophys. J.* **333**: 646.
[Press]

Efron, B. 1979. "Computers and the Theory of Statistics: Thinking the Unthinkable." *SIAM Rev.* **21**: 460–480.
[Casdagli, Theiler]

———. 1982. *The Jackknife, The Bootstrap, and Other Resampling Plans.* Philadelphia: Society for Industrial and Applied Mathematics.
[LeBaron]

Efron, B., and R. Tsibirani. 1986. "Bootstrap Methods for Standard Errors, Confidence Intervals, and Other Measures of Statistical Accuracy." *Stat. Sci.* **1**: 54–77.
[Theiler]

Elgar, S., and G. Mayer-Kress. 1989. "Observations of the Fractal Dimension of Deep- and Shallow-Water Ocean Surface Gravity Waves." *Physica D* **37**: 104–108.
[Paluš]

elHefnawy, A., G. M. Saidel, and E. N. Bruce. 1988. "CO_2 Control of the Respiratory System: Plant Dynamics and Stability Analysis." *Ann. Biomed. Engr.* **16**: 445–461.
[Rigney]

Ellner, S. 1991. "Detecting Low-Dimensional Chaos in Population Dynamics Data: A Critical Review." In *Chaos and Insect Ecology*, edited by J. A. Logan and F. P. Hain, 65–92. Blacksburg, VA: University of Virginia Press.
[Theiler]

Ellner, S., A. R. Gallant, D. McCaffery, and D. Nychka. 1991. "Convergence Rates and Data Requirements for Jacobian-Based Estimates of Lyapunov Exponents from Data." *Phys. Lett. A* **153**: 357–363.
[Kaplan]

Elman, J. L. 1990. "Finding Structure in Time." *Cog. Sci.* **14**: 179–212.
[Dirst, Mozer]

Elman, J. L., and D. Zipser. 1988. "Discovering the Hidden Structure of Speech." *J. Acoust. Soc. Am.* **83**: 1615–1625.
[Dirst, Mozer]

Elsner, J. B., and A. A. Tsonis. 1992. "Nonlinear Prediction, Chaos, and Noise." *Bull. AMS* **73**: 49–60.
[Lewis]

Engle, R. F. 1982. "Autoregressive Conditional Heteroskedasticity with Estimates of the Variance of United Kingdom Inflation." *Econometrica* **50**: 987–1007.
[LeBaron, Lewis]

Engle, R. F., T. Ito, and W. L. Lin. 1990. "Meteor Showers or Heat Waves? Heteroskedastic Intra-Daily Volatility in the Foreign Exchange Market." *Econometrica* **58**: 525–542.
[LeBaron]

Eubank, R. L. 1988. *Spline Smoothing and Nonparametric Regression.* New York: Marcel Dekker.
[Lewis]

Eubank, S. G., and D. Farmer. 1990. "An Introduction to Chaos and Randomness." In *1989 Lectures in Complex Systems*, edited by E. Jen. Santa Fe Institute Studies in the Sciences of Complexity, Lect. Vol. 1, 75–190. Redwood City, CA: Addison-Wesley.
[Casdagli]

Ezersky, A. B., and M. I. Rabinovich. 1990. "Nonlinear Wave Competition and Anisotropic Spectra of Spatio-Temporal Chaos of Faradey Ripples." *Europhys. Lett.* **13(3)**: 243–249.
[Afraimovich]

Falconer, K. 1990. *Fractal Geometry.* Chichester: Wiley.
[Casdagli]

Faraday, M. 1831. "Experimental Researches in Electricity. " *Proc. Roy. Soc. Lond.* **3**: 91–93.
[Afraimovich]

Farmer, J. D. 1981. "Spectral Broadening of Period-Doubling Bifurcation Sequences." *Phys. Rev. Lett.* **47**: 179–182.
[Theiler]

——.1982. "Chaotic Attractors of an Infinite-Dimensional Chaotic System." *Physica* **4D**: 366–393.
[Glass]

Farmer, J. D., J. Crutchfield, H. Froehling, N. Packard, and R. Shaw. 1980. "Power-Spectra and Mixing Properties of Strange Attractors." *Ann. NYAS* **357**: 453–472.
[Theiler]

Farmer, J. D., and J. J. Sidorowich. 1987. "Predicting Chaotic Time Series." *Phys. Rev. Lett.* **59(8)**: 845–848.
[Casdagli, Gershenfeld, Kostelich, Rigney]

——. 1988. "Exploiting Chaos to Predict the Future and Reduce Noise." *Evolution, Learning, and Cognition*, edited by Y. C. Lee. Singapore: World Scientific.
[Gershenfeld, Lewis, Sauer, Smith, Zhang]

————. 1991. "Optimal Shadowing and Noise Reduction." *Physica D* **47**: 373–392.
[Kantz]

Fedtke, T., ed. 1984. *B.A.C.H. Fugen der Familie Bach.* Frankfurt: Peters.
[Dirst]

Feinstone, L. J. 1987. "Minute by Minute: Efficiency, Normality, and Randomness in
Intra-Daily Asset Prices." *J. Appl. Econ.* **2**: 193–214.
[LeBaron]

Feldman, J. A., and D. H. Ballard. 1982. "Connectionist Models and Their Properties."
Cog. Sci. **6**: 205–254.
[Zhang]

Feldman, J. L., and J. D. Cowan. 1975. "Large-Scale Activity in Neural Nets. II.
A Model for the Brainstem Respiratory Oscillator." *Biol. Cyber.* **17**: 39–51.
[Rigney]

Ferguson, J. D. 1980. "Hidden Markov Analysis: An Introduction." In *Proceedings of
the Symposium on the Applications of Hidden Markov Models to Text and Speech,*
8–15. Princeton, NJ: IDA-CRD.
[Fraser]

Finnoff, W., F. Hergert, and H. G. Zimmermann. 1992. "Improving Model Selection by
Nonconvergent Methods." Unpublished manuscript, submitted to *Neural
Networks.*
[Mozer]

Fisher, J. T., J. P. Mortola, J. B. Smith, G. S. Fox, and S. Weeks. 1982. "Respiration
in Newborns: Development of the Control of Breathing." *Am. Rev. Respir. Dis.*
125: 650–657.
[Glass]

Fleming, D., ed. 1964. *The Mechanistic Conception of Life by Jacques Loeb.* Cambridge,
MA: Harvard University Press.
[Rigney]

Forte, A. 1964. "A Theory of Set-Complexes for Music." *J. Music Theory* **VIII**:
145–148.
[Dirst]

Fouad, F. M., R. C. Tarazi, C. M. Ferrario, S. Fighaly, and C. Alicandri. 1984. "Assess-
ment of Parasympathetic Control of Heart Rate by a Noninvasive Method." *Am.
J. Physiol.* **246**: H838–H842.
[Rigney]

Frasconi, P., M. Gori, and G. Soda. 1992. "Local Feedback Multilayered Networks."
Neur. Comp. **4**: 120–130.
[Mozer]

Fraser, A. M. 1989a. "Reconstructing Attractors from Scalar Time Series: A Comparison of Singular System and Redundancy Criteria." *Physica D* **34**: 391–404.
[Fraser]

————. 1989b. "Information and Entropy in Strange Attractors." *IEEE Trans. Info.
Theory* **IT-35**: 245–262.
[Gershenfeld, Paluš, Pineda]

——. 1989c. "Reconstructing Attractors from Scalar Time Series: A Comparison of Singular System and Redundancy Criteria." *Physica D* **34**: 391–404.
[Gershenfeld, Paluš]

Fraser, A. M., and A. Dimitriadis. 1993. "Forecasting Probability Densities Using Hidden Markov Models with Mixed States." This volume.
[Casdagli, Gershenfeld]

Fraser, A. M., and H. L. Swinney. 1986. "Independent Coordinates for Strange Attractors from Mutual Information." *Phys. Rev. A* **33**: 1134–1140.
[Gershenfeld, Paluš, Pineda]

Fredkin, E., and T. Toffoli. 1982. "Conservative Logic." *Intl. J. Theor. Phys.* **21**: 219–253.
[Gershenfeld]

Freeman, W. J. 1987. "Simulation of Chaotic EEG Patterns with a Dynamic Model of the Olfaction System." *Biol. Cyber.* **56**: 139–150.
[Rigney]

Freyschuss, U., and A. Melcher. 1976. "Respiratory Sinus Arrhythmia in Man: Relation to Cardiovascular Pressure." *Scand. J. Clin. Lab. Invest.* **36**: 221–229.
[Rigney]

Friedman, J. H. 1991a. "Adaptive Spline Networks." Technical Report 107, Department of Statistics, Stanford University, Stanford, CA, March.
[Lewis]

——. 1991b. "Estimating Functions of Mixed Ordinal and Categorical Variables Using Adaptive Splines." Technical Report 108, Department of Statistics, Stanford University, Stanford, CA, June.
[Lewis]

——. 1991c. "Multivariate Adaptive Regression Splines." *Ann. Stat.* **19**: 1–142. With discussion.
[Gershenfeld, Lewis]

Froehling, H., J. P. Crutchfield, D. Farmer, N. H. Packard, and R. Shaw. 1981. "On Determining the Dimension of Chaotic Flows." *Physica D* **3**: 605–617.
[Hübner]

Fucks, W. 1962. "Mathematical Analysis of Formal Structure of Music." *IRE Trans. Info. Theory* **8**: 225–228.
[Dirst]

Funahashi, K.-I. 1989. "On the Approximate Realization of Continuous Mappings by Neural Networks." *Neur. Net.* **2**: 183–192.
[Gershenfeld]

Fux, J. J. 1725. *Gradus ad Parnassum.* Reprinted as *Steps to Parnassum.* New York: W. W. Norton, 1943. Extracts are reprinted in *The Study of Fugue.* New York: Dover, 1958; reprinted in 1987.
[Dirst]

Gabura, A. J. 1970. "Music Style Analysis by Computer." In *The Computer and Music,* edited by H. B. Lincoln, 223–276. Ithaca, NY: Cornell University Press.
[Dirst]

Gallager, R. G. 1968. *Information Theory and Reliable Communication.* New York: J. Wiley.
[Paluš]

Gallez, D., and A. Babloyantz. 1993. *Phys. Lett. A*: in press.
[Glass]

Garfinkel, A, M. L. Spano, W. L. Ditto, and J. N. Weiss. 1992. "Controlling Cardiac Chaos." *Science* **257**: 1230–1235.
[Glass]

Geman, S., E. Bienenstock, and R. Doursat. 1992. "Neural Networks and the Bias/Variance Dilemma." *Neur. Comp.* **5**: 1–58.
[Casdagli, Gershenfeld, Mozer, Sauer, Wan]

Geman, S., and M. Miller. 1976. "Computer Simulation of Brainstem Respiratory Activity." *J. Appl. Physiol.* **41**: 931–938.
[Rigney]

Gencay, R., and W. D. Dechert. 1992. "An Algorithm for the n Lyapunov Exponents of an n-Dimensional Unknown Dynamical System." *Physics D* **59**:142–157.
[Gershenfeld]

Gershenfeld, N. A. 1989. "An Experimentalist's Introduction to the Observation of Dynamical Systems." In *Directions in Chaos,* edited by B.-L. Hao, Vol. 2, 310–384. Singapore: World Scientific.
[Casdagli, Gershenfeld]

——. 1992. "Dimension Measurement on High-Dimensional Systems." *Physica D* **55**: 135–154.
[Gershenfeld, Pineda]

——. 1993a. "Embedding, Expectations, and Noise." Preprint.
[Gershenfeld]

——. 1993b. "Information in Dynamics." In *Proceedings of the Workshop on Physics of Computation,* edited by Doug Matzke, 276–280. Los Alamitos, CA: IEEE Press.
[Gershenfeld]

Gershenfeld, N. A., and A. S. Weigend. 1993. "The Future of Time Series: Learning and Understanding." This volume.
[Casdagli, Dirst, Glass, Kaplan, Kostelich, Lendaris, Paluš, Sauer, Wan]

Gersho, A., and R. M. Gray. 1992. *Vector Quantization and Signal Compression.* Norwell, MA: Kluwer.
[Fraser]

Gerstein, G. L., and M. Mandelbrot. 1964. "Random Walk Models for the Spike Activity of a Single Neuron." *Biophys. J.* **4**: 41–68.
[Glass]

Gibbs, J. W. 1902. *Elementary Principles in Statistical Mechanics Developed with Especial Reference to the Rational Foundation of Thermodynamics.* New Haven, CT: Yale University Press. Republished by Dover, New York, in 1960.
[Fraser]

Giddens, D. P., and R. I. Kitney. 1985. "Neonatal Heart Rate Variability and Its Relation to Respiration." *J. Theor. Biol.* **113**: 759–780.
[Rigney]

Gingerich, O. 1992. *The Great Copernicus Chase and Other Adventures in Astronomical History.* Cambridge, MA: Sky.
[Gershenfeld]

Giona, M., F. Lentini, and V. Cimagalli. 1991. "Functional Reconstruction and Local Prediction of Chaotic Time Series." *Phys. Rev. A* **44**: 3496–3502.
[Gershenfeld]

Glass, L. 1987. "Is the Respiratory Rhythm Generated by a Limit Cycle Oscillator?" In *Concepts and Formalizations in the Control of Breathing*, edited by G. Benchetrit, P. Baconnier, and J. Demongeot, 247–263. Manchester: University Press.
[Rigney]

Glass, L., P. Hunter, and A. McCulloch, eds. 1991. *Theory of Heart: Biomechanics, Biophysics, and Nonlinear Dynamics of Cardiac Function.* New York: Springer-Verlag.
[Glass]

Glass, L., D. Kaplan, and J. Lewis. 1992. "Tests for Deterministic Dynamics in Real and Model Neural Networks." In *Proceedings of the Second Annual Conference on EEG and Nonlinear Dynamics*, edited by B. Jansen. Singapore: World Scientific.
[Glass]

Glass, L., and D. T. Kaplan. 1993. "Complex Dynamics in Physiology and Medicine." This volume.
[Casdagli]

Glass, L., and M. C. Mackey. 1979. "Pathological Conditions Resulting from Instabilities in Physiological Control Systems." *Ann. NYAS* **316**: 214–235.
[Glass]

———. 1988. *From Clocks to Chaos: The Rhythms of Life.* Princeton, NJ: Princeton University Press.
[Glass, Rigney]

Glass, L., and C. P. Maltag. 1990. "Chaos in Multi-Looped Negative Feedback Systems." *J. Theor. Biol.* **145**: 217–223.
[Glass]

Gödel, K. 1931. "Über formal unentscheibare Sätze der *Principia Mathematica* und verwandter Systeme, I." *Monatshefte f—'ur Mathematik und Physik* **38**: 173–198. An English translation of this paper is found in *On Formally Undecidable Propositions* by K. Gödel (New York: Basic Books, 1962).
[Gershenfeld]

Goldberger, A. L., D. R. Rigney, J. Mietus, E. M. Antman, and S. Greenwald. 1988. "Nonlinear Dynamics in Sudden Cardiac Death Syndrome: Heart Rate Oscillations and Bifurcations." *Experientia* **44**: 983–987.
[Rigney]

Goldberger, A. L., D. R. Rigney, and B. J. West. 1990. "Chaos and Fractals in Human Physiology." *Sci. Am.* **262**: 42–49.
[Glass, Kaplan]

Golub, G., and C. Van Loan. 1989. *Matrix Computations*, 2nd ed. Baltimore, MD: The Johns Hopkins University Press.
[Sauer]

Gooch, R. 1983. "Adaptive Pole-Zero Filtering: The Equation Error Approach." Ph.D. Dissertation, Stanford University, Stanford, CA.
[Wan]

Goodhart, C. A. E., and L. Figliuoli. 1991. "Every Minute Counts in Financial Markets." *J. Intl. Money & Fin.* **10**: 23–52.
[LeBaron]

Goodman, L. 1964. "Oscillatory Behavior of Ventilation in Resting Man." *IEEE Trans. Biomed. Engr.* **BME-11**: 82–93.
[Rigney]

Granger, C. W. J. 1992. "Forecasting Stock Market Prices: Lessons for Forecasters." *Intl. J. Forecasting* **14**: 3–24.
[Granger]

———. 1993. "Forecasting in Economics." This volume.

Granger, C. W. J., and A. P. Andersen. 1978. *An Introduction to Bilinear Time Series Models*. Gottingen: Vandenhoek and Ruprecht.
[Gershenfeld]

Granger, C. W. J., and P. Newbold. *Forecasting Economic Time Series*, 2nd ed. San Francisco: Academic Press, 1986.
[Granger]

Granger, C. W. J., and T. Terasvirta. 1992. *Modeling Nonlinear Economic Relationships*. Oxford: Oxford University Press.
[Hutchinson, Zhang]

Grassberger, P. 1983. "Generalized Dimension of Strange Attractors." *Phys. Lett. A* **97**: 227–230.
[Afraimovich, Hübner]

———. 1985. "Generalizations of the Hausdorff Dimension of Fractal Measures." *Phys. Lett. A* **107**: 101–105.
[Pineda]

———. 1986. "Do Climatic Attractors Exist?" *Nature* **323**: 609–612.
[Theiler]

———. 1988. "Finite Sample Corrections to Entropy and Dimension Estimates." *Phys. Lett A* **128**: 369–373.
[Gershenfeld, Pineda]

Grassberger, P., R. Hegger, H. Kantz, C. Schaffrath, and T. Schreiber. 1993. "On Noise Reduction Methods for Chaotic Data." *Chaos* **3(2)**: (17 pages).
[Kantz]

Grassberger, P., and I. Procaccia. 1983a. "Characterization of Strange Attractors." *Phys. Rev. Lett.* **50**: 346–349.
[Gershenfeld, Hübner, Kaplan, Paluš, Pineda]

———. 1983b. "Estimation of the Kolmogorov Entropy from a Chaotic Signal." *Phys. Rev. A* **28**: 2591–2593.
[Hübner, Theiler]

———. 1983c. "Measuring the Strangeness of Strange Attractors." *Physica D* **9**: 189–208.
[Afraimovich, Glass, Kantz, Kaplan, Paluš, Theiler]

Grassberger, P., T. Schreiber, and C. Schaffrath. 1991. "Nonlinear Time Sequence Analysis." *Intl. J. Bif. & Chaos* 1: 521–547.
[Glass, Kaplan, Kostelich, Smith]

Grebogi, C., S. M. Hammel, J. A. Yorke, and T. Sauer. 1990. "Shadowing of Physical Trajectories in Chaotic Dynamics: Containment and Refinement." *Phys. Rev. Lett.* 65: 1527.
[Kantz]

Grebogi, C., E. Ott, S. Pelikan, and J. A. Yorke. 1984. "Strange Attractors that Are Not Chaotic." *Physica D* 13: 261–268.
[Paluš]

Green, M. L., and R. Savit. 1991. "Dependent Variables in Broadband Continuous Time Series." *Physica D* 50: 521–544.
[Gershenfeld]

Grodins, F. S. 1963. *Control Theory in Biological Systems.* New York: Columbia University Press.
[Rigney]

Grodins, F. S., J. Buell, and A. Bart. 1967. "A Mathematical Analysis and Digital Simulation of the Respiratory Control System." *J. Appl. Physiol.* 22: 260–276.
[Rigney]

Guckenheimer, J. 1982. "Noise in Chaotic Systems." *Nature* 298: 358–361.
[Gershenfeld]

Gudmundsson, G. 1970. "Short-Term Variations of a Glacier-Fed River." *Tellus XXII* 3: 341–353.
[Lewis]

Guevara, M. R., L. Glass, M. C. Mackey, and A. Shrier. 1983. "Chaos in Neurobiology." *IEEE Trans. Syst., Man, & Cyber.* SMC-13: 790–798
[Glass]

Guevara, M. R., L. Glass, and A. Shrier. 1981. "Phase-Locking, Period-Doubling Bifurcations and Irregular Dynamics in Periodically Stimulated Cardiac Cells." *Science* 214: 1350–1353
[Glass]

Guevara, M. R., A. Shrier, and L. Glass. 1990. "Chaotic and Complex Cardiac Rhythms." In *Cardiac Physiology: From Cell to Bedside*, edited by D. P. Zipes and J. Jalife, 192–201. Philadelphia: W. B. Saunders.
[Glass]

Guevara, M. R., G. Ward, A. Shrier, and L. Glass. 1984. "Electrical Alternans and Period-Doubling Bifurcations." *IEEE Comp. Cardio.*: 167–170.
[Glass]

Guillemin, V., and A. Pollack. 1974. *Differential Topology.* Englewood Cliffs, NJ: Prentice-Hall.
[Gershenfeld]

Guilleminault, C., and M. Partinen, eds. 1990. *Obstructive Sleep Apnea Syndrome. Clinical Research and Treatment.* New York: Raven Press.
[Rigney]

Gutowitz, H., ed. 1990. "Cellular Automata: Theory and Practice." *Physica D* **45**: 1–483.
[Swinney]

——. 1991. *Cellular Automata, Theory and Experiment*. Cambridge, MA: MIT Press.
[Gershenfeld]

Guyton, A. C. 1991. "Blood Pressure Control—Special Role of the Kidneys and Body Fluids." *Science* **252**: 1813–1816.
[Glass]

Haken, H. 1975. "Analogy Between Higher Instabilities in Fluids and Lasers." *Phys. Lett.* **53A**: 77–78.
[Hübner]

——. 1985. "Light." In *Laser Light Dynamics*, Vol. 2. Amsterdam: North-Holland.
[Hübner]

Hammel, S. M. 1990. "A Noise Reduction Method for Chaotic Systems." *Phys. Lett. A* **148**: 421.
[Kantz]

Hammerstrom, D. 1993. "Neural Networks at Work." *IEEE Spectrum* June: 26–32.
[Gershenfeld]

Havstad, J. W., and C. L. Ehlers. 1989. "Attractor Dimension of Nonstationary Dynamical Systems from Small Data Sets." *Phys. Rev. A* **39**: 845.
[Glass]

Hayashi, H., and S. Ishizuka. 1986. "Chaos in Molluscan Neuron." In *Chaos in Biological Systems*, edited by H. Degn, A. V. Holden, and L. F. Olsen, 157–166. New York: Plenum Press.
[Glass]

Haymet, B. T., and D. I. McCloskey. 1975. "Baroreceptor and Chemoreceptor Influences on Heart Rate During the Respiratory Cycle in the Dog." *J. Physiol. (Lond.)* **245**: 699–712.
[Rigney]

He, X., and A. Lapedes. 1991. "Nonlinear Modeling and Prediction by Successive Approximations Using Radial Basis Fuctions." Technical Report LA-UR-91-1375, Los Alamos National Laboratory, Los Alamos, NM.
[Zhang]

Hellman, J. B., and R. W. Stacy. 1976. "Variation of Respiratory Sinus Arrhythmia with Age." *J. Appl. Physiol.* **41**: 734–738.
[Rigney]

Hentschel, H. G. E., and I. Procaccia. 1983. "The Infinite Number of Generalized Dimensions of Fractals and Strange Attractors." *Physica D* **8**: 435–444.
[Gershenfeld, Hübner, Pineda]

Heppner, J., C. O. Weiss, U. Hübner, and G. Schinn. 1980. "Gain in CW Laser Pumped FIR Laser Gases." *IEEE J. Quant. Electron.* **16**: 392–402.
[Hübner]

Hertz, J. A., A. S. Krogh, and R. G. Palmer. 1991. *Introduction to the Theory of Neural Computation.* Santa Fe Institute Studies in the Sciences of Complexity, Lect. Notes Vol. I. Redwood City, CA: Addison-Wesley.
[Gershenfeld]

Herz, A. V. M. 1991. "Global Analysis of Parallel Analog Networks with Retarded Feedback." *Phys. Rev. A* **44**: 1415–1418.
[Mozer]

Hewlett, W. B. 1993. Article on musical databases, to be submitted to *Comp. Music J.*
[Dirst]

Hewlett, W. B., and E. Selfridge-Field. 1989. "Databases and Musical Information." In *Computing and Musicology*, Vol. 5, 35–38. CCARH (Center for Computer-Assisted Research in the Humanities, 525 Middlefield Road, Suite 120, Menlo Park, CA 94025, USA. E-mail: xb.car@forsythe.stanford.edu.)
[Dirst]

Higgs. 1877. "Bach's 'Art of Fugure.'" In *Proceedings of the Musical Association*, Session 5.2.1877, 45–73. London.
[Dirst]

Hild, H., J. Feulner, and W. Menzel. 1992. "HARMONET: A Neural Net for Harmonizing Chorals in the Style of J. S. Bach." In *Advances in Neural Information Processing Systems*, edited by J. E. Moody, S. J. Hanson, and R. P. Lippman, vol. 4, 267–274. San Mateo, CA: Morgan Kauffman.
[Dirst]

Hiller, L. A., and L. M. Isaacson. 1957. *Illiac Suite for String Quartet.* New Music. Bryn Mawr, PA: Theodore Presser. See, also, Hiller and Isaacson (1959).
[Dirst]

——. 1959. *Experimental Music.* New York: McGraw-Hill.
[Dirst]

Hinich, M. J. 1982. "Testing For Gaussianity and Linearity of a Stationary Time Series." *J. Time Series Anal.* **3**: 169–176.
[Theiler]

Hinton, G., and S. Nowlan. 1990. "The Bootstrap Widrow-Hoff Rule as a Cluster-Formation." *Neur. Comp.* **2(3)**: 355–362.
[Zhang]

Hinton, G. E., and T. J. Sejnowski. 1986. "Learning and Relearning in Boltzmann Machines." In *Parallel Distributed Processing*, edited by D. E. Rumelhart and J. L. McClelland, volume 1. Cambridge, MA: MIT Press.
[Gershenfeld]

Hinton, G. E., and D. van Camp. 1993. "Keeping Neural Networks Simple by Minimizing the Description Length of the Weights." Preprint, Computer Science Department, University of Toronto, June 1993.
[Gershenfeld]

Hirsch, J. A., and B. Bishop. 1981. "Respiratory Sinus Arrhythmia in Humans: How Breathing Pattern Modulates Heart Rate." *Am. J. Physiol.* **241**: H620–H629.
[Rigney]

Hjellming, R. M., and R. Narayan. 1986. "Refractive Interstellar Scintillation in 1741–038." *Astrophys. J.* **310**: 768.
[Press]

Hochreiter, J. 1991. Diploma Thesis, Institute für Informatik, Technische Universität Muenchen, unpublished.
[Mozer]

Hodgkin, A. L., and A. F. Huxley. 1952. "A Quantitative Description of Membrane Current and Its Application to Conduction and Excitation in Nerve." *J. Physiol. (Lond.)* **117**: 500–544.
[Glass]

Hofstadter, D. R. 1979. *Gödel, Escher, Bach: An Eternal Golden Braid.* New York: Basic Books.
[Gershenfeld]

Hogenboom, E., W. Klische, C. O. Weiss, and A. Godone. 1985. "Instabilities of a Homogeneously Broadened Laser." *Phys. Rev. Lett.* **55**: 2571–2574.
[Hübner]

Holland, J. 1992. *Adaptation in Natural and Artificial Systems.* Cambridge, MA: MIT Press.
[Rigney]

Holzfuss, J., and U. Parlitz. 1991. "Lyapunov Exponents from Time Series." In *Proceedings of the Conference "Lyapunov Exponents," Oberwolfach 1990,* edited by L. Arnold, H. Crauel, and J. P. Eckmann. Lecture Notes in Mathematics. Berlin: Springer-Verlag.
[Kantz]

Horgan, J. D., and D. L. Lange. 1962. "Analog Computer Studies of Periodic Breathing." *IRE Trans. Biomed. Elec.* **9**: 221–228.
[Rigney]

Hornik, K., M. Stinchombe, and H. White. 1989. "Multilayer Feedforward Networks are Universal Approximators." *Neur. Net.* **2**: 359–366.
[Wan]

Hrushesky, W. J. M., D. Fader, O. Schmitt, and V. Gilbert. 1984. "The Respiratory Sinus Arrhythmia: A Measure of Cardiac Age." *Science*: 1001–1004.
[Rigney]

Hsieh, D. A. 1989. "Testing for Nonlinear Dependence in Daily Foreign Exchange Rate Changes." *J. Business* **62(3)**: 339–368.
[LeBaron, Theiler]

——. 1991. "Chaos and Nonlinear Dynamics: Application to Financial Markets." *J. Finance* **46**: 1839–1877.
[Casdagli, Theiler]

Hsü, K. J., and A. J. Hsü. 1990. "Fractal Geometry of Music: Physics of Melody." *Proc. Natl. Acad. Sci. USA* **87**: 938–941.
[Dirst]

Hu, M. J. C. 1964. "Application of the Adaline System to Weather Forecasting."
E. E. Degree Thesis. Technical Report 6775-1, Stanford Electronic Laboratories,
Stanford, CA, June.
[Gershenfeld]

Hübler, A. 1989. "Adaptive Control of Chaotic Systems." *Helv. Phys. Acta* **62**:
343–346.
[Gershenfeld]

Hübner, U., N. B. Abraham, and C. O. Weiss. 1989. "Dimensions and Entropies of
Chaotic Intensity Pulsations in a Single-Mode Far-Infrared NH_3 Laser." *Phys. Rev.
A* **40**: 6354–6365.
[Hübner, Sauer, Smith, Wan]

Hübner, U., W. Klische, N. B. Abraham, and C. O. Weiss. 1989. "Comparison of
Lorenz-Like Laser Behavior with the Lorenz Model." In *Coherence and Quantum
Optics VI*, edited by J. H. Eberly, L. Mandel, and E. Wolf, 517–520. New York &
London: Plenum Press.
[Hübner]

——.1990. "On Problems Encountered with Dimension Calculations." In *Measures of
Complexity and Chaos*, edited by N. B. Abraham, A. M. Albano, A. Passamente,
and P. E. Rapp, 133–136. New York: Plenum Press.
[Hübner]

Hübner, U., W. Klische, and C. O. Weiss. 1992. "Generalized Dimensions of Laser Attractors." *Phys. Rev. A* **45**: 2128–2130.
[Hübner]

Hübner, U., C. O. Weiss, N. B. Abraham, and D. Tang. 1993. "Lorenz-Like Chaos in
NH_3-FIR Lasers." This volume.
[Casdagli, Dirst, Gershenfeld, Kaplan, Kostelich, Paluš, Sauer, Smith, Wan]

Hunt, F., and F. Sullivan. 1986. "Efficient Algorithms for Computing Fractal Dimensions." In *Dimensions and Entropies in Chaotic Systems: Quantification of Complex Behavior*, edited by G. Mayer-Kress, 74–81. Berlin: Springer-Verlag.
[Pineda]

Husmann, H. 1938. "Die 'Kunst der Fuge' als Klavierwerk." *Bach Jahrbuch* **1938**:
1–61.
[Dirst]

Iberall, A. S., and A. C. Guyton, eds. 1973. *Regulation and Control in Physiological
Systems*. Pittsburgh, PA: Instrument Society of America.
[Rigney]

Irie, B., and S. Miyake. 1988. "Capabilities of Three-Layered Perceptrons." *Proceedings
of the IEEE Second International Conference on Neural Networks*, Vol. I, 641–647.
Conference was held in San Diego, CA, July, 1988.
[Wan]

Jackson, R. 1970. "Harmony Before and After 1910: A Computer Comparison." In *The
Computer and Music*, edited by H. B. Lincoln, 132–146. Ithaca, NY: Cornell Uni-
versity Press.
[Dirst]

Jakubith, S., H. H. Rotermund, W. Engell, A. von Oertzen, and G. Ertl. 1990. "Spatio-Temporal Concentration Patterns in a Surface Reaction: Propagating and Standing Waves, Rotating Spirals, and Turbulence." *Phys. Rev. Lett.* **65**: 3013–3016.
[Swinney]

Jannet, T. C., G. N. Kay, and L. C. Sheppard. 1990. "Automated Administration of Lidocaine for the Treatment of Ventricular Arrhythmias." *Med. Prog. Tech.* **16**: 53–59.
[Rigney]

Jenkins, G. M., and D. G. Watts. 1968. *Spectral Analysis and Its Applications.* San Francisco: Holden-Day.
[Theiler]

Jensen, M. C. 1978. "Some Anomalous Evidence Regarding Market Efficiency." *J. Fin. Econ.* **6**: 95–101.
[Zhang]

Johnson, Jr., C. 1984. "Adaptive IIR Filtering: Current Results and Open Issues." *IEEE Trans. Inform. Th.* **IT-30(2)**: 237–250.
[Wan]

Johnson, E. E., S. F. Idriss, C. Cabo, S. B. Melnick, W. M. Smith, and R. E. Ideker. 1992. "Evidence that Organization Increases During the First Minute of Ventricular Fibrillation in Pigs Mapped with Closely Spaced Electrodes." *J. Am. Coll. Card.* **19(90A)**: abstract.
[Glass]

Jordan, M. I. 1987. "Attractor Dynamics and Parallelism in a Connectionist Sequential Machine." In *Proceedings of the Eighth Annual Conference of the Cognitive Science Society*, 531–546. Hillsdale, NJ: Erlbaum.
[Mozer]

Kabal, P. 1990. "The Stability of Adaptive Minimum Mean Square Error Equalizers Using Delayed Adjustment." In *IEEE Trans. Comm.* **Com-31(3)**: 430–432.
[Wan]

Kailath, T. 1980. *Linear Systems.* Englewood Cliffs, NJ: Prentice-Hall.
[Wan]

Kalli, S., J. Gronlund, H. Ihalainen, A. Siimes, I. Valimaki, and K. Antila. 1988. "Multivariate Autoregressive Modeling of Autonomic Cardiovascular Control in Neonatal Lamb." *Comp. Biomed. Res.* **21**: 512–530.
[Rigney]

Kameoka, A., and M. Kuriyagawa. 1969. "Consonance Theory Part II: Consonance of Complex Tones and Its Calculation Method." *J. Acous. Soc. Am.* **45**: 1460–1469.
[Dirst]

Kamke, E. 1959. *Differentialgleichungen Lösungsmethoden und Lösungen.* Leipzig. (In Russian: Moscow: Nauka, 1971).
[Paluš]

Kaneko, K. 1989. "Spatio-Temporal Chaos in One- and Two-Dimensional Coupled Map Lattices" *Physica D* **37**: 60–82.
[Swinney]

Kanter, M. 1979. "Lower Bounds for Nonlinear Prediction Error in Moving-Average Processes." *Ann. Prob.* **7**: 128–138.
[Theiler]

Kantz, H. 1993. "Noise Reduction by Local Reconstruction of the Dynamics." This volume.
[Gershenfeld]

Kaplan, D. T. 1992. "Geometrical Techniques for Analyzing ECG Dynamics." *J. Electrocardiology* **24**: 77–82.
[Glass]

———. 1993. "A Geometrical Statistic for Detecting Deterministic Dynamics." This volume.
[Dirst, Gershenfeld, Paluš]

———. n.d. "Evaluating Deterministic Structure in Maps Deduced from Discrete-Time Measurements." *Intl. J. Bif. & Chaos*, in press.
[Glass, Kaplan]

Kaplan, D. T., and R. J. Cohen. 1990. "Is Fibrillation Chaos?" *Cir. Res.* **67**: 886–892.
[Glass, Theiler]

Kaplan, D. T., M. I. Furman, S. M. Pincus, S. Ryan, L. Lipsitz, and A. L. Goldberger. 1991. "Aging and the Complexity of Cardiovascular Dynamics." *Biophys. J.* **59**: 945–949.
[Glass]

Kaplan, D. T., and L. Glass. 1992. "A Direct Test for Determinism in a Time Series." *Phys. Rev. Lett.* **68(4)**: 427–430.
[Glass, Kaplan, Paluš]

———. 1993. "Coarse-Grained Embeddings of Time Series: Random Walks, Gaussian Random Processes, and Deterministic Chaos." *Physica D* **64**: 431–454.
[Kaplan]

Kaplan, D. T., and M. Talajic. 1991. "Dynamics of Heart Rate." *Chaos* **1**: 251–256.
[Glass, Rigney]

Kaplan, J., and J. Yorke. 1979. "Chaotic Behavior of Multidimensional Difference Equations." In *Functional Differential Equations and Approximation of Fixed Points*, edited by H. O. Peitgen and H. O. Walther. Lecture Notes in Mathematics, Vol. 730, 219. New York: Springer-Verlag.
[Gershenfeld]

Katok, A. B., Ya. G. Sinai, and A. M. Stepin. 1985. "Theory of Dynamical Systems and General Groups of Transformations with Invariant Measure." In *Modern Problems of Mathematics. Fundamental Research*, Vol. 2, 5–121. Moscow: VINITI Ak. Nauk SSSR.
[Afraimovich]

Katona, P. G., and F. Jih. 1975. "Respiratory Sinus Arrhythmia: Noninvasive Measure of Parasympathetic Cardiac Control." *J. Appl. Physiol.* **39**: 801–805.
[Rigney]

Keenan, D. M. 1985. "A Tukey Nonadditivity-Type Test for Time Series Nonlinearity." *Biometrika* **72**: 39–44.
[Theiler]

Kennel, M. B., R. Brown, and H. D. I. Abarbanel. 1992. "Determining Minimum Embedding Dimension Using a Geometrical Construction." *Phys. Rev. A* **45**: 3403–3411.
[Gershenfeld, Glass, Pineda]

Kennel, M. B., and S. Isabelle. 1992. "Method to Distinguish Possible Chaos from Colored Noise and to Determine Embedding Parameters." *Phys. Rev. A* **46**: 3111–3118.
[Theiler]

Kepler, J. 1619. *Harmonices mundi libri V.* Linz, Austria: Johannis Plancus.
[Dirst]

Khandokhin, P. A., Ya. I. Khanin, and I. V. Koryukin. 1988. "Bifurcations and Chaos in the Three-Level Model of a Laser with Coherent Optical Pumping." *Opt. Commun.* **65**: 367–372.
[Hübner]

Khinchin, A. I. 1957. *Mathematical Foundations of Information Theory.* New York: Dover.
[Paluš]

Khoo, M. C. K., ed. 1989. *Modeling and Parameter Estimation in Respiratory Control.* New York: Plenum Press.
[Rigney]

———. 1991. "Periodic Breathing." In *The Lung: Scientific Foundations*, edited by R. G. Crystal and J. B. West., 1419–1431. New York: Raven Press.
[Rigney]

Khoo, M. C. K., A. Gottschalk, and A. I. Pack. 1991. "Sleep-Induced Periodic Breathing and Apnea: A Theoretical Study." *J. Appl. Physiol.* **70**: 2014–2024.
[Rigney]

Khoo, M. C. K., R. E. Kronauer, K. P. Strohl, and A. S. Slutsky. 1982. "Factors Inducing Periodic Breathing in Humans: A General Model." *J. Appl. Physiol.* **53**: 644–659.
[Rigney]

Khoo, M. C. K., and V. Z. Marmarelis. 1989. "Estimation of Peripheral Chemoreflex Gain from Spontaneous Sigh Responses." *Ann. Biomed. Engr.* **17**: 557–570.
[Rigney]

Kirchheim, H. R. 1976. "Systemic Arterial Baroreceptor Reflexes." *Physiol. Rev.* **56**: 100–177.
[Rigney]

Kitney, R. I. 1981. "Modelling Respiratory Sinus Arrhythmia and Its Application to the Study of Neuropathy." In *Computing in Medicine*, edited by J. P. Paul, M. M. Jordan, M. W. Ferguson-Pell, and B. J. Andrews, 126–134, London: Macmillan.
[Rigney]

Kitney, R. I., T. Fulton, A. H. McDonald, and D. A. Linkens. 1985. "Transient Interactions Between Blood Pressure, Respiration, and Heart Rate in Man." *J. Biomed. Engr.* **7**: 217–224.
[Rigney]

Kitney, R. I., D. Linkens, A. Selman, and A. McDonald. 1982. "The Interaction Between Heart Rate and Respiration: Part II—Nonlinear Analysis Based on Computer Modeling." *Automedica* 4: 141–153.
[Rigney]

Kitney, R. I., and O. Rompelman, eds. 1981. *The Study of Heart Rate Variability.* Clarendon: Oxford University Press.
[Glass, Rigney]

Kleiger, R. E., J. P. Miller, J. T. Bigger, A. J. Moss, and the Multicenter Post-Infarction Research Group. 1987. "Decreased Heart Rate Variability and Its Association with Increased Mortality After Acute Myocardial Infarction." *Am. J. Cardiol.* **59**: 256–262.
[Glass]

Kleinfeld, D. 1986. "Sequential State Generation by Model Neural Networks." *Proc. Natl. Acad. Sci.* **83**: 9469–9473.
[Mozer]

Klische, W., and C. O. Weiss. 1985. "Instabilities and Routes to Chaos in a Homogeneously Broadened One- and Two-Mode Ring Laser." *Phys. Rev. A* **31**: 4049–4051.
[Hübner]

Knuth, D. E. 1981. *Semi-Numerical Algorithms. Art of Computer Programming*, Vol. 2, 2nd ed. Reading, MA: Addison-Wesley.
[Gershenfeld]

Kobayashi, M., and T. Musha. 1982. "$1/f$ Fluctuation of Heartbeat Period." *IEEE Trans. Biomed. Engr.* **BME-29**: 456– 457.
[Glass, Rigney]

Kobayashi, Y. 1973. "Franz Hauser und seine Bach-Handschriftensammlung." Dissertation, Göttingen University, Germany.
[Dirst]

Koepchen, H. P., S. M. Milton, and A. Trzebski, eds. 1980. *Central Interaction Between Respiratory and Cardiovascular Control Systems.* Berlin: Springer-Verlag.
[Rigney]

Kohonen, T. 1990. "The Self-Organizing Map." *Proc. IEEE* **78**: 1464–1480. (The program can be obtained from cochlea.hut.fi via anonymous ftp from the directory /pub/som_pak).
[Dirst]

Kohonen, T., P. Laine, K. Tiits, and K. Torkkola. 1991. "A Nonheuristic Automatic Composing Method." In *Music and Connectionism*, edited by P. M. Todd and D. G. Loy, 229–242. Cambridge, MA: MIT Press.
[Dirst]

Kolmogorov, A. 1941. "Interpolation und Extrapolation von stationären zufälligen Folgen." *Bull. Acad. Sci. (Nauk)* **5**: 3–14. U.S.S.R., Ser. Math.
[Gershenfeld]

. 1959. *Dokl. Akad. Nauk SSSR* **124**: 754.
[Paluš]

——. 1965. "Three Approaches to the Quantitative Definition of Information." *Prob. Infor. Trans.* **1**: 4–7.
[Gershenfeld]

Kolneder, W. 1977. *Die Kunst der Fuge: Mythen des 20. Jahrhunderts.* Wilhelmshaven: Heinrichshofen.
[Dirst]

Korenberg, M. J., and I. W. Hunter. 1990. "The Identification of Nonlinear Biological Systems: Wiener Kernel Approaches." *Ann. Biomed. Engr.* **18**: 629–654.
[Rigney]

Kostelich, E. J. 1993. "Problems in Estimating Dynamics from Data." *Physica D*, in press.
[Kostelich]

Kostelich, E. J., and D. P. Lathrop. 1993. "Times Series Prediction Using the Method of Analogues." This volume.

Kostelich, E. J., and H. L. Swinney. 1989. "Practical Considerations in Estimating Dimension from Time Series Data." *Phys. Scripta* **40**: 436–441.
[Pineda]

Kostelich, E. J., and J. A. Yorke. 1988. "Noise Reduction in Dynamical Systems." *Phys. Rev. A* **38**: 1649–1652.
[Kantz]

——. 1990. "Noise Reduction: Finding the Simplest Dynamical System Consistent with the Data." *Physica D* **41**: 183–196.
[Kostelich]

Kot, M., G. S. Sayler, and T. W. Schultz. 1992. "Complex Dynamics in a Model Microbial System." *Bull. Math. Biol.* **54(4)**: 619–648.
[Paluš]

Koza, J. R. 1993. *Genetic Programming.* Cambridge, MA: MIT Press.
[Gershenfeld]

Kramer, S., and M. Marder. 1992. "Evolution of River Networks." *Phys. Rev. Lett.* **68**: 205–208.
[Swinney]

Krasner, S., ed. 1990. *The Ubiquity of Chaos.* Washington, DC: American Association for the Advancement of Science.
[Glass]

Kroll, M. W., and K. W. Fulton. 1991. "Slope Filtered Pointwise Correlation Dimension Algorithm and Its Evaluation with Prefibrillation Heart Rate Data." *J. Electrocardiology* **24**: 97–101.
[Glass]

Kühn, R., and J. L. van Hemmen. 1991. "Self-Organizing Maps and Adaptive Filters." In *Models of Neural Networks*, edited by E. Domany, J. L. van Hemmen, and K. Schulten, 213–280. New York: Springer-Verlag.
[Mozer]

Kullback, S. 1959. *Information Theory and Statistics.* New York: Wiley.
[Paluš]

Kullback, S., and R. A. Leibler. 1951. "On Information and Sufficiency." *Ann. Math. Stat.* **22**: 79–86.
[Fraser]

Kung, S. Y. 1993. *Digital Neural Networks*. Englewood Cliffs, NJ: Prentice Hall.
[Gershenfeld]

Künsch, H. R. 1989. "The Jackknife and the Bootstrap for General Stationary Observations." *Ann. Stat.* **17**: 1217–1241.
[Theiler]

Kurths, J., and H. Herzel. 1987. "An Attractor in a Solar Time Series." *Physica D* **25**: 165–172.
[Theiler]

Labeyrie, E., J. F. Bertholon, Y. Shikata, and A. Teillac. 1987. "Ventilatory Adaptation Hysteresis for CO_2 Regulation." In *Concepts and Formalizations in the Control of Breathing*, edited by G. Benchetrit, P. Baconnier, and J. Demongeot, 69–86. Manchester: University Press.
[Rigney]

Laguarta, F., J. Pujol, R. Vilaseca, and R. Corbalan. 1988. "Instabilities in Doppler-Broadened Optically Pumped Far-Infrared Lasers." *J. Phys.* **49**: C2-409–C2-412.
[Hübner]

Laird, P., and R. Saul. 1993. "Discrete Sequence Prediction and Its Applications." *Machine Learning*: submitted.
[Gershenfeld]

Landau, I. 1979. *Adaptive Control: The Model Reference Approach*. New York: Marcel Dekker.
[Wan]

Landauer, R. 1991. "Information is Physical." *Physics Today* **44**: 23.
[Gershenfeld]

Lang, K., and G. Hinton. 1988. "A Time-Delay Neural Network Architecture for Speech Recognition." Technical Report CMU-CS-88-152, Carnegie-Mellon University, Pittsburgh, PA.
[Wan]

Lang, K. J., A. H. Waibel, and G. E. Hinton. 1990. "A Time-Delay Neural Network Architecture for Isolated Word Recognition." *Neur. Net.* **3**: 23–43.
[Gershenfeld]

Lapedes, A., and R. Farber. 1987. "Nonlinear Signal Processing Using Neural Networks." Technical Report No. LA-UR-87-2662, Los Alamos National Laboratory, Los Alamos, NM.
[Casdagli, Gershenfeld, Mozer, Wan]

Larsen, L. E., and D. O. Walter. 1970. "On Automatic Methods of Sleep Staging by EEG Spectra." *Electroencephal & Clin. Neurophysiol.* **28**: 459–467.
[Glass]

Lathrop, D. P., and E. J. Kostelich. 1989. "The Characterization of an Experimental Strange Attractor by Periodic Orbits." *Phys. Rev. A* **40**: 4028–4031.
[Kostelich]

Lawandy, N. M., and D. V. Plant. 1986. "On the Experimental Accessibility of the Self-Pulsing Regime of the Lorenz Model for Single-Mode Homogeneously Broadened Lasers." *Opt. Commun.* **59**: 55–58.
[Hübner]

LeBaron, B. 1989. "Nonlinear Dynamics and Stock Returns." *J. Business* **62(3)**: 311–337.
[Zhang]

——. 1991. "Technical Trading Rules and Regime Shifts in Foreign Exchange." Technical Paper, University of Wisconsin, Madison, WI.
[LeBaron]

——. 1992a. "Do Moving-Average Trading Rule Results Imply Nonlinearities in Foreign Exchange Markets?" Technical Report 9222, Social Systems Research Institute, University of Wisconsin, Madison, WI.
[Theiler]

——. 1992b. "Improving Forecasts Using a Volatility Index." *J. Appl. Econ.* **7**: S137–S150.
[LeBaron]

——. 1993. "Nonlinear Diagnostics and Simple Trading Rules for High-Frequency Foreign Exchange Rates." This volume.
[Gershenfeld, Zhang]

le Cun, Y. 1989. "Generalization and Network Design Strategies." In *Connectionism in Perspective*, edited by R. Pfeifer, Z. Schreter, F. Fogelman, and L. Steels. Amsterdam: North Holland.
[Gershenfeld, Wan]

le Cun, Y., B. Boser, et al. 1989. "Backpropagation Applied to Handwritten Zip Code Recognition." *Neur. Comp.* **1**: 541–551.
[Wan]

le Cun, Y. J. S. Denker, and S. A. Solla. 1990. "Optimal Brain Damage." In *Advances in Neural Information Processing Systems 2 (NIPS*89)*, edited by D. S. Touretzky, 598–605. San Mateo, CA: Morgan Kaufmann.
[Gershenfeld]

le Cun, Y., I. Kanter, and S. A. Solla. 1991. "Second-Order Properties of Error Surfaces: Learning Time and Generalization." In *Advances in Neural Information Processing Systems 3 (NIPS*90)*, edited by R. P. Lippmann, J. Moody, and D. S. Touretzky, 918–924. San Mateo, CA: Morgan Kaufmann.
[Mozer]

Lee, T.-H., H. White, and C. W. J. Granger. 1992. "Testing for Neglected Nonlinearity in Time Series Models: A Comparison of Neural Network Methods and Alternative Tests." *J. Econometrics*: in press.
[LeBaron, Theiler]

Leger, C., D. N. Politis, and J. P. Romano. 1992. "Bootstrap Technology and Applications." *Technometrics*: in press.
[Theiler]

Lehár, J., J. N. Hewitt, D. H. Roberts, and B. F. Burke. 1992. "The Radio Time Delay in the Double Quasar 0957+561." *Astrophys. J.* **384**: 453.
[Press]

Leite, J. R. R., M. Ducloy, A. Sanchez, D. Seligson, and M. S. Feld. 1977. "Laser Saturation Resonances in NH_3 Observed in the Time-Delayed Mode." *Phys. Rev. Lett.* **39**: 1469–1472.
[Hübner]

Lendaris, G. G., and A. M. Fraser. 1993. "Visual Fitting and Extrapolation." This volume.

Lequarré, J. Y. 1993. "Foreign Currency Dealing: A Brief Introduction." This volume.
[Gershenfeld, Zhang]

Levich, R. M. 1979. *The International Money Market: An Assessment of Forecasting Techniques and Market Efficiency.* Greenwich, CT: JAI Press.
[LeBaron]

———. 1981. "How to Compare Chance with Forecasting Expertise." *Euromoney* August: 61–78.
[LeBaron]

Levy, M. N., H. DeGeest, and H. Zieske. 1966. "Effects of Respiratory Center on the Heart." *Circ. Res.* **18**: 67–78.
[Rigney]

Lewis, P. A. W. 1970. "Remarks on the Theory, Computation, and Application of the Spectral Analysis of Series of Events." *J. Sound & Vibra.* **12**: 353–375.
[Rigney]

———, ed. 1972a. *Stochastic Point Processes: Statistical Analysis, Theory and Applications.* New York: Wiley.
[Rigney]

———. 1972b. "Recent Results in the Statistical Analysis of Univariate Point Processes." In *Stochastic Point Processes*, edited by P. A. W. Lewis, 1–54. New York: Wiley.
[Rigney]

Lewis, P. A. W., and B. K. Ray. 1992. "Nonlinear Modeling of Multivariate and Categorical Time Series Using Multivariate Adaptive Regression Splines." *Proceedings of the ASA Joint Meetings, Business and Economic Statistics Section*: in press.
[Lewis]

Lewis, P. A. W., B. K. Ray, and J. G. Stevens. 1993. "Modeling Time Series Using Multivariate Adaptive Regression Splines (MARS)." This volume.
[Gershenfeld]

Lewis, P. A. W., and J. G. Stevens. 1991. "Nonlinear Modeling of Time Series Using Multivariate Adaptive Regression Splines (MARS)." *J. Am. Stat. Assoc.* **87**: 864–877.
[Lewis]

———. 1992. "Semi-Multivariate Modeling of Time Series Using Adaptive Spline Threshold Autoregression (ASTAR)." *Water Res. Bull.*: in press.
[Lewis]

Lewis, T. O., and P. L. Odell. 1971. *Estimation in Linear Models.* Englewood Cliffs, NJ: Prentice-Hall.
[Press]

Li, M. Y., T. Win, C. O. Weiss, and N. R. Heckenberg. 1990. "Attractor Properties of Laser Dynamics: A Comparison of NH_3 Laser Emission with the Lorenz Model." *Opt. Commun.* **80**: 119–126.
[Hübner]

Liebert, W., K. Pawelzik, and H. G. Schuster. 1991. "Optimal Embeddings of Chaotic Attractors from Topological Considerations." *Europhys. Lett.* **14**: 521–526.
[Pineda]

Liebert, W., and H. G. Schuster. 1989. "Proper Choice of the Time Delay for the Analysis of Chaotic Time Series." *Phys. Lett. A* **142**: 107–111.
[Gershenfeld, Pineda]

Liebovitch, L. S., and T. Toth. 1989. "A Fast Algorithm to Determine Fractal Dimension by Box Counting." *Phys. Lett. A* **141**: 386–390.
[Pineda]

Lii, K. S., and M. Rosenblatt. 1982. "Deconvolution and Estimation of Transfer Function Phase and Coefficients for Non-Gaussian Linear Processes." *Ann. Stat.* **10**: 1195–1208.
[Theiler]

Lincoln, W. P., and J. Skrzypek. 1990. "Synergy of Clustering Multiple Back-Propagation Networks." In *Advances in Neural Information Processing Systems 2*, edited by D. S. Touretzky, 650–657. San Mateo, CA: Morgan Kaufmann.
[Mozer]

Linsay, P. 1991. "An Efficient Method of Forecasting Chaotic Time Series Using Linear Interpolation." *Phys. Lett. A* **153**: 353–356.
[Rigney]

Ljung, L. 1987. *System Identification: Theory for the User.* Englewood Cliffs, NJ: Prentice-Hall.
[Wan]

Loeve, M. 1977. *Probability Theory*, Vol. 2, 144. New York: Springer.
[Swinney]

Long, G., F. Ling, and J. Proakis. 1989. "The LMS Algorithm with Delayed Coefficient Adaptation." *IEEE Trans. Acoust., Speech, & Signal Proc.* **27(9)**: 1397–1459.
[Wan]

Longobardo, G. S., N. S. Cherniak, and A. P. Fishman. 1966. "Cheyne-Stokes Breathing Produced by a Model of the Human Respiratory System." *J. Appl. Physiol.* **21**: 1839–1846.
[Rigney]

Longtin, A. 1993. "Nonlinear Forecasting of Spike Trains from Sensory Neurons." *Intl. J. Bif. & Chaos*: in press.
[Glass]

Lopes, O. U., and J. F. Palmer. 1976. "Proposed Respiratory Gating Mechanism for Cardiac Slowing." *Nature* **264**: 454.
[Rigney]

Loredo, T. J. 1992. "The Promise of Bayesian Inference for Astrophysics." In *Statistical Challenges in Modern Astronomy*, edited by E. D. Feigelson and G. J. Babu. New York: Springer.
[Press]

Lorenz, E. N. 1963. "Deterministic Non-Periodic Flow." *J. Atmos. Sci.* **20**: 130–141.
[Gershenfeld, Hübner, Kostelich, Paluš, Sauer, Theiler]

Lorenz, E. N. 1989. "Computational Chaos—A Prelude to Computational Instability." *Physica D* **35**: 299–317.
[Gershenfeld]

Loy, D. G. 1991. "Connectionism and Musiconomy." In *Music and Connectionism*, edited by P. M. Todd and D. G. Loy, 20–36. Cambridge, MA: MIT Press.
[Dirst]

MacGregor, R. J. 1987. *Neural and Brain Modeling*. San Diego: Academic Press.
[Rigney]

MacGregor, R., and E. Lewis. 1977. *Neural Modeling*. New York: Plenum Press.
[Wan]

MacKay, D. 1992a. "Bayesian Interpolation." *Neur. Comp.* **4(3)**: 415–447.
[Wan]

——. 1992b. "A Practical Bayesian Framework for Backpropagation Networks." *Neur. Comp.* **4(3)**: 448–472.
[Wan]

Mackey, M. C., and L. Glass. 1977. "Oscillations and Chaos in Physiological Control Systems." *Science* **197**: 287–289.
[Rigney]

Makridakis, S., A. Andersen, R. Carbone, R. Fildes, M. Hibon, R. Lewandowski, J. Newton, E. Parzen, and R. Winkler. 1984. *The Forecasting Accuracy of Major Time Series Methods*. New York: Wiley.
[Gershenfeld]

Makridakis, S., and M. Hibon. 1979. "Accuracy of Forecasting: An Empirical Investigation." *J. Roy. Stat. Soc. A* **142**: 97–145. With discussion.
[Gershenfeld]

Malley, J. D. 1986. *Optimal Unbiased Estimation of Variance Components*. Lecture Notes in Statistics, Vol. 39. Berlin: Springer-Verlag.
[Press]

Mark, R. G., and G. B. Moody. 1989. "Automated Arrhythmia Analysis." In *Encyclopedia of Medical Devices and Instrumentation*, edited by J. G. Webster, Vol. 1, 120–130. New York: John Wiley.
[Rigney]

Markov, A. A. 1913. "An Example of Statistical Investigation in the Text of 'Eugene Onyegin' Illustrating Coupling of 'Tests' in Chains." *Proc. Acad. Sci. St. Petersburg* **7**: 153–162.
[Dirst]

Marmarelis, V. Z. 1988. "Coherence and Apparent Transfer Function Measurements for Nonlinear Physiological Systems." *Ann. Biomed. Engr.* **16**: 143–157.
[Rigney]

——. 1991. "Wiener Analysis of Nonlinear Feedback in Sensory Systems." *Ann. Biomed. Engr.* **19**: 345–382.
[Rigney]

Marteau, P. F., and H. D. I. Abarbanel. 1991. "Noise Reduction in Chaotic Time Series Using Scaled Probabilistic Methods." *J. Nonlinear Sci.* **1**: 313.
[Gershenfeld]

Martin, N. F. G., and J. W. England. 1981. *Mathematical Theory of Entropy.* London: Addison-Wesley. (In Russian: Moscow: Mir, 1988).
[Paluš]

Maxwell, H. J. 1992. "An Expert System for Harmonizing Analysis of Tonal Music." In *Understanding Music with AI: Perspectives on Music Cognition,* edited by M. Balaban, K. Ebcioglu, and O. Laske, 335–353. Cambridge, MA: MIT Press.
[Dirst]

May, R. M. 1976. "Simple Mathematical Models with Very Complicated Dynamics." *Nature* **261**: 459.
[Gershenfeld]

Mayer-Kress, G., ed. 1986. *Dimensions and Entropies in Chaotic Systems: Quantification of Complex Behavior.* Berlin: Springer-Verlag.
[Afraimovich, Glass, Paluš]

Mayer-Kress, G., F. E. Yates, L. Benton, M. Keidel, W. Tirshc, S. J. Poppl, and K. Geist. 1988. "Dimensional Analysis of Nonlinear Oscillations in Brain, Heart, and Muscle." *Math. Biosci.* **90**: 155–182.
[Glass]

Mayr, E. 1982. *The Growth of Biological Thought.* Cambridge, MA: Harvard University Press.
[Rigney]

McClelland, J. L., and J. L. Elman. 1986. "Interactive Processes in Speech Perception: The TRACE Model." In *Parallel Distributed Processing: Explorations in the Microstructure of Cognition. Volume II: Psychological and Biological Models,* edited by J. L. McClelland and D. E. Rumelhart, 58–121. Cambridge, MA: MIT Press.
[Mozer]

McEwen, J. A., and C. B. Anderson. 1975. "Modelling the Stationarity and Gaussianity of Spontaneous Electroencephalographic Activity." *IEEE Trans. Biomed. Engr.* **22**: 363–369.
[Glass]

McLeod, A. I., and W. K. Li. 1983. "Diagnostic Checking ARMA Time Series Models Using Squared-Residual Autocorrelations." *J. Time Series Anal.* **4**: 269–273.
[Theiler]

McNees, S. K. 1986. "Forecasting Accuracy of Alternative Techniques: A Comparison of U.S. Macroeconomic Forecasts." *J. Bus. & Econ. Stat.* **4**: 5–15.
[Granger]

Mees, A. I. 1991. "Dynamical Systems and Tesselations Detecting Determinism in Data." *Intl. J. Bif. & Chaos* **1**: 777–794.
[Rigney]

Meese, R. A., and A. K. Rose. 1990. "Nonlinear, Nonparametric, Nonessential Exchange Rate Estimation." *Amer. Econ. Rev.* **80(2)**: 192–196.
[LeBaron]

Melcher, A. 1976. "Respiratory Sinus Arrhythmia in Man. A Study in Heart Rate Regulating Mechanisms." *Acta Physiol. Scand. Suppl.* **435**: 1–31.
[Rigney]

———. 1980. "Carotid Baroreflex Heart Rate Control During the Active and Assisted Breathing Cycle in Man." *Acta Physiol. Scand.* **108**: 165–171.
[Rigney]

Melvin, P., and N. B. Tufillaro. 1991. "Templates and Framed Braids." *Phys. Rev. A* **44**: R3419–R3422.
[Gershenfeld]

Mendel, J. 1973. *Discrete Techniques of Parameter Estimation: The Equation Error Formulation.* New York: Marcel Dekker.
[Wan]

Meyer, L. B. 1956. *Emotion and Meaning in Music.* Chicago, IL: University of Chicago Press.
[Dirst]

Meyer, T. P., and N. H. Packard. 1992. "Local Forecasting of High-Dimensional Chaotic Dynamics." In *Nonlinear Modeling and Forecasting*, edited by M. Casdagli and S. Eubank. Santa Fe Institute Studies in the Sciences of Complexity, Proc. Vol. XII, 249–263. Reading, MA: Addison-Wesley.
[Gershenfeld]

Milhorn, H., and A. C. Guyton. 1965. "An Analog Computer Analysis of Cheyne-Stokes Breathing." *J. Appl. Physiol.* **20**: 328–333.
[Rigney]

Miller, W. T., R. S. Sutton, and P. J. Werbos. 1990. *Neural Networks for Control.* Cambridge, MA: MIT Press.
[Gershenfeld]

Milton, J. G., A. Longtin, A. Beurter, M. C. Mackey, and L. Glass. 1989. "Complex Dynamics and Bifurcations in Neurology." *J. Theor. Biol.* **138**: 129–147.
[Glass]

Mitchison, G. J., and R. M. Durbin. 1989. "Bounds on the Learning Capacity of Some Multi-Layer Networks." *Biol. Cyber.* **60**: 345–356.
[Gershenfeld]

Mizrach, B. 1989. "Multivariate Nearest-Neighbor Forecasts of EMS Exchange Rates." Technical Paper, Boston University, Chestnut Hill, MA.
[LeBaron]

Möller, M., W. Lange, F. Mitschke, N. B. Abraham, and U. Hübner. 1989. "Errors from Digitizing and Noise in Estimating Attractor Dimensions." *Phys. Lett. A* **138**: 176–182.
[Hübner]

Moloney, J. V., J. S. Uppal, and R. G. Harrison. 1986. "Origin of Chaotic Relaxation Oscillations in an Optically Pumped Molecular Laser." *Phys. Rev. Lett.* **59**: 2868–2871.
[Hübner]

Montgomery, D. R., and W. E. Dietrich. 1992. "Channel Initiation and the Problem of Landscape Scale." *Science* **255**: 826–830.
[Dirst, Swinney]

Moody, G. B. 1992. "ECG-Based Indices of Physical Activity." In *Computers in Cardiology 1992*, 403–406. Los Alamitos, CA: IEEE Computer Society Press.
[Rigney]

Moody, J. 1992. "The *Effective* Number of Parameters: An Analysis of Generalization and Regularization in Nonlinear Systems." In *Advances in Neural Information Processing Systems 4*, edited by J. E. Moody, S. J. Hanson, and R. P. Lippmann. San Mateo, CA: Morgan Kaufmann.
[Gershenfeld, Theiler, Wan]

Moody, J., and C. Darken. 1989. "Fast Learning in Networks of Locally Tuned Processing Units." *Neur. Comp.* **1(2)**: 281–294.
[Zhang]

Moore, C. 1991. "Generalized Shifts: Unpredictability and Undecidability in Dynamical Systems." *Nonlinearity* **4**: 199–230.
[Gershenfeld]

Morgan, J. N., and J. A. Sonquist. 1963. "Problems in the Analysis of Survey Data, and a Proposal." *J. Am. Stat. Assoc.* **58**: 415–434.
[Lewis]

Morgera, S. D. 1985. "Information Theoretic Complexity and Its Relation to Pattern Recognition." *IEEE Trans. Sys., Man, & Cyber.* **SMC-15 No. 5**: 608–619.
[Paluš]

Moroney, D., ed. 1989. Bach, J. S. *Die Kunst der Fuge.* München: Henle Verlag.
[Dirst]

Mortola, J. P. 1991. "Comparative Aspects of Respiratory Mechanics in Newborn Mammals." In *Basic Mechanisms of Pediatric Respiratory Disease*, edited by V. Chernick and R. B. Mellins, 80–88. Philadelphia: B. C. Decker.
[Glass]

Mortola, J. P., J. Milic-Emili, A. Nowaraj, B. Smith, G. Fox, and S. Weeks. 1984. "Muscle Pressure and Flow During Expiration in Infants." *Am. Rev. Resp. Dis.* **129**: 49–53.
[Glass]

Mozart, W. A. 1787. *Musikalisches Würfelspiel* (KV Anh. 294d), Edition 4474. Mainz: B. Schott's Söhne.
[Dirst]

Mozer, M. C. 1989. "A Focused Back-Propagation Algorithm for Temporal Pattern Recognition." *Complex Systems* **3**: 349–381.
[Mozer]

———— 1991. "Connectionist Music Composition Based in Melodic, Stylistic and Psychophysical Constraints." In *Music and Connectionism*, edited by P. M. Todd, and D. G. Loy, 195–211. Cambridge, MA: MIT Press.
[Dirst]

————. 1992. "The Induction of Multiscale Temporal Structure." In *Advances in Neural Information Processing Systems IV*, edited by J. E. Moody, S. J. Hanson, and R. P. Lippman, 275–282. San Mateo, CA: Morgan Kaufmann.
[Mozer]

————. 1993. "Neural Net Architectures for Temporal Sequence Processing." This volume.
[Dirst, Gershenfeld]

Mozer, M. C., and T. Soukup. 1991. "CONCERT: A Connectionist Composer of Erudite Tunes." In *Advances in Neural Information Processing Systems 3*, edited by R. P. Lippmann, J. Moody, and D. S. Touretzky, 789–796. San Mateo, CA: Morgan Kaufmann.
[Mozer]

Musha, T., and K. Goto. 1989. "Feature Extraction of Melody and Creation of Resembling Melodies with a Digital Filter." In *Proceedings of the First International Conference on Music Perception and Cognition*, 71–72. Kyoto.
[Dirst]

Myers, C. 1990. "Learning with Delayed Reinforcement Through Attention-Driven Buffering." Technical Report, Neural Systems Engineering Group, Department of Electrical Engineering, Imperial College of Science, Technology, and Medicine, London.
[Mozer]

Nather, R. E., D. E. Winget, J. C. Clemens, C. J. Hansen, and B. P. Hine. 1990. "The Whole Earth Telescope: A New Astronomical Instrument." *Astrophys. J.* **361**: 309–361.
[Clemens]

Neil, E., and J. F. Palmer. 1975. "Effect of Spontaneous Respiration on the Latency of Reflex Cardiac Chronotropic Responses to Baroreceptor Stimulation." *J. Physiol. (Lond.)* **247**: 16P.
[Rigney]

Niedermeyer, E., and F. Lopes da Silva. 1987. *Electroencephalography: Basic Principles, Clinical Applications and Related Fields*, 2nd ed. Baltimore: Urban & Schwarzenberg.
[Glass]

Nordström, T., and B. Svensson. 1992. "Using and Designing Massively Parallel Computers for Artificial Neural Networks." *J. Parallel & Distrib. Comp.* 14: 26.
[Afraimovich]

Normand, C., Y. Pomeau, and M. G. Velarde. 1977. "Convective Instability: A Physicist's Approach." *Rev. Mod. Phys.* **49(3)**: 581–624.
[Afraimovich]

Nottebohm, G. 1881. "J. S. Bach's letzte Fuge." *Musik-Welt* **20** (5 March, 1881): 232–236.
[Dirst]

Nowlan, S. J., and G. E. Hinton. 1992. "Simplifying Neural Networks by Soft Weight-Sharing." *Neur. Comp.* **4**: 473–493.
[Gershenfeld]

Nugent, S. T., and J. P. Finley. 1987. "Periodic Breathing in Infants—A Model Study." *IEEE Trans. Biomed. Engr.* **34**: 482–485.
[Rigney]

Nychka, D., S. Ellner, D. McCaffrey, and A. R. Gallant. 1992. "Finding Chaos in Noisy Systems." *J. Roy. Stat. Soc. B* **54(2)**: 399–426.
[Casdagli, Gershenfeld, Glass]

Ocasio, W. C., D. R. Rigney, K. P. Clark, and R. G. Mark. 1993. "bpshape.wk4: A Computer Program that Implements a Physiological Model for Analyzing the Shape of Blood Pressure Waveforms." *Comp. Meth. Prog. Biomed.*: in press.
[Rigney]

Oppenheim, A. V., and R. W. Schafer. 1989. *Discrete-Time Signal Processing.* Englewood Cliffs, NJ: Prentice Hall.
[Gershenfeld]

Orlin Grabbe, J. 1986. *International Financial Markets.* New York: Elsevier.
[Lequarré]

Osborne, A. R., A. D. Kirwin, A. Provenzale, and L. Bergamasco. 1986. "A Search for Chaotic Behavior in Large and Mesoscale Motions in the Pacific Ocean." *Physica D* **23**: 75–83.
[Theiler]

Osborne, A. R., and A. Provenzale. 1989. "Finite Correlation Dimension for Stochastic Systems with Power-Law Spectra." *Physica D* **35**: 357–381.
[Casdagli, Kaplan, Paluš, Theiler]

Ott, E., C. Grebogi, and J. A. Yorke. 1990. "Controlling Chaos." *Phys. Rev. Lett.* **64**: 1196.
[Gershenfeld]

Ouyang, Q., Z. Noszticzius, and H. L. Swinney. 1992. "Spatial Bistability of Two-Dimensional Turing Patterns in a Reaction-Diffusion System." *J. Phys. Chem.* **96**: 6773–6776.
[Swinney]

Ouyang, Q., and H. L. Swinney. 1991a. "Transition to Chemical Turbulence." *Chaos* **1**: 411–420.
[Swinney]

——. 1991b. "Transition from a Uniform State to Hexagonal and Striped Turing Patterns." *Nature* **352**: 610–612.
[Swinney]

Pack, A. I., A. Gottschalk, M. Cola, and A. Goldszmidt. 1989. "Sleep State and Periodic Respiration." In *Modeling and Parameter Estimation in Respiratory Control*, edited by M. C. K. Khoo, 181–191. New York: Plenum Press.
[Rigney]

Packard, N. H. 1990. "A Genetic Learning Algorithm for the Analysis of Complex Data." *Complex Systems* **4**: 543–572.
[Gershenfeld, Rigney]

Packard, N. H., J. P. Crutchfeld, J. D. Farmer, and R. S. Shaw. 1980. "Geometry from a Time Series." *Phys. Rev. Lett.* **45(9)**: 712–716.
[Afraimovich, Fraser, Gershenfeld, Sauer, Smith, Wan]

Palis, J., Jr., and W. de Melo. 1982. *Geometric Theory of Dynamical Systems, an Introduction.* New York: Springer.
[Paluš]

Paluš, M. 1993a. "Testing for Nonlinearity in the EEG." In *Nonlinear Dynamical Analysis of the EEG*, edited by B. H. Jansen and M. E. Brandt. Singapore: World Scientific, 100–114.
[Paluš]

——. 1993b. "Identifying and Quantifying Chaos Using Information-Theoretic Functionals." This volume.
[Gershenfeld]

——. n.d. "Kolmogorov Entropy from Time Series Using Information Functionals." *Physica D*, submitted.
[Paluš]

Parlitz, U. 1992. "Identification of True and Spurious Lyapunov Exponents from Time Series." *Intl. J. Bif. & Chaos* **2**: 155–165.
[Gershenfeld]

Parkes, J. D. 1985. *Sleep and Its Disorders.* London: W.B. Saunders.
[Rigney]

Patil, C. P., M. S. Jacobi, K. B. Saunders, and B. McA. Sayers. 1987. "Linear Segment Analysis to Quantify Irregular Breathing in Man." In *Concepts and Formalizations in the Control of Breathing*, edited by G. Benchetrit, P. Baconnier, and J. Demongeot, 325–334. Manchester: University Press.
[Rigney]

Pawelzik, K., and H. G. Schuster. 1987. "Generalized Dimensions and Entropies from a Measured Time Series." *Phys. Rev. A* **35**: 481–484.
[Hübner]

Pearlmutter, B. A. 1989. "Learning State Space Trajectories in Recurrent Neural Networks." *Neur. Comp.* **1**: 263–269.
[Mozer]

Pedly, T. J., and J. O. Kessler. 1992. "Hydrodynamic Phenomena in Suspensions of Swimming Microorganisms." *Ann. Rev. Fluid Mech.* **24**: 313–358.
[Swinney]

Petersen, K. 1989. *Ergodic Theory*, 2nd ed. Cambridge Studies in Advanced Mathematics, Vol. 2. Cambridge, MA: Cambridge University Press.
[Gershenfeld, Paluš]

Petrillo, G. A., and L. Glass. 1984. "A Theory for Phase Locking of Respiration in Cats to a Mechanical Ventillator." *Am. J. Physiol.* **246**: R311–R320.
[Rigney]

Pettis, K. W., T. A. Bailey, A. K. Jain, and R. C. Dubes. 1979. "An Intrinsic Dimensionality Estimator from Near-Neighbor Information." *IEEE Trans. Patt. Anal. & Mach. Intel.* **PAMI-1**: 25–37.
[Gershenfeld]

Pham Dinh, T., J. Demongeot, P. Baconnier, and G. Benchetrit. 1983. "Simulation of a Biological Oscillator: The Respiratory System." *J. Theor. Biol.* **103**: 113–132.
[Rigney]

Pi, H., and C. Peterson. 1993. "Finding the Embedding Dimension and Variable Dependences in Time Series." Preprint LU TP 93-4, Department of Theoretical Physics, University of Lund, March 1993. Submitted to *Neural Computation*.
[Gershenfeld]

Pick, A., and R. Langendorf. 1979. *Interpretation of Complex Arrhythmias*. Philadelphia: Lea & Febiger.
[Glass]

Pierce, J. R. 1980. *An Introduction to Information Theory: Symbols, Signals and Noise*, 2nd ed. New York: Dover. The first edition (1961) was published as *Symbols, Signals and Noise: The Nature and Process of Communication*, by Harper & Brothers.
[Dirst]

Pincus, S. M. 1992. "Approximating Markov Chains." *Proc. Natl. Acad. Sci. USA* **89**: 4432–4436.
[Glass]

Pincus, S. M., T. R. Cummins, and G. G. Haddad. 1993. "Heart Rate Control in Normal and Aborted SIDS Infants." *Am. J. Physiol.* **33**: R638–R646.
[Glass]

Pineda, F. J., and J. C. Sommerer. 1993. "Estimating Generalized Dimensions and Choosing Time Delays: A Fast Algorithm." This volume.
[Gershenfeld]

Plaut, D. C., and G. E. Hinton. 1987. "Learning Sets of Filters Using Backpropagation." *Comp. Speech & Lang.* **2**: 35–61.
[Mozer]

Poggio, T., and F. Girosi. 1990. "Networks for Approximation and Learning." *Proc. IEEE* **78(9)**: 1481–1497.
[Gershenfeld, Zhang]

Politis, D. N., and J. P. Romano. 1991. "The Stationary Bootstrap." Technical Report 365, Department of Statistics, Stanford University, Stanford, CA.
[Theiler]

Poritz, A. B. 1988. "Hidden Markov Models: A Guided Tour." In *Proceedings of the IEEE International Conference on Acoustics, Speech and Signal Processing*. Washington, DC: IEEE.
[Fraser]

Powell, M. J. D. 1987. "Radial Basis Functions for Multivariate Interpolation: A Review." In *IMA Conference on "Algorithms for the Approximation of Functions and Data,"* edited by J. C. Mason and M. G. Cox. Shrivenham: RMCS.
[Gershenfeld, Smith]

Press, W. H., B. P. Flannery, S. A. Teukolsky, and W. T. Vetterling. 1992. *Numerical Recipes in C: The Art of Scientific Computing*, 2nd ed. Cambridge: Cambridge University Press.
[Casdagli, Fraser, Gershenfeld, Paluš, Press, Theiler]

Press, W. H., and G. B. Rybicki. 1989. "Fast Algorithm for Spectral Analysis of Unevenly Spaced Data." *Astrophys. J.* **338**: 277–280.
[Rigney]

———. 1992. "Interpolation, Realization, and Reconstruction of Noisy, Irregularly Sampled Data." *Astrophys. J.* **398**: 169.
[Press]

———. 1993. "Large-Scale Linear Methods for Interpolation, Realization, and Reconstruction of Noisy, Irregularly Sampled Data." This volume.

Press, W. H., G. B. Rybicki, and J. N. Hewitt. 1992a. "The Time Delay of Gravita-
tional Lens 0957+561. I. Methodology and Analysis of Optical Photometric Data."
Astrophys. J. **385**: 404.
[Press]

———. 1992b. "The Time Delay of Gravitational Lens 0957+561. II. Analysis of Radio
Data and Combined Optical-Radio Analysis." *Astrophys. J.* **385**: 416.
[Press, Rybecki]

Priestley, M. 1981. *Spectral Analysis and Time Series.* London: Academic Press.
[Gershenfeld, Rigney]

Priestley, M. B. 1988. *Nonlinear and Non-Stationary Time Series.* New York: Academic
Press.
[Lewis, Sauer]

Principe, J. C., B. de Vries, and P. Oliveira. 1993. "The Gamma Filter—A New Class
of Adaptive IIR Filters with Restricted Feedback." *IEEE Trans. Sig. Proc.* **41**:
649–656.
[Gershenfeld]

Pritchard, W. S., and D. W. Duke. n.d. "Dimensional Analysis of No-Task Human
EEG Using the Grassberger- Procaccia Method." *Psychophysiology*: in press.
[Glass]

Provenzale, A., A. R. Osborne, and R. Soj. 1991. "Convergence of the K_2 Entropy for
Random Noises with Power-Law Spectra." *Physica D* **47**: 361–372.
[Theiler]

Pukkila, T., S. Koreisha, and A. Kallinen, 1990. "The Identification of ARMA Models."
Biometrika **77**: 537–548.
[Theiler]

Rabiner, L. R., and B. H. Juang. 1986. "An Introduction to Hidden Markov Models."
IEEE ASSP Magazine, January: 4–16.
[Gershenfeld]

Rabinovich, M. I., and M. M. Sushchik. 1990. "The Regular and Chaotic Dynamics of
Structures in Fluid Flows." *Sov. Phys. Usp.* **33(1)**: 1–35.
[Afraimovich]

Rao, C. R. 1973. *Linear Statistical Inference and Its Applications,* 2nd ed. New York:
Wiley.
[Press]

Redlich, A. N. 1993. "Redundancy Reduction as a Strategy for Unsupervised Learning." *Neur. Comp.* **5**: 289–304.
[Dirst]

Renyi, A. 1970. *Probability Theory.* Amsterdam: North-Holland.
[Pineda]

Rico-Martinez, R., I. G. Kevrekidis, and R. A. Adomaitis. 1993. "Noninvertibility in
Neural Networks." In *Proceedings of ICNN, San Francisco, 1993,* 382–386. Piscat-
away, NJ: IEEE Press.
[Gershenfeld]

Riggs, D. S. 1963. *The Mathematical Approach to Physiological Problems.* Baltimore,
MD: Williams & Wilkins.
[Rigney]

———. 1970. *Control Theory and Physiological Feedback Mechanisms.* Baltimore, MD:
Williams & Wilkins.
[Rigney]

Rigney, D. R., A. L. Goldberger, W. C. Ocasio, Y. Ichimaru, G. B. Moody, and
R. G. Mark. 1993. "Multi-Channel Physiological Data: Description and Analysis."
This volume.
[Casdagli, Gershenfeld, Rigney]

Rigney, D. R., W. C. Ocasio, K. P. Clark, J. Y. Wei, and A. L. Goldberger. 1992. "Deterministic Mechanism for Chaos and Oscillations in Heart Rate and Blood Pressure." *Circulation* **86**: 1–659.
[Rigney]

———. 1993. "Forecasting Beat-to-Beat Heart Rate and Blood Pressure Fluctuations
Using a Nonlinear Physiological Model." Submitted.
[Rigney]

Ring, M. B. 1991. "Incremental Development of Complex Behaviors Through Automatic Construction of Sensory-Motor Hierarchies." In *Machine Learning: Proceedings of the Eighth International Workshop,* edited by L. Birnbaum and G. Collins,
343–347. San Mateo, CA: Morgan Kaufmann.
[Mozer]

Rissanen, J. 1986. "Stochastic Complexity and Modeling." *Ann. Stat.* **14**: 1080–1100.
[Gershenfeld]

———. 1987. "Stochastic Complexity." *J. Roy. Stat. Soc. B* **49**: 223–239. With discussion: 252–265.
[Gershenfeld]

———. 1989. *Stochastic Complexity and Statistical Inquiry.* Singapore: World Scientific.
[Fraser]

Rissanen, J., and G. G. Langdon. 1981. "Universal Modeling and Coding." *IEEE Trans.
Info. Theory.* **IT-27**: 12–23.
[Gershenfeld]

Ritchie, R. G., E. A. Ernst, B. L. Pate, J. P. Pearson, and L. C. Sheppard. 1990. "Au-
tomatic Control of Anesthetic Delivery and Ventilation During Surgery." *Med.
Prog. Tech.* **16**: 61–67.
[Rigney]

Robinson, A. J., and F. Fallside. 1987. "The Utility-Driven Dynamic Error Propagation Network." Technical Report CUED/F-INFENG/TR.1, Department of Engineering, Cambridge University, Cambridge, MA.
[Mozer]

Rogowski, Z., I. Gath, and E. Bental. 1981. "On the Prediction of Epileptic Seizures." *Biol. Cyber.* **42**: 9–15.
[Rigney]

Rosenblatt, F. 1962. *Principles of Neurodynamics.* New York: Spartan.
[Mozer]

Rössler, O. E. 1976. "An Equation for Continuous Chaos." *Phys. Lett. A* **57**: 397–398.
[Kostelich, Paluš, Theiler]

Rubin, D. M. 1992. "Use of Forecasting Signatures to Help Distinguish Periodicity, Randomness, and Chaos in Ripples and Other Spatial Patterns." *Chaos* **2**: 525–535.
[Casdagli, Theiler]

Rubio, J. E. 1972. "A New Mathematical Model of the Respiratory Center." *Bull. Math. Biophys.* **34**: 467–481.
[Rigney]

Ruelle, D. 1990. "Deterministic Chaos: The Science and the Fiction." *Proc. Roy. Soc. London A.* **427**: 241–248.
[Afraimovich, Casdagli, Pineda]

Ruelle, D., and J. P. Eckmann. 1985. "Ergodic Theory of Chaos and Strange Attractors." *Rev. Mod. Phys.* **57**: 617–656.
[Gershenfeld]

Rumelhart, D. E., R. Durbin, R. Golden, and Y. Chauvin. 1993. "Backpropagation: The Basic Theory." In *Backpropagation: Theory, Architectures and Applications*, edited by Y. Chauvin and D. E. Rumelhart. Hillsdale, NJ: Lawrence Erlbaum.
[Casdagli, Gershenfeld, Mozer]

Rumelhart, D. E., G. E. Hinton, and R. J. Williams. 1986a. "Learning Internal Representations by Error Propagation." In *Parallel Distributed Processing: Explorations in the Microstructure of Cognition. Volume I: Foundations*, edited by D. E. Rumelhart and J. L. McClelland, 318–362. Cambridge, MA: MIT Press/Bradford Books.
[Gershenfeld, Mozer, Wan, Zhang]

———.1986b. "Learning Representations by Back-Propagating Errors." *Nature* **323**: 533–536.
[Gershenfeld]

Russell, D. A., J. D. Hanson, and E. Ott. 1980. "Dimension of Strange Attractors." *Phys. Rev. Lett.* **45**: 1175.
[Gershenfeld]

Ryan, J. C., and N. M. Lawandy. 1987. "Instabilities in a Three-Level Coherently Pumped Laser." *Opt. Commun.* **64**: 54–58.
[Hübner]

Sakamoto, Y., M. Ishiguro, and G. Kitagawa. 1986. *Akaike Information Criterion Statistics.* Dordrecht: D. Reidel.
[Gershenfeld]

Samet, H. 1992. *The Design and Analysis of Spatial Data Structures.* Reading, MA: Addison-Wesley.
[Pineda]

Sano, M., and Y. Sawada. 1985. "Measurement of the Lyapunov Spectrum from a Chaotic Time Series." *Phys. Rev. Lett.* **55**: 1082–1085.
[Glass, Kantz]

Sauer, M., C. Grebogi, E. Ott, T. Sauer, and J. A. Yorke. 1992. "Estimating Correlation Dimension from a Time Series." Preprint, University of Maryland.
[Pineda]

Sauer, T. 1992. "A Noise Reduction Method for Signals from Nonlinear Systems." *Physica D* **58**: 193–201.
[Gershenfeld, Kantz, Kostelich, Sauer]

———. 1993. "Time Series Prediction Using Delay Coordinate Embedding." This volume.
[Casdagli, Gershenfeld, Smith]

Sauer, T., J. A. Yorke, and M. Casdagli. 1991. "Embedology." *J. Stat. Phys.* **65(3/4)**: 579–616.
[Afraimovich, Fraser, Gershenfeld, Glass, Kantz, Kaplan, Kostelich, Sauer, Smith]

Saul, J. P., R. D. Berger, P. Albrecht, S. P. Stein, M. H. Chen, and R. J. Cohen. 1991. "Transfer Function Analysis of the Circulation: Unique Insights into Cardiovascular Regulation." *Am. J. Physiol.* **261**: H1231–H1245.
[Rigney]

Saul, J. P., R. D. Berger, M. H. Chen, and R. J. Cohen. 1989. "Transfer Function Analysis of Autonomic Regulation. II. Respiratory Sinus Arrhythmia." *Am. J. Physiol.* **256**: H153–H161.
[Rigney]

Saund, E. 1989. "Dimensionality-Reduction Using Connectionist Networks." *IEEE Transactions on Pattern Analysis and Machine Intelligence (T-PAMI)* **11**: 304–314.
[Gershenfeld]

Scheinkman, J. A., and B. LeBaron. 1989. "Nonlinear Dynamics and Stock Returns." *J. Business* **62**: 311–338.
[Theiler]

Schmidhuber, J. 1992a. "A Fixed-Size Storage $O(n^3)$ Time Complexity Learning Algorithm for Fully Recurrent Continually Running Networks." *Neur. Comp.* **4**: 243–248.
[Mozer]

———. 1992b. "Learning Unambiguous Reduced Sequence Descriptions." In *Advances in Neural Information Processing Systems IV*, edited by J. E. Moody, S. J. Hanson, and R. P. Lippman, 291–298. San Mateo, CA: Morgan Kaufmann.
[Mozer]

Schmidhuber, J., D. Prelinger, and M. C. Mozer. 1993. "Continuous History Compression." In the proceedings of the International Workshop on Neural Networks, held in Aachen, Germany, in press.
[Mozer]

Schmieder, W. 1990. *Thematisch-systematisches Verzeichnis der musikalischen Werke von Johann Sebastian Bach: Bach-Werke-Verzeichnis (BWV)*. Wiesbaden: Breitkopf und Härtel.
[Dirst]

Schottstaedt, W. G. 1989 "Automatic Counterpoint." In *Current Directions in Computer Music Research*, edited by M. V. Matthews and J. R. Pierce, 215–224. Cambridge, MA: MIT Press.
[Dirst]

Schreiber, T. 1992. "An Extremely Simple Nonlinear Noise-Reduction Algorithm." *Phys. Rev. E* **47**: 2401–2404.
[Kantz]

Schreiber, T., and P. Grassberger. 1991. "A Simple Noise Reduction Method for Real Data." *Phys. Lett. A* **160**: 411.
[Kantz]

Schroeder, W. A. 1982. *Missouri Water Atlas*. Columbia: University of Missouri Press.
[Swinney]

Schulenberg, D. 1992. *The Keyboard Music of J. S. Bach*. New York: Schirmer.
[Dirst]

Schuster, A. 1898. "On the Investigation of Hidden Periodicities with Applications to a Supposed 26-Day Period of Meteorological Phenomena." *Terr. Mag.* **3**: 13–41.
[Gershenfeld]

Schuster, H. G. 1988. *Deterministic Chaos: An Introduction*, 2nd rev. ed. Weinheim: Physik.
[Casdagli, Paluš]

Schwan, H. P., ed. 1969. *Biological Engineering*. New York: McGraw-Hill.
[Rigney]

Schwanauer, S. M. 1993. "A Learning Machine for Tonal Composition." In *Machine Models of Music*, edited by S. M. Schwanauer, and D. A. Levitt, 512–532. Cambridge, MA: MIT Press.
[Dirst]

Schwanauer, S. M., and D. A. Levitt, eds. 1993. *Machine Models of Music*, 533–538. Cambridge, MA: MIT Press.
[Dirst]

Schwartz, E. I. 1992. "Where Neural Networks are Already at Work: Putting AI to Work in the Markets." *Bus. Week* November 2: 136–137.
[Gershenfeld]

Schwarz, G. 1978. "Estimating the Dimension of a Model." *Ann. Stat.* **6**: 461–464.
[Theiler]

Segel, L. A. 1984. *Modeling Dynamic Phenomena in Molecular and Cellular Biology*. Cambridge: Cambridge University Press.
[Rigney]

Selman, A., A. McDonald, R. Kitney, and D. Linkens. 1982. "The Interaction Between Heart Rate and Respiration. Part I—Experimental Studies in Man." *Automedica* **4**: 131–139.
[Rigney]

Serre, T., J. R. Buchler, and M. J. Goupil. 1991. "Predicting White Dwarf Light Curves." In *White Dwarfs*, edited by G. Vauclair and E. Sion, 175. Boston: Kluwer. [Clemens]

Shannon, C. E., and W. Weaver. 1964. *The Mathematical Theory of Communication.* Urbana: University of Illinois Press. [Paluš]

Shaw, R. S. 1981. "Strange Attractors, Chaotic Behavior and Information Flow." *Z. Naturforsch.* **36A**: 80–112. [Gershenfeld, Paluš]

Shepard, R. N. 1982a. "Geometrical Approximations to the Structure of Musical Pitch." *Psych. Rev.* **89**: 305–333. [Dirst]

———. 1982b. "Structural Representations of Musical Pitch." In *The Psychology of Music*, edited by D. Deutsch, 343–390. New York: Academic Press. [Dirst]

Sheppard, L. C. 1980. "Computer Control of the Infusion of Vasoactive Drugs." *Ann. Biomed. Engr.* **8**: 431–444. [Rigney]

———. 1989. "Automation of the Infusion of Drugs Using Feedback Control." *J. Cardiothor. Anes.* **3**: 1–3. [Rigney]

Shumway, R. H. 1988. *Applied Statistical Time Series Analysis.* Englewood Cliffs, NJ: Prentice Hall. [Zhang]

Shynk, J. 1989. "Adaptive IIR Filtering." *IEEE ASSP Magazine* **6(2)**: 4–21. [Wan]

Silverman, B. W. 1985. "Some Aspects of the Spline Smoothing Approach to Non-Parametric Regression Curve Fitting." *J. Roy. Stat. Soc. (Ser B)* **47**: 1–52. [Lewis]

Silverman, B. W. 1986. *Density Estimation for Statistics and Data Analysis.* New York: Chapman and Hall. [Gershenfeld]

Simon, H. A., and R. K. Sumner. 1968. "Pattern in Music." In *Formal Representation of Human Judgement*, edited B. Kleinmuntz. New York: John Wiley & Sons. Reprinted in *Machine Models of Music*, edited by S. M. Schwanauer and D. A. Levitt, 83–110. Cambridge, MA: MIT Press, 1993. [Dirst]

Sinai, Ya. G. 1959. *Dokl. Akad. Nauk SSSR* **12**: 768. [Paluš]

———. 1976. *Introduction to Ergodic Theory.* Princeton: Princeton University Press. [Paluš]

Singer, A. 1990. "Implementations of Artificial Neural Networks on the Connection Machine." *Parallel Computing* **14**: 305. [Zhang]

Sirovich, L. 1987. "Turbulence and the Dynamics of Coherent Structures." *Qtr. Appl. Math.* **XLV**: 561–590.
[Swinney]

Skarda, C. A., and W. J. Freeman. 1987. "How Brains Make Chaos in Order to Make Sense of the World." *Behav. Brain Sci.* **10**: 161–195.
[Glass]

Skinner, J. E., A. L. Goldberger, G. Mayer-Kress, and R. E. Ideker. 1990. "Chaos in the Heart: Implications for Clinical Cardiology." *Biotechnology* **8**: 1018–1024.
[Glass]

Skinner, J. E., C. M. Pratt, and T. Vybiral. 1993. "Reduction in the Correlation Dimension of Heartbeat Intervals Precedes Imminent Ventribular Fibrillation in Human Subjects." *Am. Heart J.* **125**: 731–743.
[Glass]

Skilling, J., D. Robinson, and S. Gull. 1990. "Probabilistic Displays." In *Maximum Entropy and Bayesian Methods, Laramie, 1990*, edited by W. Grandy and L. Schick, 365–368. Dordrecht: Kluwer.
[Wan]

Smith, J. M., E. Clancy, C. R. Valeri, J. N. Ruskin, and R. J. Cohen. 1988. "Electrical Aternans and Cardiac Electrical Stability." *Circulation* **77(1)**: 110.
[Glass]

Smith, L. A. 1988. "Intrinsic Limits on Dimension Calculations." *Phys. Lett. A* **133**: 283–288.
[Pineda]

——. 1990. "Quantifying Chaos with Predictive Flows and Maps: Locating Unstable Periodic Orbits." in *Measurés of Complexity and Chaos*, edited by N. B. Abraham, A. M. Albano, A. P. Pasamante, and R. E. Rapp. NATO ASI Series. New York: Plenum.
[Smith]

——. 1992. "Identification and Prediction of Low-Dimensional Dynamics." *Physica D* **58**: 50–76.
[Smith, Theiler]

——. 1993. "Does a Meeting in Sante Fe Imply Chaos?" This volume.
[Casdagli, Gershenfeld]

Smith, L. A., K. Godfrey, P. Fox, and K. Warwick. 1991. "A New Technique for Fault Detection in Multi-Sensor Probes." In *Control 91*, Vol. 1, 1062. IEE Conference Publication 332.
[Smith]

Smith, P. L. 1979. "Splines as a Useful and Convenient Statistical Tool." *Amer. Stat.* **33(2)**: 57–62.
[Lewis]

Smith, R. L. 1992. "Optimal Estimation of Fractal Dimension." In *Nonlinear Modeling and Forecasting*, edited by M. Casdagli and S. Eubank. Santa Fe Institute Studies in the Sciences of Complexity, Proc. Vol. XII, 115–135. Reading, MA: Addison-Wesley.
[Pineda]

Smolensky, P. 1990. "Tensor Product Variable Binding and the Representation of Symbolic Structures in Connectionist Networks." *Art. Int.* **46**: 159–216.
[Mozer]

Smolensky, P., M. C. Mozer, and D. E. Rumelhart, eds. 1994. *Mathematical Perspectives on Neural Networks.* Hillsdale, NJ: Lawrence Erlbaum.
[Gershenfeld]

Solomonoff, R. J. 1964. "A Formal Theory of Induction Inference, Parts I and II." *Information & Control* **7**: 1–22, 221–254.
[Gershenfeld]

Sompolinsky, H., and I. Kanter. 1986. "Temporal Association in Asymmetric Neural Networks." *Phys. Rev. Lett.* **57**: 2861–2864.
[Mozer]

Sorenson, H. W. 1970. "Least-Squares Estimation: From Gauss to Kalman." *IEEE Spectrum*: 63–68.
[Fraser]

Sparrow, C. 1982. *The Lorenz Equations: Bifurcations, Chaos, and Strange Attractors.* Applied Mathematical Sciences, Vol. 41. New York: Springer-Verlag.
[Hübner]

Stearns, S. 1981. "Error Surfaces or Recursive Adaptive Filters." *IEEE Trans. Circ. Sys.* **CAS-28(6)**: 603–606.
[Wan]

Stevens, J. E. 1991. "An Investigation of Multivariate Adaptive Regression Splines for Modeling and Analysis of Univariate and Semi-Multivariate Time Series Systems." Ph.D. Thesis, Naval Postgraduate School, Monterey, CA.
[Lewis]

Stewart, G. W. 1973. *Introduction to Matrix Computations.* New York: Academic Press.
[Kostelich]

Stornetta, W. S., T. Hogg, and B. A. Huberman. 1988. "A Dynamical Approach to Temporal Pattern Processing." In *Neural Information Processing Systems*, 750–759. New York: American Institute of Physics.
[Mozer]

Stuart, A., and J. K. Ord. 1987. *Kendall's Advanced Theory of Statistics*, 5th ed., Vol. 1. New York: Griffin. Previous editions published as *The Advanced Theory of Statistics* by M. Kendall and A. Stuart.
[Press]

Subba Rao, T. 1992. "Analysis of Nonlinear Time Series (and Chaos) by Bispectral Methods." In *Nonlinear Modeling and Forecasting*, edited by M. Casdagli and S. Eubank. Santa Fe Institute Studies in the Sciences of Complexity, Proc. Vol. XII, 199–226. Reading, MA: Addison-Wesley.
[Gershenfeld]

Subba Rao, T., and M. M. Gabr. 1980. "A Test For Linearity of Stationary Time Series." *J. Time Series Anal.* **1**: 145–158.
[Theiler]

Sugihara, G., and R. M. May. 1990. "Nonlinear Forecasting as a Way of Distinguishing Chaos from Measurement Error in Time Series." *Nature* **344**: 734–741.
[Glass, Kaplan, Rigney, Smith]

Sussman, G. J., and J. Wisdom. 1988. "Numerical Evidence that the Motion of Pluto is Chaotic." *Science* **241**: 433–437.
[Gershenfeld]

———. 1992. "Chaotic Evolution of the Solar System." *Science* **257** (1992): 56–62.
[Gershenfeld]

Svarer, C., L. K. Hansen, and J. Larsen. 1993. "On Design and Evaluation of Tapped-Delay Neural Network Architectures." In *IEEE International Conference on Neural Networks, San Francisco (March 1993)*, 46–51. Piscataway, NJ: IEEE Service Center.
[Gershenfeld]

Swift, J., and P. C. Hohenberg. 1977. "Hydrodynamic Fluctuations at the Convective Instability." *Phys. Rev.* **15**: 319–328.
[Afraimovich]

Swinney, H. L. 1993. "Spatio-Temporal Patterns: Observations and Analysis." This volume.
[Gershenfeld]

Takens, F. 1981. "Detecting Strange Attractors in Turbulence." In *Dynamical Systems and Turbulence*, edited by D. A. Rand and L.-S. Young. Lecture Notes in Mathematics, Vol. 898, 336–381. Warwick, 1980. Berlin: Springer-Verlag.
[Afraimovich, Fraser, Gershenfeld, Kostelich, Paluš, Pineda, Sauer, Smith, Wan, Zhang]

———. 1984. *Estimating the Dimension of an Attractor.* Lecture Notes in Mathematics, Vol. 1125, 99–106. Berlin: Springer-Verlag.
[Pineda]

———. 1985. In *Nonlinear Dynamics and Turbulence*, edited by G. T. Barenblatt, G. Iooss, and D. Joseph. London: Pitman.
[Hübner]

Tam, W. Y., and H. L. Swinney. 1990. "Spatiotemporal Patterns in a One-Dimensional Open Reaction-Diffusion System." *Physica D* **46**: 10–22.
[Swinney]

Tam, W. Y., J. A. Vastano, H. L. Swinney, and W. Horsthemke. 1988. "Regular and Chaotic Chemical Spatiotemporal Patterns." *Phys. Rev. Lett.* **61**: 2163–2166.
[Swinney]

Tamaki, N., T. Yasuda, R. H. Moore, J. B. Gill, C. A. Boucher, A. M. Hutter, H. K. Gold, and H. W. Strauss. 1988. "Continuous Monitoring of Left Ventricular Function by an Ambulatory Radionuclide Detector in Patients with Coronary Artery Disease." *J. Amer. Coll. Card.* **12**: 669–679.
[Rigney]

Tang, D. Y., M. Y. Li, and C. O. Weiss. 1991. "Field Dynamics of a Single-Mode Laser." *Phys. Rev. A* **44**: 7597–7604.
[Hübner]

Tang, D. Y., and C. O. Weiss. 1992. "Phase Dynamics of a Detuned Single-Mode Laser." *Appl. Phys. B* **54**: 2548–2553.
[Hübner]

Tang, D. Y., C. O. Weiss, E. Roldan, and G. J. de Valcarcel. 1992. "Deviation from Lorenz-Type Dynamics of an NH$_3$ Ring Laser." *Opt. Commun.* **89**: 47–53.
[Hübner]

Tank, D. W., and J. J. Hopfield. 1987. "Neural Computation by Concentrating Information in Time." *Proc. Natl. Acad. Sci.* **84**: 1896–1900.
[Mozer]

Taylor, S. J. 1986. *Modelling Financial Time Series.* New York: Wiley.
[Lewis]

——. 1992. "Rewards Available to Currency Futures Speculators: Compensation for Risk or Evidence of Inefficient Pricing?" *Econ. Rec.* **68** (Supplement): 105–116.
[LeBaron]

Temam, R. 1988. *Infinite-Dimensional Dynamical Systems in Mechanics and Physics.* Applied Mathematical Sciences, Vol. 68. Berlin: Springer-Verlag.
[Gershenfeld]

Termonia, Y., and Z. Alexandrowicz. 1983. "Fractal Dimension of Strange Attractors from Radius Versus." *Phys. Rev. Lett.* **51(14)** : 1265–1268.
[Afraimovich]

Theiler, J. 1986. "Spurious Dimension from Correlation Algorithm Applied to Time Series Data." *Phys. Rev. A* **34**: 2427–2432.
[Paluš]

——. 1987. "Efficient Algorithm for Estimating the Correlation Dimension from a Set of Discrete Points." *Phys. Rev. A* **36**: 4456–4462.
[Pineda]

——. 1988. "Quantifying Chaos: Practical Estimation of the Correlation Dimension." Ph.D. Thesis, California Institute of Technology, Pasadena, CA.
[Theiler]

——. 1990. "Estimating Fractal Dimension." *J. Opt. Soc. Am. A* **7(6)**: 1055–1073.
[Gershenfeld, Kaplan, Theiler]

——. 1991. "Some Comments on the Correlation Dimension of $1/f^\alpha$ Noise." *Phys. Lett. A* **155(8,9)**: 480–493.
[Gershenfeld, Kaplan, Theiler]

Theiler, J., and S. Eubank. 1992. "Don't Bleach Chaotic Data." Technical Report LA-UR-92-1575, Los Alamos National Laboratory, Los Alamos, NM.
[Kaplan, Theiler]

Theiler, J., S. Eubank, A. Longtin, B. Galdrikian, and J. D. Farmer. 1992a. "Testing for Nonlinearity in Time Series: The Method of Surrogate Data." *Physica D* **58**: 77–94.
[Paluš, Smith, Theiler]

Theiler, J., B. Galdrikian, A. Longtin, S. Eubank, and J. D. Farmer. 1992b. "Using Surrogate Data to Detect Nonlinearity in Time Series." In *Nonlinear Modeling and Forecasting*, edited by M. Casdagli and S. Eubank. Santa Fe Institute Studies in the Sciences of Complexity, Proc. Vol. XII, 163–188. Reading, MA: Addison-Wesley.
[Glass, Kaplan, Theiler]

Theiler, J., P. S. Linsay, and D. M. Rubin. 1993. "Detecting Nonlinearity in Data with Long Coherence Times." This volume.
[Dirst, Gershenfeld, LeBaron, Paluš]

Thisted, R., and B. Efron. 1987. "Did Shakespeare Write a Newly Discovered Poem?" *Biometrika* **74**: 445–455.
[Dirst]

Tiao, G. C., and R. S. Tsay. 1989. "Model Specification in Multivariate Time Series." *J. Roy. Stat. Assoc. (Ser B)* **51**: 157–213.
[LeBaron, Lewis]

Todd, P. M. 1988. "A Sequential Network Design for Musical Applications." In *Proceedings of the 1988 Connectionist Models Summer School*, edited by D. S. Touretzky, G. E. Hinton, and T. J Sejnowski, 76–84. San Mateo, CA: Morgan Kaufmann.
[Dirst]

———. 1989. "A Connectionist Approach to Algorithmic Composition." *Comp. Music J.* **13 (4)**: 27–43. Reprinted in *Music and Connectionism,* edited by P. M. Todd, and D. G. Loy, 179–194. Cambridge, MA: MIT Press.
[Dirst]

Todd, P. M., and D. G. Loy, eds. 1991. *Music and Connectionism.* Cambridge, MA: MIT Press.
[Dirst]

Tong, H. 1983. *Threshold Models in Nonlinear Time Series Analysis.* Berlin: Springer-Verlag.
[Lewis]

———. 1990. *Nonlinear Time Series Analysis: A Dynamical Systems Approach.* Oxford: Oxford University Press.
[Casdagli, Gershenfeld, Lewis, Rigney, Sauer, Theiler]

Tong, H., and K. S. Lim. 1980. "Threshold Autoregression, Limit Cycles and Cyclical Data." *J. Roy. Stat. Soc. B* **42**: 245–292.
[Gershenfeld]

Tong, H., B. Thanoon, and G. Gudmundsson. 1985. "Threshold Time Series Modeling of Two Icelandic Riverflow Systems." *Water Res. Bull.* **21**: 651–660.
[Lewis]

Tovey, D. F. 1931. *A Companion to "The Art of Fugue."* London: Oxford University Press.
[Dirst]

Tresser, C. 1981. "Modeles Simples de Transitions vers la Turbulence." Ph.D. Thesis, University of Nice.
[Hübner]

Trunk, G. V. 1968. "Representation and Analysis of Signals: Statistical Estimation of Intrinsic Dimensionality and Parameter Identification." *General Systems* **13**:49–76.
[Gershenfeld]

Tsay, R. S. 1986. "Nonlinearity Tests for Time Series." *Biometrika* **73(2)**: 461–466.
[Theiler]

——. 1989. "Testing and Modeling Threshold Autoregressive Processes." *J. Am. Stat. Assoc.* **84**: 231–240.
[Lewis]

——. 1991. "Nonlinear Time Series Analysis: Diagnostics and Modelling." *Stat. Sinica* **1**: 431–451.
[Theiler]

——. 1992. "Model Checking via Parametric Bootstraps in Time Series Analysis." *Appl. Stat.* **41**: 1–15.
[Theiler]

Tsonis, A. A., and J. B. Elsner. 1992. "Nonlinear Prediction as a Way of Distinguishing Chaos from Random Fractal Sequences." *Nature* **359**: 217–220.
[Rigney]

Tufillaro, N. B., R. Ramshankar, and J. P. Gollub. 1989. "Order-Disorder Transition in Capillary Ripples." *Phys. Rev. Lett.* **62**: 422–425.
[Swinney]

Tukey, J. W. 1977. *Exploratory Data Analysis.* Reading, MA: Addison-Wesley.
[Dirst, Gershenfeld]

Turing, A. M. 1936. "On Computable Numbers, with an Application to the Entscheidungsproblem." *Proc. London Math. Soc.* **42**: 230–265.
[Gershenfeld]

Turjanmaa V., S. Kalli, M. Sydanmaa, and A. Uusitalo. 1990. "Short-Term Variability of Systolic Blood Pressure and Heart Rate in Normotensive Subjects." *Clin. Physiol.* **10**: 389–401.
[Rigney]

Tyssedal, J. S., and D. Tjostheim. 1988. "An Autoregressive Model with Suddenly Changing Parameters and an Application to Stock Market Prices." *Appl. Stat.* **37**: 353–369.
[Lewis]

Ulam, S. 1957. "The Scottish Book: A Collection of Mathematical Problems." Unpublished manuscript. See also the special issue on Dr. Ulam: *Los Alamos Science* **15** (1987).
[Gershenfeld]

Unnikrishnan, K. P., J. J. Hopfield, and D. W. Tank. 1991. "Connected-Digit Speaker-Dependent Speech Recognition Using a Neural Network with Time-Delayed Connections." *IEEE Trans. Signal Proc.* **39**: 698–713.
[Mozer]

Unno, W., Y. Osaki, H. Ando, and H. Shibahashi. 1979. *Nonradial Oscillations of Stars.* Tokyo: University of Tokyo Press.
[Clemens]

Uppal, J. S., R. G. Harrison, and J. V. Moloney. 1987. "Gain, Dispersion, and Emission Characteristics of Three-Level Molecular Laser Amplifier and Oscillator Systems." *Phys. Rev. A* **36**: 4823–4834.
[Hübner]

van der Pol, B., and J. van der Mark. 1928. "The Heartbeat Considered as a Relaxation Oscillator and an Electrical Model of the Heart." *Phil. Mag.* **6**: 763–775.
[Glass]

Vanderriest, C., J. Schneider, G. Herpe, M. Chevreton, M. Moles, and G. Wlerick. 1989. "The Value of the Time Delay $\Delta T(A, B)$ for the 'Double' Quasar 0957+561 from Optical Photometric Monitoring." *Astron. & Astrophys.* **215**: 1.
[Press]

Városi, F. 1988. "Efficient Use of Disk Storage for Computing Fractal Dimension." Unpublished manuscript, University of Maryland.
[Pineda]

Vastano, J. A., T. Russo, and H. L. Swinney. 1990. "Bifurcation to Spatially Induced Chaos in a Reaction-Diffusion System." *Physica D* **46**: 23–42.
[Swinney]

Volterra, V. 1959. *Theory of Functionals and of Integral and Integro-Differential Equations.* New York: Dover.
[Gershenfeld]

Voss, R. F., and J. Clarke. 1978. "'1/f Noise' in Music: Music from 1/f Noise." *J. Acous. Soc. Am.* **63**: 258–263.
[Dirst]

Wada, T., H. Akaike, and E. Kato. 1986. "Autoregressive Models Provide Stochastic Descriptions of Homeostatic Processes in the Body." *Nippon Jinzo Gakkai Shi* **28**: 263–268.
[Rigney]

Waibel, A. 1989. "Modular Construction of Time-Delay Neural Networks for Speech Recognition." *Neur. Comp.* **1(1)**: 39–46.
[Wan]

Waibel, A., T. Hanazawa, G. Hinton, K. Shikano, and K. Lang. 1989. "Phoneme Recognition Using Time-Delay Neural Networks." *IEEE Trans. Acoust., Speech, & Signal Proc.* **37(3)**: 328–339.
[Mozer, Wan]

Waldrop, M. M. 1992. *Complexity: The Emerging Science at the Edge of Order and Chaos.* New York: Simon & Schuster.
[Rigney]

Wallace, C. S., and D. M. Boulton. 1968. "An Information Measure for Classification." *Comp. J.* **11**: 185–195.
[Gershenfeld]

Walters, P. 1982. *An Introduction to Ergodic Theory.* Berlin: Springer.
[Paluš]

Wan, E. 1990a. "Temporal Backpropagation for FIR Neural Networks." In the proceedings of the International Joint Conference on Neural Networks, held in San Diego, CA, 575–580.
[Wan]

———. 1990b. "Temporal Backpropagation: An Efficient Algorithm for Finite Impulse Response Neural Networks." *Proceedings of the 1990 Connectionist Models Summer School*, 131–140. San Mateo, CA: Morgan Kaufmann.
[Wan]

———. 1993. "Times Series Prediction Using a Connectionist Network with Internal Delay Lines." This volume.
[Casdagli, Gershenfeld, Kostelich, Mozer, Sauer, Smith]

Watrous, R. L., and L. Shastri. 1987. "Learning Acoustic Features from Speech Data Using Connectionist Networks." In *Proceedings of the Ninth Annual Conference of the Cognitive Science Society*, 518–530. Hillsdale, NJ: Erlbaum.
[Mozer]

Wegman, E. J., and I. W. Wright. 1983. "Splines in Statistics." *J. Am. Stat. Assoc.* **78**: 351–365.
[Lewis]

Wei, S., and W. William. 1990. *Time Series Analysis: Univariate and Multivariate Methods.* Reading, MA: Addison-Wesley.
[Wan]

Weigend, A. S. 1991 "Connectionist Architectures for Time Series Prediction of Dynamic Systems." Ph.D. Thesis, Stanford University.
[Gershenfeld, Wan]

———. 1993. "Book Review of Hertz, Krogh and Palmer, *Introduction to the Theory of Neural Computation.*" *Art. Int.* **62(1)**: in press.
[Casdagli]

———. n.d. "The Effective Number of Hidden Units." Preprint, University of Colorado at Boulder, in preparation.
[Gershenfeld]

Weigend, A. S., and M. Dirst. 1993. "Programming Bach: J. S. Bach's Neural Network." Preprint, University of Colorado at Boulder, in preparation.
[Dirst]

Weigend, A. S., and N. A. Gershenfeld, eds. 1993. *Time Series Prediction: Forecasting the Future and Understanding the Past.* Santa Fe Institute Studies in the Sciences of Complexity, Proc. Vol. XV. Reading, MA: Addison-Wesley.
[Casdagli, Gershenfeld]

Weigend, A. S., B. A. Huberman, and D. E. Rumelhart. 1990. "Predicting the Future: A Connectionist Approach." *Intl. J. Neur. Sys.* **1**: 193–209.
[Casdagli, Dirst, Gershenfeld, Kaplan, Lewis, Mozer, Rigney, Theiler, Zhang]

———. 1992. "Predicting Sunspots and Exchange Rates with Connectionist Networks." In *Nonlinear Modeling and Forecasting*, edited by M. Casdagli and S. Eubank. Santa Fe Institute Studies in the Sciences of Complexity, Proc. Vol. XII, 395–432. Redwood City, CA: Addison-Wesley.
[Gershenfeld, LeBaron, Mozer, Wan]

Weigend, A. S., and D. E. Rumelhart. 1991a. "The Effective Dimension of the Space of Hidden Units." In *Proceedings of International Joint Conference on Neural Networks, Singapore,* 2069–2074. Piscataway, NY: IEEE Service Center.
[Dirst, Smith]

——. 1991b. "Generalizations Through Minimal Networks with Applications to Forecasting." In *INTERFACE '91—23rd Symposium on the Interface: Computing Science and Statistics,* edited by E. M. Keramidas, 362–370. Conference held in Seattle, WA, in April 1991. Interface Foundation of North America.
[Casdagli, Gershenfeld, Wan]

Weiss, C. O., N. B. Abraham, and U. Hübner. 1988. "Homoclinic and Heteroclinic Chaos in a Single-Mode Laser." *Phys. Rev. Lett.* **61**: 1587–1590.
[Hübner]

Weiss, C. O., and J. Brock. 1986. "Evidence for Lorenz-Type Chaos in a Laser." *Phys. Rev. Lett.* **57**: 2804–2806.
[Hübner]

Weiss, C. O., and W. Klische. 1984. "On the Observability of Lorenz Instabilities in Lasers." *Opt. Commun.* **51**: 47–48.
[Hübner]

Weiss, C. O., W. Klische, N. B. Abraham, and U. Hübner. 1989. "Comparison of NH_3 Laser Dynamics with the Extended Lorenz Model." *Appl. Phys. B* **49**: 211–215.
[Hübner]

Weiss, C. O., W. Klische, P. S. Ering, and M. Cooper. 1985. "Instabilities and Chaos of a Single Mode NH_3 Ring Laser." *Opt. Commun.* **52**: 405–408.
[Hübner]

Werbos, P. 1974. "Beyond Regression: New Tools for Prediction and Analysis in the Behavioral Sciences." Ph.D. Thesis, Harvard University, Cambridge, MA.
[Gershenfeld, Wan]

——. 1988. "Generalization of Backpropagation with Application to a Recurrent Gas Market Model." *Neur. Net.* **1**: 339–356.
[Wan]

White, D. A., and D. A. Sofge, eds. 1992. *Handbook of Intelligent Control.* Van Nostrand Reinhold.
[Casdagli, Gershenfeld]

White, H. 1990. "Connectionist Nonparametric Regression: Multilayer Feedforward Networks Can Learn Arbitrary Mappings." *Neur. Net.* **3**: 535–549.
[Gershenfeld]

Whitney, H. 1936. "Differentiable Manifolds." *Ann. Math.* **37**: 645.
[Paluš]

Whittle, P. 1963. *Prediction and Regulation by Linear Least-Squares Methods.* London: The English Universities Press.
[Rigney]

Widrow, B., and M. E. Hoff. 1960. "Adaptive Switching Circuits." In *1960 IRE WESCON Convention Record,* Vol. 4, 96–104. New York: IRE.
[Gershenfeld]

Widrow, B., and S. D. Stearns. 1985. *Adaptive Signal Processing*. Englewood Cliffs, NJ: Prentice Hall.
[Mozer, Wan]

Wiener, N. 1949. *The Extrapolation, Interpolation and Smoothing of Stationary Time Series with Engineering Applications*. New York: Wiley.
[Gershenfeld]

Williams, R. N. 1991. *Adaptive Data Compression*. Norwell, MA: Kluwer.
[Fraser]

Williams, R., and D. Zipser. 1989. "A Learning Algorithm for Continually Running Fully Recurrent Neural Networks." *Neur. Comp.* **1**: 270–280.
[Mozer, Wan]

Winget, D. E. 1988. "Asteroseismology of Compact Objects" In *IAU Colloquium No. 123, Advances in Helio- and Asteroseismology*, edited by J. Christensen-Dalsgaard and S. Frandsen, 231. Dordrecht: Reidel.
[Clemens]

Winget, D. E., R. E. Nather, J. C. Clemens, J. Provencal, S. J. Kleinman, P. A. Bradley, M. A. Wood, C. F. Claver, M. L. Fruch, A. D. Graver, B. P. Hine, C. J. Hansen, G. Fontaine, N. Achilleos, D. T. Wicteramasinghe, T. M. K. Mavar, S. Seetha, B. N. Asholca, D. O'Donoghue, B. Warner, D. W. Kurtz, D. A. Buckley, J. Brickhill, G. Vanclair, N. Dolez, M. Chevreton, M. A. Barstow, J. E. Solheim, A. Kanaan, S. O. Kepler, G. W. Henry, and S. D. Kawaler. 1991. "Asteroseismology of the DOV star PG1159-035 with the Whole Earth Telescope." *Astrophys. J.* **378**: 326–378.
[Clemens]

Winston, P. H. 1984. *Artificial Intelligence*, 2nd ed. Reading, MA: Addison-Wesley.
[Zhang]

Wolf, A., J. B. Swift, H. L. Swinney, and J. A. Vastano. 1985. "Determining Lyapunov Exponents from a Time Series." *Physica D* **16**: 285–317.
[Glass, Kantz, Kaplan, Paluš]

Wolff, C. 1975. "The Last Fugue: Unfinished?" *Curr. Music.* **19**: 71–77.
[Dirst]

Womack, B. F. 1971. "The Analysis of Respiratory Sinus Arrhythmia Using Spectral Analysis and Digital Filtering." *IEEE Trans. Biomed. Engr.* **BME-18**: 399–409.
[Rigney]

Wright, J. J., R. R. Kydd, and A. A. Sergejew. 1990. "Autoregression Models of EEG." *Biol. Cyber.* **62**: 201–210.
[Rigney]

Yamashiro, S. M. 1989. "Distinguishing Random from Chaotic Breathing Pattern." In *Modeling and Parameter Estimation in Respiratory Control*, edited by M. C. K. Khoo, 137–145. New York: Plenum.
[Rigney]

Yates, F. E. 1981. "Analysis of Endocrine Signals: The Engineering and Physics of Biochemical Communication Systems." *Biol. Reproduc.* **24**: 73–94.
[Rigney]

Yates, F. E., A. Garfinkel, D. O. Walter, and G. B. Yates, eds. 1987. *Self-Organizing Systems: The Emergence of Order.* New York: Plenum.
[Rigney]

Yip, K. M.-K. 1991. "Understanding Complex Dynamics by Visual and Symbolic Reasoning." *Art. Intel.* **51**: 179–221.
[Gershenfeld]

Yule, G. 1927. "On a Method of Investigating Periodicity in Disturbed Series with Special Reference to Wolfer's Sunspot Numbers." *Phil. Trans. Roy. Soc. London* **A 226**: 267–298.
[Casdagli, Gershenfeld, Wan]

Zeghlache, H., and P. Mandel. 1985. "Influence of Detuning on the Properties of Laser Equations." *J. Opt. Soc. Am.* B **2**: 18–22.
[Hübner]

Zeng, X., R. Eykholt, and R. A. Pielke. 1991. "Estimating the Lyapunov-Exponent Spectrum from Short Time Series of Low Precision." *Phys. Rev. Lett.* **66**: 3229–3232.
[Glass]

Zhang, X., and J. Hutchinson. 1993. "Dumb Algorithms on Fast Machines: Practical Issues in Nonlinear Time Series Prediction." This volume.
[Casdagli, Dirst, Gershenfeld, Kaplan Mozer, Zhang]

Zhang, X., M. McKenna, J. Mesirov, and D. Waltz. 1990. "The Backpropagation Algorithm on Grid and Hypercube Architectures." *Parallel Comp.* **14**: 317–327.
[Zhang]

Zhong, F., R. Ecke, and V. Steinberg. 1991. "Asymmetric Modes and the Transition to Vortex Structures in Rotating Rayleigh-Bénard Convection." *Phys. Rev. Lett.* **67**: 2473–2476.
[Swinney]

Ziehn, B. 1894. "Gehört die 'unvollendete' Back-Fuge zur 'Kunst der Fuge' oder nicht?" In *Allgemeine Musik-Zeitung*, 435–437.
[Dirst]

Ziman, J. M. 1979. *Models of Disorder.* Cambridge: Cambridge University Press.
[Afraimovich]

Index

A

Abelson, H., 18
Abraham, N. B., 387
activation function, 27
adaptive regression splines
    see multivariate adaptive regression
    splines
adaptive spline threshold autoregressive
(ASTAR) model, 296, 303, 310, 315
Afraimovich, V. S., 539, 549
Akaike, H., 15, 31, 267, 437, 448
algorithmic information theory, 60
analysis by synthesis, 164
anaphylactic shock, 109
Anderson, T. W., 437
apnea, 109, 116, 118, 121
architecture, 25, 28, 34, 244, 251, 256
ARMA model
    see autoregressive moving average
    (ARMA) model
artificial intelligence, 3, 221, 348
asteroseismology, 141-142
attractor, 32, 36, 326
auto-association, 57, 160
autocorrelation coefficients, 12, 85, 87, 90,
94, 222, 431
autoprediction, 251
autoregressive conditional heteroskedastic
(ARCH) model, 314, 467
autoregressive hidden Markov models
(ARHMMs), 268
autoregressive (AR) models, 13, 128, 165,
196, 303, 433
autoregressive moving average (ARMA)
model, 13, 15, 198, 212-213, 225, 433,
436, 438, 447, 451, 453
    vs. Fourier transform, 438
average root mean squared errors
(ARMSE), 536

B

Bach, J. S., 6, 153
Bachrach, J., 253-254
backpropagation, 28, 196, 198, 202, 222,
253
bacteria, 566

Ballard, D. H., 258
barriers to transport, 356
Barron, A. R., 33
Baum, L. E., 265, 270
Bayesian, 19, 35
Beethoven, Ludwig von, 167
Bellman, R. E., 298
Belousov-Zhabotinsky reaction, 560
Bénard convection, 78, 540
Bengio, Y., 256
Bernstein, L., 166
Bharucha, J. J., 167
bias, 26, 198
bias-variance dilemma, 56, 183, 348
bifurcation, 128, 515
bifurcation diagram of the Lorenz model,
76
binary tree, 51, 287
Bingham, S., 370
bit-interleave, 376
block entropy, 47
blood oxygen, 105, 525
blood pressure, 109, 125, 513
Bollerslev, T., 459
bootstrapping, 361, 416, 434, 437, 473,
522
bottom up, 164
box-counting algorithm, 51, 367, 369, 370,
375
BPTT, 253, 256
Bradley, E., 18, 61
Brahe, T., 18
brain activity, 525
brain-style computation
    see neural networks
Brock, J., 79
Brock, W. A., 45, 435, 464, 466, 473
Brock-Dechert-Scheinkman (BDS) test,
45, 464
Broomhead, D. S., 31, 184, 330
Brown, P. F., 159
Bulirsch-Stoer method, 413
Buntine, W. L., 19

C

Canadian lynx data, 298
Cannon, W. B., 106

capacity of a network, 29
capacity dimension, 369
capillary ripples, 553
Caprile, B., 230
Casdagli, M. C., 23, 32, 179, 267, 315, 327, 332, 347, 446, 523
categorical time series as predictor variables (CASTAR), 308, 311
Cawley, R., 478, 480
cell growth, 513
cellular automata, 567
central pattern generators, 126
chaos, 127, 220, 265, 365, 383, 387, 404, 421, 435, 452, 513, 527, 541, 564
    conditions for, 78
    Lorenz-like, 82, 89
    low-dimensional, 366
    spatio-temporal, 553, 559
    also see deterministic chaos
chaotic dynamics, 513
characterization methods, 42, 45
Chatfield, C., 20, 221
chicken pox, 522
Chomsky, N., 167
chord, 160
Chui, C. K., 567
Clemens, J. C., 6, 139
clustering, 163
coarse grained flow averages, 417
Cohen, R. J., 435, 522
coherence effects, 104
coherence time, 429, 431, 452
coherent structures, 560
cointegration, 534-535
collapses, 10, 325
computational complexity, 51, 370
conditional densities, 267
configuration space, 20
confinement, 356
Connection Machine, 219, 222
connectionist models, 19, 25, 56
    also see neural networks
connectionist networks
    see neural networks
Connor, J., 254
constrained static network, 200
context, 165

context-free grammars, 37, 359
convolution filter, 12
correlation dimension, 85, 88, 96, 100, 367, 369, 415, 475, 486, 547, 550, 555
correlation integral, 87, 93-94, 100, 369, 375, 435, 442, 486
coupled map lattices, 567
covariance matrix, 58, 161, 268, 390
Cover, T. M., 29
Craven, P., 302
critical Rayleigh number R_c, 75
cross-correlation function (CCF), 362
cross correlations, 468-469
cross validation, 35, 210, 239
Crutchfield, J. P., 33
currency exchange, 2, 39, 131, 133-134, 257, 309, 426, 457
curse of dimensionality, 33, 165, 298
Curtiss, P., 244

D

D_q
    see generalized dimensions
Darwinism, 107
data-rich, 18, 164
Data Set A, 73, 175, 189, 195, 283-284, 296, 315, 319, 323, 325, 347, 354, 367, 380, 387, 405, 415, 419, 475
Data Set B, 105, 308, 347, 415, 425, 519
Data Set C, 39, 131, 219, 243, 257, 260 309, 415, 426, 457
Data Set D, 38, 219, 222, 265, 279, 347, 356, 367, 383, 387, 406-407, 415, 421-422, 439
Data Set E, 139, 387, 429
Data Set F, 151, 219
data sets
    Canadian lynx, 298
    riverflow, 304
    sea surface temperature, 298, 304
    sleep stage, 105, 109, 308, 313, 513
    sunspot, 2, 55, 298, 304
    white dwarf stars, 139, 406, 439
    also see Data Sets A–F, and physiological data
data sonification, 42

databases, 385
Davies, C. T. M., 477
Davies, M. E., 478
de Vries, B., 244, 248, 254
Dechert, W. D., 45
decoding layer, 57
delay coordinate embedding, 175, 178, 246, 254
    filtered, 23, 179
delay vectors, 266, 351
Dempster, A. P., 270
deterministic chaos, 16, 109, 349, 358, 388, 391, 399, 415, 432, 476
    also see chaos
deterministic modeling, 56, 347, 349, 352, 451, 522
deterministic vs. stochastic models, 53, 55-56, 349
detuning, 84, 90
Deutsche mark, 135
    also see currency exchange
Diebold, F. X., 39
diffeomorphism, 45, 266
difference representation, 157
differenced time series, 362
differential equations, 4, 177
digital imaging systems, 558
dimension, 20, 109, 486, 520, 526, 544
    of the attractor, 175
    spectra, 100
Dimitriadis, A., 37, 359
direct forecasts, 36, 327-328
direct vs. iterated prediction, 36, 185
Dirst, M., 37, 151
disorder, 550
Domowitz, I., 459
downsample, 180
dstool, 18
Durbin, R., 29
DVS plots
    see deterministic vs. stochastic
    models
dynamical systems, 3, 539, 541, 546, 557

E

Ebcioglu, K., 163
Eckmann, J.-P., 176, 284

economics, 7, 529, 533
EEG
    see electroencephalograms
efficient market hypothesis, 39
Efron, B., 158, 434, 473
electrocardiograms (ECGs), 514, 521
electroencephalograms (EEGs), 55, 364, 514, 525
electronic brokeraging, 133
Ellner, S., 435, 448
Elman, J. L., 167, 246, 251, 253
embedding, 20, 24, 30, 45, 52, 175-176, 178-179, 219, 353, 367, 374, 477
    autoregressive models, 30
    by expectation values, 25, 366
    delay space, 246, 254
    phase plane, 517
    spatial, 539
    also see overembedding
embedding dimension, 88, 326, 351
embedology, 177, 265, 267
    also see state-space reconstruction
encoding, 57, 221
    also see representation
end-use models, 534
Engle, R. F., 314, 467
entrainment, 328
entropy, 46, 49, 85, 88, 159, 367, 387-388, 401, 405, 526, 544, 547, 556, 557
    Komogorov-Sinai, 49, 387, 401, 476,547
Entscheidungsproblem, 60
epileptic seizures, 109
equation-error adaptation, 205
equilibrium model, 148
ergodic process, 431
error backpropagation, 28
error-correction model, 535
error bars, 8, 10, 34, 211
exchange rates
    see currency exchange and
    Data Set C
expectation-maximization algorithm (EM), 265, 270, 281
exchange rates
    see currency exchange rates
expectations, 166, 529

expectations, 166, 529
expert system, 163
exploratory data analysis, 42
exponential trace memory, 247
external shocks, 2, 106

F

$f(\alpha)$ spectrum of singularities, 557
Fallside, F., 253
false nearest-neighbors approach, 52, 372
Farber, R., 25
Farmer, J. D., 31, 185, 284, 330, 477, 520
feedforward networks, 29, 359
Ferguson, J. D., 269
filtered delay coordinate, 23, 175, 179-180, 182
finite impulse response filter, 12
Finnoff, W., 261
FIR network, 200
fitting
    see training
fixed mass techniques, 369
flexibility, 59, 348
fluid dynamics, 73
forecasting
    see time series prediction
forward-backward algorithm, 274, 276 279
Fourier transform, 12, 92, 142, 149, 159, 435-436, 438, 566
    algorithm, 452
    discrete, 442
    fast, 397, 442
    vs. ARMA model, 438
fractal delay coordinate embedding prevalence theorem, 178
fractal dimension, 430, 547
Frasconi, P., 254
Fraser, A. M., 32, 45, 51, 62, 265, 267, 359, 367, 371-373, 380, 387-388, 403
frequency locking, 564
Friedman, J. H., 31, 297, 301, 315
Froehling, H., 87
ftp, 7, 41, 70, 569
fugue, 6, 153, 167, 237
function approximation, 178

fundamental vs. technical analysis, 135

G

g-mode pulsations, 146
Gabura, A. J., 160
gamma memory, 248, 250
Gaussian error model, 36, 353
Gaussian processes, 418, 433
Geman, S., 56, 184, 259, 348
generalized correlation integral, 97
generalized cross-validation criterion, 302
generalized dimensions, 47, 85, 97, 367-368, 557
Gershenfeld, N., 1, 51, 349, 354, 356, 368
Gilligan's Island, 154
Giona, M., 32
Glass, L., 399, 418, 513-514
global predictors, 328
global radial basis function prediction, 33, 334
Gödel, K., 60
goodness-of-fit criterion, 315
Granger, C. W. J., 31, 529
Grassberger, P., 45, 87, 97, 368, 377, 415, 434, 478, 482, 486, 520, 549
Grebogi, C., 393, 408, 477

H

Haken, H., 73, 77
halting problem, 60
Hammel, S. M., 478
harmony, 160
Hausdorff dimension, 99
Haussler, D., 29
heart rate, 105-106, 109, 112, 121, 123-126, 425, 522-525
Hentschel, H. G. E., 47, 97, 371
Herz, A. V. M., 254
Herzel, H., 434
hidden filter hidden Markov models (HFHMMS), 273
hidden Markov models (HMM), 37, 39, 62, 166, 265, 268, 269
    autoregressive, 268
    hidden filter, 273
hidden state, 265
hidden units, 26, 161

Hild, H., 163
Hinton, G. E., 29, 34, 57, 199, 246
histograms, 39, 158-159
HMM
    see hidden Markov models
Hodgkin, A. L., 515
Hopfield, J. J., 249, 253
Hsieh, D. A., 467
Hsu, G. H., 478, 480
Huberman, B. A., 57
Hübner, U., 73
Hutchinson, J., 39, 244, 262, 359
Huxley, A. F., 515

I

I memory
    see memory taxonomy
incomplete data, 270
indeterministic, 432
    also see stochastic
indirect characterization, 53
infinite impulse response filter (IIR), 13
information dimension, 51, 367, 369, 375
information theory, 29, 45, 388
information-theoretic functionals, 387
interactions
    number of, 29
    order of, 29, 348
interpolation, 181
intervention analysis, 308
iterative forecasts, 36, 327-328

J

Jenkins, G. M., 436
joint entropy, 47
joint mutual information, 48
Jones, R., 184
Jordan, M. I., 247, 254

K

k-means clustering, 229
k-nearest-neighbor algorithms, 32, 55,
369-370, 446
Kahle, E., 171
Kalman filter, 20, 214, 268, 536

Kaplan, D. T., 52, 399, 418, 435, 513,
517, 522
Kaplan-Yorke conjecture, 97
Kennel, M. B., 372, 435, 520
Kepler, J., 18
Khinchin, A. I., 388
King, G. P., 31, 184
KL distance
    see Kullback-Leibler distance
Kohonen, T., 162, 165
Kolmogorov, A. N., 11, 401
Kolmogorov-Sinai entropy, 387, 401,
476, 547
Kostelich, E. J., 283, 371, 476-477
Koza, J., 29
Kullback, S., 272, 388
Kullback-Leibler distance, 272
kurtosis, 463

L

lacunarity, 368
lag time, 55, 351
Lang, K., 34, 199
Lapedes, A., 25, 244, 246, 350
lasers, 73, 78, 380
    also see Data Set A
lead time, 55, 351
learning, 19, 253
"learning" approach, 2
LeBaron, B., 40, 457, 473
le Cun, Y., 57, 202, 258
Leibler, R. A., 272
Lempel-Ziv algorithm, 159
leptokurtosis, 461
Lequarré, J., 4, 131
Levich, R. M., 470
Lewis, P. A. W., 31, 296, 308
Liebert, W., 372, 375
Liebovitch, L.S., 369
Lii, K. S., 433
likelihood, 10, 211
Lim, K. S., 17
linear correlations, 43
linear models, 16, 347
linear predictive coding, 15
linear redundancy, 390-391
linear structure, 43

Ljung-Box-Pierce test, 463
local expansion rates, 50, 294
local linear approximation, 3, 353
local linear models, 11, 16-17, 31, 175, 359
local predictors, 332
local radial basis function prediction, 334
logistic map, 16, 452
long coherence time, 429, 444, 452
long-term behavior, 2, 14, 208, 210
long-term prediction, 228
Lorenz, E. N., 10, 73, 283-284
Lorenz system, 10, 21, 73, 75, 77, 82, 85, 187, 406, 420, 453
Lorenz-Haken model, 73-74, 78, 83
Lorenz-like chaos, 82, 89
Lorenz-like pulsations, 102
low-dimensional chaos, 56, 366
low-dimensional determinism, 34, 419
low-dimensionality, 31, 366
low-pass embedding, 23, 32, 179
low-pass linear filter, 442
Lowe, D., 330
Loy, D. G., 154
Lyapunov exponents, 49, 59, 97, 109, 176, 179, 293, 416, 430, 452, 475-476, 487, 520, 544, 547, 557
    spatial, 556

M
MIT/BIH Polysomnographic Database, 111
machine learning, 3, 348
machine with a mechanism, 107
Mackey, M. C., 514
marginal redundancy, 390
Markov, A. A., 159
Markov models, 165
Markov process, 267, 269
MARS program
    see multivariate autoregression splines
maximal Lyapunov exponent, 487
maximum likelihood, 36
maximum norm, 88
Maxwell, H. J., 161
Maxwell's Demon, 45

May, R. M., 16, 522
Mayer-Kress, G., 387
MDL
    see minimum description length
Mean Squared Error (MSE)
    see normalized mean squared error
measles, 522
medical data
    see physiological data
melody, 158
memory, 247-253
    delay-line, 246
    gamma, 248
    TIS, 251-252
    transformed input, 251
memory taxonomy, 253
metaphor of a machine, 106
method of analogues, 283
method of delays, 327
metric, 37, 165
metric entropy, 387
Meyer, L. B., 166
minimum description length, 60, 267
Mitchinson, G. J., 29
mixture of Gaussians, 37
model vs. system, 431
models
    ARCH, 314, 467
    ARMA, 213, 433, 436, 438, 447, 451, 453
    ASTAR, 296, 303, 310, 315
    autoregressive, 13, 128, 165, 196, 268, 303, 433
    CASTAR, 308, 311
    deterministic, 53, 56, 347, 349
    deterministic vs. stochastic (DVS), 53, 55-56
    end-use, 534
    equilibrium, 148
    error-correction, 535
    filters, 12
    fitted, 124
    Gaussian error, 353
    hidden Markov, 37, 61, 166, 265, 268-269, 273
    linear, 11, 16-17, 31, 446
    Lorenz-Haken, 73, 83

models (continued)
  Markov, 159
  MARS, 31, 297, 299, 316
  moving average, 11
  multiscale integration, 254
  nonlinear, 16-17, 348, 449
  recursive partitioning, 300
  regime-switching, 535
  regression, 15, 296
  SMASTAR, 304
  state-space, 10, 24, 214, 219, 221
  stochastic, 2, 53, 56, 347, 353, 532
  system-independent, 108
  tapped delay-line, 20, 246
  threshold autoregressive (TAR), 17, 298, 304, 311
  TI-delay, 254
  TI-exponential memory, 254
  Volterra series, 17
moderate-dimensional dynamics, 421
Moody, J., 31, 229, 438
moving average (MA), 11, 198, 433
Mozart, W. A., 152
Mozer, M. C., 29, 165, 166, 243, 247, 254, 256
multilayer perceptrons, 25-26, 236, 240
multiscale integration model, 254
multivariate analysis, 107, 111, 362
multivariate autoregression splines (MARS), 31, 296, 299, 316
musical texts, 151
mutual information, 48, 58, 367, 388, 557, 159

N

Nason, J. M., 39
Navier-Stokes equations, 75
Neal, R., 249
nearest-neighbor predictor, 333
nearest neighbors, 32, 55, 164, 333
network dimensions, 48, 209
networks, 57, 200
  feedforward, 29, 359
  recurrent, 166, 252, 359
  time-delay neural (TDNN) 34, 199
  also see neural networks

neural networks, 18-19, 25, 160, 165, 197, 209, 243, 252, 256, 316, 359
  and machine learning, 29
  vs. autoregressive models, 26
  vs. polynomial expansion, 29
neural systems, 527
Newton, I., 18
NH$_3$ laser, 74
noise, 2, 7, 13, 23, 93, 110, 125, 145, 183, 356, 358, 417, 430, 477, 489, 524, 531
  also see white noise
noise reduction, 44, 337, 475-476, 484
noise terms, 353
nonlinear diagnostics, 464
nonlinear dynamics, 515, 527
nonlinear models, 10, 16-17, 348, 449
nonlinearity, 58, 108, 388, 391, 439, 446, 522
nonstationarity, 108, 408, 453, 523, 531
normalized mean squared error (NMSE), 63, 70, 190, 209, 470
Nowlan, S. J., 57
Nychka, D., 527
Nyquist frequency, 420

O

O memory
  see memory taxonomy
observability, 266
observables, 21, 177, 543
Onegin, E., 159
Oregonator kinetics, 560
orthogenesis, 107
orthogonal decomposition, 32, 567
Osborne, A. R., 415
out-of-sample, 19
  also see test set
outliers, 352
  also see robust error
output-error adaptation, 206
overembedding, 478
overfitting, 19, 240, 334, 348, 366, 437
overquantization, 404, 411
overtraining
  see overfitting

P

Packard, N., 20, 29, 176, 542
Paluš, M., 51, 387
parallel distributed processing, 25
    also see neural networks
partial autocorrelation, 159
pattern formation, 540, 560, 564
pattern recognition, 244
Pawelzik, K., 97
Pearlmutter, B. A., 255
Pepys, S., 152
period doubling, 82-83
periodic breathing, 118, 121, 123, 127
periodic oscillations, 564
periodic signals, 450
periodogram, 42
Pesin identity, 49
phase plane embedding, 517
    also see phase portrait
phase portraits, 43, 85, 87, 353
physiological data, 105, 513, 523, 525, 527
    blood, 105, 109, 125, 513, 525
    brain activity, 525
    chicken pox, 522
    electrocardiograms (ECGs), 514, 521
    electroencephalograms (EEGs), 54, 364, 514, 525
    epileptic seizures, 109
    measles, 522
    sudden death, 109
    also see apnea, heart rate, periodic breathing, respiration, and sleep apnea
Pincus, S. M., 527
Pineda, F. J., 51, 367
planetary motion, 18
plane-wave condition, 79
plateau, 88, 94
Plaut, D. C., 246
Poggio, T., 229
Poincaré, H., 18
Poincaré map, 475, 488, 518
point predictions, 34
Politis, D. N., 434
polynomial curve fitting, 19

polynomial regression, 29
polyphonic structure, 162
power spectra, 3, 12, 16, 43-44, 85, 92
Prandtl number, 76
prediction function, 178
prediction horizon, 49-50, 186
prediction time, 55
predictive accuracy, 350
    also see error bars
predictor profile, 332
Press, W., 4, 493
Priestley, M., 53
principal components, 26, 161
Principe, J. C., 35, 244, 248, 254
Procaccia, I., 45, 47, 87, 97, 368, 371, 415, 486, 520, 549
Provenzale, A., 415, 435
pseudocode, 377
pulse trains, 85
Pushkin, A., 159

Q

quadratic map, 16
quadratic prediction, 334
quantization, 32, 50, 328, 333, 412
quasi-periodicity, 564
quicksort algorithm, 376

R

Rabinovich, M. I., 539
radial basis functions, 33, 229, 240, 315, 328, 370
random walk, 418
randomness and order, 152
Rayleigh-Bénard convection, 73-75
Rayleigh number R, 75
reactive, 107
real-time recurrent learning (RTRL), 253, 256
reconstructed space, 265
    also see state space
reconstruction, 323-324, 326-327, 331-332, 368
recurrent networks, 37, 166, 252, 359
recursion (Box-Jenkins procedure), 15
recursive partitioning, 299
redundancy, 48, 159, 388, 390, 407, 411

regime-switching models, 535
regression problem, 15, 296
regular grammars, 37, 359
reinjection, 284
Renyi, A., 368
Renyi dimensions
    see generalized dimensions
Renyi entropy of second order, K_2, 85
representation, 10, 156, 178, 221
resampling, 434
respiration, 105, 513, 525
    also see periodic breathing, and
    apnea
return maps, 102, 518
returns, 461
Rigney, D., 105
Rissanen, J., 29, 268
riverflow data, 304
Robinson, A. J., 253
robust error, 55, 63, 352
root-mean-squared error, 228
Rosenblatt, M., 433
Rössler system, 284, 406, 413
rotating-wave approximation, 77
RTRL
    see real-time recurrent learning
Rubin, D. M., 361, 450
Ruelle, D. M., 20, 176, 284, 368
Rumelhart, D. E., 28, 37, 57, 211,
253, 334, 348
run-length encoding, 157, 237
runaway extensions, 328
Runge-Kutta method, 85

S

saddle orbits, 284
Samet, H., 385
sample autocorrelation, 436
sampling rate, 326
Sanger, T., 154
saturation, 326
Sauer, T., 20, 32, 175, 179, 267, 284, 342,
368, 478, 480, 483, 489
Saund, E., 37, 57
scaling region, 88-89
Scargle, J., 447
Scheinkman, A., 45

Schmidhuber, J., 255
Schreiber, T., 478, 482, 487
Schuster, A., 42
Schuster, H. G., 97, 375
Schwanauer, S. M., 163
Schwarz, G., 437
Schwarz-Rissanen criterion, 302, 310, 315
sea surface temperature data, 298, 304
Sejnowski, T., 29
self-organizing feature maps, 162
semiclassical laser equations, 77
semi-multivariate threshold autoregressive
models
    see threshold autoregressive models
Shakespeare, W., 158
Shannon, C. E., 45, 388
Shaw, R. S., 45, 403
Shepard, R. N., 162, 165
shift correction, 88
short-term memory, 250
short-term prediction, 2, 35, 226
Sidorowich, J. J., 31, 185, 330, 477
sigmoid, 27, 251
signature, 404
simple current network (SRN), 253
single delay reconstruction, 331
singular value decomposition (SVD), 32,
175, 182, 290
Skinner, J. E., 520, 526
sleep apnea, 110, 313, 360
    also see apnea
sleep stage data, 105, 109, 308, 313, 513
SMASTAR, 311
Smith, J. M., 516
Smith, L. A., 32, 323, 368, 435
Smith, P. L., 301
Smith, R. L., 368
Smolensky, P., 29, 258
solar radio pulsations, 435
solution manifold, 20
spatial disorder, 539
spatial embedding, 539
spatial Lyapunov exponents, 556
spatio-temporal analogy, 542
spatio-temporal chaos, 61, 553, 559
spatio-temporal patterns, 557-558
spiral chaotic pulsations, 82

spot currency trading, 133
spot deals, 132
squashing function
    see sigmoid
SRN
    see simple recurrent network
state space, 10, 24, 30-31, 177, 213-214
219, 221
state-space reconstruction, 3, 19, 21, 53,
265-266, 351, 353
static, 198
stationarity, 6, 326, 453
statistical inference, 61
statistical physics, 539
statistics and neural networks, 29, 348
stellar oscillations, 140
stochastic influences, 26, 55, 62, 349
stochastic models, 2, 347, 353, 532
stochastic point process, 113
stochastic systems theory, 265
stock market, 4, 39, 529-530
    also see trading rules
strange attractors, 77, 97, 430
strange nonchaotic attractor, 393
structural pattern recognition, 25, 244
subcritical Hopf bifurcation, 77, 85
subset selection, 41, 353
sudden death, 109
Sugihara, G., 349, 522
Sullivan, F., 370
sunspot data, 2, 55, 298, 304
supercritical bifurcations, 562
surprise, 160
surrogate, 42, 429
surrogate data, 44, 397, 416, 419, 421, 425
    FT vs. ARMA, 438
Swift-Hohenberg equation, 541
Swinney, H. L., 45, 51, 367, 371-372, 380,
557, 560
Swiss franc, 4, 135
    also see currency exchange
system vs. model, 431

T
Takens, F., 20, 176, 179, 226, 266-267,
284, 327, 368, 542, 548
Takens theorem, 196, 368, 403, 543

Tam, W. Y., 560
Tang, D. Y., 102
tangent manifold, 179
Tank, D. W., 249, 253
tapped delay-line memory, 20, 246
TDNN
    see time-delay neural network
teacher-forcing, 35, 205
temporal backpropagation, 199, 202-203,
214
temporal pattern recognition, 25, 61, 244
test set, 19, 351
Theiler, J., 341-342, 368, 370-371, 397-
398, 415, 429, 435, 473, 522
thematic structure, 163
theory of nonradial pulsations, 146
Thisted, R., 158
three-body problem, 18
threshold autoregressive (TAR) model,
17, 298, 303-304, 311
TI memory
    see memory taxonomy
time-delay embedding
    see embedding
time-delay neural network (TDNN), 199-
200, 254, 256
time delays, 93, 367
time invariance, 11
times series
    definition of, 2, 430
    forecasting goals, 2
    prediction, 2, 11, 367
TIS memory, 251-252
    also see memory taxonomy
temporal sequence processing, 243
TO memory
    see memory taxonomy
Todd, P. M., 154, 166-167
Tong, H., 17, 298, 303-304, 348, 435
topological entropy, 547
TOS memory
    see memory taxonomy
Toth, T., 369
trading rules, 457, 470
training, 19, 210, 240, 245, 261
training set, 182, 210, 351
trajectory, recurrence, 417

transaction costs, 457, 470
transformed input and state, 251
truncated spline functions, 301
Tsay, R. S., 435, 466
Turing, A. M., 60

U

underfitting, 348
undersampling, 286
"understanding" approach, 2
unfolding in time, 201
univariate analysis, 363
universality, 558
unstable periodic orbits, 557
update extensions, 328
upsampling, 181
U.S. dollar, 4, 135
    also see currency exchange

V

variable encoding, 258
variable predictability, 294
Városi, F., 370
verisimilitude, 128
visual fitting of data, 10, 62, 319
volatility, 2, 41, 162
Volterra series, 17

W

Wahba, G., 302
Waibel, A., 34, 199, 246, 254
Wan, E., 34, 195, 246, 254

Watrous, R. L., 256
Watts, D. G., 436
wavelet transforms, 567
Weigend, A. S., 1, 19, 32, 211, 244, 259,
304, 334, 347, 354, 356, 438, 446
weighted regression, 185
weights, 19, 25, 197
Weiss, C. O., 79
Werbos, P., 28, 61
white dwarf stars, 139, 406, 439
white noise, 13, 43, 405, 415, 486
Whole Earth Telescope, 6, 139
Widrow, B., 25, 258
Wiener, N., 11
Williams, R., 253
Wold decomposition, 432
Wolf, A., 487, 520
Wolf sunspot data
    see sunspot data

Y

Yip, K., 18
Yorke, J. A., 179, 477
Yule, G., 2, 350
Yule-Walker equations, 13, 437

Z

Zeghlache, H., 79
Zhang, X., 39, 219, 244, 260, 262, 359
Zheleznyak, A. L., 539
Zipser, D., 253